Developing Microsoft Dynamics GP Business Applications

Build dynamic, mission-critical applications with this hands-on guide

Leslie Vail

[PACKT] PUBLISHING enterprise
professional expertise distilled

BIRMINGHAM - MUMBAI

Developing Microsoft Dynamics GP Business Applications

First published: December 2012

Production Reference: 1191212

Published by Packt Publishing Ltd.
Livery Place
35 Livery Street
Birmingham B3 2PB, UK.

ISBN 978-1-84968-026-4

www.packtpub.com

Cover Image by Artie Ng (artherng@yahoo.com.au)

Credits

Author
Leslie Vail

Reviewers
Mohammad R. Daoud
Frank Hamelly
Vaidhyanathan Mohan
Jivtesh Singh

Acquisition Editor
Rashmi Phadnis

Lead Technical Editors
Susmita Panda
Dayan Hyames

Technical Editors
Arun Nadar
Jalasha D'costa
Prashant Salvi

Copy Editors
Insiya Morbiwala
Aditya Nair
Alfida Paiva

Project Coordinator
Arshad Sopariwala

Proofreaders
Maria Gould
Sandra Hopper

Indexer
Hemangini Bari

Graphics
Valentina D'silva
Aditi Gajjar

Production Coordinators
Conidon Miranda
Pooja Chiplunkar

Cover Work
Conidon Miranda

About the Author

Leslie Vail is a CPA and has been working as a Microsoft Dynamics GP Consultant for nearly 20 years. She began with Version 1.0 in 1993. During this period she completed numerous implementations, conversions, and custom-development projects. She has been a Session Leader at many partner and customer-technical conferences, and conducts training classes throughout Northern and Central America.

Leslie has been a Microsoft Dynamics GP Microsoft **Most Valuable Professional** (**MVP**) since 2007. She is recognized throughout the industry for her product expertise and contributions to the Dynamics community. She is the Principal of ASCI, Inc., a consulting firm located in Dallas, TX.

As a **Microsoft Certified Trainer** (**MCT**), she serves as a **Subject Matter Expert** (**SME**) for the Microsoft **Assessments and Certification Exams** (**ACE**) team. She is a member of the US MCT Advisory Council, and has been listed as one of the Microsoft Dynamics Top 100 most influential people by DynamicsWorld. She is one of the top contributors to the Microsoft Dynamics GP Newsgroup and the Dynamics Community forum. Leslie maintains the popular Dynamics Confessor Blogspot blog (`http://dynamicsconfessions.blogspot.com/`).

Leslie has reviewed and developed Microsoft Courseware, coauthored the book *Confessions of a Dynamics GP Consultant* published by *Accolade Publications, Inc.*, and has been the Technical Editor of several books dedicated to Microsoft Dynamics GP.

Leslie provides implementation and consulting services for companies ranging from a family office to a multinational manufacturing firm. She is a Microsoft Certified IT Professional in Microsoft Dynamics GP Applications and Microsoft Dynamics GP Installation & Configuration, as well as a Microsoft Certified **Database Administrator** (**DBA**).

She holds a Microsoft Certified Technology Specialist certification in Dexterity, Modifier with VBA, Integration Manager, Report Writer, HR/Payroll, Financials, Inventory and Order Processing, FRx Report Designer, SQL Server 2000, SQL Server 2008, and Microsoft XP Professional.

A skilled developer, Leslie uses Dexterity, Modifier with VBA, Integration Manager, and eConnect to provide custom solutions to her clients. She is a **Certified Integration Developer (CID)**, a Dexterity CID, a Dynamics Tools CID, and a Dexterity Certified Systems Engineer.

Her training proficiency spans the entire Microsoft Dynamics GP product line. She is an experienced trainer and gives classes for Dexterity, Financials, Inventory & Order Processing, HR/Payroll, Integration Manager, Modifier with VBA, FRx Report Designer, SQL Server Reporting Services, Report Writer, Crystal Reports, SmartList Builder, Excel Report Builder, Integrated Excel Reports, Extender, and System Manager.

Prior to working with Microsoft Dynamics GP, Leslie was the Tax Director for a large financial institution; before that, she worked for one of the original "Big Eight" accounting firms as a Senior Tax Accountant.

Acknowledgement

First and foremost, I would like to thank Kerry George for asking me to write this book in the first place. Your unwavering encouragement kept me pressing on one page at a time. I thank the army of kind people at Packt Publishing for your tireless help and support. You were unbelievably patient and calming throughout the process. I couldn't have done it without you.

I would especially like to thank my editors, Jalasha D'costa and Prashant Salvi, for making the book actually flow smoothly. You corrected so many little things and gave me such superb suggestions. People reading this book will think I have excellent grammar and punctuation skills because of you. They are of course wrong, but they won't be reading this acknowledgement, so it won't be found out.

To my reviewers Mohammad R. Daoud, Jivtesh Singh, Vaidhyanathan Mohan, and Frank Hamelly, I thank you for your valuable time, ideas, and insights. You are this author's secret weapon. You helped me close up the holes in my content and kept me honest when I skipped over things. This is a much better, more complete book because of you and your willingness to help and advise me. I was truly blessed the day you said "yes" when Packt asked you to review this book.

I thank David Musgrave, the worldwide wizard of Dexterity, and his trusty sidekick, Mariano Gomez, for their unending motivation and friendship. I have learned so much from you two guys over the years. The development community relies on your expertise and your willingness to share your knowledge, to move forward. I hope that outpouring of knowledge never stops, because when I use it it makes me look smart.

A special shout goes out to Diane Bilyeu for putting up with me when I spent all night at her kitchen table writing Chapter One. Diane has taken the gesture of a nod and a smile to a whole new level, and I appreciate it immensely.

To my friends, family, colleagues, and clients, can you believe it's finally over? At last you will not have to listen to my unending chatter about this book. Of course, the unending chatter will continue, but the topic will be different.

Last but not least, I thank everyone on the Dynamics community forum. Anyone who has ever asked a question, or suggested an answer, and even those who lurk without posting, I thank you. While researching content for this book, I found so many answers from community postings. Keep up the good work! I'll see you online.

About the Reviewers

Mohammad R. Daoud has been working as a Microsoft Dynamics GP Consultant since 2004. His began his career by working with Dynamics GP Version 7.5; he studied every single detail of the application's technicalities and did a lot of successful implementations that included functional consultations, analysis, and custom development projects. He holds a graduate degree in Computer Science and is currently pursuing an MBA degree in Accounting.

In January 2007 he was nominated to receive the MVP certificate, and was certified in April 2008 for his online contributions to the Dynamics community (Dynamics GP newsgroups, forums, user groups, and his blog `http://mohdaoud.com`). He was listed as one of Microsoft Dynamics's Top 100 most influential people in 2009 by DynamicsWorld (`http://dynamicsworld.co.uk/Top-100-List.php`)

Mohammad's certificates include:

- Microsoft Most Valuable Professional (MVP)
- Microsoft Certified Trainer (MCT)
- Microsoft Certified Technology Specialist (MCTS)
- Microsoft Certified Professional (MCP)
- Microsoft Certified IT Professional - Dynamics (MCITP)
- Microsoft Certified Technology Specialist (MCTS)

Mohammad has successfully completed the following Microsoft Certification Exams:

- Microsoft Dynamics GP 2010 Installation & Configuration
- Managing Microsoft Dynamics Implementations
- Microsoft Dynamics GP 10.0 Inventory & Order Processing
- Microsoft Dynamics GP 10.0 Financials
- Microsoft Dynamics GP 10.0 Installation & Configuration
- Microsoft SQL Server 2005 - Implementation and Maintenance

- Microsoft Dynamics GP 9.0 Financials
- Microsoft Dynamics GP 9.0 Inventory & Order Processing
- Microsoft Dynamics GP 9.0 Modifier with VBA
- Microsoft Dynamics GP 9.0 Report Writer
- Microsoft Dynamics GP 9.0 Installation & Configuration
- Installing, Configuring, and Administering Microsoft SQL Server 2000
- Designing and Implementing Databases with Microsoft SQL Server 2000

Frank Hamelly is a business and technology professional with over 25 years of experience in implementing and supporting various ERP systems and business process reengineering initiatives across all organizational areas and across various industries, for small, mid-sized, and even Fortune 500 companies. He has held numerous positions in accounting, finance, customer service, and information systems. His application experience includes SAP R/3, Baan, Fourth Shift, Peachtree, MAS 90/200, Quickbooks, and Microsoft Dynamics ERP. His industry experience includes manufacturing, telecom, aerospace, life sciences, utilities, and media.

Frank holds a degree in Business Administration with an Accounting major from the University of Pittsburgh. He is a **Microsoft Certified Professional** (**MCP**), **Microsoft Certified IT Professional** (**MCITP**), **Microsoft Certified Trainer** (**MCT**), and has been named an MVP every year since 2008. He writes articles for Dynamics-related websites and is a regular speaker at Microsoft Dynamics Convergence, MSDynamicsWorld's Decisions virtual conference. He is also a trainer and presenter for **Great Plains Users Group** (**GPUG**), and has reviewed a number of books written by fellow MVPs.

Frank is the author of the GP2themax blog, with 5,000 visits per month.

Frank is also the owner at NOVA Solutions, LLC, located in the Raleigh-Durham-Chapel Hill area of North Carolina. He is also the reviewer of the book *Dynamics GP Cookbook, Mark Polino, Packt Publishing*.

Vaidhyanathan Mohan is a Microsoft Dynamics GP consultant and an enthusiast of all related technologies. He started his career as a Microsoft Dynamics GP developer, gained invaluable experience with that, and became a consultant who now handles implementation, analysis, development, and administration of Microsoft Dynamics GP.

Vaidhyanathan possesses a Bachelor's degree in Mathematics and a Master's degree in Computer Applications. He is an active blogger who blogs about Microsoft Dynamics GP and related technologies. His blog, *Dynamics GP – Learn & Discuss*, has been recognized and added to Microsoft's Community Blog List. His active presence on many Dynamics GP forums is felt as he answers users' queries and shares his knowledge.

I sincerely thank my parents, Mohan and Vijaya, for their unconditional love and sacrifices in molding me. I thank my brother, Karthikeyan Mohan, from my heart. Without his care and inspiration, I would be nothing.

I thank all my peers, who were instrumental in building my career and experience. To name a few, Rajesh Hari, Geeth, Subhash, Suresha, Ravindranath, Sajeesh KA, David Musgrave, Mariano Gomez, Mark Polino, and the entire GP community.

I thank Jimmy Grewal and Prem Nair, for their tremendous guidance and support.

Finally, I thank my wife Rajeswari and my daughter Sreenidhi. They are the meaning to my life. Their love and understanding will forever drive me to learn and achieve more.

Jivtesh Singh is a Dynamics GP consultant, Systems Implementer, and has been associated with Microsoft technologies since the launch of the Microsoft .NET framework. Jivtesh has over 10 years of experience in the development and maintenance of enterprise software using best practices of coding, refactoring and usage of design patterns, and test-driven development.

Jivtesh recently built a Kinect interface to control Microsoft Dynamics GP 2010 R2 Business Analyzer with gestures. Later, he built a part of the GP Future demo for the Convergence GP Keynote event.

Jivtesh has set up a custom search engine directory for the Dynamics GP blog at `www.gpwindow.com` to help with easier access of Dynamics GP resources for the GP community. With MVP Mark Polino, he has also set up a Dynamics GP product directory at `www.dynamicsgpproducts.com`.

Jivtesh's accomplishments include:

- His blog on Dynamics GP – `www.jivtesh.com`
- Jivtesh's custom search engine for GP blogs – `www.gpwindow.com`
- Dynamics GP products website – `www.dynamicsgpproducts.com`

www.PacktPub.com

Support files, eBooks, discount offers and more

You might want to visit www.PacktPub.com for support files and downloads related to your book.

Did you know that Packt offers eBook versions of every book published, with PDF and ePub files available? You can upgrade to the eBook version at www.PacktPub.com and as a print book customer, you are entitled to a discount on the eBook copy. Get in touch with us at service@packtpub.com for more details.

At www.PacktPub.com, you can also read a collection of free technical articles, sign up for a range of free newsletters and receive exclusive discounts and offers on Packt books and eBooks.

PACKTLIB

http://PacktLib.PacktPub.com

Do you need instant solutions to your IT questions? PacktLib is Packt's online digital book library. Here, you can access, read and search across Packt's entire library of books.

Why Subscribe?

- Fully searchable across every book published by Packt
- Copy and paste, print and bookmark content
- On demand and accessible via web browser

Free Access for Packt account holders

If you have an account with Packt at www.PacktPub.com, you can use this to access PacktLib today and view nine entirely free books. Simply use your login credentials for immediate access.

Instant Updates on New Packt Books

Get notified! Find out when new books are published by following @PacktEnterprise on Twitter, or the Packt Enterprise Facebook page.

Table of Contents

Preface **1**

Chapter 1: Microsoft Dynamics GP Architecture **9**

The native user interface **10**
Horizontal 11
Vertical 11
Dexterity overview **12**
Resource Explorer 13
sanScript scripting language 14
Extensive function library 15
Structured exception handling 15
Integrated source code control 15
Built-in Report Writer 15
COM support 16
Graphical forms designer 16
Debugging tools 16
Dexterity design 17
Components of the Dynamics GP application **17**
Start your engines! 18
The launch file (Dynamics.set) 19
The preferences file (Dex.ini) 21
SQLLogSQLStmt 22
SQLLogODBCMessages 23
SQLLogAllODBCMessages 23
Synchronize 23
Workstation=WINDOWS 23
Workstation2 24
OLEPath 25
RememberUser 26
ShowAdvancedMacroMenu 27
ExportOneLineBody 28

The Dexterity Runtime Engine (Dynamics.exe) 31
SQL table and procedure names **32**
Original table-naming convention 33
00000 – Master tables 36
40000 – Setup tables 36
50000 – Temp tables 36
60000 – Relation or Cross Reference tables 36
70000 – Report Options tables 37
80000 – Posting Journal Reprint tables 37
90000 – Miscellaneous tables 38
10000, 20000, and 30000 - Work, Open, and History Transaction tables 38
Stored procedures 39
Current table-naming convention 41
What you see – the User Interface (UI) **43**
Push buttons 44
Note button (record level) 45
Printer icon 45
Zoom fields 45
Lookup button 46
Browse buttons 46
Sort-by List 46
Note button (window level) 47
Additional window elements **48**
E-mail Link 48
Map Link 49
Quantity alert button 50
Multicurrency button 50
Show Details 50
Information button 51
Expansion arrow 52
Help button 53
Summary **54**
Chapter 2: Integrating Application Fundamentals **55**
Defining the project **56**
Changing a window's look or behavior 56
Changing current functionality 58
Creating new functionality 58
Exchanging data between systems 58
Storing additional data 58
Types of integrations 59

Overview of available tools **60**
Dexterity 60
Capabilities of Dexterity 61
Limitations 63
Developer skills required 63
End-user prerequisites 63
Visual Studio Tools for Dynamics GP (VS Tools) 63
Capabilities of VS Tools 64
Developer skills required 64
End-user prerequisites 64
Modifier with VBA (Visual Basic for Applications) 65
Capabilities of Modifier with VBA 67
Developer skills required 67
End-user prerequisites 67
Continuum 68
Capabilities of Continuum 68
Developer skills required 69
End-user prerequisites 69
Extender / eXtender Enterprise 69
Capabilities of Extender and eXtender Enterprise 69
Developer skills required 70
End-user prerequisites 70
DDE \ ODBC \ ADO \ OLE Automation 70
DDE 71
ODBC 71
ADO 71
OLE Automation 72
Capabilities of DDE \ ODBC \ ADO \ OLE Automation 72
End-user prerequisites 72
Integration Manager 73
Capabilities of Integration Manager 73
Developer skills required 74
End-user prerequisites 74
Table Import 74
Capabilities of Table Import 75
Developer skills required 76
End-user prerequisites 76
eConnect 76
Capabilities of eConnect 78
Developer skills required 79
End-user prerequisites 79
Web services 79
Capabilities of web services 80
Developer skills required 80
End user prerequisites 81

Modifying the user interface **81**
Dexterity 81
VS Tools 83
WinForm properties 84
WinForm control properties 84
Modifier with VBA 87
Extender / eXtender Enterprise 88
Forms 89
Detail forms 90
Windows 90
Detail windows 91
Notes 93
Changing or adding functionality **94**
Dexterity 94
Form events 94
Window events 95
Field events 95
Scrolling window events 96
Triggers 96
Five triggers in Dexterity 97
VS Tools 98
Form events 100
Window events 100
Scrolling window events 102
Field events 105
Procedure events 106
Function events 106
Modifier with VBA 107
Window events 108
Modal dialog events 108
Field events 109
Scrolling window events 110
Report events 112
Band events 112
Continuum 113
eXtender Enterprise 114
Adding information not previously collected **114**
DUOS 114
Summary **116**
Chapter 3: Getting Started with Dexterity **117**
Overview of the development process **117**
Installing the software 118
Preparing your development environment 118

Developing the application 118
Creating the chunk file 119
Delivering the final product 120
Preparing the development environment 121
Installing Dexterity and the SDK 121
Modifying the Dex.ini file 122
Creating the development dictionary 123
Moving to test mode 124
Dynamics GP desktop 126
Modifying user security 126
Installing DexSense 129
Installing the Support Debugging Tool (SDT) 130
Blast off! 130
Overview of Dexterity 130
Components of Dexterity 131
Resources and their relationships 132
DataType 133
Format 133
Field 134
Composite 138
Table 139
Form and window 140
Scrolling window 142
Navigating the Resource Explorer 148
Worksets 149
Summary 150
Chapter 4: Building the User Interface 151
Overview 152
Workset 153
Base resources 155
Data types 155
Format 157
Fields 158
Creating tables and keys 159
Tables 160
Customer Contact Master 160
Contact Phone Master 163
Table naming conventions 164
Table options 165
Types of tables 166

Creating forms and windows 167
Maintenance form and window creation 167
Attaching tables 169
Setting window properties 170
Removing window fields 171
Adding fields to the window 173
New scrolling window 180
Lookup form and window creation 181
Window fields 182
Scrolling windows 185
Working with window fields 188
Adding static text 188
Column headings 189
Summary 190
Chapter 5: sanScript – Making It Work 191
Introduction to sanScript 192
Scripts 192
Syntax rules 194
Script flow 194
Script naming conventions 195
Table operations 196
get 198
change 198
remove 199
save table 199
release table 199
copy to table 199
copy from table 200
Creating a record 200
Retrieving a record 202
Customer 202
Customer zoom 203
Browse buttons 206
Updating a record 209
Deleting a record 209
Ranges 210
Setting a range 210
Creating a virtual key 214
range where 218
Scrolling windows 220
Big and Small Line item 222
BrowseOnly windows 224
Lookup windows 225

Editable windows 228
Line events 229
AddsAllowed windows 230
Triggers 230
Form trigger 232
Form trigger registration 232
Form trigger considerations 233
Cross-dictionary considerations 233
Focus trigger 233
Focus trigger registration 234
Focus trigger considerations 235
Cross-dictionary considerations 235
Database trigger 235
Database trigger registration 235
Database trigger considerations 237
Cross-dictionary considerations 237
Procedure trigger 237
Procedure trigger registration 237
Procedure trigger considerations 238
Cross-dictionary considerations 239
Function trigger 239
Function trigger registration 239
Function trigger considerations 240
Cross-dictionary considerations 240
Create your form trigger! 240
Processing procedure 240
Summary 243
Chapter 6: Deploying a Dexterity Solution 245
System requirements 246
General requirements 246
Feature-specific requirements 247
Minding versions and builds 251
Table creation routines 253
Using the SQL Maintenance window 255
Building a utilities window 257
Automatically creating the tables upon launch 263
Completing the application 263
Forms and windows 263
Linking your prompts 264
Linking your lookups 264
Adding tool tips 264
Hyperspacing your lookup buttons 265
Linking your formats 265
Setting your tab order 266
Complying with user interface standards 267

Tables 267
Reports 267
Referential diagnostics 268
Linked prompts 269
Table relationships validation 271
Creating the chunk file 272
Extracting resources 273
Transfer dictionary module 275
Testing in a multi-dictionary environment 281
Chunk doesn't unchunk 281
Testing tools and techniques 282
Additional resources available 283
Distributing the completed application 284
Sending the chunk 284
Windows Installer services 285
Summary 285
Chapter 7: Creating Customizations with Modifier 287
Overview of Modifier 287
Two tools in one! 288
Modifying windows and window fields 288
Launching Modifier 289
The window properties 291
Size 292
Opening position 292
The tab sequence 295
The window layout 298
Modifying the General Entry window 304
Adding and modifying window fields 307
Adding fields to the scrolling window 309
Modifying static text 313
Adding or changing graphic elements 319
Changing global resources 325
Pictures and native pictures 325
Strings 328
Formats 328
Data types 332
Messages 333
Summary 334
Chapter 8: Creating Customizations with VBA 335
VBA overview 336
Components 336

Objects 338
Properties 339
Methods 342
Events 343
UserForms 345
Modules 346
Debugging 346
Setting options 348
Windows and window fields 348
Creating the Summary button 349
Creating the Go To button 350
Adding objects to the project 352
Adding the Vendor Maintenance window 353
Adding additional windows and window fields 354
Using methods and properties 354
Setting field values 356
Cross-dictionary access 358
Referencing the Collections module 358
Scrolling windows 361
Adding a scrolling window to the project 362
Grid events 362
Line got focus 362
Line lost focus 363
Line change 363
Filtering records 363
BeforeLinePopulate 364
Fun with dialogs 367
BeforeModalDialog 368
AfterModalDialog 370
The Dynamic User Object Store 370
Architecture 371
Declaring the objects 372
Retrieving data 372
Saving data 373
Deleting data 374
Deploying a Modifier/VBA customization 378
Creating package files 378
Limitations of packages 380
Editing packages 380
Known issues with Windows 7 381
Summary 383

Chapter 9: Code-free Customization 385
Overview of tools 385
SmartList Builder 386
Excel Report Builder 386
Drill-Down Builder 387
Extender 387
SmartList Builder 387
Getting Started with SmartList Builder 388
Importing the templates 390
Creating a SmartList object 392
Adding tables 394
Fields 397
Field options 400
Currency fields 402
Date fields 402
Integer and long integer fields 402
String fields 402
Calculated fields 403
Calculated field 1: QTY Available for Sale 405
Calculated field 2: List of On Hand QTY 406
Calculated field 3: CONSTANT 2 407
Calculated field 4: CONSTANT 4 408
Restrictions 408
Go Tos 410
Go To: Item maintenance 412
Go To: Item transaction inquiry 413
Granting security to a SmartList Builder object 417
Excel Report Builder 419
Drill Down Builder 422
Extender 429
Overview 429
Extender editions 429
Extender Standard 429
eXtender Enterprise 430
Working with Extender 430
Summary 444
Chapter 10: Creating Customizations with VS Tools 445
Architecture 446
Dexterity Shell 446
Dexterity Bridge 447
Application assemblies 447
Add-ins folder 447

Installing VS Tools **448**
Download it 448
Run the installation 450
Vendor Quick Entry project **451**
Creating the new project 451
Adding the new window 453
Window controls 454
Button 455
TextBox 457
Label 459
RadioButton and GroupBox 460
ComboBox 460
Adding window controls **460**
TextBox controls and properties 460
Label controls and properties 461
Button controls and properties 462
RadioButton and GroupBox 464
ComboBox 465
Accessing dictionary resources **465**
Referencing the application assembly 466
Referencing the namespace 467
Building dictionary assemblies **468**
Dictionary Assembly Generator (DAG) 468
Using the DAG 468
Creating the AddIn assembly 470
Opening your window **471**
Code the action 472
Building and testing your assembly 474
Table operations **474**
Creating a record 475
Retrieving a record 477
Updating a record 478
Deleting a record 478
Clearing the window **480**
Working with ranges **480**
Building and deploying the application **483**
Dynamics GP 2013 consideration **484**
Summary **485**
Chapter 11: Upgrading Customizations **487**
Using the SDK **487**
Script changes 489

Data model changes 491
New tables 492
Deleted tables 492
New columns 492
Deleted columns 493
New indexes 493
Deleted indexes 493
Different data types 494
Different segments 494
Different index columns 494
New RW relations 495
Deleted RW relations 495
Table changes 495
Table changes – summary 496
Table changes – detail 497
Form changes 498
Dexterity **500**
Setting up generic source code control 500
Installing Dexterity Source Code Control Server (DSCCS) 501
Configuring the DSCCS 504
Resolving validation errors 506
Checking in the old dictionary 507
Checking in the old dictionary to start the new project 510
Creating the new development dictionary 511
Making changes to your code 514
Data type changes 514
Field changes 516
Procedure or function changes 516
Table changes 516
Functionality changes 517
Completing the update 517
Converting the data 517
Recreating alternate forms and reports 518
Updating forms and report dictionaries 518
GP 2013 considerations 518
Testing your application and building the update chunk 519
Modifier with VBA **519**
Modifier 520
VBA 520
Environment changes 521
Window changes 522
Report changes 524
GP 2013 considerations 524

Extender and Builder(s) **525**
Extender 525
SmartList Builder and Excel Report Builder 526
Visual Studio Tools (VS Tools) **526**
Downloading the application 526
Opening the solution 527
Rebuilding application assemblies 529
Updating references to the assemblies 531
Fixing the code 532
Building the new solution 532
GP 2013 considerations 533
Summary **534**
Index **563**

Preface

Microsoft Dynamics GP (Dynamics GP) is an exceptional Enterprise Resource Planning system used throughout the world. Released in 1993, it was the first in its class to deliver a solution that was completely re-written to capitalize on the Windows operating system. Dynamics GP was designed and built so that outside developers could easily enhance its functionality by writing add-ons, or so called third-party applications.

In the beginning, you had only one development tool, Dexterity. Today there are so many tools available making it difficult to choose which one is right for you and your project. This book can help you decide by giving you the chance to work with seven of the most popular tools available today.

What this book covers

Chapter 1, Microsoft Dynamics GP Architecture, includes a description of the components making up Dynamics GP and how they interact. This chapter also provides a description of the Dynamics GP table naming conventions, data flow, and elements of the user interface.

Chapter 2, Integrating Application Fundamentals, includes a short description of how an integrating application works and a brief overview of various tools you can use to build an integration. You will learn which tools can modify the user interface or change the functionality of Dynamics GP.

The tools discussed include:

- Dexterity
- Visual Studio Tools for Dynamics GP (VS Tools)
- Modifier with Visual Basic for Applications (VBA)
- Continuum
- Extender and eXtender Enterprise
- Dynamic Data Exchange (DDE)
- Open Database Connectivity (ODBC)
- ActiveX Data Objects (ADO)
- Object Linking and Embedding (OLE) Automation
- Integration Manager
- Table Import
- Web Services

Chapter 3, Getting Started with Dexterity, will teach you how to prepare your development environment for creating an integrating application with Dexterity. You will learn about the various components making up Dexterity and how they work together.

Chapter 4, Building the User Interface, introduces you to your first hands-on exercise in which you will build the user interface for your project. Additionally you will create the tables, forms, windows, and base resources you need for your sample integration.

Chapter 5, sanScript – Making It Work, will teach you the basics of the sanScript language and will continue your integration project. You will bring life to the window you created in *Chapter 4* by adding code to object events.

Chapter 6, Deploying a Dexterity Solution, will help you finalize your Dexterity integration project by creating deployment scripts and building your chunk file. You will then install this chunk file and see your program run in the multi-dictionary environment of Dynamics GP.

Chapter 7, Creating Customizations with Modifier, will help you start a new project using the Modifier. In this project you will modify the user interface you created in *Chapter 4*, as well as numerous alterations to native Dynamics GP windows.

Chapter 8, Creating Customizations with VBA, covers how VBA is used in Dynamics GP. The Dynamics GP object model, VBA object properties, and object events are covered in this chapter. Several projects are included in this chapter that will put code behind the modifications that you created in *Chapter 7* using the Modifier tool.

Chapter 9, Code-free Customizations, shows you how you can create some very functional customizations using no code at all! You will create a customization using each of the tools listed as follows:

- SmartList Builder
- Excel Report Builder
- Drill Down Builder
- Extender

Chapter 10, Creating Customizations with VS Tools, will help you use Visual Studio along with the Software Development Kit for Visual Studio Tools for Dynamics GP (VS Tools). This chapter starts with how to install VS Tools, and ends with your having created a VS Tools add-in for Dynamics GP. The various VS Tools components are described as you work through your project.

Chapter 11, Upgrading Customizations, shows that once you complete your customization you need to know how to keep it up-to-date so that it will work with future releases of Dynamics GP. This chapter guides you through what to do, and how to do it. Included in this chapter are projects in which you will take a Dexterity project and a VS Tools project through an actual upgrade. The following customization types are covered:

- Dexterity
- Modifier with VBA
- Extender
- SmartList Builder
- Report Builder
- VS Tools

Chapter 12, The Wrap Up, asks you the question – so where do you go from here? In addition, it provides a run down on what we covered as well as information on books, blogs, websites, and training that can help you further your skills as a Dynamics GP developer.

Chapter 12, The Wrap Up, is not present in the book but is available as a free download from the following link: `http://www.packtpub.com/sites/default/files/downloads/0264EN_12_The_Wrap_Up.pdf`.

Appendix A, Dexterity Control Types, provides more complete information about the control types supported by Dexterity. Here you'll find images of the control types as well as a brief description of their characteristics and usage. *Appendix A* is not present in the book but is available as a free download from the following link: `http://www.packtpub.com/sites/default/files/downloads/0264EN_AppA_Dexterity_Control_Types.pdf`.

Appendix B, Event Matrix, provides a matrix of the window events across Dexterity, VS Tools, and VBA. Use this matrix to compare how the events of each tool aligns with the others. *Appendix B*, is not present in the book but is available as a free download from the following link: `http://www.packtpub.com/sites/default/files/downloads/0264EN_AppB_Event_Matrix.pdf`

What you need for this book

To complete all of the projects in this book you need to have a basic understanding of the functionality of Dynamics GP. You also need to have had a little experience with some (any) computer programming language to help you understand the concepts of scripting.

In addition, you need the following software:

- Dynamics GP 2010 with the following modules installed, registered, and working:
    - Modifier (installs automatically with Dynamics GP system files)
    - SmartList Builder
    - Extender
    - Fixed Assets
- The current build of the Support Debugging Tool for Dynamics GP 2010.
- Dynamics GP 2010 installation media. You will need this to install Dexterity 2010 and the Dynamics GP Software Development Kit.
- The current build of DexSense for Dexterity 2010.
- Microsoft Visual Studio version 2005 or later installed and working.
- Software Development Kit for Visual Studio Tools for Dynamics GP 2010.

Information on how to obtain the Support Debugging Tool, DexSense, and the Software Development Kit for Visual Studio Tools is provided within the chapter that the tool is discussed.

Who this book is for

This book is for a developer who is just starting to work with Dynamics GP and is looking for a good introduction to what's going on under the covers of the application as well as the customization tools available to enhance it.

This book will introduce you to several of the tools available and give you a chance to work with each one of them. You will get a feel for what it's like to use the tool, and perhaps learn which of the tools you may like to explore further.

This book includes step-by-step instructions for completing small integration projects using each of the tools listed as follows:

- Dexterity
- Modifier with VBA
- Extender
- SmartList Builder
- Excel Report Builder
- Drill Down Builder
- Visual Studio Tools for Dynamics GP

The intent of these projects is not to create complete working applications, nor to make you an expert on any of these applications, but rather to provide insight into how the tool works. Think of it as a plate of appetizers for tasting each of the tools included.

Conventions

In this book, you will find a number of styles of text that distinguish between different kinds of information. Here are some examples of these styles, and an explanation of their meaning.

Code words in text are shown as follows: "You can register a function trigger against any dictionary in the application using the function `Trigger_RegisterFunctionByName()`".

A block of code is set as follows:

```
range clear table RM_Customer_MSTR;

range table RM_Customer_MSTR where
physicalname('State' of table RM_Customer_MSTR) + "'= IL'";
fill window Customer_Lookup_Scroll;

range clear table RM_Customer_MSTR;
```

New terms and **important words** are shown in bold. Words that you see on the screen, in menus or dialog boxes for example, appear in the text like this: "Your new **Additional** menu will be on the **Customer Maintenance** window".

[Warnings or important notes appear in a box like this.]

[Tips and tricks appear like this.]

Reader feedback

Feedback from our readers is always welcome. Let us know what you think about this book—what you liked or may have disliked. Reader feedback is important for us to develop titles that you really get the most out of.

To send us general feedback, simply send an e-mail to `feedback@packtpub.com`, and mention the book title through the subject of your message.

If there is a topic that you have expertise in and you are interested in either writing or contributing to a book, see our author guide on `www.packtpub.com/authors`.

Customer support

Now that you are the proud owner of a Packt book, we have a number of things to help you to get the most from your purchase.

Downloading the example code

You can download the example code files for all Packt books you have purchased from your account at `http://www.packtpub.com`. If you purchased this book elsewhere, you can visit `http://www.packtpub.com/support` and register to have the files e-mailed directly to you.

Downloading the color images of this book

We also provide you a PDF file that has color images of the screenshots/diagrams used in this book. The color images will help you better understand the changes in the output.You can download this file from `http://www.packtpub.com/sites/default/files/downloads/0264EN_graphics.pdf`

Errata

Although we have taken every care to ensure the accuracy of our content, mistakes do happen. If you find a mistake in one of our books—maybe a mistake in the text or the code—we would be grateful if you would report this to us. By doing so, you can save other readers from frustration and help us improve subsequent versions of this book. If you find any errata, please report them by visiting `http://www.packtpub.com/support`, selecting your book, clicking on the **errata submission form** link, and entering the details of your errata. Once your errata are verified, your submission will be accepted and the errata will be uploaded to our website, or added to any list of existing errata, under the Errata section of that title.

Piracy

Piracy of copyright material on the Internet is an ongoing problem across all media. At Packt, we take the protection of our copyright and licenses very seriously. If you come across any illegal copies of our works, in any form, on the Internet, please provide us with the location address or website name immediately so that we can pursue a remedy.

Please contact us at `copyright@packtpub.com` with a link to the suspected pirated material.

We appreciate your help in protecting our authors, and our ability to bring you valuable content.

Questions

You can contact us at `questions@packtpub.com` if you are having a problem with any aspect of the book, and we will do our best to address it.

1
Microsoft Dynamics GP Architecture

You can develop applications that seamlessly integrate with Microsoft Dynamics GP, so that the user will not be able to tell when your code is running versus the original Dynamics GP code. Creating such integrating applications first requires that you have an intimate knowledge of Dynamics GP's architecture. By architecture, we mean the design that allows the product to evolve with technological advancements and embrace the increasing demands for performance and capacity. This lasting design is a framework of interdependent components that weave together to create the application we know today. Understanding the components of Dynamics GP and how they work together will prepare you to design a solution that will evolve alongside Dynamics GP as it meets the demands of this ever-changing market.

This chapter is a 10,000 foot view of Dynamics GP, how it was built, why it was built that way, and how this affects your application. Upon completion of this chapter, you will understand the components that make up Dynamics GP and how your application fits into that architecture. This chapter will provide an overview of the following topics:

- The native user interface
- Dexterity overview
- Components of the Dynamics GP application
- SQL table and procedure names
- What you see - the user interface (UI)

The native user interface

In developing an application exposed to the user, you must give careful consideration to the user's experience. By using objects that mimic the native user interface, the user will see your windows as if they were the native windows of the core application, and in many ways they will be. When Dynamics GP was released back in 1993, it was platform and database independent. It would run on both Macintosh and Windows clients. It would run on a Faircom Server as handily as it would run on an NT Server. In the beginning, Dynamics GP supported three database management systems: Btrieve (later named PSQL 2000), FairCom's c-tree Plus, and Microsoft SQL Server (starting with Dynamics GP release 3.15).

While today Dynamics GP only supports Microsoft SQL Server (SQL), be mindful that Dexterity continues to support the legacy databases. In building your own solution, these other databases are available for you to exploit. For instance, many developers use the c-tree database to create local temp tables. They are faster to access than SQL tables and easier to create. As an example, the On-Line Field Descriptions are held in c-tree tables on the workstation (`OLFD001.dat` and `OLFD001.idx`), as well as the AutoComplete data (`AutoCmpl.dat` and `AutoCmpl.idx`). Dynamics GP has a unique architecture that lends itself to functional enhancements created by third-party developers. In fact, third-party add-ons were encouraged from day one in order to increase the software's appeal in different industries. From the beginning, the plan was for Dynamics GP to provide the base, the foundation, and for developers to build on that base to create cohesive integrating applications. New products are released daily that expand and complement the core functionality of Dynamics GP. Today there are upwards of 5,000 third-party products that have been registered as Dynamics GP add-ons.

Customizations are typically called *add-on* or *third-party* products because they supplement Dynamics GP functionality. These add-on or third-party solutions are developed and distributed by Microsoft Partners worldwide.

While customization opportunities abound, the source code, screen designs, and report constructions are protected from permanent changes. Because of the distinctive architecture of Dynamics GP, a developer can neither see nor modify the original code, so the business logic is not directly altered. You can build a completely integrated application with no access to the source code. In other words, you as the developer cannot hurt the original code by creating add-on modules or a new version of a window or report.

Add-on solutions are broadly grouped into two categories:

- Horizontal
- Vertical

Horizontal

A horizontal add-on solution supplements one of the existing functions of Dynamics GP or adds a new function that is not targeted at a specific industry. Horizontal solutions can be used in nearly any industry and have a broad appeal. For example, an application that monitors user activity to log the user out after a certain period of inactivity, or an application that allows you to find a specific record using multiple search criteria, would be examples of *horizontal* solutions. Horizontal solutions work together to make Dynamics GP a feature-rich product with near-unlimited options.

A sampling of available horizontal solutions would include the following:

- Collections management
- Sorted lookups
- Commissions management
- Login management
- Advanced password controls
- Cross-company period close controls
- Tools for finding field and table information
- Navigation tools for finding windows
- Search tools for locating reports
- Helpers for designing user security
- Task schedulers
- Pop-up notes

Vertical

A vertical add-on expands Dynamics GP's functionality to focus on a specific industry. Specialized applications for healthcare, manufacturing, publishing, life sciences, banking, and so on are examples of vertical add-ons. These are also known as industry-specific oriented solutions. Some vertical solutions morph Dynamics GP into looking like a program that was designed from the beginning to satisfy the razor-sharp needs of a specific trade. This ability is part of what makes Dynamics GP so fascinating to work with. You can take a generalized Enterprise Resource Planning (ERP) system and turn it into what looks and feels like it was a customized solution from the ground up.

A sampling of available vertical solutions would include the following:

- Field service
- Project management
- Construction
- Education
- Government
- Healthcare
- Retail
- Professional services
- Not for profit

To find out more information about existing solutions, try Microsoft's Solution Finder at `http://pinpoint.microsoft.com/`. Solution Finder is an online tool you can use to search for an existing third-party application to fit your business need. Another excellent source of information is ISV-Central at `http://www.isv-central.com/search/products.aspx`.

ISV Central is a community resource designed to share product information between partners and customers.

Both sites will provide you with product overviews as supplied by the software publisher (no independent analysis here), links to the website of the developer, and information as to whether the software is certified for Microsoft Dynamics (CfMD). CfMD products have met Microsoft's highest standard for partner-developed products. Having the CfMD logo beside your solution is a very high achievement indeed.

Looking for an existing product is one of the steps often overlooked by even the best programmers. It's always a good idea to make sure someone else hasn't already written the application before you get too involved in coding it!

Dexterity overview

Dexterity is more than a programming language; it is a comprehensive Integrated Development Environment (IDE). Dexterity is the tool that was used to create Dynamics GP in the first place. This proprietary tool was created by Great Plains Software, Inc. in the late 1980s. While most other financial application vendors were using toolsets developed in Visual Basic, Dexterity was written in C++ with the .NET elements written in C#. This is significant because Visual Basic is only available from one vendor, Microsoft. On the other hand, C++ is a standards-based programming language. As a standards-based language, you can purchase it from any number of vendors.

C++ is compatible with C and other languages, thereby offering portability and modularity. Being a standards-based language means there is an actual ISO standard defining the C++ architecture. You can purchase a copy of the latest currently published C++ standard at `http://goo.gl/DP72P`.

This section describes Dexterity in general and the features comprising it.

Resource Explorer

The Resource Explorer is the *home page* of the development environment. You design the user interface, build the code that defines the business logic, and create reports, all from this single interface. In short, it is from within the Resource Explorer that you create every aspect of your Dexterity application. The Resource Explorer is illustrated in the following screenshot:

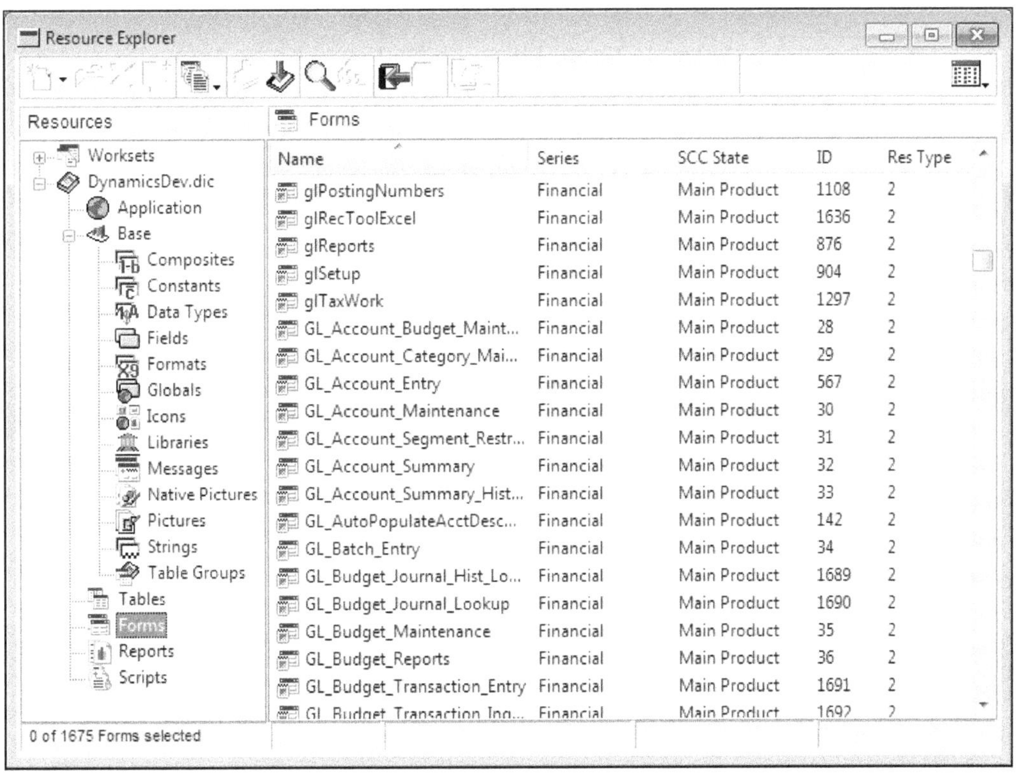

Resources are stored in a file with a `.dic` extension called a dictionary. Each dictionary file includes all of the resources necessary to deliver a complete application. We'll talk about Dexterity resources in more detail in *Chapter 3, Getting started with Dexterity*, but in general, resources include:

- Application Information - the product's name and build number
- Base components:
    - Composites - a group of fields that form a single data type
    - Constants - an alias for another value used in scripts for readability
    - Data types - the type of information displayed or stored in a field
    - Fields - a single element of data
    - Formats - defines how a data type is displayed
    - Globals - fields that are available to any script at any time
    - Icons - a picture resource
    - Libraries - resource libraries and COM type libraries
    - Messages - predefined string values with an associated numeric reference used in scripts allowing for translation and substitution
    - Native Pictures - images that can be displayed on only one platform
    - Pictures - images that can be displayed across platforms
    - Strings - every occurrence of a string used in a dictionary
    - Table Groups - logical collections of related tables
- Tables - a defined group of global fields
- Forms - a collection of windows, scripts, menus, commands, and related tables
- Reports - displaying or printing of a result set from a database query
- Function scripts - global routines using parameters to pass data that always return a value
- Procedure scripts - global routines using parameters to pass data

sanScript scripting language

sanScript is the internal programming language you will use to write the code that provides business logic to the application. Scripts are attached to object events such as opening or closing a window, changing the value of a field, selecting a push button, picking a menu item, and so on. Your scripts run when the user causes the object event to occur. This behavior is what makes Dexterity an event-driven language.

Extensive function library

The function library contains hundreds of specialized functions encompassing 45 different areas that allow you to implement key functionality in the application. Actions such as adding an auto-complete value to a specified field or adding an item to a list view field are examples of functions included in the function library.

Structured exception handling

Dexterity includes a built-in process you can use to recover from and document errors that occur while your application is running. Exception handling allows you to save the current state of execution at a specified point and then switch the execution to a subroutine known as an **exception handler**. The exception handler attempts to resolve the exception and then returns execution to the previously saved position.

Dexterity uses the `throw` and `catch` method to handle exceptions. The exception is raised, or thrown, to the exception handler. The `catch` method is the exception handler taking over code execution.

Integrated source code control

Dexterity supports integration with Microsoft Visual SourceSafe and Team Foundation Server. A generic provider is also included if you want to use another source code control program. The Generic provider uses a collection of text files to implement source control. You will be setting up and using the Generic provider in *Chapter 11, Upgrading Customizations*.

Built-in Report Writer

A single-pass, banded report writer is integrated into Dexterity. You can create simple text reports and lists or you can create presentation-quality documents with graphics and colors. However, since it only passes through the data once, a single-pass report writer has some pretty big limitations. For example, if you need to calculate a sum from which you can derive a percentage that you can show in the detail band of the report, you will need to write your own function to create the sum. Also, you cannot create the simple *page X of Y* notation.

To help with some of the limitations of a single-pass report writer, hundreds of *user-defined functions* are available to you for retrieving or manipulating data in a manner that you otherwise could not. For example, the `RW_AccountNumber` and `RW_AccountDescription` functions return the account number and account description of an account according to the account index you pass to the function.

The downside is that you won't find these user-defined functions documented in the software manuals. You will need to search the Knowledge Base in CustomerSource or PartnerSource to find instructions on how to use several of the functions. Using simply RW in your search criteria will retrieve many articles with instructions for several user-defined functions.

Report Writer also includes 21 built-in (well-documented) system functions that you can use to create **calculated fields**. Functions such as `MONTH_NAME`, used for pulling the name of the month out of a date value, and `STRIP`, for removing trailing spaces from a string field, are both examples of system functions.

As a developer, you can write your own user-defined functions to use with the Report Writer in order to solve problems similar to the derived percentage issue discussed earlier.

COM support

Component Object Model (**COM**) automation is a method you can use to access features and objects across different applications and languages. An object is something that can be acted upon, such as a button or a window. When using COM, one application can cause the other application to do something by following the rules described in an **Application Programming Interface** (**API**). Dynamics GP can provide access to its objects, which can be used by other programs. When it provides access to its objects, it is acting as an automation server. Dynamics GP can also access objects exposed by other applications. When Dynamics GP accesses objects on another automation server, it is acting as an automation client. The thing that allows access to objects is a server, and the one that is accessing the objects is a client. With Dexterity's COM support, Dynamics GP can act as both.

Graphical forms designer

The Graphical forms designer is a WYSIWYG (What You See Is What You Get) tool used to layout and design the visual components of the application. You will create all of the components of the user interface using the Graphical forms designer.

Debugging tools

Dexterity includes a source-level script debugger that you can use for interactive debugging of your application. These built-in tools provide the ability for you to set breakpoints in any script, apply conditions to break points, access scripts that are currently in the call stack, step through scripts one line at a time, display and set the values of fields and variables, and otherwise examine the state of your application at breakpoint.

In addition to the internal debugging tools, a Dexterity-based suite of tools called the **Support Debugging Tool** is available from Microsoft that allows you to examine, with precision, the specific series of events that led up to an error in your application. With the Support Debugging Tool, you can control and monitor all aspects of script execution. And best of all, it's free! Support for this tool is located at the Support Debugging Tool Portal at `http://aka.ms/SDT`.

Dexterity design

Dexterity was ahead of its time, and is still unequaled in the development world. As we said earlier, Dexterity is both platform independent and database independent. Only the runtime engine is dependent on the platform. Standalone applications you develop using Dexterity can run on the Macintosh as well as the PC. Dynamics GP itself, however, no longer supports the Macintosh.

Dexterity currently supports three different databases Pervasive Software's P.SQL 2000 (formerly known as Btrieve), FairCom Corporation's c-tree Plus, and Microsoft's SQL Server. Switching between platforms or databases does not require any changes in the business logic code. If you're using the Microsoft SQL database, you can even use Dexterity to create and call SQL Stored Procedures.

The unique design of Dexterity keeps the technology piece of the application separate from the business logic. This strategy provides a means for Dynamics GP to update its technology without changing its code. For instance, Dynamics GP updated from a 16-bit application to a 32-bit application without re-writing the business logic. Beginning with version 7.0, Dexterity added COM (Component Object Model) support. Again, no code changes were required.

Components of the Dynamics GP application

As discussed earlier, if you are going to build solutions that integrate with Dynamics GP, it is critical that you know how it works. By "know how it works" we don't mean you should learn the intricacies of how to run the general ledger, but rather understand what makes this program tick. Learn how data is exchanged and how you can interact with that exchange. You need to understand the table structure and how to discover the location of the data you seek.

Remember, Dynamics GP is essentially just another database program. When working with any database application, you need to know your CRUD!

- C = Create a Record
- R = Read a Record
- U = Update a Record
- D = Delete a Record

Start your engines!

Let's find out what it takes to launch the application - what happens when a user double-clicks on the Dynamics GP program icon?

If you look at the properties of the Dynamics GP shortcut, you'll see the expected executable file (`Dynamics.exe`) and something extra. The extra is a reference to the `Dynamics.set` file. The full shortcut on my machine looks like this:

```
"C:\Program Files (x86)\Microsoft Dynamics\GP2010\Dynamics.exe"
Dynamics.set
```

There is no path designation for the `Dynamics.set` file because that file is in the same location as the runtime engine. If you place the `Dynamics.set` file in a different location, you would also need to provide the full path to its location.

For example, if you placed the `Dynamics.set` file in the folder `C:\LaunchFile`, the properties of your shortcut would look like this:

```
"C:\Program Files (x86)\Microsoft Dynamics\GP2010\Dynamics.exe" "C:\
LaunchFile\Dynamics.set"
```

Without the `Dynamics.set` parameter, the Dynamics runtime engine will launch and you will get the following window:

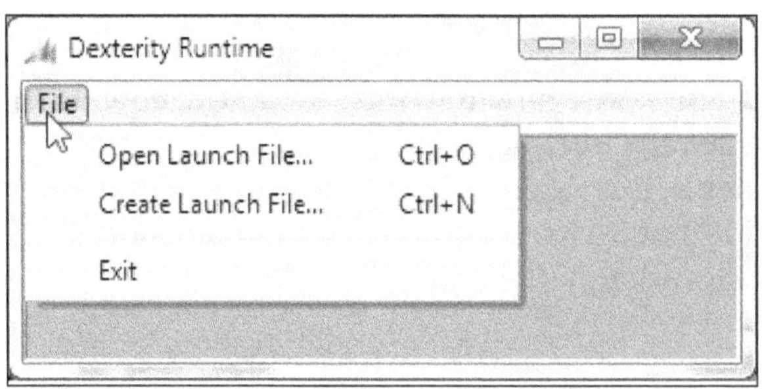

Selecting the **File** menu will provide you with the opportunity to either **Open Launch File...** or **Create Launch File....** Do not select **Create Launch File...** as the file that is created is not usable. Instead, open the `Dynamics.set` file, which is Dynamics GP's proper launch file.

We have just identified two of the files involved in getting Dynamics GP to run. These two files are components of Dexterity. The runtime environment consists of seven main components:

- The **launch** file (`Dynamics.set`) defines each of the components included when Dynamics GP loads
- The **Preferences** file (`Dex.ini`) stores settings such as OLE pathname and the location of the help files
- The Dexterity **Runtime engine** (`Dynamics.exe`) interprets the resources in the dictionary to present a functioning application
- The Application **dictionary** (`Dynamics.dic`) holds all resources in an application
- The **Forms** dictionary (`Forms.dic`) stores forms modified by the user
- The **Reports** dictionary (`Reports.dic`) stores reports modified by the user
- The Microsoft **SQL Server** databases hold all of the data elements of the system and the companies

The launch file (Dynamics.set)

As stated earlier, the `Dynamics.set` file is also known as the launch file. It defines each of the components included when Dynamics GP loads. Several pieces of information are necessary for a successful launch. The combination of information in the `Dynamics.set` file and the `Dex.ini` file identify each of these components. From the `Dynamics.set` file we get the following information:

- The total number of dictionaries used by the Dynamics GP application when launched from this workstation
- The product ID assigned to each dictionary
- Each product's name
- The name of each set of dictionary locations in this `Dynamics.set` file
- The path to and the name of the product's core Application dictionary
- The path to and the name of the product's modified Forms dictionary
- The path to and the name of the product's modified Reports dictionary

The following screenshot shows the various elements of a `Dynamics.set` file:

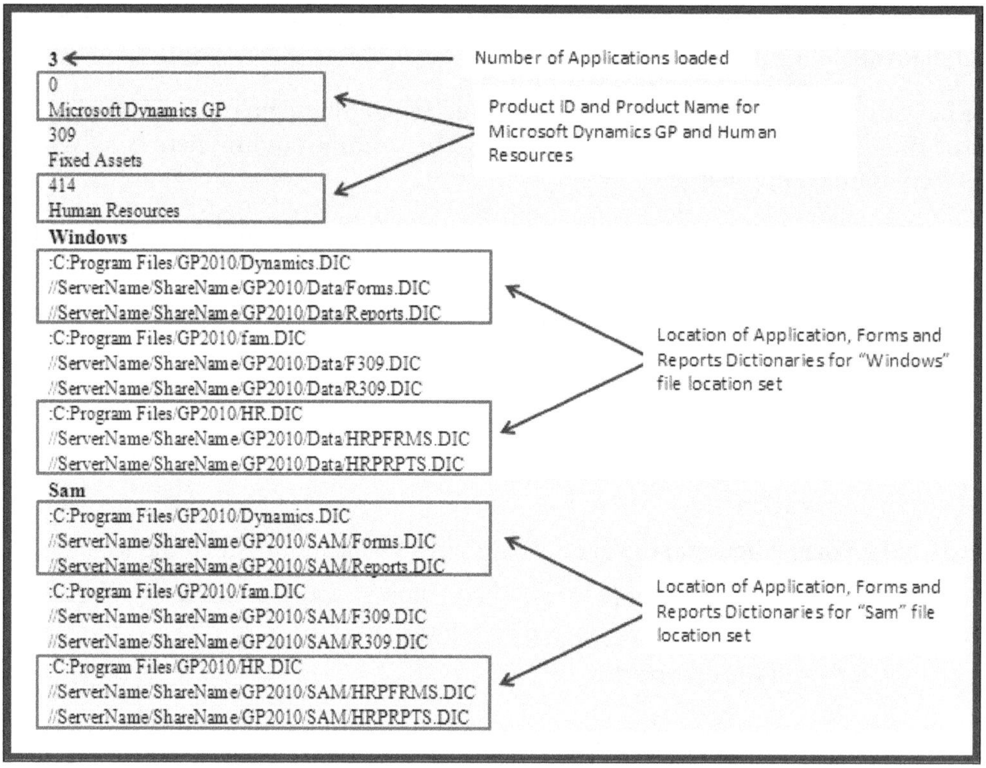

The path locations must be listed in the same order that they appear in the top section of the file. The elements should be in the following order:

- Application dictionary
- Modified Forms dictionary
- Modified Reports dictionary

The `Dynamics.set` file illustrated earlier contains two dictionary location IDs, *Windows* and *Sam*. The `Workstation2=` switch in the `Dex.ini` file indicates which set of paths should be used by the workstation. We'll talk about the `Dex.ini` file in the next section.

The product ID is assigned to the developer by Microsoft; no two products can have the same ID. The dictionary names are determined by you, the developer, but can be changed by the user without ill effect. Any time you refer to an application dictionary in your code, you identify it by the product ID, not the name. The `Dynamics.set` file is a simple text file and can be read and edited using Notepad.

The dictionaries are accessed by Dynamics GP in the order that they are listed in the `Dynamics.set` file. The products may be listed in any order with the exception that the `Dynamics.dic` file must always be listed first, and always has the product ID of zero.

Dynamics GP may respond differently if the order of the products is changed. Be aware of this possibility as you test your add-on application.

If there is a deviation in the order across different products' dictionaries, Dynamics GP will not launch. If there is a deviation in the order of the dictionaries for the same product (Application dictionary, Forms dictionary, Reports dictionary), Dynamics GP will launch, but upon accessing the resources, an error message will display. Study the position of products and dictionaries in the `Dynamics.set` file carefully if unusual errors are presented at launch.

The preferences file (Dex.ini)

The `Dex.ini` file lives in the `Data` folder inside of the `GP2010` folder. In releases 9 and before, the `Dex.ini` file was located in the same folder as the `Dynamics.exe` file.

The location of the `Dex.ini` file in a default 32-bit installation is as follows: `C:\Program Files\Microsoft Dynamics\GP2010\Data\Dex.ini`.

The location of the `Dex.ini` file in a default 64-bit installation is as follows: `C:\Program Files (x86)\Microsoft Dynamics\GP2010\Data\Dex.ini`.

The `Dex.ini` file is also known as the *Preferences* file. It works the same as all other `.ini` files in that it stores settings that are later read by the application in performing some task. Like the `Dynamics.set` file, you can read and edit the `Dex.ini` file using Notepad, as it is also a simple text file.

The `Dex.ini` file is divided into sections by a bracketed word and each section contains zero or more lines of *switches* that control Dynamics's behavior. Some switches are mandatory and are typically added by the application itself; other switches are optional according to user preferences; still other switches are included by the application that do nothing at all. The bracketed section names should never be changed, but you have a wide variety of switches to choose from.

An excerpt of a typical `Dex.ini` file is shown in the following screenshot. Explanations will be included for each of the `Dex.ini` switches enclosed in a box. An extensive list of `Dex.ini` switches is available at `http://dynamicsconfessions.blogspot.com/2010/07/dexini-switches-now-available-to.html`.

SQLLogSQLStmt

When set to `TRUE`, the SQLLogSQLStmt switch causes the system to create (or append to) a text file named `DEXSQL.LOG` and logs all SQL statements being sent to the SQL Server. This file is created in the same folder that holds the `Dex.ini` file; the `DEXSQL.LOG` file is used to troubleshoot errors in the application.

SQLLogODBCMessages

When set to TRUE, the SQLLogODBCMessages switch causes the system to create (or append to) a text file named `DEXSQL.LOG` and logs all ODBC messages returned from the SQL Server. This file is created in the same folder that holds the `Dex.ini` file. The `DEXSQL.LOG` is used to troubleshoot errors in the application.

SQLLogAllODBCMessages

The SQLLogAllODBCMessages switch is one of those settings that appears in the `Dex.ini` file that does nothing at all. The only setting you need for ODBC logging is `SQLLogODBCMessages=TRUE`. Nearly every technology support person will tell you that this has to be set to TRUE if you want to include ODBC messages in the `Dexsql.log` and that is just wrong. You can delete this setting from the `Dex.ini` file if you want.

Synchronize

The Synchronize switch indicates whether Dynamics Utilities has aligned the `Dynamics.dic` to the **account framework**. The account framework was defined when Dynamics GP was initially installed. `FALSE` signifies that the `Dynamics.dic` has been synchronized. `TRUE` indicates that synchronization is needed. You use the Dynamics Utilities application to accomplish synchronization. The account framework is stored in the `DYNAMICS` database and is not consistent across installations.

Workstation=WINDOWS

The Workstation switch was used to read the Location Translation table (`DYNAMICS.dbo.SY03600`) to determine if any string substitutions needed to be made to the pathnames listed inside the `Dynamics.set` file. You will find this entry in every `Dex.ini` file, but Dynamics GP will launch just fine without it. Although **WINDOWS** is populated as the default setting on windows workstations, the entry was not dependent on the operating system. You could have changed it to CLARK and as long as you had CLARK defined in the Location Translation table, the program would have been just fine. The Location Translation table exists in the `DYNAMICS` database as table `SY03600`.

Workstation2

The Workstation2 switch identifies which set of dictionary locations listed in the `Dynamics.set` file should be used by the runtime engine. Though rarely implemented, the `Dynamics.set` file could contain several sets of file locations. Revisiting our earlier `Dynamics.set` file (refer to the following screenshot), we see that there are two sets of dictionary locations defined: **Windows** and **Sam**. If `Workstation2=Sam` were used in the `Dex.ini` file, any modified reports would come from the `Reports.dic` file at `//ServerName/ShareName/GP2010/SAM/Reports.dic`. Conversely, if the `Dex.ini` file setting was `Workstation2=Windows`, modified reports would come from the `Reports.dic` file at `//ServerName/ShareName/GP2010/Data/Reports.dic`.

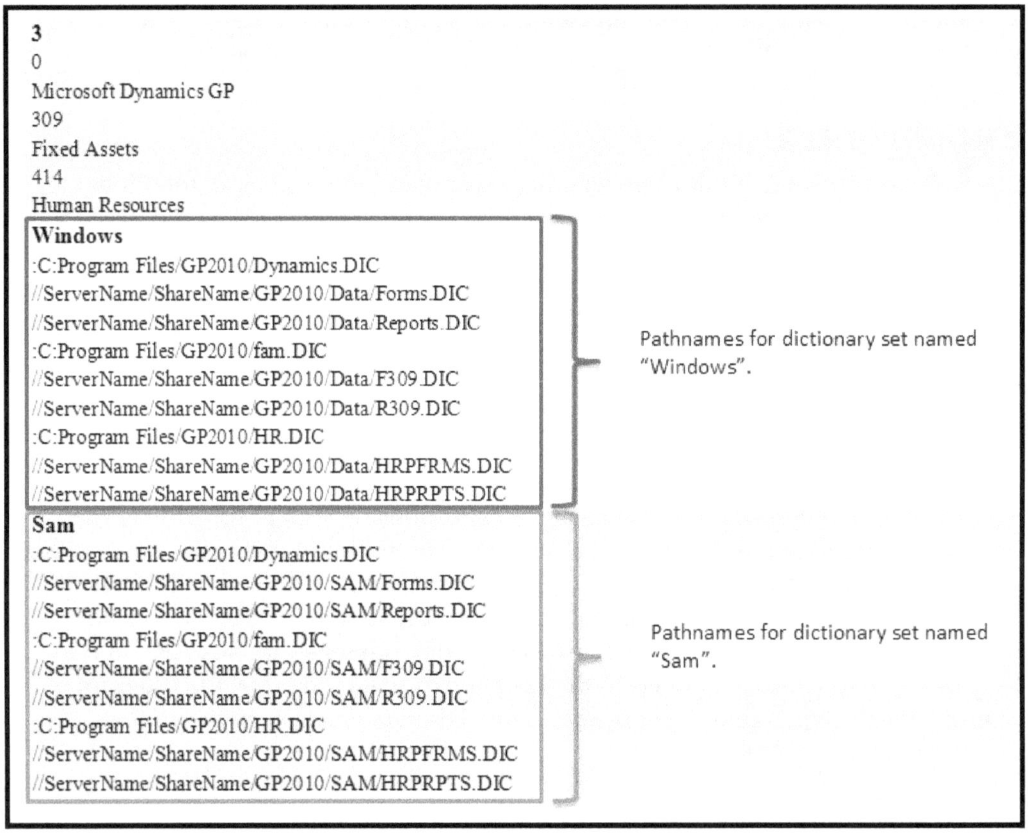

3
0
Microsoft Dynamics GP
309
Fixed Assets
414
Human Resources
Windows
:C:Program Files/GP2010/Dynamics.DIC
//ServerName/ShareName/GP2010/Data/Forms.DIC
//ServerName/ShareName/GP2010/Data/Reports.DIC
:C:Program Files/GP2010/fam.DIC
//ServerName/ShareName/GP2010/Data/F309.DIC
//ServerName/ShareName/GP2010/Data/R309.DIC
:C:Program Files/GP2010/HR.DIC
//ServerName/ShareName/GP2010/Data/HRPFRMS.DIC
//ServerName/ShareName/GP2010/Data/HRPRPTS.DIC
Sam
:C:Program Files/GP2010/Dynamics.DIC
//ServerName/ShareName/GP2010/SAM/Forms.DIC
//ServerName/ShareName/GP2010/SAM/Reports.DIC
:C:Program Files/GP2010/fam.DIC
//ServerName/ShareName/GP2010/SAM/F309.DIC
//ServerName/ShareName/GP2010/SAM/R309.DIC
:C:Program Files/GP2010/HR.DIC
//ServerName/ShareName/GP2010/SAM/HRPFRMS.DIC
//ServerName/ShareName/GP2010/SAM/HRPRPTS.DIC

The Workstation2 switch is required in order to launch Dynamics GP. If it is missing, you will be presented with the following dialog box asking you to select a dictionary location ID. Notice that the system reads the `Dynamics.set` file in order to determine the available choices.

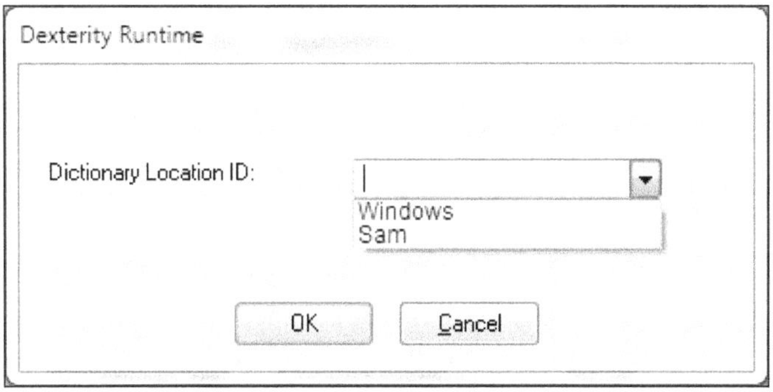

Upon selecting a Dictionary Location ID, Dynamics GP will launch normally. The default setting for the Workstation2 switch is **Windows**.

OLEPath

The OLEPath switch is used to indicate where the OLE object container is located. A user can attach and/or link documents to certain record types in Dynamics GP. The ability to take advantage of this functionality is dependent on the *OLEPath* switch properly pointing to the location of the OLE container. This is an often-overlooked switch. It is either absent, and therefore no OLE support exists, or it is improperly set. You should include some logic in your application to check for this switch if you need to interact with OLE objects.

If the switch is missing, the user is not even aware that OLE attachments are a possibility because the visual queue is missing from the window, as shown in the following screenshot:

If the switch is set to a local path (that is the default), then only one workstation can access the objects. If each user on the system should have the ability to read the OLE objects, then OLEPath should point to a shared location.

If the shared location is a mapped network drive, then each user's profile on each machine they use to access Dynamics GP must include the same drive mapping. While drive mapping may involve a little more diligence in creating the appropriate network login scripts, its advantage is that the directory structure necessary to store OLE notes will be created automatically by the application the first time the OLE container is accessed.

If the shared location is indicated by a Uniform Naming Convention (UNC) path then the full directory structure supporting OLE notes for each company must exist before the workstation attempts to attach or access OLE objects.

That's a lot of rules for a simple switch, but this switch opens up a very powerful feature in Dynamics GP that should be exploited.

To illustrate what the full directory structure looks like, let's assume your installation has two company databases, `DB1` and `DB2`. Let's further assume that you want to store the OLE Notes in `\\ServerName\ShareName\GP2010\Data\` and you want to use a UNC reference for the *OLEPath* switch in the `Dex.ini` file. Since the note folders will not be created automatically if you use the UNC reference, you must create the folders yourself. The following structures are needed:

```
\\ServerName\ShareName\GP2010\Data\DB1\OLENotes\
```

```
\\ServerName\ShareName\GP2010\Data\DB2\OLENotes\
```

If you create these folders manually, your UNC reference in the `Dex.ini` file will work.

RememberUser

The RememberUser switch indicates whether the checkbox on the **Welcome to Microsoft Dynamics GP** window (the Welcome window) is available.

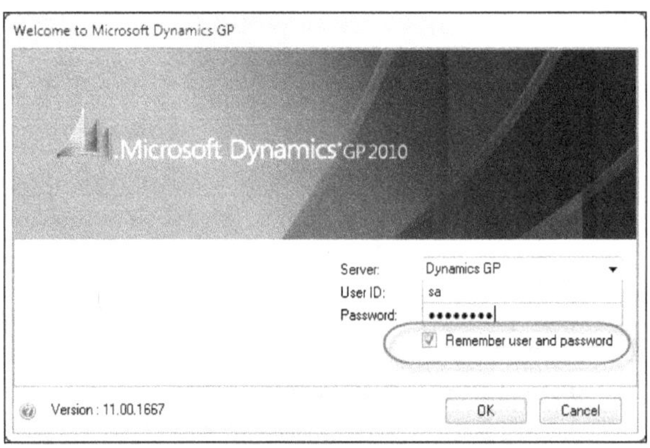

If the setting is TRUE, then the checkbox on the Welcome window will be active and you can click on it and mark it. If it is set to FALSE, then the field will be disabled.

Your ability to click in the box, however, does not change whether the user and password are in fact remembered. If the option **Remember the user and password** has not been selected in the **System Preferences** window, the fact that you checked the box on the Welcome window has no impact on system behavior. When you close Dynamics GP, the Dex.ini setting is changed to FALSE if the System Preferences object has not been checked. It's odd behavior, but it's good to know that changing the Dex.ini setting doesn't cause the user and password to be stored.

ShowAdvancedMacroMenu

When the ShowAdvancedMacroMenu switch is set to TRUE, an **Advanced** menu item is revealed under **Microsoft Dynamics GP | Tools | Macro** that contains several additional utilities which you can use to work with Macros. Refer to the following screenshot:

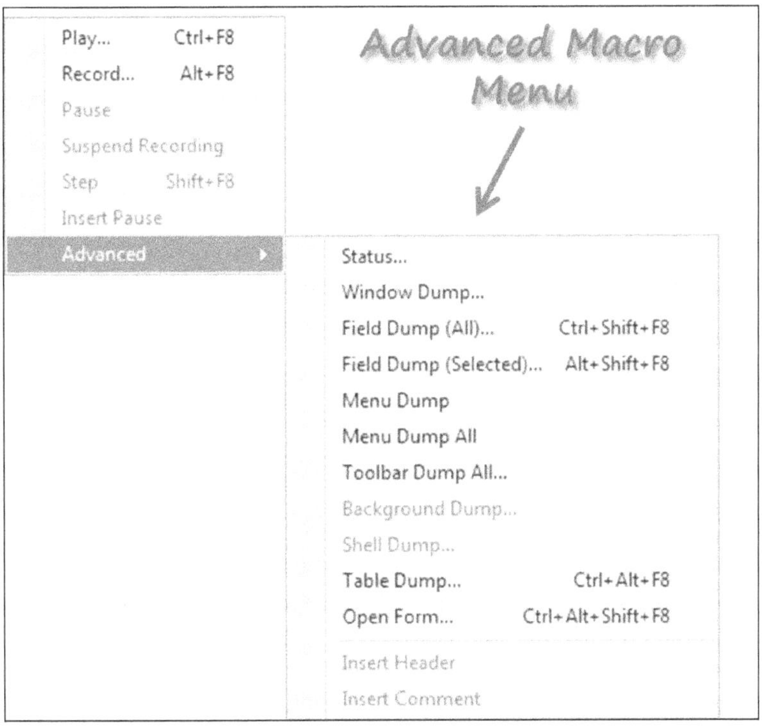

Since the macro language built into Dynamics GP is used extensively in testing, it is definitely something you should learn to use. `ShowAdvancedMacroMenu = TRUE` should be automatically added every time you set up your Dynamics GP environment.

One of the handiest Advanced Macro utilities gives you the ability to jump to a specific line in a macro. This is especially helpful if your macro should fail (imagine!). The dialog that comes up tells you what line it failed on; if it's line 587, it could be a long day trying to count down to that line. The Macro Status tool lets you jump to that specific line to inspect your macro.

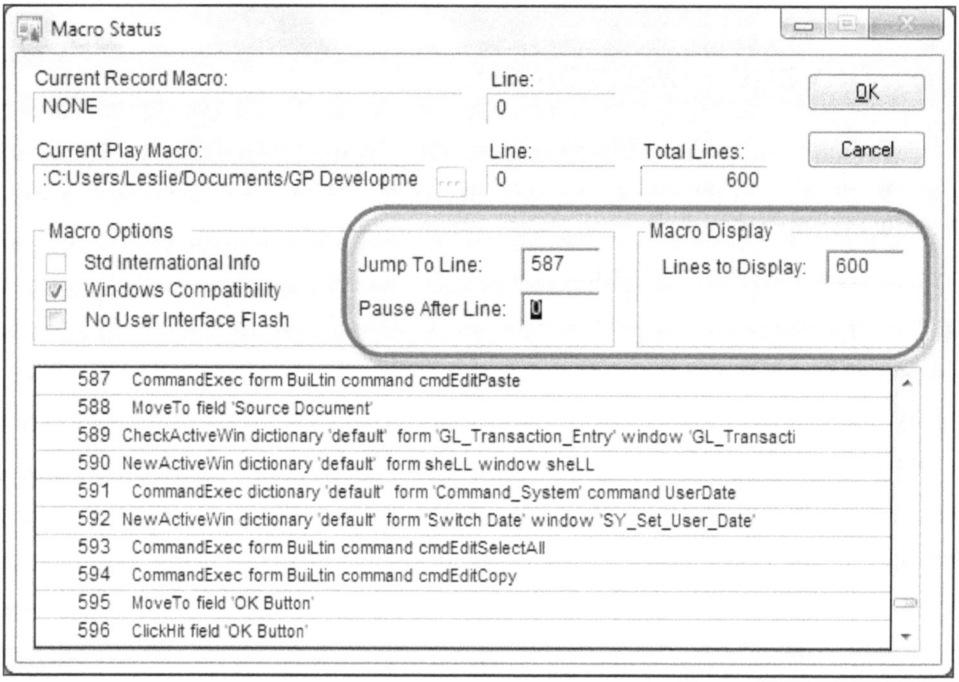

ExportOneLineBody

When the ExportOneLineBody switch is set to `TRUE`, any fields in the body of a report in Report Writer come out as a single line when printing the report to a file. It isn't displayed on the window like that, but in the file it is. This is very helpful if you want to save a report to a spreadsheet-friendly format, such as a tab-delimited file for instance. The Report Writer layout window is only as wide as the paper size set under **Printer Setup**. It is very easy to run out of space if you have to set the fields horizontally across the layout. An old trick to get more space was to load print drivers for a plotter because plotter paper was so much wider. With this switch in the `Dex.ini` file, you can simply list the fields vertically and they will come out as a single row once the report is printed to a file.

There are some benefits to using Report Writer to export data. One, it doesn't require that you have access to a SQL tool, and two, you can make use of data in temporary tables, and you can embed Excel formulas into the report.

The following screenshot shows the Report Writer view of a short sales report. Notice that the last four lines of the screenshot are really Excel formulas. When the report is printed to the screen, it looks much like the layout, only with data. However, if you look at the file, all of the fields come out in a single row per record. If you open this file using Excel with the R1C1 reference style, the hardcoded fields become valid Excel formulas.

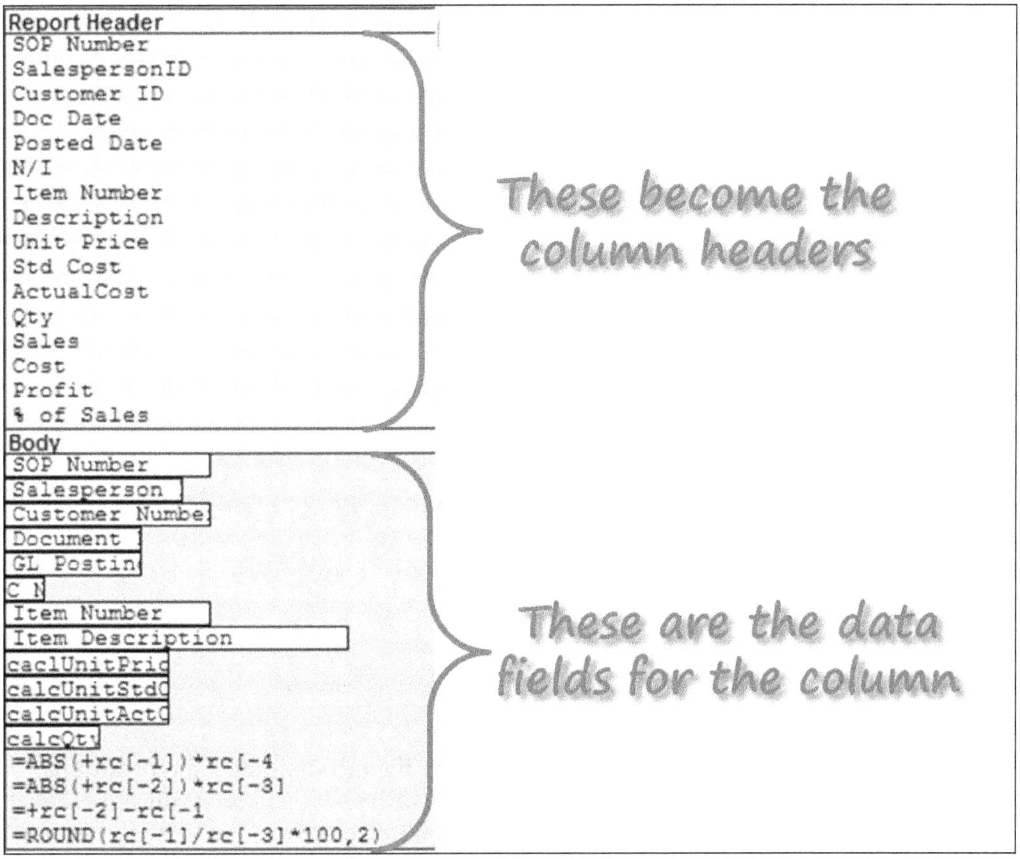

The following screenshot represents what the report looks like when printed to the screen. The screen output follows the same form as the report layout.

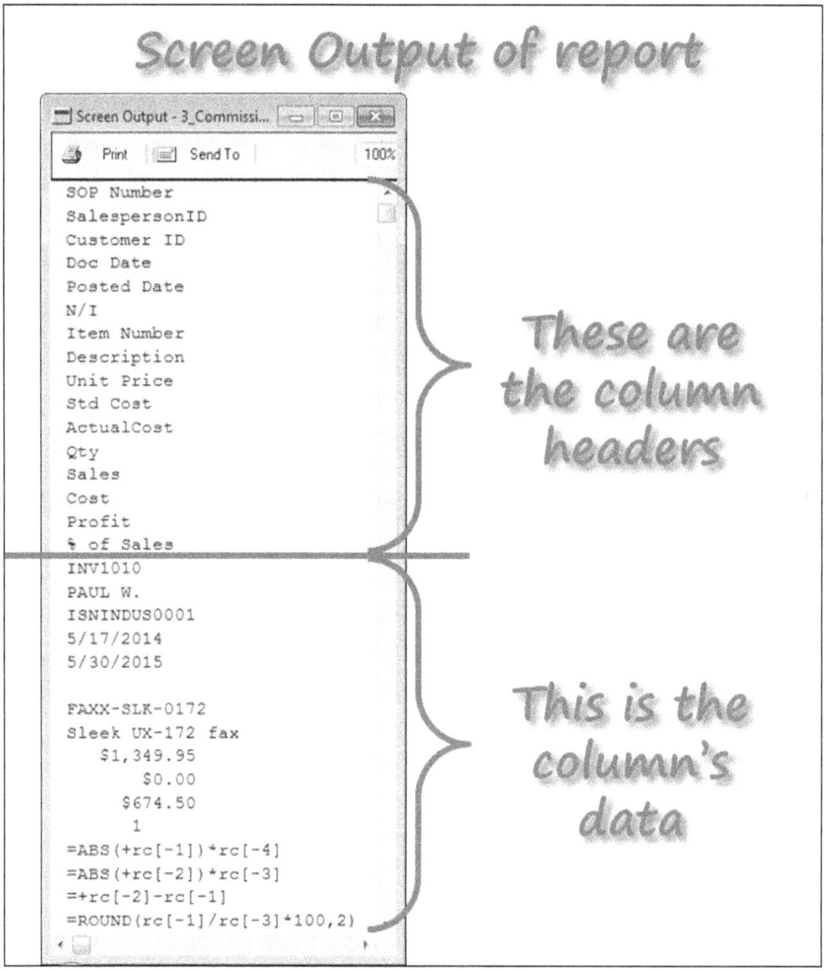

The following spreadsheet is the result of opening the tab-delimited file produced by the report. Notice how the formula `=ABS(+rc[-2])*rc[-3]` coded into the report layout resolves to the formula `=ABS(+L2)*K2` in cell `N2`.

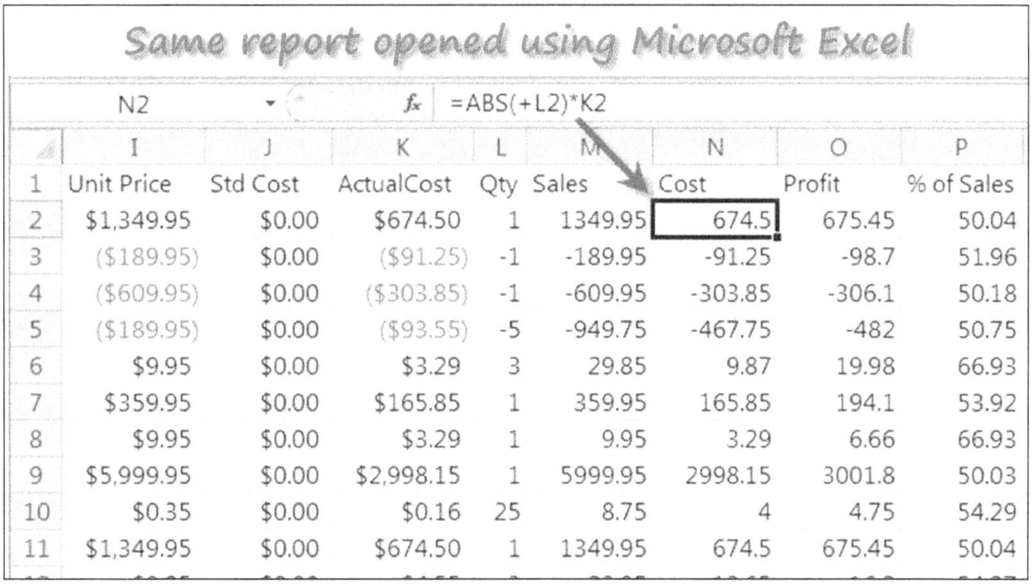

As you can see, these simple `.ini` switches can be used to your advantage when working with Dynamics GP. It's worth your time to find them and engage them to help cut down the time you spend navigating through the complexities of the application. Even better, develop some of your own `.ini` switches to store settings or reveal functionality for your application.

The Dexterity Runtime Engine (Dynamics.exe)

The `Dynamics.exe` file is the runtime engine. Information from the `Dynamics.set` and `Dex.ini` files locate the `Dynamics.dic` file. The `Dynamics.dic` file contains all of the resources and scripting that results in the functioning application when interpreted by the runtime engine. Resources include all of the fields, forms, windows, tables, scripts, reports, buttons, boxes, formats, strings, procedures, functions, and so on. In short, everything, that is the application, you have come to know is Dynamics GP.

Two additional dictionary files, `Reports.dic` and `Forms.dic`, are generated if the user launches the Report Writer (for reports) or the Modifier (for forms). Remember, only user-created (or modified) resources live in these dictionaries. All of the original resources remain in the `Dynamics.dic`.

The following diagram presents the components of Dynamics GP if only the core modules were loaded. In reality, a typical Dynamics GP installation includes nearly 20 dictionaries. We have seen implementations that are comprised of over 40 dictionaries. Each module has its own modified forms and reports dictionaries. You determine what these dictionaries, if created, are named when you build your application's chunk file. The chunk file is a self-installing data dictionary file that extracts to become the application dictionary of your customization.

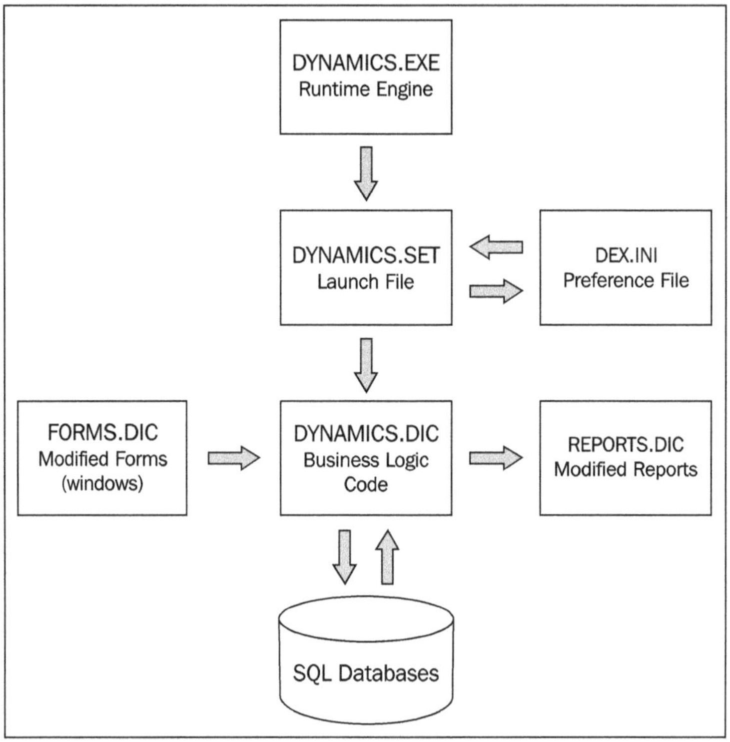

SQL table and procedure names

A typical installation of Dynamics GP adds roughly 500 tables to the `DYNAMICS` database and each company can generate another 2,000 tables. In addition to the tables, there are over 400 views and 20,000 stored procedures per company. That's right, 20,000. This is a big data model to navigate! What makes it even more challenging is that there are no foreign keys defined for the tables. Without foreign keys you cannot easily discern which tables are related to which other tables. As you can see in the following screenshot, a quick scan of SQL Server Management Studio doesn't yield much more information:

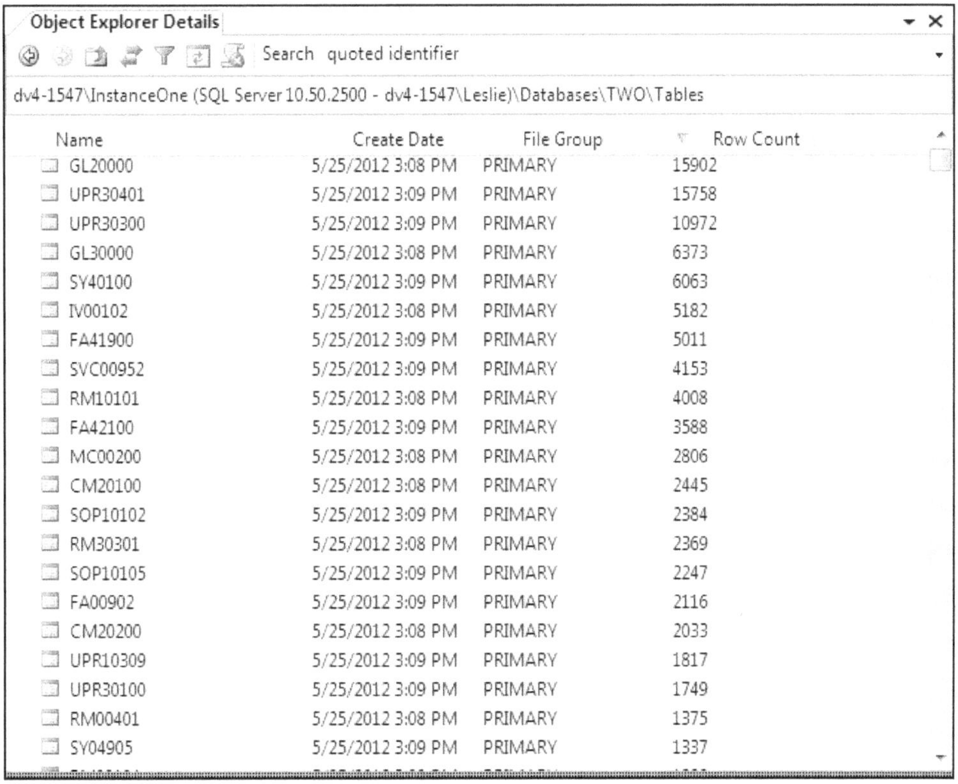

It is no surprise when clients tell me they are very confused by the seemingly cryptic table names. We're so used to those legacy table names that we don't appreciate what a wonder it must be for a new user or developer.

Any effort that involves manipulating data or creating reports will sooner or later result in the question *How do we know which table to use?* This query spawns much gnashing of teeth and all sorts of unusual behavior. So, what do the table names mean?

Original table-naming convention

There is actually a very good naming convention for Dynamics GP data tables. However, this is only a convention. While you will not be arrested by the table-naming police for not following the convention, adhering to a standard makes it easier on everybody. The published naming conventions were followed by the Dynamics GP programmers pretty diligently, but not so much by many third-party developers. Here are the basics: the first two or three characters will indicate the module name, and the remaining numbers indicate the type of table.

Some of the more popular module prefixes are listed in the following table:

Prefix	Module
AA	Analytical Accounting
AF	Advanced Financial Analysis
AHR	Advanced Human Resources
APR	Advanced Payroll
ASI	SmartList Favorites
BM	Bill of Materials
CM	Cash Management (Bank Rec)
DD	Direct Deposit
DTA	Multi-dimensional Analysis
ECM	Enhanced Commitment Management
EDCML	Multilingual Checks
EDCVAT	VAT Daybook
EHW	Employee Health and Wellness
ENC	Encumbrance Management
ERB	Excel ReportBuilder
EXT	Extender
FA	Fixed Assets
GL	General Ledger
HR	Human Resources
IV	Inventory
IVC	Invoicing (NOT Sales Order Processing)
LK	Linked Transactions
MC	Multicurrency
ME	EFT (Electronic Funds Transfer)
MX	Electronic Signatures / Audit Trails

Prefix	Module
PA	Project Accounting
PM	Payables Management (Accounts Payable)
POP	Purchase Order Processing
RM	Receivables Management (Accounts Receivable)
RVLP	Payables Document Management
SLB	SmartList Builder
SOP	Sales Order Processing
SVC	Field Service
SY	System or Company
UPR	US Payroll (Canadian payroll is CPR)
WDC	Field-Level Security (Advanced Security)

After the prefix, the number indicates the table type. Knowing these numbers will help you zero in on the correct table. The following table sets out the numbering convention used to indicate the table type.

Developers often put their company's initials at the beginning of the table name. For instance, table `WDC41101` is the **Advanced Security Setup** table, which is part of the **Advanced Security** module. The Advanced Security module was developed by **Winthrop Dexterity Consultants (WDC)**.

Table number	Description	Abbreviation
00000	Master tables	MSTR
10000	Work tables	WORK
20000	Open tables	OPEN
30000	History tables	HIST
40000	Setup tables	SETP
50000	Temp tables	TEMP
60000	Relation tables	REL

Table number	Description	Abbreviation
70000	Report Options tables	ROPT
80000	Posting Journal Reprint tables	REPRINT
90000	A mixed bag. There is no consistency in this group	

An explanation of how data flows through the WORK, OPEN, and HIST tables deserves its own section; the remaining table types are covered next.

00000 – Master tables

Master tables are mostly what you find under the "Cards" area. These are your Customers, Vendors, Inventory Items, GL Accounts, and the like. For instance, the information you see on the **Customer Maintenance** window is stored in the `RM00101` table.

40000 – Setup tables

Setup tables include choices you have made to initiate a module. For instance, the information entered on the **Payables Management Setup** window is stored in the `PM40100` table.

50000 – Temp tables

Temp tables are tables that are used temporarily by the system and the records in those tables can normally be deleted without issue. For instance, the Net Profit Temporary table or `AF50000` table is used by the Advanced Financial Analysis module to hold the Net Profit amount that will be used on one of the financial statements. Once the statement has been printed, the number is irrelevant.

60000 – Relation or Cross Reference tables

Relational tables are used to store information that spans more than one module. For instance, the SOP/POP link table (`SOP60100` table) holds the information about POP documents linked to SOP documents. Another example is the Sales Customer Item Cross Reference (`SOP60300` table). This table stores how customer item numbers relate to regular item numbers. There are not that many tables in the 60000 range.

70000 – Report Options tables

Report Options tables contain all of the information you enter in any of the report options windows that defines what information you would like to appear on a particular report. The following screenshot comes from the **Payables Trial Balance Report Options** window; information recorded on this window would be stored in the Report Options series of tables (`PM70500`).

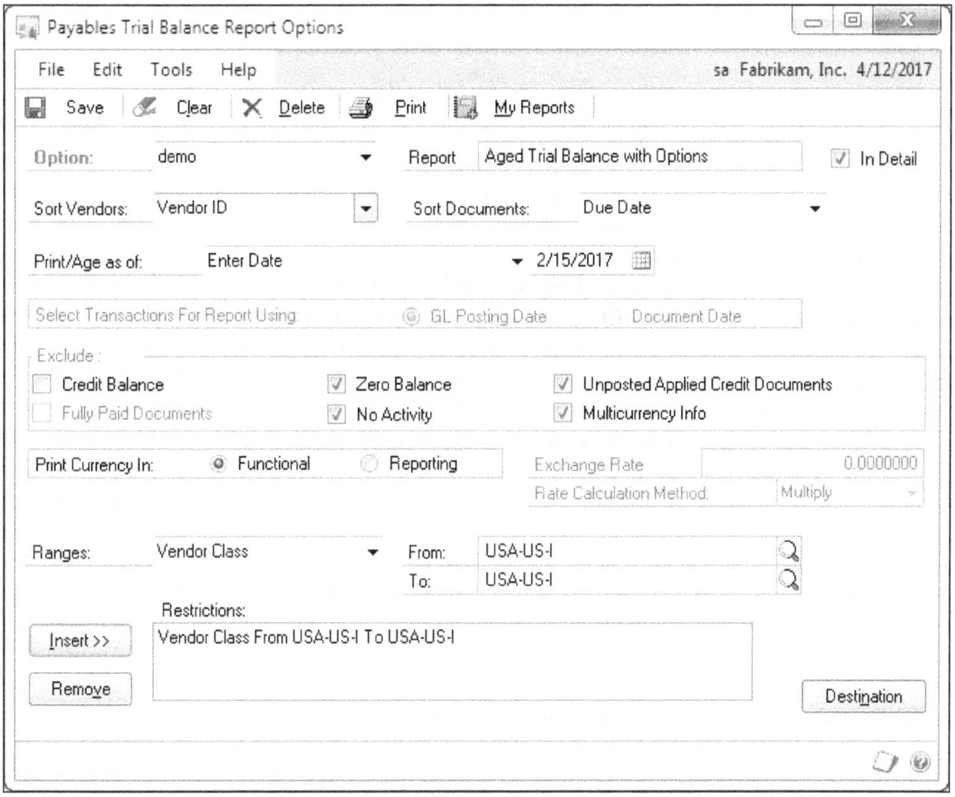

80000 – Posting Journal Reprint tables

Posting Journal Reprint tables contain all of the information you need to reprint the posting journals. So, don't feel like you MUST print all of those reports the system generates after posting a transaction. The data is waiting for you in these tables, should you ever need to reprint them.

90000 – Miscellaneous tables

The Miscellaneous tables group is a mixed bag of information. To give you an idea of what kind of information is included in the 90000 tables, take a look at the following table. There you can see the **Display names** of several 90000 tables.

Physical name	Display name
CM90000	CM Transmission Log
CM90001	Checkbook EFT Log
CN90000	Collections - User Preferences
ERB90100	Data Connection Products
ERB90200	Data Connection Series
ERB90300	Data Connections
ERB90400	Data Connection Restrictions
ERB90450	Data Connection Restriction Values
PA91301	PA Contract Segment Override Header
PA91304	PA Contract Segment Override Detail
PA92301	PA Contract Template Seg Override Header
PA92304	PA Contract Template Seg Override Detail
PDK90003	PDK File Error Log
PDK90100	PDK Security
PP900000	Deferral Opened Periods
PTO90000	PTO Pending Master Conversion
SE90001	Account Rollups Account List Accelerator
SE988977	Account Rollups Options Columns
SLB90000	Third Party GoTo Types
SY90000	SY_User_Object_Store (Dynamics User Object Store)
SY90100	Default Chart of Accounts

10000, 20000, and 30000 - Work, Open, and History Transaction tables

If your customization will be interacting with transactions, you need to understand how the transactions flow through the transaction tables. Generally, there are three phases to a transaction: **Work**, **Open**, and **History**. How transactions move through each of these phases varies by module, with no two modules working exactly the same way, but the concepts are the same.

With a broad brush we will paint the Work phase of a transaction as an unposted transaction. In my world, the term *posted* means to be committed to the ledger. Once a transaction is posted, it cannot be deleted nor can it be unposted. Before it is posted, you can still work on it. Therefore, work transactions are not posted. Transactions in the Work phase are stored in the 10000 tables. For example, if you were looking for an unposted inventory adjustment, you would start by looking in the `IV10000` table.

The next phase of a transaction is the Open phase. What we know about an Open transaction is that it has been posted, but it is not yet in history. Examples of Open transactions include a vendor invoice that has not been paid by the company or an amount due from a customer as an account receivable. Often, transactions that are in the Open phase are called *outstanding* transactions. Open transactions are stored in the 20000 tables. If you were hunting down an unpaid vendor invoice, you would look in the `PM20000` table.

The final phase of a transaction is the History phase. A transaction in history has been closed and settled. No amounts are outstanding, nothing is unapplied, and it is a fully completed transaction. How a transaction moves to history is different for each module; whether a transaction can come out of history and move back into the Open phase also varies by module. What we do know about a transaction in history is that it has been posted and is no longer considered open. History transactions are stored in the 30000 tables. A void sales order would be found in the `SOP30200` table.

Victoria Yudin has some great information on the popular tables from each module at `http://victoriayudin.com/gp-tables/`.

You can find a more thorough explanation of how transactions flow through the Work, Open, and History tables at `http://tinyurl.com/d5townx`.

Stored procedures

Whenever Dexterity is used to create a SQL table, a number of stored procedures are automatically created and are used to optimize database performance when performing table operations. The names of most of the auto-generated stored procedures that apply to normal table operations typically begin with `zDP_`, followed by the table's physical name and a suffix that indicates the purpose of the stored procedure. The following table lists the purpose and numbers of these procedures:

Suffix	Purpose	Quantity
F	First record	One per key
L	Last record	One per key
N	Next record	One per key

Suffix	Purpose	Quantity
SD	Delete a record	One per table
SI	Insert a record	One per table
SS	Select a record	One per key
UN	Unpositioned Next	One per non-unique key

For example, the following table shows that the stored procedures are auto-generated for the PM Class Master file (`PM00100`). This table has three keys; keys 1 and 2 are unique, key 3 is not.

Stored procedure	Function
`zDP_PM00100F_1`	Get the first record using key 1
`zDP_PM00100F_2`	Get the first record using key 2
`zDP_PM00100F_3`	Get the first record using key 3
`zDP_PM00100L_1`	Get the last record using key 1
`zDP_PM00100L_2`	Get the last record using key 2
`zDP_PM00100L_3`	Get the last record using key 3
`zDP_PM00100N_1`	Get the next record using key 1
`zDP_PM00100N_2`	Get the next record using key 2
`zDP_PM00100N_3`	Get the next record using key 3
`zDP_PM00100SD`	Delete a record from the table
`zDP_PM00100SI`	Insert a record into the table
`zDP_PM00100SS_1`	Select a record using key 1
`zDP_PM00100SS_2`	Select a record using key 2
`zDP_PM00100SS_3`	Select a record using key 3
`zDP_PM00100UN_3`	Unpositioned next for key 3 which is a non-unique key. Keys 1 and 2 are unique keys

Other automatically generated stored procedure prefixes include:

Prefix	Purpose
`Smxxx`	System Manager stored procedures found in the DYNAMICS database
`GLxxx`	General Ledger
`glpxxx`	General Ledger Posting
`glpmc`	General Ledger Multicurrency Posting
`Duxxx`	Microsoft Dynamics GP Utilities
`Frlxxx`	Microsoft FRx

Current table-naming convention

Back in the Dark Ages, we had only eight characters to work with for naming tables and fields. Who needed more than eight characters, right? By following the strict but sensible naming convention described in the previous section, thousands of integrating products have been created including tens of thousands of tables. Using the old naming convention, very few duplicate table names have cropped up.

The new table naming convention includes real words and abbreviations that we all can understand instead of the eight characters we were limited to before. Sounds good, but can you imagine working with an application containing 3,000 tables that are not named in any consistent manner? As with most things, when it comes to naming database tables, consistency is a virtue.

While there is still no table-naming-convention police that will visit your office, we think you will find the table-naming convention described next to be a best practice. This new naming convention is both easy to understand and easy to implement. Dexterity programmers will recognize this as the table's **Technical Name**.

Table names will comprise of a module abbreviation, followed by a term that describes the contents of the table, followed by a subtype abbreviation (if appropriate), and then by a main type abbreviation:

```
MODULE_Contents_SUB_MAIN
```

So the General Ledger chart of accounts would be translated to:

```
GL_Account_MSTR
```

The common module abbreviations are the same ones described in the *Original table-naming convention* section. It's from this table that you would get the GL portion of the previous table name.

The following table shows common subtable abbreviations and the type of subtable. For the Asset Financial Detail Master table of the Fixed Assets module, the new physical name would be `FA_Financial_DTL_MSTR`. The DTL portion of the name would come from the following table:

SUB table abbreviation	Subtype
ADDR	Address
BHDR	Batch Header
DTL	Detail
HDR	Header
HTAX	Tax Header
LINE	Line Item
LTAX	Line Item Tax
SERL	Serial/Lot Number
DIST	Account Distributions

The following table shows common Main table abbreviations. For the Invoicing Transaction History table, the physical name would be `IVC_HDR_HIST`. The subtype of HDR comes from the previous table and the Main table type of HIST comes from the following table:

MAIN table abbreviation	Main type
MSTR	Master tables
WORK	Work tables
OPEN	Open tables
HIST	History tables
SETP	Setup tables
TEMP	Temp tables
REL	Relation tables
ROPT	Report Options tables
REPRINT	Posting Journal Reprint tables

Although SQL allows a table name of 128 characters, the Dexterity limit is 80 characters. Adhering to the most restrictive case of 80 characters will help ensure the portability of integrating applications. As a side note, if we need more than 80 characters for a table name, perhaps we need to reconsider our approach.

The previous tables certainly do not represent the exhaustive list of all possible table abbreviations, but they will get you off to a good start. Your fellow developers will thank you for your consistency. Don't forget to add your company's initials at the beginning of the name; this will both help reduce duplicate table names, and make your module's tables easier to spot.

What you see – the user interface (UI)

If you intend to create a user experience, you need to know the design standards so that you can build your interface to appear and behave exactly like the native Dynamics GP windows. Changing from the Dynamics GP native windows to your windows must be seamless. The "look and feel" must be identical.

Just how tall is that **Save** button? How wide is it supposed to be? What is expected from a scrolling window or a lookup window? Have you provided for the addition of a record note or a Linked Lookup? Can you add records "on the fly"? Did you include scroll buttons and expansion buttons?

Not only must your application be bug free, it also needs to be free of distractions such as fat buttons and strange icons. Fortunately, you do not have to guess the answers to the earlier questions. Dynamics GP has each of the properties spec'd out for you. In Appendix F of the Dexterity Basics training manual, the user interface guidelines are covered in detail. Depending on the enhancement plan you've signed up for, you can download the Dexterity Basics training manual from `http://tinyurl.com/btpbnk7`. Contact your partner regarding how to acquire the Dexterity Basics training manual if you cannot download it.

For now, let's review some of the more common window elements that users expect to find on a window. The following image shows the **Vendor Maintenance** window with some of its window elements highlighted.

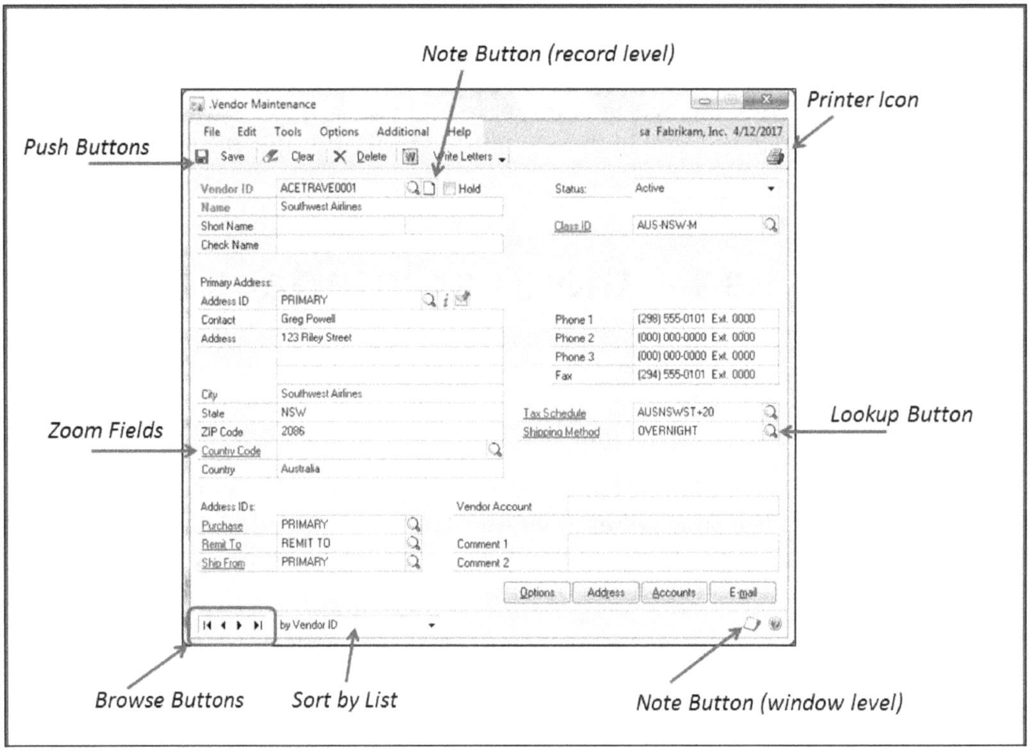

A description of each of the highlighted controls is as follows:

Push buttons

Push buttons appear in the window's control area. Push buttons define actions that you can perform on the record displayed. Depending on the type of window, the buttons expected are as follows:

Window type	Buttons
Maintenance	Save, Clear, Delete
Transaction	Save, Delete, Void, Post
Inquiry	OK, Redisplay
Non-modal Dialog	OK, Cancel
Modal Dialog	OK

Note button (record level)

Record-level notes are big text fields that will hold up to 32,000 characters. The note icon will change appearance to indicate the presence of a note (as shown in the following screenshot). Record-level notes pertaining to all records are typically centrally stored in the company database in the Record Notes Master table (`SY03900`).

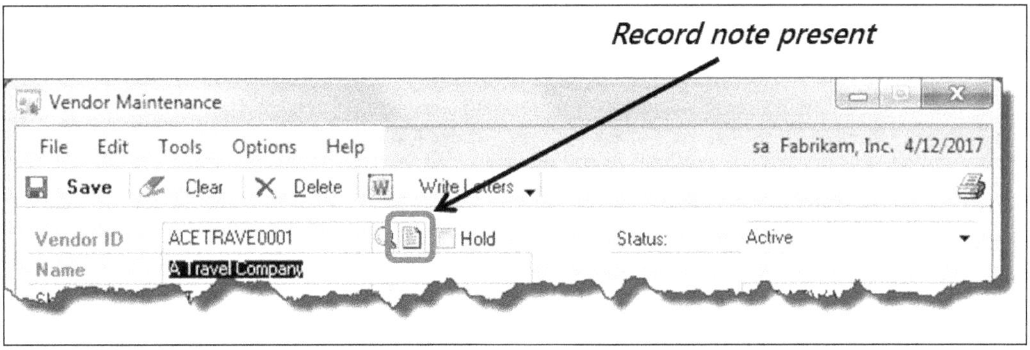

Printer icon

For any given window, a button should typically be available that will print a list report containing the contents of the window. This is the same report that prints from the **File** | **Print** command when navigating the user interface.

Zoom fields

These fields allow a user to drill down to a lower level of detail from the current window. You can identify a zoom field by the blue underlined prompt. When the mouse passes over a zoom field it turns into a zoom pointer. If the field is a master record, the zoom typically goes to a setup or maintenance window for that item. If the zoom field is a financial field, the zoom navigation is normally from a summary level to a more detailed level.

The following screenshot highlights the zoom pointer and the blue underlined prompt:

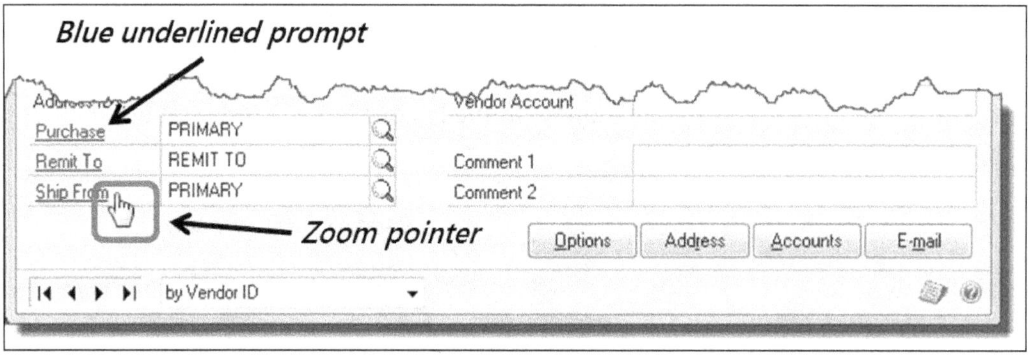

Lookup button

Pressing a lookup button should open a lookup window that contains a listing of records. You select a record on the lookup window and that value will be returned to the originating window. For example, the **Lookup** button next to the **Customer Number** on the **Customer Maintenance** window will open the **Customers Lookup** window. Selecting a customer from the lookup list will bring the selected customer's information back into the **Customer Maintenance** window.

Browse buttons

Browse buttons provide a means of navigation between records in the table. The left-most button will take you to the first record in the table. The right-most button will take you to the last record in the table. The two inner buttons will navigate forward or backward one record at a time.

Sort-by List

Next to the browse buttons, you will normally find a list field that is used to sort by a stated criteria . There is usually one record listed for each key in the table. Choosing a specific sort-by will set the table so that the records scroll according to the key you have selected, and the records will be listed in that order when displayed in the lookup window.

The following screenshot highlights the sort-by list:

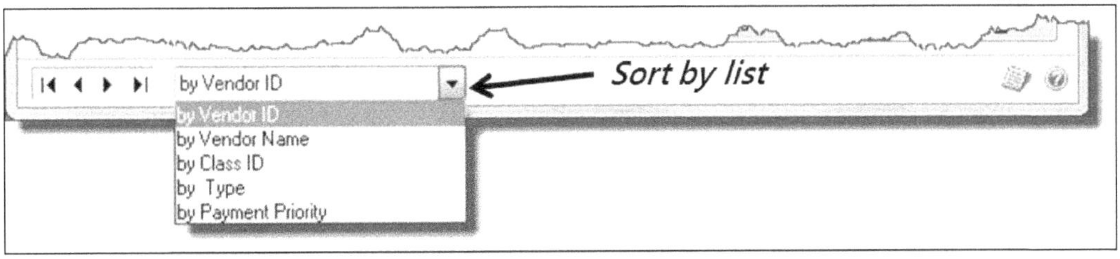

Note button (window level)

Window-level notes provide a place to store information regarding a window. The notes are stored by company so that two companies do not share the same set of window notes even though they each have a similar window. For example, each company has a **Vendor Maintenance** window, but the two windows do not retrieve the same notes.

Window notes provide an excellent place for the user to document instructions on the proper way to fill out a window, to document naming conventions such as Vendor, Customer IDs, and so on. While this element may not directly relate to the functionality of your application, it's one of the elements a Dynamics GP user has come to expect. The icon changes to an image with lines on it when a window note is present.

The following screenshot highlights what the icon looks like when a note is present:

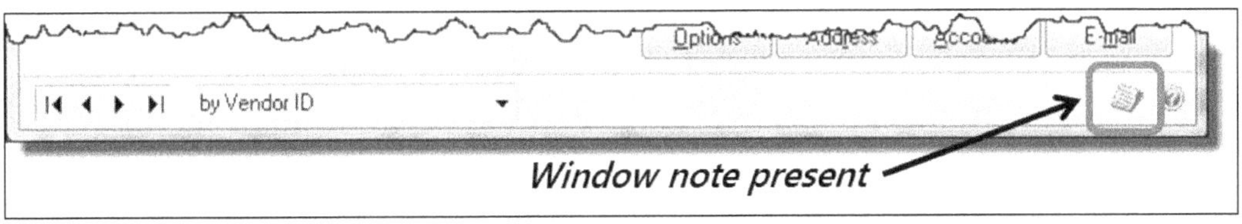

Additional window elements

The following screenshot shows the **Sales Transaction Entry** window with several window elements highlighted:

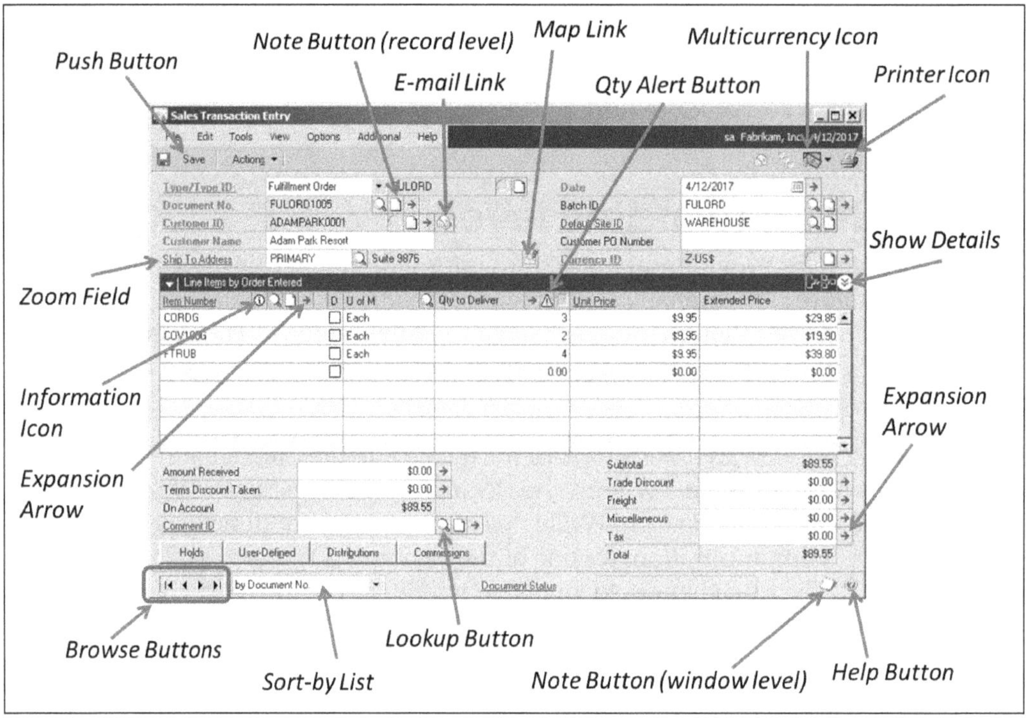

A description of the selected controls is as follows:

E-mail Link

Selecting the E-mail Link button opens the **Sales E-mail Detail Entry** window as shown in the following screenshot:

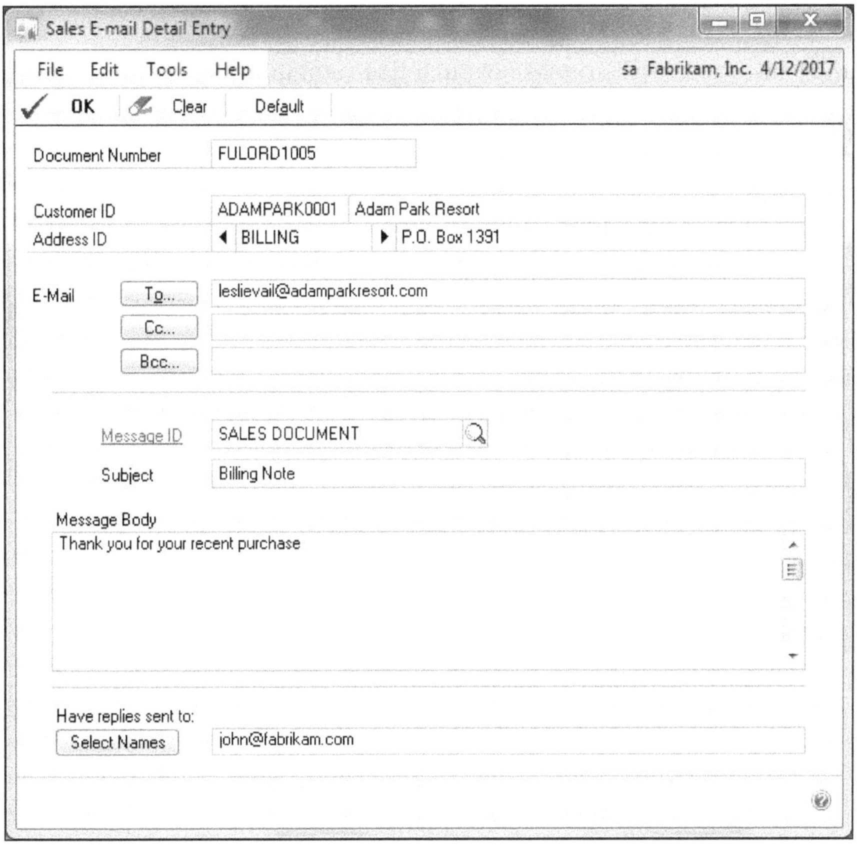

Map Link

Pushing the **Map Link** button will launch Bing Maps and throw it the address listed on the window. This is a very handy feature. If you use an address in your application, it should include this behavior, as it is an expected piece of functionality.

You do not have to use Bing Maps if you don't want to. Mariano Gomez has an article on his blog regarding how to change the default map services. Use this URL to access the article: `http://tinyurl.com/c6vosoy`.

Quantity alert button

The quantity alert icon will appear next to a line-item quantity if you still have a process to complete regarding that quantity amount. For instance, if you have not fully allocated the quantity on an invoice, the icon will show up on that line. By selecting that line and then pushing the **Quantity Alert** button, you will be greeted with a message telling you what the system is unhappy about.

Multicurrency button

By pressing the multicurrency button, you can switch the display between the originating and functional currency. You can also modify the reporting currency rate if applicable. The following screenshot shows you what the window looks like when the multicurrency button is pressed:

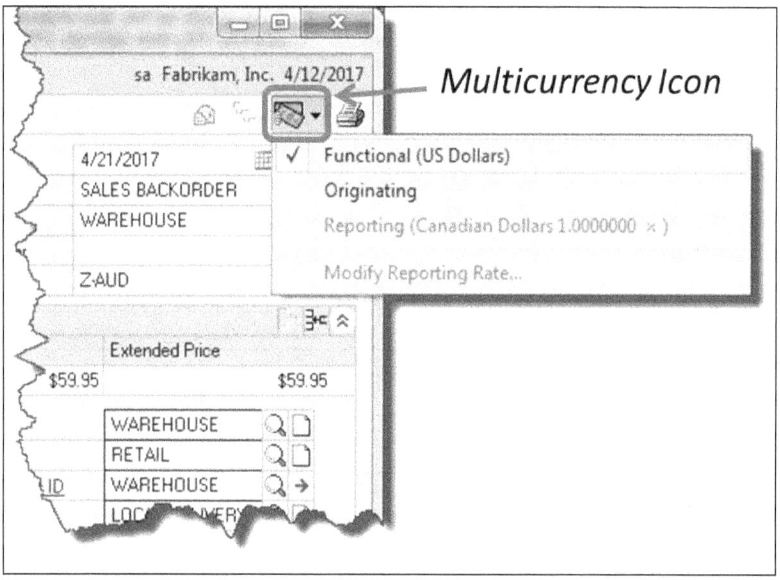

Show Details

The **Show Details** button always shows additional information about the line item selected. These buttons are used to expand and shrink the view of a scrolling window. Sometimes the grid view is just expanded to show more lines of information for a single record. Other times, you'll see a larger area that doesn't look anything like a scrolling window.

In the case of the **Sales Transaction Entry** window, much more than just another row of information is revealed, as shown in the following screenshot. Lately, the **Show Details** view has been used in this way to deliver information using a more attractive layout.

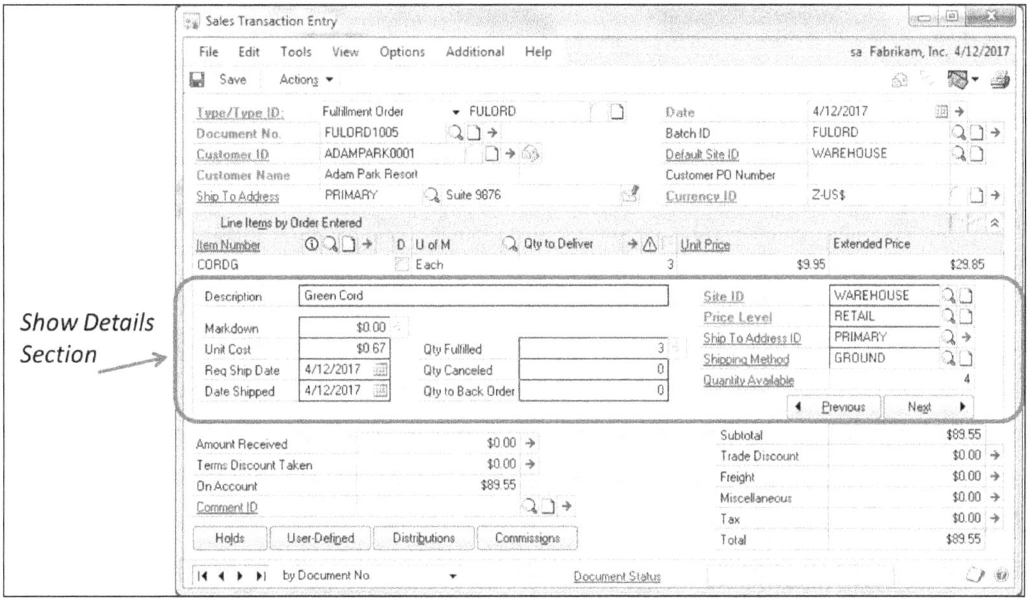

Information button

This button is more fully described as the "line-item information" button. It's used on the **Sales Transaction Entry** and **Purchase Order Entry** windows. This button is identified in the following screenshot:

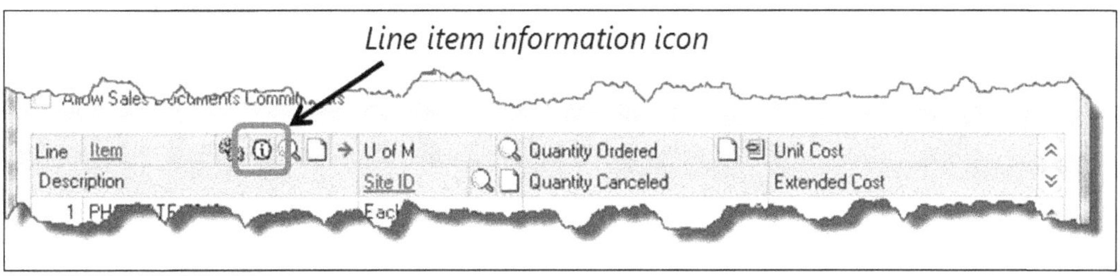

Pressing this button on the **Sales Transaction Entry** window or the **Purchase Order Entry** window opens a **Sales Quantity Status** window with information pertaining to the line item selected. The following is a screenshot of both the **Sales Quantity Status** window and the **Purchasing Quantity Status** window.

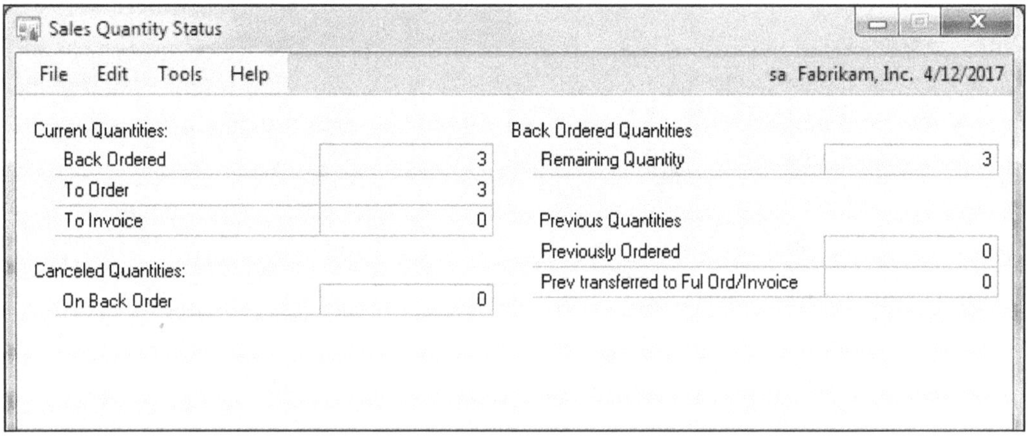

Expansion arrow

Expansion arrows can reveal a whole host of information. You can take a straightforward-looking window such as the **Sales Transaction Entry** window and stuff a massive amount of information in it via the use of expansion arrow buttons. Indeed, the **Sales Transaction Entry** window contains thirteen expansion arrow buttons.

If you need a lot more information about a particular record and want to keep your user interface clean, expansion arrow buttons are a good solution for you.

The following screenshot shows the window that you open when you press the **Item Detail** expansion arrow on the **Sales Transaction Entry** window:

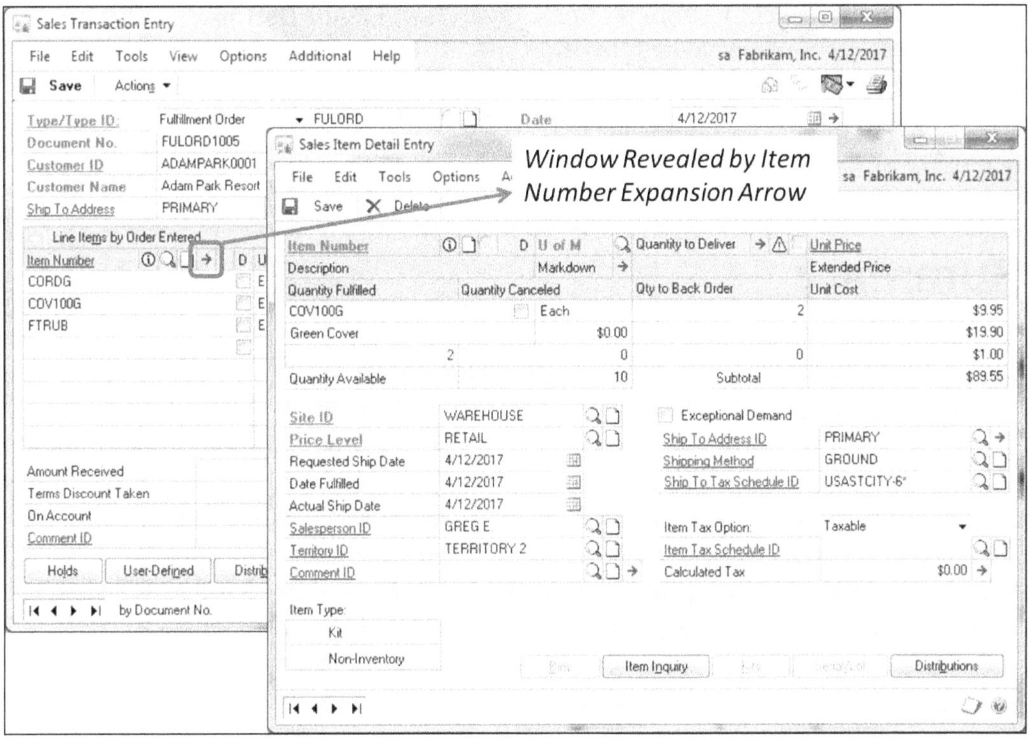

Help button

By pressing the Help button in the lower right-hand corner of the window (or *F1*), you launch a context-sensitive help screen, as shown in the following screenshot:

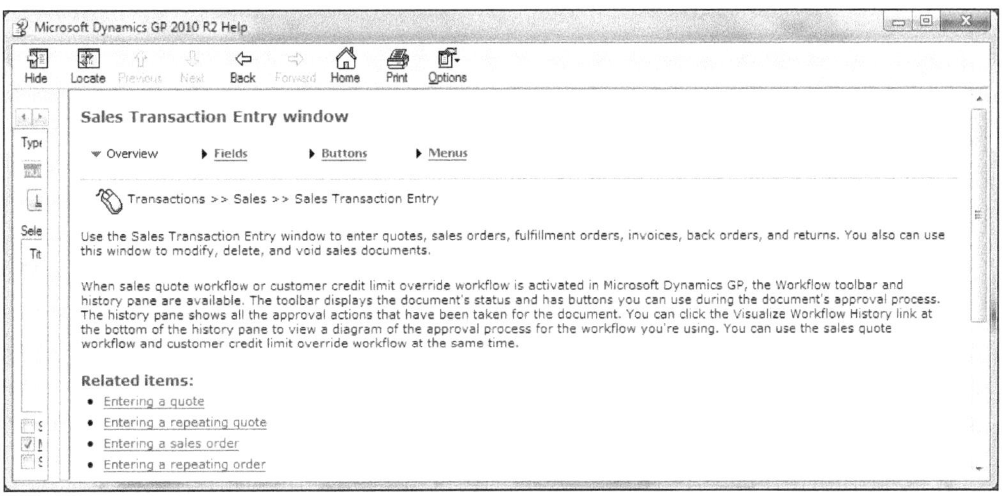

The **Help** window includes four sections: **Overview**, **Fields**, **Buttons**, and **Menus**. Each of these sections reveals information specific to the selected window. For example, the **Fields** section of the **Sales Transaction Entry** window defines, in alphabetical order, each field that appears on the window.

Your help file should also be context sensitive and include similar information.

Summary

In this chapter, we learned that Dynamics GP was designed as a platform- and database-independent application. We were introduced to the foundation that Dynamics GP is built upon and how from the beginning, developers have been enriching its functionality by creating a wide array of integrating applications. We explored the features of the Dexterity toolset and how we can create our own custom applications using the very toolset Dynamics GP was written with.

We learned about the seven main components of the Dynamics GP runtime environment and how they interact. We know the details of what happens when Dynamics GP is launched. We also have a new collection of switches we can use with the `Dex.ini` file to achieve more functionality or change the behavior of Dynamics GP.

The table-naming conventions have been demystified as well as the general flow of transactions through those tables. We know the purpose and naming conventions for the auto-generated stored procedures and that they aid in optimizing the performance of table operations. Finally, many of the standard window elements were introduced, such as zooms and note buttons.

In the next chapter, we will learn the fundamentals of integrating applications.

2

Integrating Application Fundamentals

This chapter will present a series of questions developers should ask themselves, and answer, before beginning development. You will learn about the various tools available for customizing Dynamics GP along with the skills required to use them. We will also go over any additional products the end user may need to implement your solution. Finally, you will learn what you should have in your development-tool arsenal and where to get it.

Key topics in this chapter include:

- Defining the project
- Overview of available tools
- Modifying the user interface
- Changing or extending functionality
- Adding information not previously collected

Defining the project

Before you start installing software and designing windows, you need a plan. It's almost time to search for your flowcharting template, but first you need to answer this query: Just what *are* you trying to accomplish? Let's begin to define the project by responding to some fundamental questions:

- Do you want to change the way a window looks or behaves?
- Do you want to change or extend current Dynamics GP functionality?
- Do you need to create brand-new functionality?
- Do you need to exchange data between dissimilar systems?
- Are you just trying to store some additional static data?

Changing a window's look or behavior

Maybe the field prompts are wrong or the tabbing order is unacceptable. Perhaps there are not enough, or even too many, fields on the window. The window may need additional navigation options such as menus or buttons. You might want a field to be created only if certain criteria are met. For instance, if your customer were a reseller, the **Tax Schedule ID** field would not be required; but if your customer were not a reseller, the **Tax Schedule ID** field would be required. Maybe more visual cues should be present on the window, such as a red/green light indicator on the **Customer Maintenance** window to represent their payment pattern similar to the following screenshot:

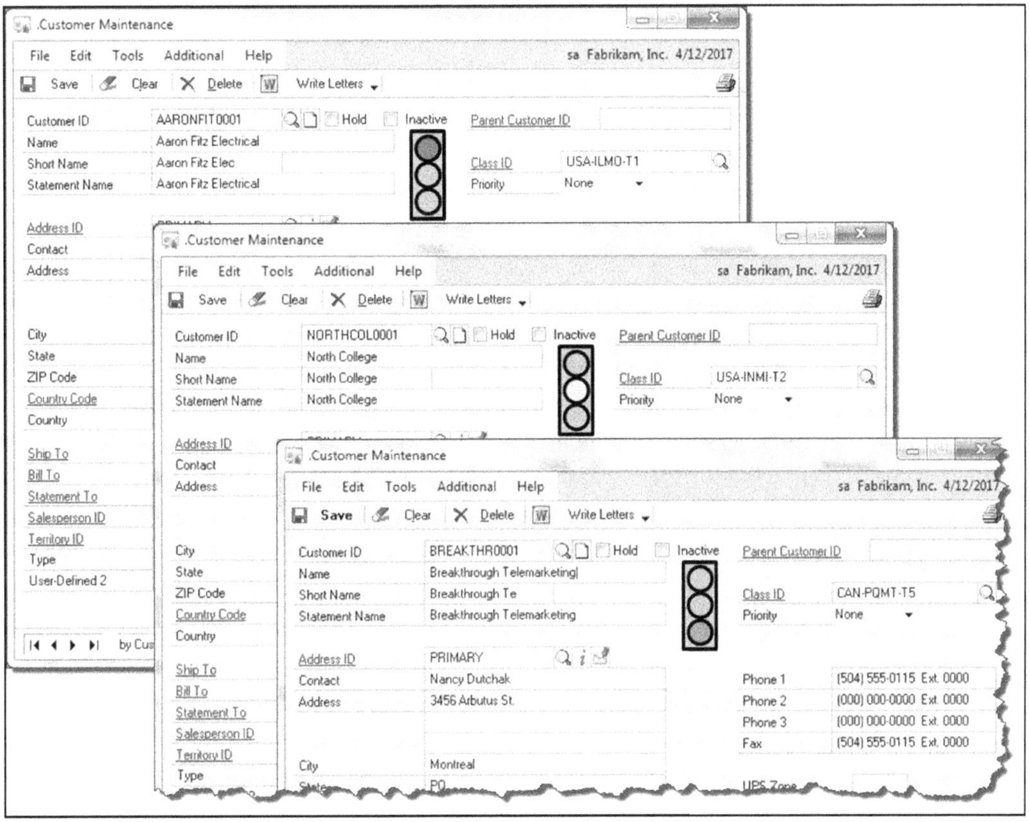

You may want a more obvious cue if a record note exists, like a bigger icon or an icon that flashes! You might want to see the quantity of a stock item available in the **Sales Order Processing** lookup window. This list could go on forever.

Changing current functionality

Many times Dynamics GP is just missing a little something with the way it processes certain transaction types. For example, perhaps you would like to be warned if you are entering a **Payables Transaction** for a vendor with an outstanding purchase order, or you need a receivables document to move to a history table automatically when you pay it, instead of having to run a monthly routine.

This list *does* go on forever.

Creating new functionality

This category is filled with things that Dynamics GP doesn't do at all. As we discussed in *Chapter 1, Microsoft Dynamics GP Architecture*, Dynamics GP was built with these types of customizations in mind. Dynamics GP constructed the foundation, and developers like you make its functionality boundless. Many of our vertical solutions are present here; applications for running mining operations, restaurants, and retail stores have been developed for Dynamics GP. The unique needs of a myriad of industries have been satisfied by third-party applications.

Exchanging data between systems

Very often, you accumulate detailed information in a different system of record and you need to import it into Dynamics GP. Sometimes you need to export information out of Dynamics GP in order to update another system.

For instance, **Point of Sale** (**POS**) systems update the general ledger with daily sales, and payroll services send weekly payroll details to upload into the general ledger. Vendors send new price sheets that cause adjustments in the list price. These list price changes need to update the website as well as the accounting system. A constant stream of data flows back and forth every day and it needs to update other systems, or be updated itself. The goal is to take the information from the point of original entry, and electronically place it wherever it needs to go. We only want to touch the data once; dual entry needs to be eliminated. Our aim is to have only one version of the truth.

Storing additional data

Quite often the fields available for user customization, so called user-defined fields, are far too few. This problem is nearly universal when it comes to the inventory.

Take, for example, a company that trades in high-end audio equipment. For a preamp they may need to know the distortion percentage, the number and types of inputs, outputs available, and so on.

For speakers, they need a completely different set of information, such as sizes and types of drivers. Yes, there is much more information that needs to be at a salesperson's fingertips than the part number and the price.

All of that additional information needs a place where you can enter it, and a table to call home.

Types of integrations

At the end of the day, there are generally two types of integrations:

Database-level integrations include tasks such as the following:

- Importing data into Dynamics GP
- Exporting data out of Dynamics GP
- Storing additional data in new tables
- Synchronizing data between Dynamics GP and peripheral systems

User-interface-level integrations include tasks such as the following:

- Adding entirely new windows
- Adding fields or controls to an existing window
- Adding navigation items to the home page
- Adding new menus
- Changing field locations on a window
- Changing a window's tab order

A single customization often involves both types of integrations.

OK, so now that the interrogation is complete, it's time to find that template and start flowcharting!

Overview of available tools

Once you have defined what you are trying to do, you need to find a tool that will do it. Fortunately, there are many tools available to help you accomplish your goal.

Each tool does something a little different from the others. Which tool to select may be obvious, but often there are several ways to achieve the same result. A hybrid approach may be necessary when no single tool will go the distance.

The several tools at your disposal, which we will discuss in this section, include:

- Dexterity
- Visual Studio Tools for Dynamics GP (VS Tools)
- Modifier with VBA (Visual Basic for Applications)
- Continuum
- Extender / eXtender Enterprise
- DDE / ODBC / ADO / OLE Automation
- Integration Manager
- Table Import
- eConnect
- Web services

The decision of which tool to use requires careful consideration of the type of customization, the capabilities of the tool, the skills of the developer, and any prerequisites or licensing requirements imposed on the end user.

Dexterity

Dexterity is a complete **Integrated Development Environment** (**IDE**) and the native language of Dynamics GP. You can create the tightest, most seamless integrations using Dexterity. All of the resources of a Dexterity integration are stored in a dictionary file. By resources we mean the business logic, fields, windows, table definitions, push buttons, and so on. The runtime engine interprets these resources and presents a single application to the user.

The extensibility of Dynamics GP allows multiple Dexterity integrations to run at the same time, yet appear to the end user as a single program running alone. This unique environment is referred to as a **multidictionary** environment. The following diagram illustrates a multidictionary environment where several dictionaries are functioning together to create the end-user application.

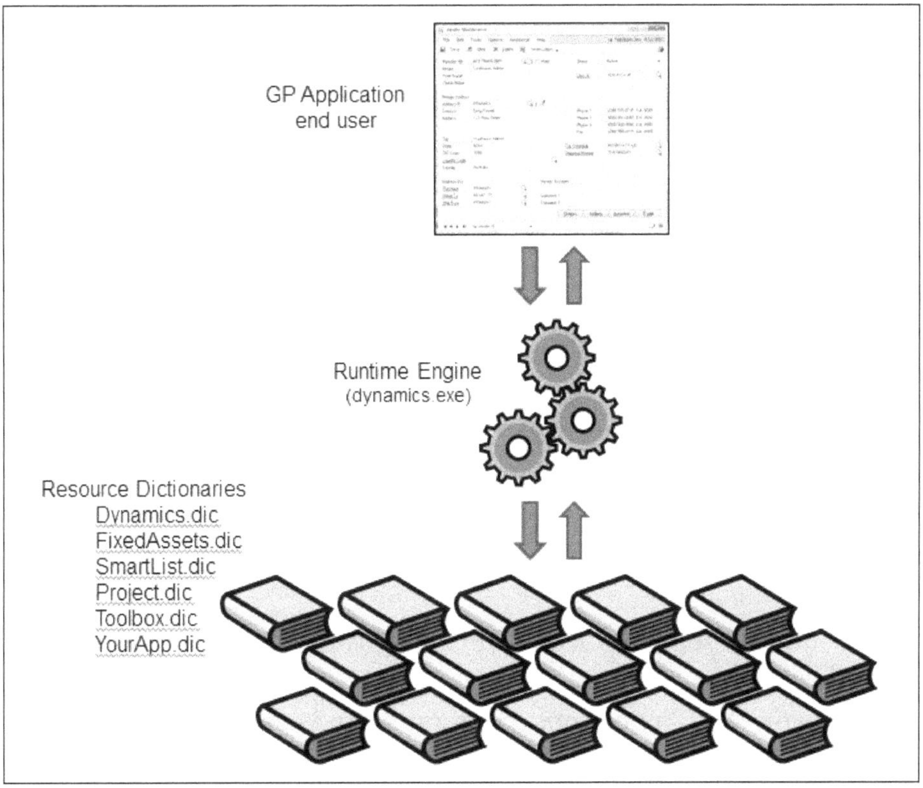

Dexterity Multidictionary Environment

If you need your application to have access to all of the resources in the Dynamics dictionary and behave exactly as Dynamics GP, Dexterity is your tool.

Capabilities of Dexterity

Using Dexterity, you can perform the following:

- Access and manipulate all of the resources exposed by the `Dynamics.dic` file.
- Create new custom windows using the built-in WYSIWYG graphical forms designer in a style indistinguishable from the native Dynamics GP windows.

- Create your own version of an existing window to use in place of the original Dynamics GP window (an Alternate Window).
- Create new reports using the built-in report writer.
- Create your own version of an existing report to use in place of the original Dynamics GP report (an Alternate Report).
- Use the sanScript scripting language to create business logic that extends existing Dynamics GP functionality or creates wholly new functionality. Scripts can respond to user actions such as pushing a button, changing a field, or closing a window.
- Use Dexterity triggers to watch for events such as opening a window or tabbing off a field. You can trigger off events in any customization written in Dexterity; you are not limited to just the `Dynamics.dic` file. When the trigger fires, it runs a procedure written by you.
- Use the extensive library of over 1,000 pre-written functions to perform otherwise complicated tasks.
- Use the integrated debugging tools to debug your application, even when your application is running in multidictionary mode.
- Create structured error handling.
- Call the same procedures used by Dynamics GP to execute subroutines.
- Create SQL tables and automatically generate stored procedures to handle table operations.
- Include the resources you create in the Dynamics GP security model without writing a line of code.
- Access the .NET Framework, web services, and any other features made available by other applications through **Component Object Model** (**COM**).
- Provide access to end-user customization tools such as Report Writer, Modifier with VBA, and the Import utility.
- Create your own toolbar.
- Create navigation to your application from the homepage.
- Create a navigation list exposing your data.

Even though Dynamics GP is not open source, all of these capabilities are available because Dynamics GP was written to embrace the creation of integrating applications. In short, if you use Dexterity to create your customization, it can do anything Dynamics GP can do.

Limitations

As robust as it is, Dexterity does have its limitations:

- You cannot modify a form in a third-party dictionary, meaning you cannot modify a Dexterity window that lives in a dictionary other than the `dynamics.dic` file
- Dexterity does not support Unicode; so if you need support for Chinese, Japanese, and Korean hieroglyphs, you have a problem to solve
- Dexterity cannot access fields added using the Modifier tool; therefore, you cannot attach sanScript code to those fields
- Dexterity does not support dynamically loading images from a database

Developer skills required

The following skills are generally necessary to develop an integration using Dexterity:

- Thorough knowledge of the sanScript scripting language
- Experience working with the data model of Dynamics GP
- Experience navigating the user interface of Dynamics GP
- A proficiency in database design
- An understanding of SQL Server
- An understanding of SQL Server stored procedures

End-user prerequisites

There are no end-user prerequisites. The customization results in a `.cnk` file that any user could drop into the folder containing `dynamics.exe`. The next time Dynamics GP is launched, the `.cnk` file gets extracted, adds the appropriate settings to the `Dynamics.set` file, and becomes a `.dic` dictionary file.

Visual Studio Tools for Dynamics GP (VS Tools)

The Dynamics GP development community was elated when **VS Tools** was released. Finally, a non-Dexterity programmer had a real opportunity to customize Dynamics GP! The world was buzzing with rumors that Dynamics GP was being rewritten in .NET, and Dexterity would soon be a thing of the past. Well, that didn't happen; but VS Tools really opened the floodgates for customizations. It turns out that there are actually more .NET programmers than Dexterity programmers! Imagine that.

VS Tools is the .NET **API** (**Application Programmer's Interface**) for Dynamics GP. By creating your customization in managed code (a .NET assembly), you can use the exhaustive list of features provided by the .NET Framework against the resources in Dexterity-created dictionaries.

Capabilities of VS Tools

Using VS Tools, you can perform the following:

- Create WinForm Windows
- Access tables, table fields, and table buffers
- Access Original, Alternate, and Modified windows
- Access window fields and global variables
- Access fields created using the Modifier tool
- Execute Dexterity commands, functions, and procedures
- Use the runtime engine's trigger system to respond to events such as a window opening or a field changing

Developer skills required

The following are the skills the developer requires:

- Knowledge of the .NET Framework
- Experience creating applications with Microsoft Visual Studio
- Understanding of a .NET language, like Visual Basic.NET or C#
- Experience working with the data model of Dynamics GP to understand which tables to access
- Familiarity with Dexterity is helpful if you want to invoke Dexterity functions or procedures
- Experience navigating the user interface of Dynamics GP to understand which events to register

End-user prerequisites

There are no end user prerequisites. The customization results in a `.dll` file that any user could drop into the `AddIns` folder inside the main Dynamics GP folder. Using a default installation, the `AddIns` folder exists here for a 64-bit machine: `C:\Program Files (x86)\Microsoft Dynamics\GP2010\AddIns` and for a 32-bit machine it is located here: `C:\Program Files\Microsoft Dynamics\GP2010\AddIns`.

Modifier with VBA (Visual Basic for Applications)

Modifier with VBA is actually two tools: **Modifier** and **VBA**. You cannot separate them, but they each do very different things. Modifier is largely a stripped-down version of the graphical forms designer found in the Dexterity tool. Using Modifier, you can make changes to existing windows, such as adding or removing fields, moving things around, or changing the tab order. Changes made with Modifier are stored in a separate dictionary. Modifications to forms in the `Dynamics.dic` file are stored in the `Forms.dic` file. Modified windows are not part of, nor do they change, the original resources in the application dictionary. Each application creates its own dictionary for storing changes made with Modifier. For example, the Fixed Asset module's modified windows are stored in the `F309.dic` file, while modified windows of the Human Resources module are stored in the `HRPFRMS.dic` file. The filenames that store modified forms are defined in the `Dynamics.set` file.

The following screenshot shows the **Vendor Maintenance** window as it appears in the Modifier. To add new fields from the **PM_Vendor_MSTR** table to the window, simply select a field from the toolbar and then drag it onto the window.

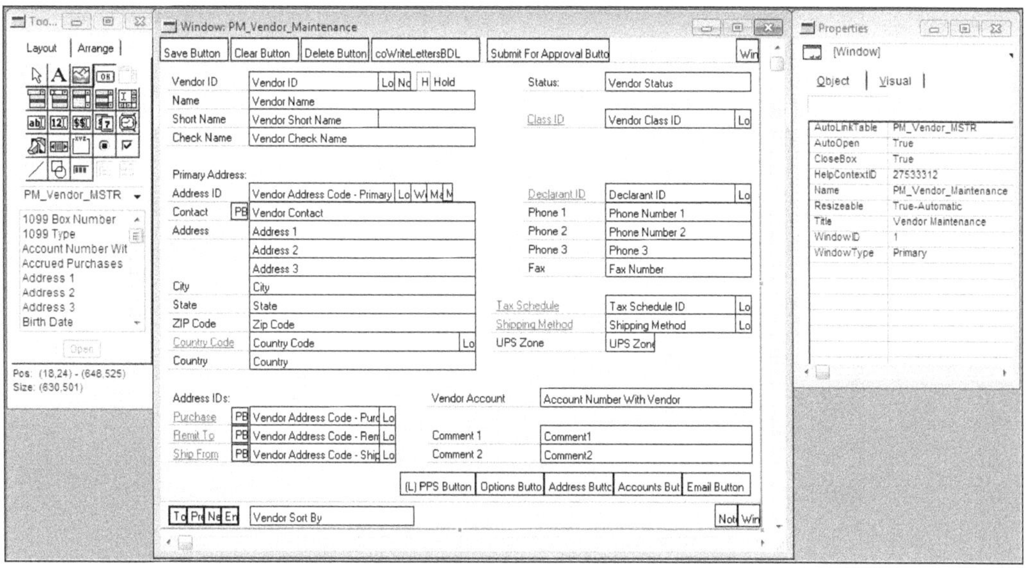

New fields created using Modifier have no life. You can use Modifier to put the button on the window, but you cannot make it do anything when pushed. The VBA side of Modifier with VBA provides the means for you to attach code to new or existing fields. The code is VBA code, not sanScript (Dexterity) code. Once you attach some VBA to the new button's `change` event, it has life. Modifier is aware of any window created with Dexterity no matter which application dictionary (`.dic` file) houses it. This fact makes Modifier an attractive tool to use when you need to add a button to a third-party window with the least amount of effort.

Modifier is largely an end-user tool; VBA code runs as clear text and is easily modified by others. You need to keep this in mind if you develop your integrations using Modifier with VBA. Protect your customizations!

Use the **Protection** tab in the **Project Properties** window to lock your VBA project from viewing:

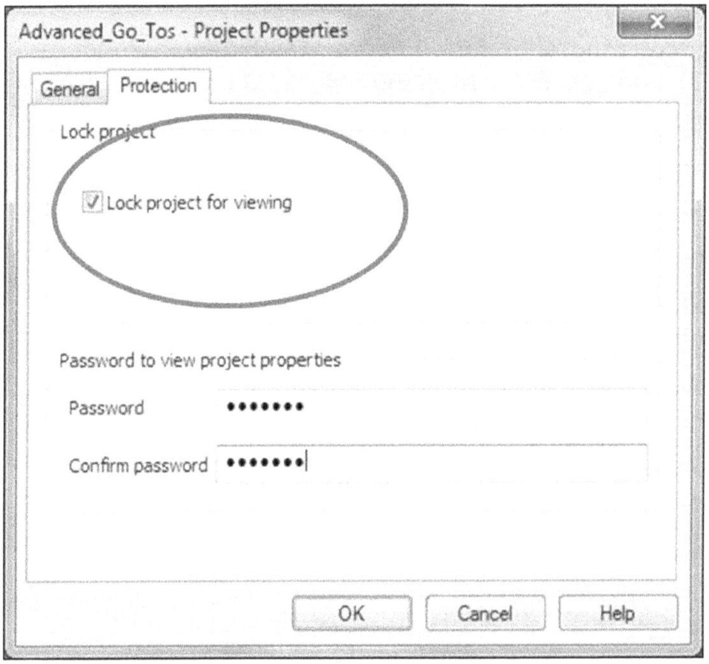

Capabilities of Modifier with VBA

The following is a list of capabilities of Modifier with VBA:

- Modify the appearance of existing windows
- Add new objects and fields to windows
- Change the tab order on a window
- Add additional business logic to windows and reports using VBA
- Add new VBA forms
- Modify windows and reports of other integrating applications created in Dexterity
- Modifier with VBA is COM compliant

Developer skills required

Different skills are required for different actions using Modifier with VBA. The primary difference is that the window changes themselves do not involve any coding, nor require any knowledge of the data model. Once you start adding VBA code to the project, developer skills become necessary.

- Knowledge of VBA is necessary for adding event code or VBA forms
- Experience navigating the user interface of Dynamics GP is helpful
- Knowledge of SQL commands is needed as the project becomes more complex
- Developer skills are not required to make window changes using the Modifier

End-user prerequisites

You, as the developer, will create a package file once the customization is complete. Any end user can import this package file, provided the Customization Site License or Modifier with VBA module is registered. Customizations created with Modifier and VBA need to be installed on each user's workstation rather than on a network share. Depending on the number of workstations, deployment and maintenance can be inconvenient.

Continuum

Continuum provides the COM API for Dynamics GP. Continuum, such as VS Tools, takes advantage of the extensive triggering system of the runtime engine. You identify which Dexterity events your application wants to be notified of, such as opening a window or changing a field, and when that event occurs, your application is notified and can execute a procedure in response to the event.

This tool, although no longer being developed, is a favorite among Visual Basic and Delphi programmers. One of the beauties of Continuum is that it comes with add-in wizards that walk you through each step required for creating an integration using a point-and-click interface. Using the wizards, you choose the windows, fields, and controls you want to integrate with, and then it generates the Visual Basic (or Delphi) code needed to complete the integration. Continuum for Visual Basic takes full advantage of **Object Linking and Embedding Automation** "**OLE Automation**" to keep data fields and buttons completely synchronized. What could be better than that?

You use a special subset of sanScript with Continuum; that subset is documented in the sanScript Supplement that comes with the Continuum. The Continuum API provides you a means to write pass-through sanScript that will execute against the Dynamics GP application. The files you need to set up a Continuum project, as well as the Continuum API Guide and Continuum sanScript Supplement, are located inside the `Tools\Continuum` folder of the installation DVD for Dynamics GP.

Capabilities of Continuum

The following are the capabilities of Continuum:

- Create a form-level integration to add additional business logic to Dynamics GP using pass-through sanScript
- Create a process integration that reacts to Dynamics GP events
- Create a database-level integration using pass-through sanScript to register triggers in Dynamics GP
- Access Dynamics GP resources using COM

Developer skills required

The following are the skills the developer requires:

- Experience navigating the user interface of Dynamics GP to understand which events to register
- Experience with whichever COM-capable development tool you selected to create the integrating application (such as Visual Basic .NET)
- Experience creating applications with Microsoft Visual Basic or Delphi is helpful
- Knowledge of the Dynamics GP dictionary resources
- Familiarity with the sanScript statements used by Continuum

End-user prerequisites

There are no end-user prerequisites.

Extender / eXtender Enterprise

Extender is an integration tool that gives you the ability to add additional data entry windows to Dynamics GP. If you have the Enterprise version of the product, you can add new functionality using scripts. **eXtender Enterprise** is as close to point-and-click Dexterity as you can get. The Enterprise edition is available only from eOne Solutions; it is not sold by Microsoft as part of Dynamics GP.

Extender allows for complete custom applications to be built by end users without a single line of code being written. The new Extender application you create comes automatically linked with all the features of Dynamics GP: SmartList objects, drill downs, advanced lookups, notes, as well as unique Extender features such as detail note tracking, e-mail merges, user-defined searches, imports, and more.

Capabilities of Extender and eXtender Enterprise

The following are the capabilities of Extender and eXtender Enterprise:

- Create new data-entry windows
- Link additional windows to any form in any dictionary
- Create new dialog windows

- Create new note windows with expanded functionality
- Create new menus
- Automate macro execution
- Create conditional windows using criteria set by the developer
- Create SQL views of the data collected in the new windows
- Import information into the new Extender data-entry windows
- Apply templates to set default values for fields
- Embed business logic behind Extender data-entry windows (eXtender Enterprise only)

Developer skills required

The following are the skills the developer requires:

- Experience navigating the user interface of Dynamics GP
- Development experience using sanScript is necessary for using the scripting functionality of the eXtender Enterprise edition
- None required for using the Extender module purchased from Dynamics GP; this tool is predominantly an end-user tool

End-user prerequisites

The user needs to have purchased the Extender module or eXtender Enterprise, depending on whether scripting is required.

DDE \ ODBC \ ADO \ OLE Automation

DDE (**Dynamic Data Exchange**), **ODBC** (**Open Database Connectivity**), **ADO** (**Active Data Objects**), and **OLE Automation** (**Object Linking and Embedding Automation**), are industry standards for accessing and exchanging data.

DDE

The primary function of DDE is to allow applications to share data. It uses the **Windows Messaging Layer** functionality within Windows to allow two running applications to share the same data. DDE is a method of asynchronous communication called a **DDE Conversation**, which allows one program to communicate with another program. The client application sends a Windows message to the server application. Windows holds the message and sends it to the server application when the server application is ready to process it. The client application can continue processing; it does not need to wait for a response from the other applications.

The only thing DDE can do is transmit data. It only appears to control another application if the other application can treat data as a command. DDE is lean and clean from the programmer's point of view; there are no `.dlls` required and it doesn't interrupt processing.

ODBC

ODBC is a technique used for accessing database information using drivers. The drivers provide a universal middleware layer that uses a standard set of commands to communicate with a **Database Management System** (**DBMS**). The driver then translates those standard commands into the correct instructions for the specific DBMS. Using ODBC, you can write programs that access data without knowing how the database is implemented. ODBC drivers are considered **OLE Providers**.

ADO

ADO is a collection of COM objects used for accessing data sources. Like ODBC, ADO is a language-neutral object model that exposes data raised by an underlying OLE Provider. ODBC drivers are the most commonly used OLE Providers.

OLE Automation

OLE Automation is a mechanism that allows for the exchange of data between applications. It provides an infrastructure whereby applications, called **automation controllers**, can access and manipulate shared automation objects that are exposed by **automation servers**. Automation controllers are also known as **automation clients**. This concept of client and server is also used in DDE conversations. Dynamics GP can be either an automation server or an automation client.

Capabilities of DDE \ ODBC \ ADO \ OLE Automation

There are some subtle differences between these applications, but we're going to group them together for the purpose of customizing Dynamics GP. These tools have the following capabilities:

- Create new data records
- Read existing records
- Update existing records
- Delete existing records
- DDE is capable of asynchronous communication

None of these tools provide a means to modify the user interface of Dynamics GP.

Developer skills required

The following are the skills the developer requires:

- Experience with the chosen development tool used to access the data.
- If using OLE Automation, knowledge of how to call the objects is required.
- If using DDE, knowledge of the supported DDE commands is required.
- Experience working with the data model of Dynamics GP is essential. No data validation is provided, so it is up to you, the developer, to ensure the correct tables are being accessed.
- An understanding of how to create a database connection string is helpful.
- ODBC and ADO require no specific developer skills.

End-user prerequisites

There are no end-user prerequisites. It's possible the end user may require an access license to the database if the database is being accessed outside of the Dynamics GP user interface.

Integration Manager

Integration Manager is an end-user tool used to import data into Dynamics GP. This tool could be used for a one-time import to convert data from an old system, or a periodic import of data in an ongoing manner. **VBScript** can be added to many events of an integration, allowing for some very complex logic to be executed. Integration Manager has a user-friendly interface for mapping data from a wide array of sources.

The following screenshot shows the destination mapping window for a **Payables Transaction** where you would identify the source for the data being imported. Most fields have a default value that will be used if the source data does not include a value for that specific field.

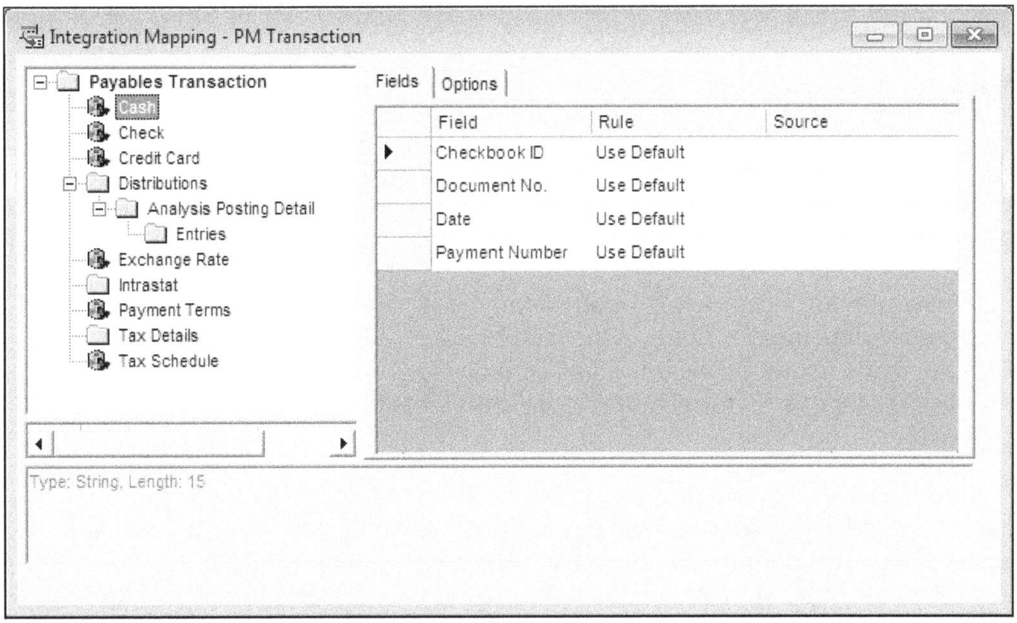

Capabilities of Integration Manager

The following is a list of the capabilities of Integration Manager:

- Import transactions with full business-logic validation
- Import or update master files with full business-logic validation
- Use VBScript at distinct points in the integration to expand functionality

Integration Manager is restricted to a predefined set of import destinations. New destinations cannot be added by the user.

Developer skills required

The following are the skills the developer requires:

- No knowledge of the database is required; this is an end-user tool
- VBScript can be used, so a proficiency in VBScript would be helpful to fully exploit the capabilities of this tool
- Integration Manager can be automated using **Windows Scheduler**, so experience with the Windows Scheduler is a plus

End-user prerequisites

The end user needs to own the Integration Manager module. This module is not included with any of the Dynamics GP Business Ready Licensing models.

Table Import

Table Import is a utility that comes with Dynamics GP and can be used to import data into any Dynamics GP table, regardless of the dictionary. Table Import does only one thing, but it does it very well and very fast. Think of Table Import as the data-slam method of importing data.

You must provide a properly formatted comma-or tab-delimited text file, and then map the file data into the fields in the target table. No translation is available, no links are made to other tables, and you cannot use ODBC. No validation against the business logic is performed. It is a direct-to-table import that you use at your own risk.

A word of caution: although there is nothing stopping you from importing transaction data, you should tread lightly in this area. Remember, you have *no* business logic validation with the Table Import utility. You would most likely use this utility for populating master file tables, especially tables without an adapter that you can use with Integration Manager. For example, you might use Table Import to bring in account categories or sales tax details.

The Dynamics GP Software Development Kit includes information that will help you use this tool to import master file data as well as a limited number of transaction types.

> Table Import can only add data; it cannot update data.

The following screenshot shows the **Table Import Definition** window. The data you want to import is in the source file. The destination table is where the data will land in Dynamics GP. The scrolling window lists all of the fields in the destination table and how they are mapped to the fields in the source file.

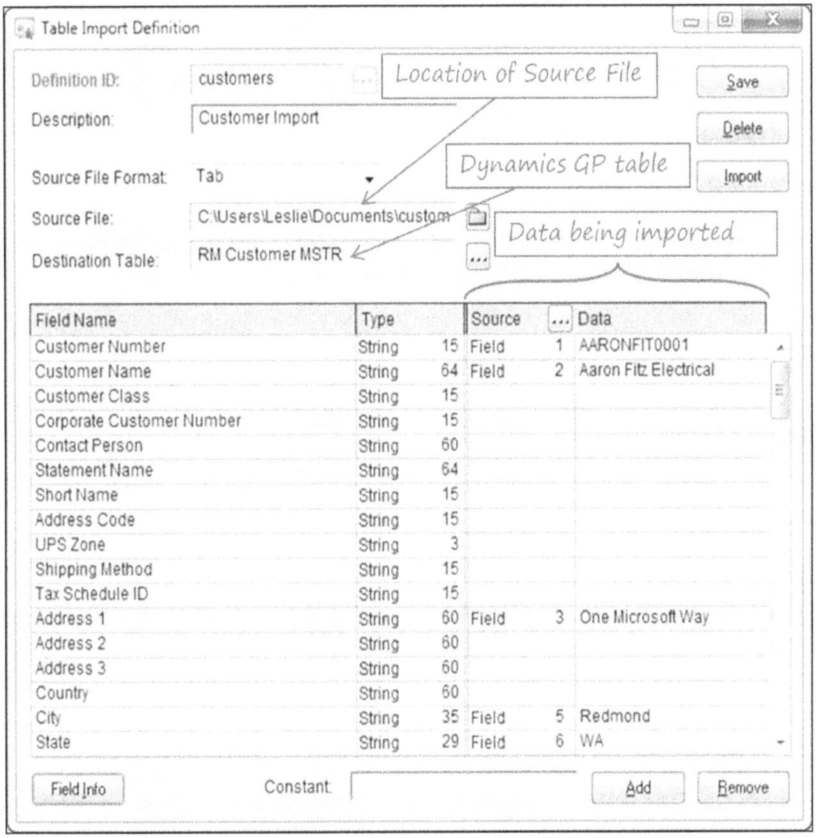

Capabilities of Table Import

The following are the capabilities of Table Import:

- Import new data into any table in the database
- Import values into a multi-select listbox field

> If you use SQL Server Management Studio to look at a multi-select listbox field, such as the DSPLKUPS (display in lookups) field in the GL00100 table, you will see <Binary data> as the field value. If you use Table Import, you can use a 32-character string field made up of T and F characters and populate the database field correctly. Integration Manager does not support this level of integration with multi-select listboxes.

Developer skills required

The skill required by the developer is as follows:

- As no validation is done on the imported data, an in-depth knowledge of the Dynamics GP data model is required

End-user prerequisites

None; the end user simply opens the **Table Import Definition** window, selects the appropriate **Definition ID**, and then executes the import.

eConnect

At its root, the **eConnect** product is a collection of encrypted, stored procedures, and the methods to access them. These stored procedures allow external applications to create, retrieve, update, delete, and void Dynamics GP documents. Unlike the Table Import utility, eConnect uses the Dynamics GP business logic to validate all transactions.

eConnect has become the integration tool of choice across the development world because it can interface with nearly anything. It contains a rich set of interoperability tools that you can use to extend the functionality of Dynamics GP. You can create stored procedures to run before or after the eConnect document exchange. As a bonus, eConnect is very fast.

The principal components and interfaces of eConnect include a .NET-managed code assembly, a Microsoft BizTalk **AIC** (**Application Integration Component**), and **MSMQ** (**Microsoft Message Queuing**) services. When you install eConnect, it creates a **WCF** (**Windows Communication Foundation**) service named **eConnect for Microsoft Dynamics GP 2010 Integration Service** (**eConnect Integration Service**). This integration service replaces the eConnect COM+ object that was used in eConnect versions prior to Dynamics GP 2010.

eConnect uses specially formatted **XML** (**Extensible Markup Language**) documents to integrate with Dynamics GP; therefore, your application must be able to create or consume XML documents. Business rules are enforced during the integration ensuring all of the necessary tables are updated with each imported item. Over 200 integration points are supported by eConnect. The required XML document schemas are provided in the eConnect documentation.

By using eConnect, external applications such as CRM programs, point-of-sale systems, web services, warehouse management systems, and other legacy applications can interact with Dynamics GP. These external applications can import or update master files and import or void transactions. The following diagram illustrates the different layers and components incorporated with eConnect.

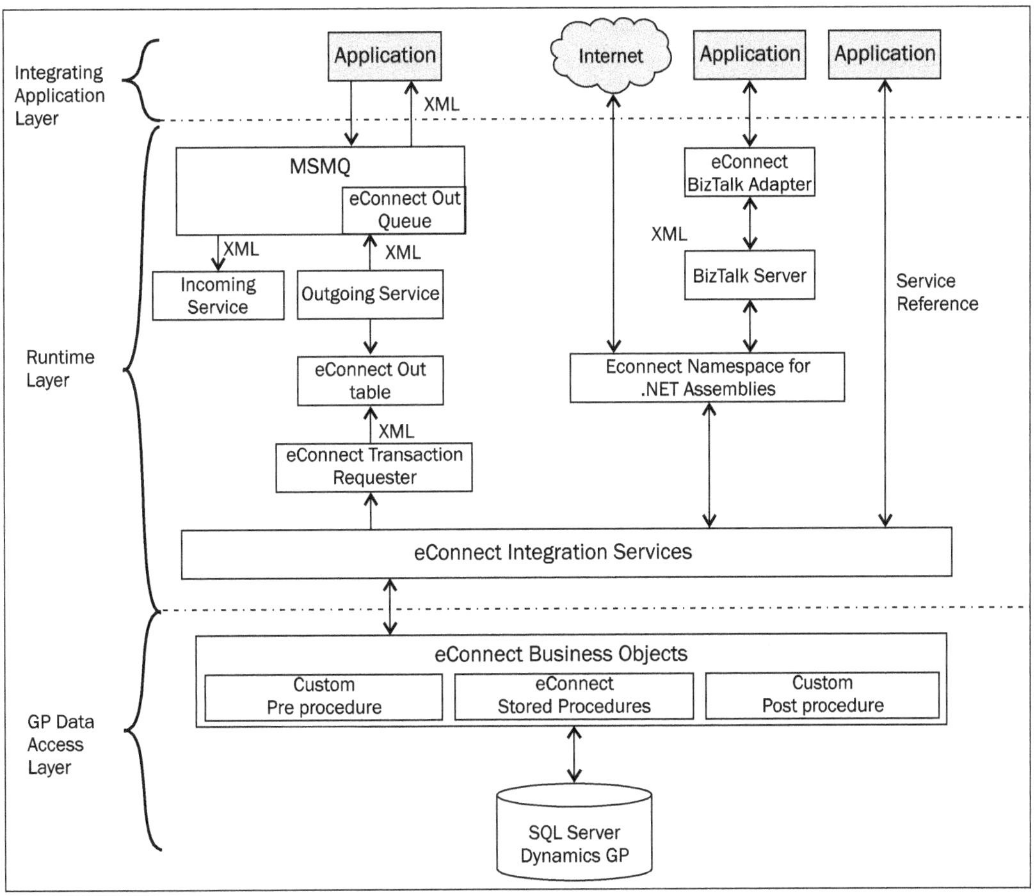

eConnect configuration

In order to communicate with eConnect Integration Service, you need to include the eConnect assembly and namespace in your development project. However, if you use a Service Reference to interact with eConnect, you do not need to add the Dynamics GP eConnect assembly or namespace to your development project. With a Service Reference, your application will be able to work directly with the eConnect Integration Service.

By using MSMQ services, eConnect integrations can run without user interaction. The MSMQ services do not post the transactions however, so you can't automate the entire process. There is a tool available from Microsoft, the `autopost.dll` file, that uses the Continuum API to call the Dynamics GP posting procedures. By using this tool, you can post imported transactions automatically instead of requiring the user to complete the post. On the downside, in order for the `autopost.dll` file to work Dynamics GP must be running. Since Dynamics GP must be running, a user license is consumed.

There is an application available from eOne Solutions, called **SmartConnect**, that provides a point-and-click interface to eConnect. The SmartConnect application transforms eConnect into a drag-and-drop end-user tool without losing the development capabilities of eConnect.

SmartConnect ships with a shared web service that allows for data to be pushed into Dynamics GP from external applications such as Access or Excel. You can also fully automate posting without the need for `autopost.dll`. Using the web service, you can update Dynamics GP without requiring a Dynamics GP login; thus you still have all of your user licenses available.

Capabilities of eConnect

The following are the capabilities of eConnect:

- eConnect allows external applications to access real-time Dynamics GP data
- eConnect provides tools that you can use to create, retrieve, update, delete, and void a wide array of Dynamics GP documents
- Imported data is validated against the Dynamics GP business logic
- eConnect is compatible with Visual Basic objects, stored procedures, BizTalk Server AIC adapters, COM integration, MSMQ, XML document exchange, and other industry-standard technologies

Developer skills required

The following are the skills the developer requires:

- Expertise working with XML documents and schemas is a must
- Experience with the selected programming application you're using to create the integration is necessary
- Knowledge of Windows Communication Foundation is needed if you are using a Service Reference
- Familiarity with Dynamics GP documents and operations

End-user prerequisites

To use an eConnect integration, the eConnect runtime components must be installed. In order to use the private message queues for the incoming and outgoing services, Windows Message Queuing must be installed and operating. In order to use the eConnect BizTalk adapter, BizTalk 2006 or later must be installed.

Web services

Web services use XML documents and acts as middleware between eConnect and Dynamics GP. There is no user interface with web services; you design the user interface with your tool of choice and then use the web services to manipulate the data. In order to use web services, the type of transaction you want to execute must be listed as one supported by this tool.

Web services for Dynamics GP integrates with web-based applications using the industry standards of XML, **SOAP** (**Simple Object Access Protocol**), **WSDL** (**Web Services Description Language**) and **UDDI** (**Universal Description, Discovery, and Integration**) on an Internet protocol across a network.

XML provides the language that you can use between different platforms and development tools. XML uses tags to define, transmit, validate, and interpret the data. A tag is written between angled brackets and marks the start point and end point of each logical unit or element of an XML document. The following is an example of a simple XML tag:

```
<email>
  <to>Leslie Vail</to>
  <from>Superman</from>
  <body>Hello World!</body>
</email>
```

SOAP is a messaging protocol you can use to encode request and response messages. SOAP messages are independent of any operating system or programming language, so they can be transported using popular Internet protocols such as **SMTP** (**Simple Mail Transfer Protocol**), **MIME** (**Multipurpose Internet Mail Extensions**), **IMAP** (**Internet Message Access Protocol**), and HTTP.

WSDL is the language used to describe what the web service does and provides information on how to interface with it.

UDDI is used for listing which specific services are available. UDDI listings are written in WSDL.

Capabilities of web services

The following are the capabilities of web services:

- Allows integration with external applications
- Customers
- Vendors
- General Ledger accounts
- Sales Order documents
- Purchase Order documents
- Receivable transactions
- Payable transactions
- General Ledger transactions
- Through the web service, an external application can create, retrieve, update, delete, and void supported Dynamics GP documents
- Provides a means to interact with Dynamics GP using a web browser

Developer skills required

The following are the skills required by the developer:

- Expertise working with XML documents and schemas is a must
- Experience with the selected programming application you select to create the integration
- Familiarity with Dynamics GP documents and operations
- Knowledge of the Dynamics Security Service
- Knowledge of Windows Communication Foundation
- .NET programming abilities

End user prerequisites

A web service solution will need to be installed by qualified IT personnel; after that, there are no end-user requirements.

Modifying the user interface

Don't like what you see? Change it. Despite the number of hours the fine people at Microsoft spent designing the user interface of Dynamics GP, sometimes changes are necessary. The changes may be purely cosmetic, such as adding a company logo to the window, or they may bring with them additional functionality, such as creating an XML document of the displayed record.

The question at hand is, which of the tools that we just discussed can you use to make a change to the user interface. Tools having the capability of modifying the user interface include:

- Dexterity
- VS Tools
- Modifier with VBA
- Extender and eXtender Enterprise

Dexterity

Using Dexterity you can change any existing window by creating an Alternate Form (window) to be used in place of the original. You start with the original window as a base and then make modifications to suit your needs.

> Modifications should only be additive; removing any of the existing controls will break the Dynamics GP application.

A word of caution: Alternate Forms and Reports can be very expensive for you to maintain. With each release of Dynamics GP, you run the risk of having to re-create the Alternate Form or Report from scratch. It is for this reason that using Alternate Forms and Reports should be avoided if at all possible.

You can also create completely new windows with Dexterity, thereby creating an interface specifically for your integrating application. This is normally the road you will travel down rather than modifying existing Dynamics GP resources.

During the development of a Dexterity integration, you will actually be working with all of the resources in the core application dictionary of Dynamics GP. The core application dictionary is named `Dynamics.dic` and contains nearly 60,000 resources that you can use.

The core modules that comprise the `Dynamics.dic` dictionary include:

- System Manager
- General Ledger
- Payables Management (Accounts Payable)
- Receivables Management (Accounts Receivable)
- Bank Reconciliation
- Purchase Order Processing
- Sales Order Processing
- Inventory

In Dexterity, windows are included in **Forms**. A Form can contain any number of windows that work together for a common purpose. For example, the **RM_Customer_Maintenance** form includes four other windows in addition to the **RM_Customer_Maintenance** window.

The following screenshot shows the **Form Definition** window for the **RM_Customer_Maintenance** form as seen in Dexterity. Note the additional four windows:

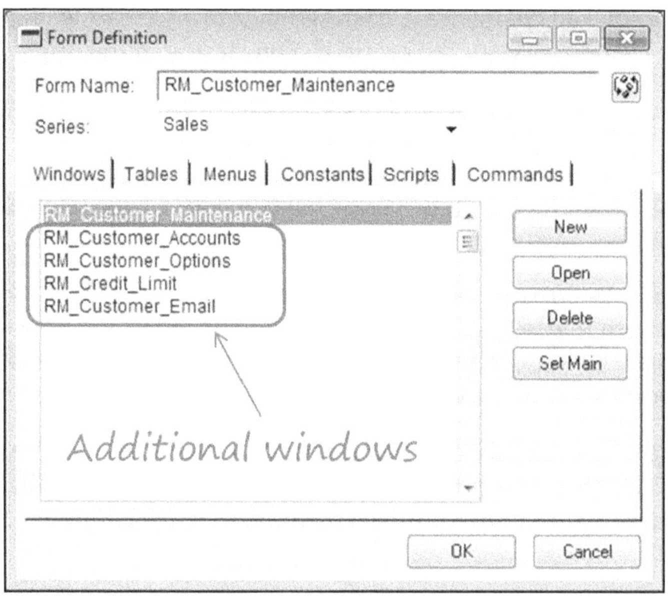

`Dynamics.dic` includes over 1,600 Forms. Forms contain windows and windows contain window objects; window objects are the push buttons, pictures, fields, and such that appear on a window. You have access to all existing window objects and can add one or more of them to a window that you create. Window objects include:

- Static text (field prompts)
- Pictures
- Push buttons
- Drop-down lists
- Comboboxes
- Listboxes
- Multi-select listboxes
- Horizontal listboxes
- Scrolling windows
- Visual switches
- Radio groups
- Lines and shapes
- Progress indicators
- Tree views
- List views
- String, integer, currency, date, and time fields
- Big text fields

Any resource you create becomes what is known as a **third-party resource** and eventually ends up in a separate dictionary that is your integrating application.

Adding sanScript code to the objects on the window will cause your program to spring into action when the user pushes a button.

VS Tools

Using VS Tools, you can add a new window but you cannot change any existing windows. WinForms are used to create new windows for add-ins developed using VS Tools. As such, you can use any WinForm control in a window design.

WinForm properties

In order to more closely match the look of a window created in Dexterity, VS Tools has added additional WinForm properties to its windows:

- `AutoSetDexColors`: Set it to **True** to match the color of the Dynamics GP window
- `ControlArea`: Set it to **True** to create a band at the top of the WinForm where the **Save**, **Clear**, and **Delete** buttons are normally placed
- `StatusArea`: Set it to **True** to draw a divider line across the bottom of the WinForm where you would normally place the browse, window note, and help buttons, and sort-by fields

WinForm control properties

In addition to the extra WinForm properties for the windows, VS Tools added additional properties to, or updated the characteristics of, several controls to more closely match the look of Dynamics GP window objects.

The VS Tools controls that were modified to better integrate with Dynamics GP include:

- Buttons
- TextBoxes
- Labels
- ComboBoxes
- ListBoxes

Buttons

Buttons include the new property `ButtonType` with the following values:

- **Standard**: There are no Dynamics GP specific appearance changes to the Standard button type for VS Tools.

  The following screenshot shows two Standard buttons:

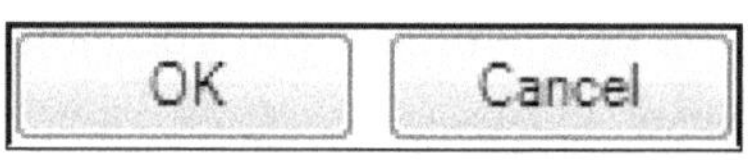

- **Toolbar**: The Toolbar button type has a flat appearance, often has a picture on it, appears highlighted when the mouse is placed over it, changes to a darker color when pushed, and you would typically place it in the window's control area. The printer button found in the upper right-hand corner of most windows is a Toolbar button.

  The following screenshot shows the printer Toolbar button in the window control area:

- **ToolbarWithSeparator**: The ToolbarWithSeparator button type is similar to a Toolbar button with the addition of a vertical separator line drawn to the right of each button. The standard **Save**, **Delete**, and **Cancel** buttons that you place in the control area at the top of a window are examples of ToolbarWithSeparator buttons.

  The following screenshot shows three ToolbarWithSeparator buttons:

- **StatusArea**: The StatusArea button type displays only graphics, has a 3D border drawn around it, and a tooltip describing the button's purpose when the mouse is placed over it. The window note and help buttons at the bottom of a window are examples of StatusArea buttons.

  The following screenshot shows a StatusArea button.

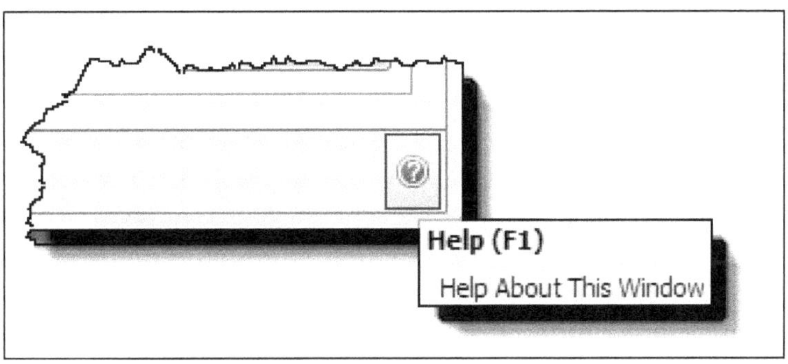

- **Field**: A field button type displays only graphics and is placed next to other controls on the form. The lookup button, record notes button, and expansion arrow button are examples of Field buttons.

  The following screenshot shows two Field buttons: the lookup button and the record notes button.

Pictures for button controls are provided in the `.png` files included with VS Tools. These pictures match the images found on Dynamics GP windows and buttons.

TextBoxes

You would use a TextBox control to display string, integer, currency, date, and time values. The TextBox control includes the new property `AutoSetDexColors`; when set to **True**, the TextBox will match the colors of Dynamics GP.

Labels

Label controls in VS Tools correspond to the field prompts in Dynamics GP. To mimic the traditional underline on Dynamics GP prompts, the `LinkField` property was added to the Label control. The `LinkField` property is set to the name of the corresponding TextBox on the window. A convenient drop-down list of control names makes it easier to select the correct field.

ComboBoxes

You would use a ComboBox control to imitate the drop-down list or combobox in Dynamics GP. The ComboBox control includes the new property `AutoSetDexColors`; when set to **True**, the ComboBox will match the colors of Dynamics GP. The following screenshot shows a VS Tools ComboBox control being used as a Dynamics GP drop-down list.

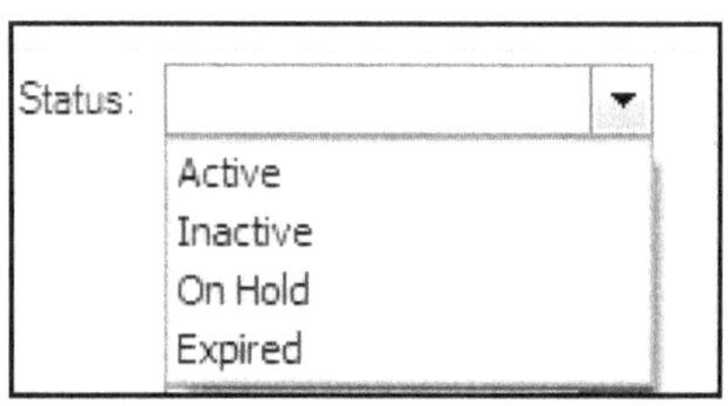

ListBoxes

You would use a ListBox control in place of the list box or multiselect list box of Dynamics GP.

Modifier with VBA

Using Modifier you can make extensive changes to existing windows in Dynamics GP, but you cannot add a new window. You can change any window created with Dexterity; you are not limited to just the windows in the `dynamics.dic` dictionary.

Modifier is an end-user tool, but may be appropriate if, for example, changes would otherwise result in the creation of an alternate window using Dexterity. Since Modifier is a stripped-down version of Dexterity, the graphical design tools in Modifier are a subset of the corresponding tools found in Dexterity.

The following are some of the changes you can make to a window with this tool; we cover Modifier in more detail in *Chapter 7, Creating Customizations with Modifier*:

- Add graphics such as a company logo to a window
- Change field prompts or add other static text to the window
- Move objects around on the window
- Move objects off the window
- Drag additional fields onto the window from the AutoLinked table (if any)
- Make fields required
- Hide fields

The following are several types of new fields you can add to a window:

- Push button
- Drop-down list
- String field
- Integer field
- Currency field
- Date field

Modifier does not alter the original window; it creates a copy of the window and you make changes to the copy. Modified windows are stored in a separate forms dictionary that cannot include any sanScript code. Each application dictionary installed creates a separate forms dictionary in which to store that application's modified forms. As you recall, the location of the forms dictionary is dictated by the `Dynamics.set` file.

Extender / eXtender Enterprise

Using Extender or eXtender Enterprise, you can add a new independent window or attach a new window to an existing Dynamics GP window. Although you can create a new window, that window comes with a predefined layout. You do not have the freedom of designing the window from scratch.

You create new windows using the **Extender Windows** window. The following screenshot shows this window being used to define a new window named **Vendor Extra**. The new window's name is determined by the **Description** field. Note that you can only put 15 fields on an Extender window. You can include as many windows as you like, but they can each contain only 15 fields.

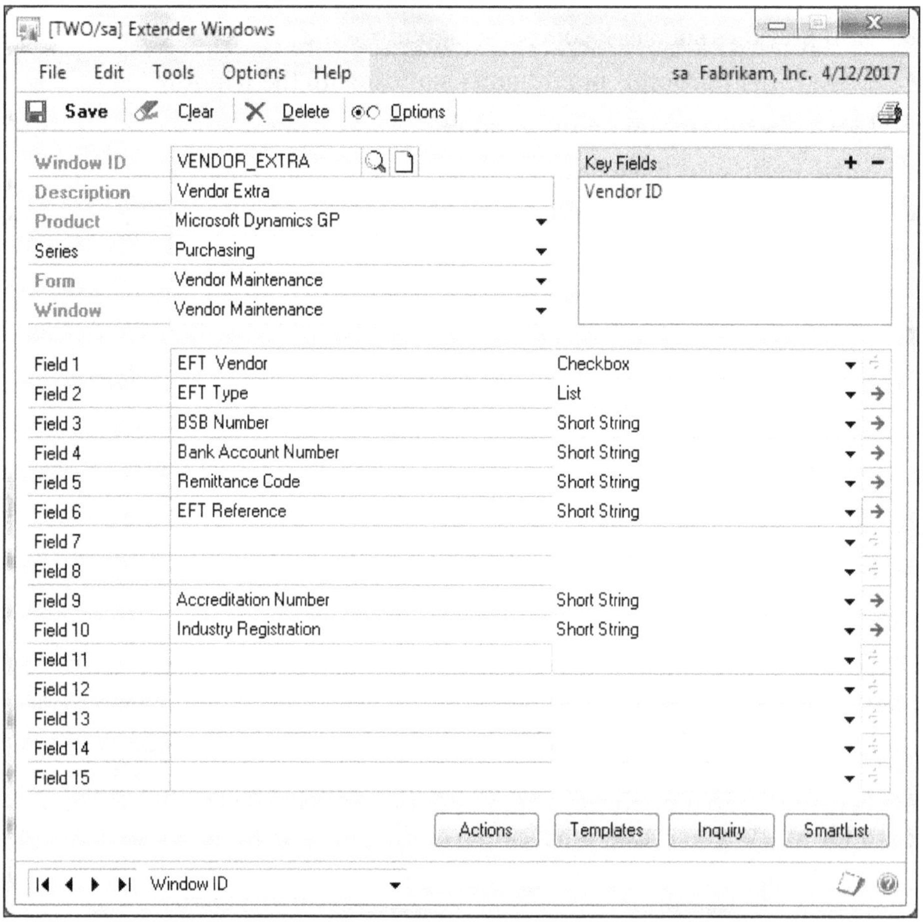

The following screenshot shows what the **Vendor Extra** window defined earlier looks like when opened in Dynamics GP.

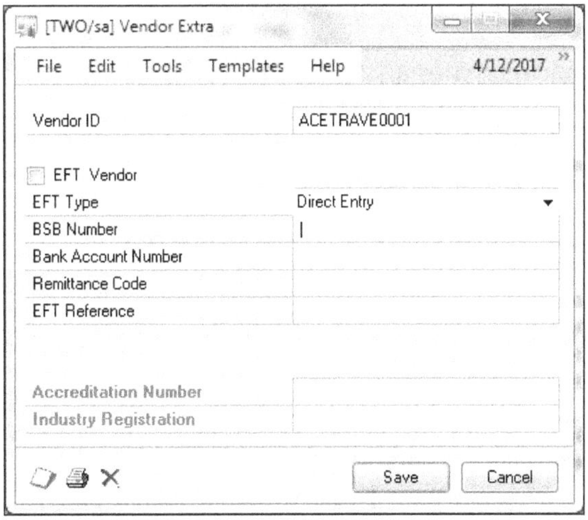

Using Extender or eXtender Enterprise, you can add several different types of data-entry windows. These are explained as follows:

Forms

An **Extender form** creates a standalone data-entry window that is not attached to an existing Dynamics GP window. You can use Extender forms to create new master records for information not stored anywhere else in Dynamics GP. For instance, you could create a master file describing vacation destinations.

An Extender form contains two columns of 12 fields each. Our new **Vacation Destination** window is shown in the following screenshot:

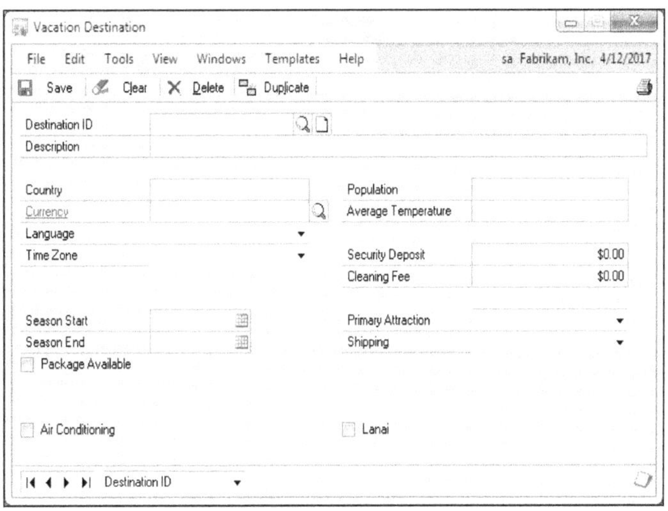

Detail forms

An **Extender Detail form** creates a standalone scrolling window similar to an invoice-entry screen. You can use it for any data that has multiple line items associated with a header record. For example, if the header record is a travel agency, the line items could be trips booked by that travel agency.

Each Extender Detail form can contain two columns of six fields each per header record. Each line item can contain 12 fields and may be two rows tall.

Our new **Travel Agencies** window is shown in the following screenshot:

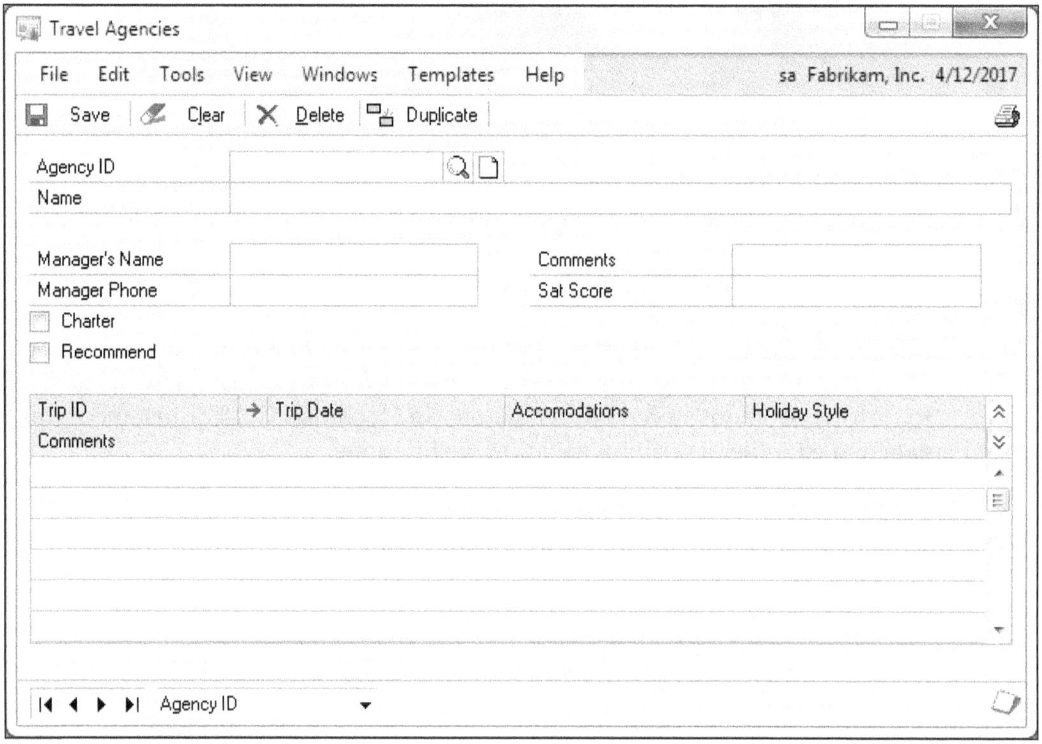

Windows

An **Extender window** creates a window that is attached to any existing Dynamics GP window. You would use Extender windows to link additional data fields to an existing record. Use an Extender window when you need to track more information than the Dynamics GP window can hold.

For example, if your inventory consists of tomato plants, you may need to track information such as days to harvest, fruit size, and hardiness. Each Extender window can contain up to 15 fields that are presented in a single column. You can attach numerous Extender windows to a single Dynamics GP window, but each Extender window is limited to 15 fields. You can also attach an Extender window to a scrolling window such as a purchase order or sales order line item.

The following screenshot shows an Extender window that supplements item information:

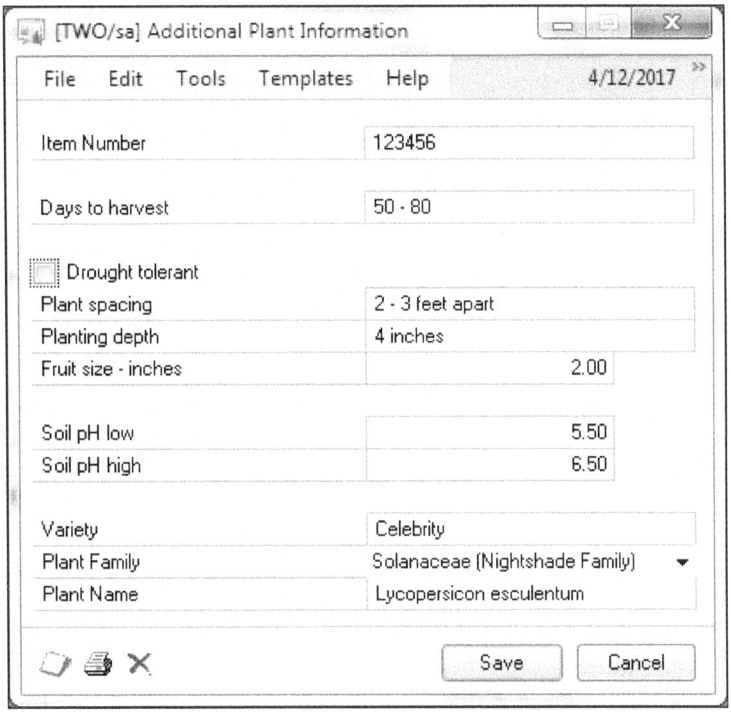

Detail windows

An **Extender Detail window** creates a window that can track multiple line items for a single Dynamics GP window record. A Detail window looks similar to a scrolling window. Using our tomato plant as an example, you could use a Detail window to list pests and problems inherent to this variety of tomato.

You can attach numerous Detail windows to a single Dynamics GP window. For our tomato plant, we could use a second Detail window to list Pruning and Staking methods.

The following screenshot shows two Detail windows that are attached to the Item Maintenance window:

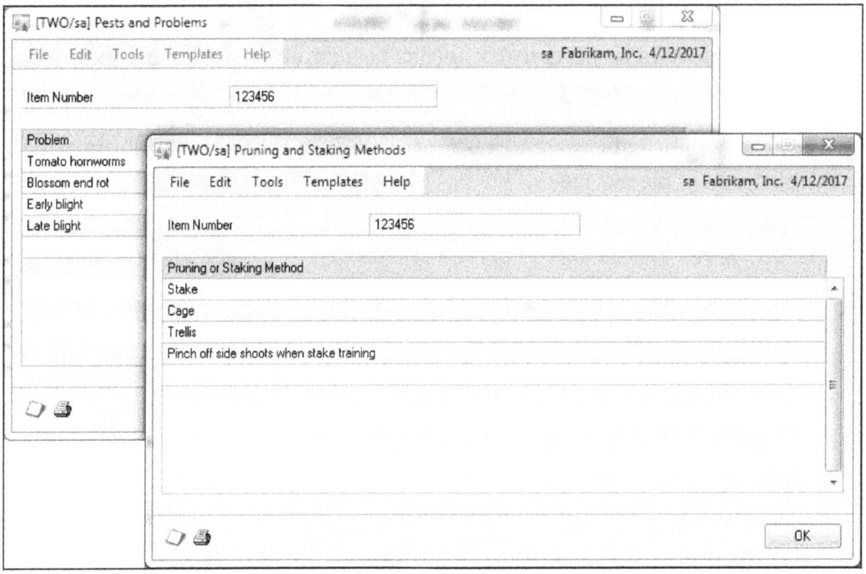

Each Detail window can contain ten fields. These fields can be distributed between two rows. You can also attach a Detail window to a scrolling window, such as an invoice line item.

The following screenshot shows a Detail window displaying two rows per line:

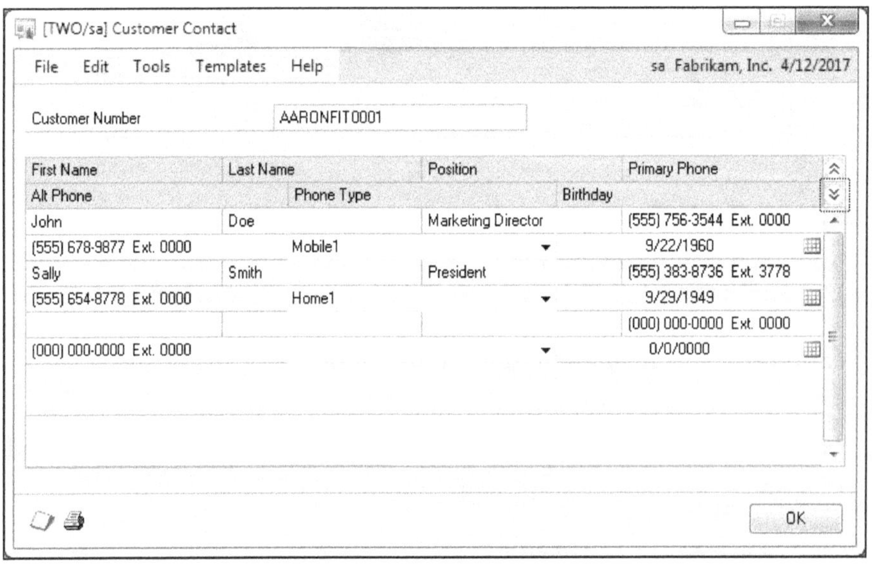

Notes

An **Extender Notes** window allows you to create multiple notes for a window record. The notes can be categorized and are stamped with the user ID, time, and date when the note was created. A Dynamics GP Notes window allows only a single note as a continuous text field, without the ability to categorize or date-stamp it.

The following screenshot shows an Extender Note created for the Employee Maintenance window:

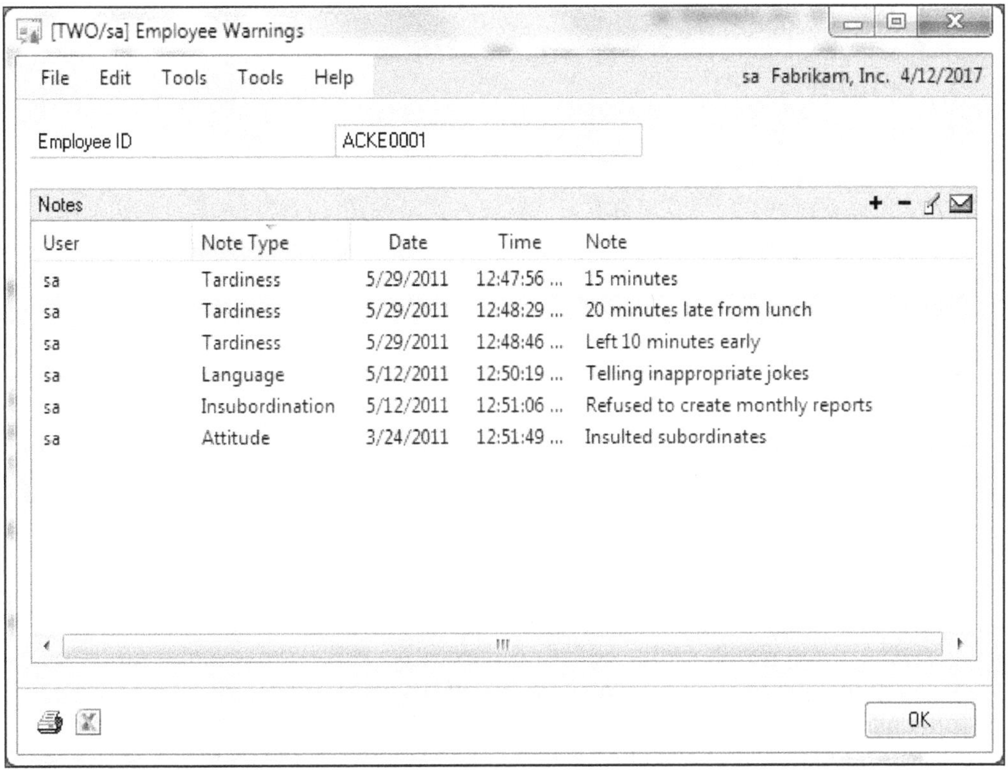

Notice that the notes are categorized into different types and include the user ID, date, and timestamp with each note.

You cannot modify existing Dynamics GP windows with Extender or eXtender Enterprise.

Changing or adding functionality

What's the point of creating an integrating application if you can't sneak in some extra functionality? Building a user interface is nice, but the real power is in making those new buttons *do* something when they are pushed. Enter the tools you can use to change the way Dynamics GP behaves. You may want to change the way the software currently accomplishes something, or add a brand-new talent to its list of abilities.

The tools highlighted in this section are capable of adding to or altering the Dynamics GP business logic:

- Dexterity
- VS Tools
- Modifier with VBA
- Continuum
- eXtender Enterprise

Dexterity

Dexterity uses the sanScript scripting language to create Dynamics GP logic.

sanScript is an event-driven language. When the events occur, whether initiated by the user or through code, scripts attached to those events are executed. Scripts can be attached to forms, windows, fields, and scrolling windows.

For the forms and windows that you have created, sanScript code will be attached directly to the object events. For forms and windows that you did not create, you will use triggers to watch for the events' occurrences. When the event occurs, you can run a global procedure known as **Trigger Processing Procedure** to execute custom business logic.

Dexterity events are used by other tools, so it is important to understand when they occur. The primary events are listed as follows:

Form events

The form events are as follows:

- **Pre**: The `Pre` event occurs when the form is opened.
- **Post**: The `Post` event occurs when the form is closed.

Window events

The windows events are as follows:

- **Pre**: The `Pre` event occurs when the window is opened.
- **Activate**: The `Activate` event occurs each time the window gains focus. The `Activate` event occurs *after* the window Pre event when the window is opened.
- **Post**: The `Post` event occurs when the window is closed.
- **Print**: The `Print` event occurs when the Print item is selected from the **File** menu. The **Print** menu selection will only be active if a script is attached to the window's `Print` event.
- **ContextMenu**: The `ContextMenu` event occurs when the right mouse button is clicked by the user. The mouse pointer must be on the window and not over a field, in order for this event to occur. This event enables scripts to create the right-click menu seen in most other Windows applications.

Field events

The field events are as follows:

- **Pre**: The `Pre` event occurs when the field is entered.
- **Change**: The `Change` event occurs when the field is exited and there has been a change in the value of the field. For toggle fields such as push buttons and visual switches, the `Change` event occurs when the user clicks on the field.
- **Post**: The `Post` event occurs when the field is exited. The field `Post` event occurs *after* the field's `Change` event.
- **MouseEnter**: The `MouseEnter` event occurs when the mouse pointer moves over the field.
- **MouseExit**: The `MouseExit` event occurs when the mouse pointer moves off the field.
- **ContextMenu**: The `ContextMenu` event occurs when the right mouse button is clicked by the user while the mouse pointer is over the field. This event allows for a field-specific menu to be displayed when the user right-clicks on the field.

Scrolling window events

Dexterity can only evaluate one record at a time; each line of a scrolling window represents a single record. Therefore, each time a line is populated with data and is displayed on the screen, the line events occur. When a scrolling window is initially filled, the line events run over and over again until the scrolling window is filled.

The scrolling window events are as follows:

- **LineFill**: The `LineFill` event occurs each time a new line is displayed in the scrolling window and each time focus is moved to an existing line that already contains data. When the scrolling window is initially filled, the `LineFill` event occurs repeatedly until the scrolling window is completely full. The `LineFill` event occurs *before* the `LinePre` event.
- **LinePre**: The `LinePre` event occurs when focus moves into the line. The `LinePre` event occurs *after* the `LineFill` event.
- **LineChange**: The `LineChange` event occurs when focus leaves the line and there has been a change to one of the fields on the line. A script placed on this event is typically used to save the line record to the table.
- **LinePost**: The `LinePost` event occurs when focus leaves the line.
- **LineDelete**: The `LineDelete` event occurs when the **Delete Row** item is selected from the **Edit** menu. The **Delete Row** menu selection will only be active if a script is attached to the `LineDelete` event.
- **LineInsert**: The `LineInsert` event occurs when the **Insert Row** item is selected from the **Edit** menu. The **Insert Row** menu selection will only be active if a script is attached to the **LineInsert** event.
- **ContextMenu**: The **ContextMenu** event occurs when the user right-clicks on the line. This event allows for a line-specific (not record-specific) menu to be displayed when the user right-clicks on the line.

Triggers

When you need to alter the business logic of Dynamics GP, you will use triggers. A trigger monitors one of the previously defined events, and when that event occurs the trigger is activated. Once activated, your scripts are called. What's more is that you get to decide whether your script runs before or after the Dynamics GP script. Often, if your script runs before, you can cancel the Dynamics GP script so that it does not run at all.

For example, let's assume your application is storing additional information about a vendor. If the vendor is deleted by Dynamics GP, you want to delete your data too. You need to register triggers in order to be notified that the user has selected to delete a vendor, and whether the delete was successful. Alternatively, perhaps you do not want to allow the vendor to be deleted under certain circumstances. When the user selects to delete the vendor, your code can cancel the delete operation, thereby changing the Dynamics GP business logic.

You can register a trigger against the events in any dictionary; you are not limited to the events of the `Dynamics.dic` dictionary.

Triggers give us the ability to interact with other dictionaries without needing the source code of the other application.

> It is important to note that Dexterity triggers can only *see* operations performed by Dexterity. Any operations performed by SQL-stored procedures are unknown to Dexterity. This is true even if the SQL-stored procedure was called by a Dexterity script.

Five triggers in Dexterity

The following are the five triggers in Dexterity:

- Database trigger
- Focus trigger
- Form trigger
- Function trigger
- Procedure trigger

More information on each trigger type is provided as follows:

- **Database trigger**: A Database trigger is activated by *successful* table operations. The table operation must be performed by sanScript code. Table operations performed by SQL-stored procedures will not activate the trigger. Table operations that you can monitor include:
    - Reading a record without locking it
    - Reading a record and placing a lock on it
    - Adding a new record
    - Deleting an existing record
    - Updating an existing record

- **Focus trigger**: A Focus trigger is activated when the specified focus event occurs. Available focus events are summarized in the following table:

Object	Event
Field	Pre, Change, Post, or ContextMenu (Note: only the change event applies to toggle fields such as push buttons and visual switches)
Form	Pre or Post
Window	Pre, Post, Activate, Print, or ContextMenu
Scrolling Window	LineFill, LinePre, LineChange, LinePost, LineInsert, LineDelete, or ContextMenu
Menu Item	Change
Script Command	Change (Note: script commands are typically used for navigation)

- **Form trigger**: A Form trigger is activated when the specified form opens. Form triggers are used to add menu items to the **Additional** menu on an existing form. You will often use an **Additional** menu item to provide navigation to your custom windows.
- **Function trigger**: A Function trigger is activated when the specified function is called.
- **Procedure trigger**: A Procedure trigger is activated when the specified procedure is called.

VS Tools

Using .NET assemblies, VS Tools can both alter existing business logic and create new functionality in Dynamics GP. We learned earlier that you can use VS Tools to create a user interface with WinForms that have been enhanced so that they more closely match Dynamics GP. In fact, it's difficult to tell the difference between a well-formatted WinForm and a native Dexterity window.

Next to Dexterity, VS Tools is the richest toolset we have for creating integrating applications. For some jobs, such as accessing resources in third-party dictionaries, it is arguably superior.

You can access any dictionary's resources by including a reference to that dictionary's application assembly in your VS Tools project. Application assemblies for the `Dynamics.dic` file and the other integrating applications included with Dynamics GP are bundled with VS Tools.

If your customization requires integration with a different third-party dictionary, you can create your own application assembly using the **Dictionary Assembly Generator** (**DAG**). The DAG utility ships with VS Tools.

Dictionary resources exposed to VS Tools through the application assembly include:

- Forms
- Windows
- Alternate windows
- Modified windows
- Window fields
- Tables
- Table fields
- Global variables
- Commands
- Functions
- Procedures

In addition to its capability of accessing dictionary resources, VS Tools can also monitor events similar to those we discussed in the previous section. When a monitored event is activated, your custom code is executed to carry out whatever functions are necessary for your customization.

The code that executes is called an **event handler**. You register the event handler against the event you want to monitor and it will run your code when the event occurs. Conceptually, this architecture is very similar to what we do in Dexterity. Registration is required in both tools and the VS Tools event handler takes the place of Dexterity's Trigger Processing Procedure.

Similar to Dexterity, you can run the event handler either before or after the Dynamics GP code for that event. If you run it before, you can often cancel the Dynamics GP scripts and take control of the application.

The following events are available in VS Tools; notice how similar they are to the Dexterity events:

Form events

The following is a list of Form events:

- **OpenBeforeOriginal**: The `OpenBeforeOriginal` event occurs when the form is opened, but *before* the Dynamics GP form Pre Script executes.
- **OpenAfterOriginal**: The `OpenAfterOriginal` event occurs when the form is opened, but *after* the Dynamics GP form Pre Script executes.
- **CloseBeforeOriginal**: The `CloseBeforeOriginal` event occurs when the form is closed, but *before* the Dynamics GP form Post Script executes.
- **CloseAfterOriginal**: The `CloseAfterOriginal` event occurs when the form is closed, but *after* the Dynamics GP form Post Script executes.

Window events

The following is a list of window events:

- **OpenBeforeOriginal**: The `OpenBeforeOriginal` event occurs when the window is opened, but *before* the Dynamics GP window `Pre` script executes.
- **OpenAfterOriginal**: The `OpenAfterOriginal` event occurs when the window is opened, but *after* the Dynamics GP window `Pre` script executes.
- **ActivateBeforeOriginal**: The `ActivateBeforeOriginal` event occurs when the window gains focus, but *before* the Dynamics GP window `Activate` script executes.
- **ActivateAfterOriginal**: The `ActivateAfterOriginal` event occurs when the window gains focus, but *after* the Dynamics GP window `Activate` script executes.
- **CloseBeforeOriginal**: The `CloseBeforeOriginal` event occurs when the window is closed, but *before* the Dynamics GP window `Post` script executes.
- **CloseAfterOriginal**: The `CloseAfterOriginal` event occurs when the window is closed, but *after* the Dynamics GP window `Post` script executes.
- **PrintBeforeOriginal**: The `PrintBeforeOriginal` event occurs when the user selects **Print** from the **File** menu, but *before* the Dynamics GP `Print` script executes.

- **PrintAfterOriginal**: The `PrintAfterOriginal` event occurs when the user selects **Print** from the **File** menu, but *after* the Dynamics GP `Print` script executes.
- **BeforeModalDialog**: The `BeforeModalDialog` event occurs when Dynamics GP has opened a modal dialog box, but before it has been presented to the user. This event is often used to programmatically answer the dialog, thereby preventing the user from ever seeing it, or to change the text of the dialog to a more understandable message.

  You can have some fun with this event by changing the message on a dialog box to something more clever than the original. Something like **This will delete all open transaction records from the database, do you want to continue?** and then you only present them with the **OK** button. This sort of message always wakes people up.

  However, the more mainstream use of this function is to make dialogs less of a nuisance. For example, if you create a new transaction and type in a new batch ID, the following modal dialog is displayed:

  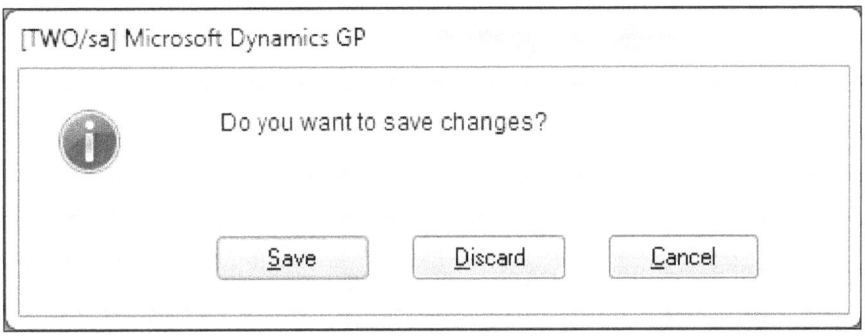

  Your event handler code for `BeforeModalDialog` could push the **Add** button so the user never even sees the question.

- **AfterModalDialog**: The `AfterModalDialog` event occurs when the user has acted on a modal dialog displayed by a window. The event handler code can evaluate the user's response to the dialog and use that information for further processing. For example, if you needed to add a record to your customization every time a new Batch ID is added, you would use the `AfterModalDialog` event to find out if the user did indeed want to add the new Batch ID.

Scrolling window events

The following is a list of scrolling window events:

- **LineFillBeforeOriginal**: The `LineFillBeforeOriginal` event occurs before the Dynamics GP `LineFill` event. This event is activated before the Dynamics GP `LineFill` script executes. The `LineFill` event occurs before the `LinePre` event. Event handler code for this event can cancel the Dynamics GP `LineFill` event.

  The events execute in this order:

  1. VS Tools `LineFillBeforeOriginal`
  2. Dexterity `LineFill`
  3. Dexterity `LinePre`

- **LineFillAfterOriginal**: The `LineFillAfterOriginal` event occurs after the Dynamics GP `LineFill` event. This event is activated after the Dynamics GP `LineFill` script executes. The `LineFill` event occurs before the `LinePre` event.

  The events execute in this order:

  1. Dexterity `LineFill`
  2. VS Tools `LineFillAfterOriginal`
  3. Dexterity `LinePre`

- **LineEnterBeforeOriginal**: The `LineEnterBeforeOriginal` event occurs when focus moves into a line on the scrolling window, but before the Dynamics GP `LinePre` event occurs. Event handler code for this event can cancel the Dynamics GP `LinePre` event.

  The events execute in this order:

  1. VS Tools `LineEnterBeforeOriginal`
  2. Dexterity `LinePre`

- **LineEnterAfterOriginal**: The `LineEnterAfterOriginal` event occurs when focus moves into a line on the scrolling window, but after the Dynamics GP `LinePre` event occurs.

  The events execute in this order:

  1. Dexterity `LinePre`
  2. VS Tools `LineEnterBeforeOriginal`

- **LineChangeBeforeOriginal**: The `LineChangeBeforeOriginal` event occurs when focus moves out of a line on the scrolling window and a field on that line has changed. This event occurs before the Dynamics GP `LineChange` event. Event handler code for this event can cancel the Dynamics GP `LineChange` event.

  The events execute in this order:

  1. VS Tools `LineChangeBeforeOriginal`
  2. Dexterity `LineChange`

- **LineChangeAfterOriginal**: The `LineChangeAfterOriginal` event occurs when focus moves out of a line on the scrolling window and a field on that line has changed. This event occurs after the Dynamics GP `LineChange` event.

  The events execute in this order:

  1. Dexterity `LineChange`
  2. VS Tools `LineChangeAfterOriginal`

- **LineLeaveBeforeOriginal**: The `LineLeaveBeforeOriginal` event occurs when focus moves out of a line on the scrolling window. This event occurs before the Dynamics GP `LinePost` event. Event handler code for this event can cancel the Dynamics GP `LinePost` event.

  The events execute in this order:

  1. VS Tools `LineLeaveBeforeOriginal`
  2. Dexterity `LinePost`

- **LineLeaveAfterOriginal**: The `LineLeaveAfterOriginal` event occurs when focus moves out of a line on the scrolling window. This event occurs after the Dynamics GP `LinePost` event.

  The events execute in this order:

  1. Dexterity `LinePost`
  2. VS Tools `LineLeaveAfterOriginal`

- **LineInsertBeforeOriginal**: The `LineInsertBeforeOriginal` event occurs when the user selects Insert Row from the Edit menu. This event occurs before the Dynamics GP `LineInsert` event. Event handler code for this event can cancel the Dynamics GP `LineInsert` event.

  The events execute in this order:

  1. VS Tools `LineInsertBeforeOriginal`
  2. Dexterity `LineInsert`

- **LineInsertAfterOriginal**: The `LineInsertAfterOriginal` event occurs when the user selects Insert Row from the Edit menu. This event occurs after the Dynamics GP `LineInsert` event.

  The events execute in this order:

  1. Dexterity `LineInsert`
  2. VS Tools `LineInsertAfterOriginal`

- **LineDeleteBeforeOriginal**: The `LineDeleteBeforeOriginal` event occurs when the user selects **Delete Row** from the **Edit** menu. This event occurs before the Dynamics GP `LineDelete` event. Event handler code for this event can cancel the Dynamics GP `LineDelete` event.

  The events execute in this order:

  1. VS Tools `LineDeleteBeforeOriginal`
  2. Dexterity `LineDelete`

- **LineDeleteAfterOriginal**: The `LineDeleteAfterOriginal` event occurs when the user selects **Delete Row** from the **Edit** menu. This event occurs after the Dynamics GP `LineDelete` event.

  The events execute in this order:

  1. Dexterity `LineDelete`
  2. VS Tools `LineDeleteAfterOriginal`

To wrap it up, the Line navigation events execute in the following order:

1. VS Tools `LineFillBeforeOriginal`
2. Dexterity `LineFill`
3. VS Tools `LineFillAfterOriginal`
4. VS Tools `LineEnterBeforeOriginal`

5. Dexterity `LinePre`
6. VS Tools `LineEnterAfterOriginal`
7. VS Tools `LineChangeBeforeOriginal`
8. Dexterity `LineChange`
9. VS Tools `LineChangeAfterOriginal`
10. VS Tools `LineLeaveBeforeOriginal`
11. Dexterity `LinePost`
12. VS Tools `LineLeaveAfterOriginal`

Field events

The order of execution for Field events work in much the same way as the Line events covered earlier. However, be aware that each line of a scrolling window also includes fields, and the field events occur for those scrolling window fields just as they do everywhere else. The Line events are simply an accessory unique to scrolling windows. Let's go over the Field events available, using VS Tools:

- **Change**: This event occurs when the value of the field changes, whether changed by the user or by Dynamics GP code, and the user attempts to leave the field.
- **ClickBeforeOriginal**: This event applies to push buttons and visual switches. The `ClickBeforeOriginal` event occurs when the user clicks on the button or switch. This event occurs before the Dynamics GP field Change event. Event handler code for this event can cancel the Dynamics GP `Change` event.
- **ClickAfterOriginal**: This event applies to push buttons and visual switches. The `ClickAfterOriginal` event occurs when the user clicks on the button or switch. This event occurs after the Dynamics GP field `Change` event.
- **EnterBeforeOriginal**: The `EnterBeforeOriginal` event occurs when focus moves into a field, but before the field Pre event occurs. Event handler code for this event can cancel the Dynamics GP `Pre` event.
- **EnterAfterOriginal**: The `EnterAfterOriginal` event occurs when focus moves into a field. This event occurs after the Dynamics GP field `Pre` event occurs.
- **LeaveBeforeOriginal**: The `LeaveBeforeOriginal` event occurs when focus moves out of a field, but before the field Post event occurs. Event handler code for this event can cancel the Dynamics GP `Post` event.
- **LeaveAfterOriginal**: The `LeaveAfterOriginal` event occurs when focus moves out of a field after the Dynamics GP field `Post` event occurs.

- **ValidateBeforeOriginal**: When a field changes (other than a push button or visual switch) and focus moves out of the field, the Validate in CIT event occurs. The Validate in CIT event can be invoked using the `ForceValidate()` method so that the Validate event will occur when focus leaves the field, whether the field has changed or not. At this point, any VS Tools validation code attached to that event will execute. This event occurs before the fields' Validate in CIT event. Event handler code for this event can cancel the Validate in CIT event. In Dexterity terms, this is similar to the `TRIGGER_REGISTER_FOCUS/TRIGGER_BEFORE_ORIGINAL` event.
- **ValidateAfterOriginal**: When a field changes (other than a push button or visual switch) and focus moves out of the field, the `Validate` event occurs. The `Validate` event can be invoked using the `ForceValidate()` method so that the `Validate` event will occur when focus leaves the field, whether the field has changed or not. At this point, any VS Tools validation code attached to that event will execute. This event occurs after the fields' `Validate` event.

Procedure events

The procedure events are as follows:

- **InvokeBeforeOriginal**: This event occurs when a Dynamics GP procedure is invoked, but before the procedure runs. The event handler can access the values of any parameters passed into the procedure and can change the values of the `out` and `inout` parameters.
- **InvokeAfterOriginal**: This event occurs when a Dynamics GP procedure is invoked after the procedure has run. The event handler can access the values of any parameters passed into the procedure and can change the values of the `out` and `inout` parameters.

Function events

The function events are as follows:

- **InvokeBeforeOriginal**: This event occurs when a Dynamics GP function is invoked, but before the function runs. The event handler can access the values of any parameters passed into the function as well as the return value. The event handler can change the values of `out` and `inout` parameters.
- **InvokeAfterOriginal**: This event occurs when a Dynamics GP function is invoked after the function has run. The event handler can access the values of any parameters passed into the function as well as the return value. The event handler can change the values of the `out` and `inout` parameters.

We'll see in the next section that VBA events follow this same pattern.

Modifier with VBA

Modifier with VBA is two tools in one. Using the Modifier component, you can make extensive changes to the user interface, as we discussed earlier. Using the VBA component, you can make those changes come to life.

One of the changes you can make using Modifier is the creation of new fields on a window. These new fields are not tied to a table, nor do they have any script behind them to make them operational. When we push the new button on the window, nothing happens. Using the VBA component, you can attach VBA code behind those buttons. Now, when the button is pushed, those new fields can be populated.

VBA is a powerful language and you can use it to create some sophisticated business logic changes, but it does not create a running application on its own. If extensive changes are necessary, you would be better served using Dexterity or VS Tools to create a new application. VBA was not intended to be used as a tool that significantly alters the nature of the application.

One big advantage that VBA holds over other tools is that it is blind to the dictionary construct, so it is simple to modify windows in any third-party dictionary. You can easily put buttons on a window in fixed assets and use VBA to cause that button to open a window in Grant Management.

In order to manipulate data, VBA makes an external connection to the database. Any data validation must be provided by you as the developer.

VBA can modify the business logic by running event procedures upon the activation of a VBA event. Event procedures can run either *before* or *after* the Dynamics GP event script executes.

For example, when a window first opens, the following events occur in this order:

1. Dexterity Form `Pre` script
2. VBA `BeforeOpen` window event
3. Dexterity window `Pre` script
4. VBA `AfterOpen` window event
5. VBA `BeforeGotFocus` field event
6. Dexterity field `Pre` script
7. VBA `AfterGotFocus` field event
8. VBA `BeforeActivate` window event
9. Dexterity `Activate` window event

VBA events are similar to those used in VS Tools; they also correspond to Dexterity events. VBA events are explained in the following sections.

Window events

The window events are as follows:

- **BeforeOpen**: The `BeforeOpen` window event occurs as the window opens, but before the Dynamics GP window `Pre` script executes. Therefore, the Dynamics GP `Pre` script could override any field presets that were made by the VBA code. VBA event procedure code can cause the window to open invisibly.
- **AfterOpen**: The `AfterOpen` window event occurs as the window opens, but *after* the Dynamics GP window `Pre` script executes. In this event, your VBA event procedure code can override any values set by the Dynamics GP `Pre` script.
- **BeforeClose**: The `BeforeClose` window event occurs as the window closes, but before the Dynamics GP window `Post` script executes.
- **AfterClose**: The `AfterClose` window event occurs as the window closes, but before the Dynamics GP window `Post` script executes. You can use the `AfterClose` event to update values in your integrating application.
- **BeforeActivate**: The `BeforeActivate` event occurs when focus comes to the window, but before the Dynamics GP window `Activate` script executes. Keep in mind that the `Activate` event also occurs each time a window is opened.
- **AfterActivate**: The `AfterActivate` event occurs when focus comes to the window after the Dynamics GP window `Activate` script executes. Keep in mind that the `Activate` event also occurs each time a window is opened.

Modal dialog events

Modal dialog events work in the sameway in VBA as they do in VS Tools. A modal dialog is a window you must dismiss before you can do anything else in the application. The dialog may be informational and just have an **OK** button on it, or it could be an **ask dialog** with up to three buttons on it.

For example, whenever you enter a new batch ID on a transaction, a modal dialog pops up and asks you if you want to create the batch. The following modal dialog is displayed:

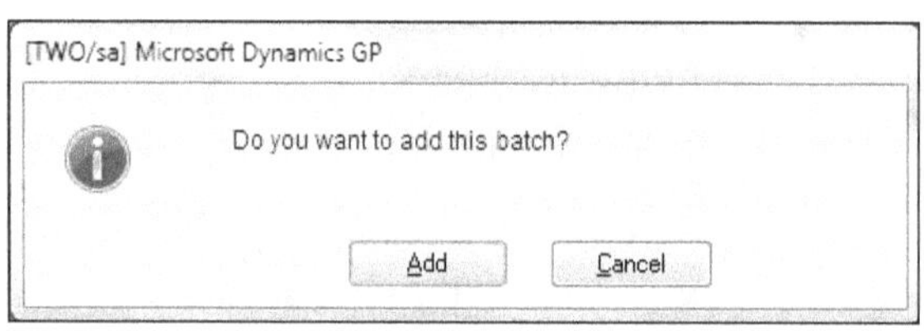

Until you dismiss this dialog, you will not be able to do anything else in the application.

The following are modal dialog events:

- **BeforeModalDialog**: The `BeforeModalDialog` event occurs when Dynamics GP raises a modal dialog, but before it has been displayed to the user. You can programmatically dismiss the dialog by answering the question or pressing the **OK** button, as appropriate. For the dialog box shown here, a VBA event procedure could push the **Add** button and the user would never see this dialog.

  Instead of dismissing the dialog, the event procedure could change the text of the dialog displayed. This event can be loads of fun because you can check the User ID and then "personalize" the message for that user. Instead of the dialog shown earlier, what would happen if the dialog looked like the one shown next? Imagine the possibilities!

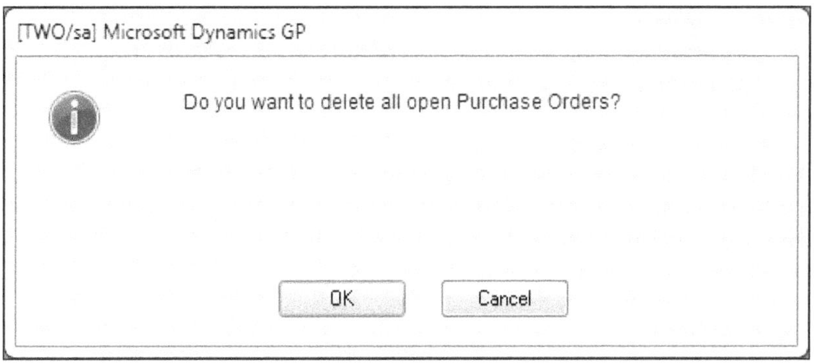

- **AfterModalDialog**: The `AfterModalDialog` event occurs when Dynamics GP raises a modal dialog and the user dismisses it. Use this event to determine how the user answered the dialog.

Field events

The field events are as follows:

- **BeforeGotFocus**: The `BeforeGotFocus` event occurs when the user enters a field. This event occurs before the Dynamics GP `Pre` event for this field.

  The VBA event procedure can prevent the Dynamics GP field `Pre` script from executing by setting the `CancelLogic` parameter to `True`. If the `Pre` script does not execute, then any event procedure attached to the `AfterGotFocus` event will not execute.

- **AfterGotFocus**: The `AfterGotFocus` event occurs when the user enters a field. This event occurs after the Dynamics GP `Pre` event for this field.
- **BeforeUserChanged**: The `BeforeUserChanged` event occurs when the user has changed the data in a field and attempts to leave the field. For push buttons and visual switches, the event occurs when the user clicks on the push button or visual switch. This event occurs before the Dynamics GP `Change` event.

  The VBA event procedure can prevent the Dynamics GP field `Change` script from executing by setting the `CancelLogic` parameter to `True`. If the `Change` script does not execute, any event procedure attached to the `AfterUserChanged` event will not execute.

- **AfterUserChanged**: The `AfterUserChanged` event occurs when the user has changed the data in a field and attempts to leave the field. For push buttons and visual switches, the event occurs when the user clicks on the push button or visual switch. This event occurs after the Dynamics GP `Change` event.
- **BeforeLostFocus**: The `BeforeLostFocus` event occurs when the user attempts to leave the field. This event occurs before the Dynamics GP `Post` event.

  The VBA event procedure can prevent the Dynamics GP field `Post` script from executing by setting the `CancelLogic` parameter to `True`. If the `Post` script does not execute, any event procedure attached to the `AfterLostFocus` event will not execute.

- **AfterLostFocus**: The `AfterLostFocus` event occurs when the user attempts to leave the field. This event occurs after the Dynamics GP `Post` event.

Scrolling window events

The scrolling window events are as follows:

- **BeforeLineGotFocus**: The `BeforeLineGofFocus` event occurs when the user enters a line on the scrolling window. This event occurs before the Dynamics GP `LinePre` event.

  The VBA event procedure cannot prevent the GP `LinePre` event from occurring.
- **AfterLineGotFocus**: The `AfterLineGofFocus` event occurs when the user enters a line on the scrolling window. This event occurs after the Dynamics GP `LinePre` event.

- **BeforeLineChange**: The `BeforeLineChange` event occurs when the user attempts to leave a line on the scrolling window and one of the fields on the line has changed. This event occurs before the Dynamics GP `LineChange` event.

  The VBA event procedure cannot prevent the GP `LineChange` event from occurring.

- **AfterLineChange**: The `AfterLineChange` event occurs when the user attempts to leave a line on the scrolling window and one of the fields on the line has changed. This event occurs after the Dynamics GP `LineChange` event.
- **BeforeLinePopulate**: The `BeforeLinePopulate` event occurs before the Dynamics GP `LineFill` event. This event is activated before the Dynamics GP `LineFill` script executes. The `LineFill` event occurs each time a new line is displayed in the scrolling window and each time focus is moved to an existing line that already contains data. When the scrolling window is initially filled, the `LineFill` event occurs repeatedly until the scrolling window is completely full. The `LineFill` event occurs before the `LinePre` event.

  The `BeforeLinePopulate` event is commonly used to filter out any records you do not want to display on the grid.

- **AfterLinePopulate**: The `AfterLinePopulate` event occurs after the Dynamics GP `LineFill` event. This event is activated after the Dynamics GP `LineFill` script executes. The `LineFill` event occurs each time a new line is displayed in the scrolling window and each time focus is moved to an existing line that already contains data. When the scrolling window is initially filled, the `LineFill` event occurs repeatedly until the scrolling window is completely full. The `LineFill` event occurs before the `LinePre` event.
- **BeforeLineLostFocus**: `The BeforeLineLostFocus` event occurs when the user exits a line on the scrolling window. This event occurs before the Dynamics GP `LinePost` event.
- **AfterLineLostFocus**: The `AfterLineLostFocus` event occurs when the user exits a line on the scrolling window. This event occurs after the Dynamics GP `LinePost` event.

Report events

In addition to windows, VBA can interact with Report Writer reports to set or retrieve values of a field and to include new fields added with the Modifier to a report.

The report events are as follows:

- **Start**: The `Start` event occurs once at the beginning of a report just before the report starts to print. No data values are available during this event. The `Start` event is typically used to populate report legends, open an ADO connection to SQL Server, and to initialize any module-level variables you use in the report.

  Legends are fields used to display information that is passed to the report when it is printed. Legend fields are string fields that are normally used to display the sorting and restriction options used in the report. Legend fields can be used in calculated fields.

- **End**: The `End` event occurs once after the report prints. The `End` event is often used to close an ADO connection to SQL Server.

Band events

Report Writer is a single-pass, banded report writer. Each report contains several sections; these sections are called bands. Each band has a corresponding band event. When a report prints, each band event occurs just before the data in the band prints. The following screenshot shows the bands that make up a report; reports may contain several Additional Header and Additional Footer sections.

The `BeforePH` event occurs before the `BeforeRH` event. Likewise, the `BeforePF` event occurs after the `BeforeRF` event. Therefore, on the first page of the report, the page header is printed before the report header. On the last page of the report, the report footer is printed before the page footer.

The band events are as follows:

- **BeforePH** (**Page Header**): The `BeforePH` event occurs just before any data in the page header prints. The page header prints at the top of every page, including the first page. Therefore, the `BeforePH` event occurs first and then the `BeforeRH` event occurs.
- **BeforeRH** (**Report Header**): The `BeforeRH` event occurs only on the first page of the report before any items in the report header print. The `BeforeRH` event occurs after the `BeforePH` event.
- **BeforeAH** (**Additional Header**): `TheBeforeAH` event occurs before items in the report's additional header print. If the report uses multiple additional headers, a `BeforeAH` event occurs for each additional header.
- **BeforeBody** (**Line Items**): The `BeforeBody` event occurs before each line item in the body section prints. If there are five body records, the `BeforeBody` event will occur five times.
- **BeforeAF** (**Additional Footer**): The `BeforeAF` event occurs before items in the report's additional footer print. If the report uses multiple additional footers, a `BeforeAF` event occurs for each additional footer.
- **BeforeRF** (**Report Footer**): The `BeforeRF` event occurs only on the last page of the report and before any items in the report footer print. The `BeforeRF` event occurs first, and then the `BeforePF` event occurs.
- **BeforePF** (**Page Footer**): The `BeforePF` event occurs just before data in the page footer prints. The page footer prints at the bottom of every page, including the last page. Therefore, the `BeforePF` event occurs after the `BeforeRF` event.

Continuum

The Continuum API allows any COM-compliant development environment to interact with Dynamics GP.

Designed originally to be used with Visual Basic and Delphi, it gives you the ability to access the OLE layer of Dynamics GP and pass commands through the application at runtime.

eXtender Enterprise

Using eXtender Enterprise you can attach sanScript to newly created forms, thereby adding functionality or modifying existing functionality. Procedures added through eXtender Enterprise can create brand-new modules to meet specific needs.

eXtender Enterprise is itself a hybrid because much of the code is built in to the application and you will not have to write those procedures. Using this, you can create some pretty impressive changes in a very short period of time, across the tables.

Sometimes modifying the user interface and business logic is not necessary because the only thing you need to do is interact with the database itself. Many tools can access a SQL database, and we have three choices:

- eConnect and SmartConnect
- Integration Manager
- Continuum

Adding information not previously collected

If you just need to add some additional static information to records in the database, several of the tools we have discussed can satisfy that need. The best tool for that sort of thing is Extender. That's all good and fine, but if you don't happen to own the Extender module, that's not a good answer.

This section explores how you can store additional data using the **DUOS** tables.

DUOS

DUOS stands for **Dynamic User Object Store**. The DUOS table is stored in the company database and has the following characteristics.

Table information:

Display name	Technical name	Physical name	Series
SY_User_Object_Store	SY_User_Object_Store	SY90000	Company

Field information:

Field name	Physical name	Storage type	Field Position
GPS_Reserved	GPS_RESERVED	Long Integer	1
ObjectType	ObjectType	String	5
ObjectID	ObjectID	String	37
PropertyName	PropertyName	String	99
PropertyValue	PropertyValue	String	131

Key Information:

Key name	DUOS_Key1	DUOS_Key2
Segment1	ObjectType	ObjectType
Segment2	ObjectID	PropertyName
Segment3	PropertyName	PropertyValue

The DUOS tables were constructed so that a user could store additional information, which did not have anywhere else to live, in the tables. They were designed to be used with the Modifier with VBA module so that extra fields added to a window could be stored without requiring you to create additional SQL tables. The DUOS table is misunderstood or unknown by most, so they do not get much attention out there in Dynamics GP land; but that stops now. You can either write to it directly from outside of Dynamics GP, or use VBA to programmatically access the table. The VBA developer's guide will walk you through the entire process. The VBA developer's guide is installed with Dynamics GP. Using the default settings, you can find it here for a 64-bit machine: `C:\Program Files (x86)\Microsoft Dynamics\GP2010\Documentation\ VBADevelopersGuide.pdf`. Find it here for a 32-bit machine: `C:\Program Files\Microsoft Dynamics\GP2010\Documentation\ VBADevelopersGuide.pdf`.

Summary

Choosing a development tool is influenced by four prominent considerations:

- Type of integration
- Capabilities of the toolset
- Skills of the developer
- Prerequisites of licensing to the end user

In this chapter we described how the following ten development tools fit into the considerations just mentioned:

- Dexterity
- VS Tools
- Modifier with VBA
- Continuum
- Extender and eXtender Enterprise
- DDE, ODBC, and ADO,and OLE Automation
- Integration Manager
- Table Import
- eConnect
- Web services

We looked at which of the tools would modify the user interface, add functionality, or just access the database directly. At the end of the day, the solution needs to meet the specification, be supportable, affordable, and upgradable.

The next chapter is more of a hands-on case study. We are going to use a couple of techniques discussed in this chapter to create a small integrating application using Dexterity.

3

Getting Started with Dexterity

In the next few chapters, you will create your first integrating application using Dexterity. As you develop this application, we will see the key concepts of Dexterity in action. You can create a standalone application using Dexterity, but our focus is on creating an application that will integrate with Dynamics GP.

In this chapter, you will set up your development environment, explore the different components that make up Dexterity, and learn how to navigate the Dexterity Resource Explorer.

The key topics in this chapter include:

- Overview of the development process
- Preparing the development environment
- Overview of Dexterity
- Navigating the Resource Explorer

Overview of the development process

Developing an integrating application using Dexterity involves several steps. This section will provide an overview of those steps.

Installing the software

The first step in the development process seems somewhat obvious, but there are several items involved. The first step is to install the software that you'll use to create your customization. The software you will install to develop an integrated application using Dexterity includes the following:

- Dynamics GP
- Dexterity
- Dynamics GP Software Development Kit
- DexSense
- Support Debugging Tool

Preparing your development environment

After you install the software, you need to prepare your development environment. Preparing the development environment involves the following steps:

- Creating a folder in which to develop your application
- Making a copy of the `Dynamics.dic` file from your production environment and copying it to your development folder
- Renaming the development copy of the `Dynamics.dic` file to something appropriate for your project

Developing the application

As you create your user interface and business logic code, you will repeatedly switch back and forth between test mode (also called Debug mode) and tools mode in order to test your application. When you're in test mode, you're actually running Dynamics GP from your development dictionary. While in tool mode, you can use the source debugging tools and the script editor to perfect your code.

While developing your application, you will, no doubt, create additional forms, windows, reports, procedures, and so on. These objects are each considered to be a **resource** that is added to the development dictionary. Every resource in the dictionary has a resource ID. Dexterity assigns the resource ID automatically, and you cannot change it. The resources you create will have a resource ID greater than or equal to 22000.

Creating the chunk file

At the end of the development process, you will use the Dexterity Utilities application to create a **Chunk** or `.cnk` file. During this procedure, you will extract all of the resources that you created from the development dictionary. The extracted resources create what is known as an **Extract** dictionary. Additionally, if you modified any existing Dynamics GP form or report, you will need to manually transfer those resources to your Extract dictionary. A modified form or report is known as an **Alternate** form or report.

This extracted dictionary will be used to create a self-extracting Chunk file. This Chunk file becomes your final product dictionary. You will add product information to the Extract dictionary, such as:

- Your application's name
- Your application's delivered dictionary name
- The names for your form and report dictionaries
- Version and build information
- Installation scripts
- The Compression method (whether or not to strip out the source code)

The development environment is graphically depicted in the following diagram:

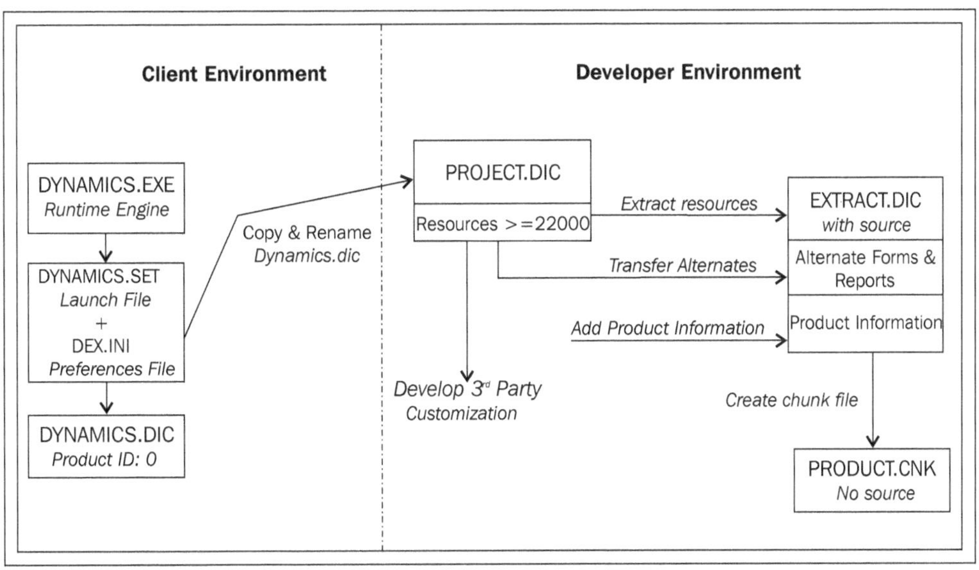

Delivering the final product

After testing and debugging and testing some more, and testing more after that, it's at last time to deliver your final application. You have a couple of different choices on how to deliver your product.

The easiest method is to just ship the `.cnk` file and instruct your user to copy the `.cnk` file into the Dynamics GP application folder and then launch Dynamics GP. The `.cnk` file will automatically expand and become your `.dic` file according to the information you provided during the chunking process.

Alternatively, you can create a Windows Installer file using the WiX installer template that is included with Dexterity. You can find the template with instructions in the `\Microsoft Dexterity\Dex 11.0\Samples\InstallerDexterity\Samples` folder.

The following illustration depicts the client environment once your application has been deployed:

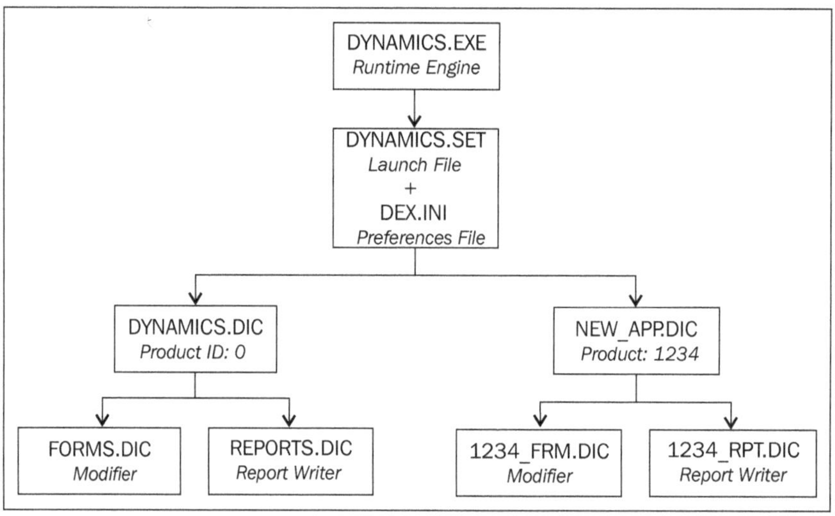

Preparing the development environment

Before you start programming, you need to set up your environment. As you gain more experience working with Dexterity, you will develop your own favorite ways to do things. For now, we will create a good basic starting point. At last, you get to install the software!

We are presuming that a test installation of Dynamics GP and Microsoft SQL is already set up and running on the workstation. The test installation of Dynamics GP should not be connected to the production copy of Dynamics GP. A test company is not sufficient; the entire environment should be for testing. We will focus on installing the Dexterity program, the Dynamics GP Software Development Kit (SDK), and other software you need for establishing your development environment.

After installing the software, you'll create the development dictionary, modify the `Dex.ini` file, and make some necessary changes to Dynamics GP security. With the security changes made, you will make sure that you can switch into test mode, and then you'll be ready to start creating your application.

Installing Dexterity and the SDK

The Dexterity installation files are located in the Dynamics GP installation media. You can download the installation DVD from the locations listed as follows:

From PartnerSource: `http://tinyurl.com/76z68gy`

From CustomerSource: `http://tinyurl.com/7vpm69q`

Alternatively, search for "Product release downloads Dynamics GP 2010 R2" and you will be directed to the correct site.

To install Dexterity, right-click on the file `\Tools\Dex\setup.exe` and select **Run as Administrator**. If you are not installing the software on a computer that has UAC (User Account Control) enabled, you can just double-click on the `setup.exe` file. Accept all of the defaults when prompted, and just press **Next**.

To install the SDK, run the following file:

`\Tools\SDK\Dynamics GP\Microsoft_DynamicsGP11_SDK_x86_en-us.msi`

Again, accept all of the defaults when prompted and press **Next**.

Modifying the Dex.ini file

Since your integrating application needs to be connected to the Dynamics GP data tables, copy the `Dex.ini` file from the Dynamics GP installation to the Dexterity installation. The following pathnames will presume that you have accepted all of the defaults during the installation:

- For a 32-bit machine, copy the `Dex.ini` file

  from `C:\Program Files\Microsoft Dynamics\GP2010\Data\Dex.ini`

  to `C:\Program Files\Microsoft Dexterity\Dex 11.0\Data\Dex.ini`

- For a 64-bit machine, copy the `Dex.ini` file

  from `C:\Program Files (86)\Microsoft Dynamics\GP2010\Data\Dex.ini`

  to `C:\Program Files (86)\Microsoft Dexterity\Dex 11.0\Data\Dex.ini`

It's OK to overwrite the existing file. Open the `Dex.ini` file in the `Dex 11.0\Data` folder and add the following line at the bottom of the file:

```
ShowResIDs=TRUE
```

This switch adds a column named **ID** to the **Resource Explorer**, as shown in the following screenshot. The **ID** column displays the resource number of each item. All of the existing items will have resource IDs less than 22,000; the resources you create will have IDs that start at 22,000. Sorting the **ID** column is a quick way to find the resources you created.

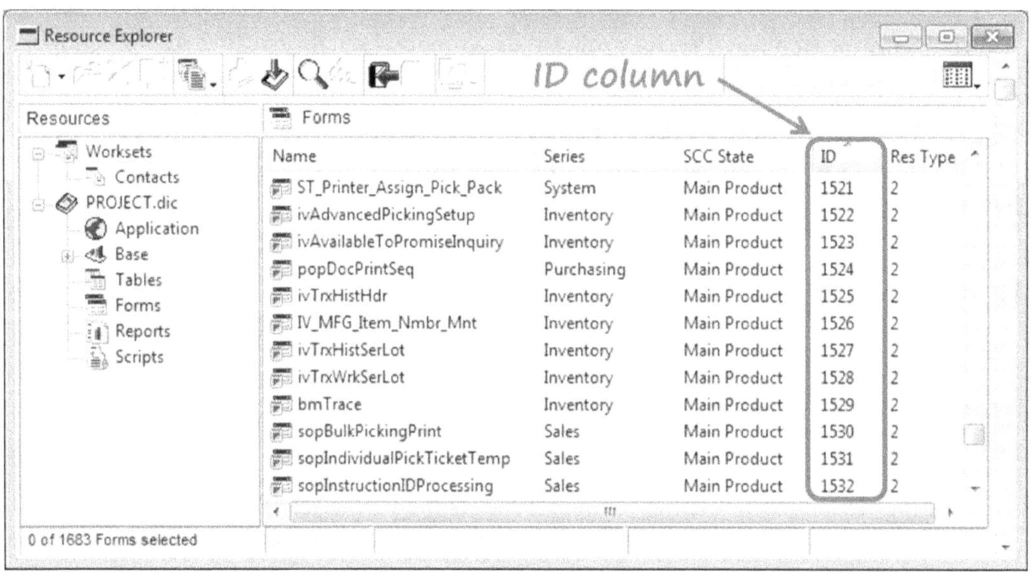

You also need to change the `DexHelpPath=` setting so that it points to the location of the Dexterity help file. If you do not change it, nothing will happen in Dexterity when you press *F1*.

If you accepted the default settings during the installation, you can modify the `DexHelpPath=` setting as follows:

- For a 32-bit machine, modify it to `DexHelpPath=C:\Program Files\Microsoft Dexterity\Dex 11.0\`.
- For a 64-bit machine, modify it to `DexHelpPath=C:\Program Files (86)\Microsoft Dexterity\Dex 11.0\`.

Creating the development dictionary

You will develop your application in a copy of the `Dynamics.dic` file that was used in production. Never open the production dictionary; instead, copy it. Follow these steps to create your development dictionary:

1. Create a folder to hold your development dictionary. Let's create a folder named `DEXDEVELOP` inside your `My Documents` folder.
2. Copy the `Dynamics.dic` file from the production installation of Dynamics GP to the `DEXDEVELOP` folder.

   If you used the default installation paths for Dynamics GP, the `Dynamics.dic` file used in production will be located in the folders indicated as follows:

   For a 32-bit machine:

   `C:\Program Files\Microsoft Dynamics\GP2010\Dynamics.dic.`

   For a 64-bit machine:

   `C:\Program Files (86)\Microsoft Dynamics\GP2010\Dynamics.dic.`

3. Rename the `Dynamics.dic` file in the `DEXDEVELOP` folder to `PROJECT.dic`.

Your development dictionary is ready to go. Launch Dexterity from **All Programs | Microsoft Dexterity | Dex 11.0 | Dexterity**, and open the `PROJECT.dic` dictionary from the `DEXDEVELOP` folder inside of your `My Documents` folder.

The Dexterity **Resource Explorer** window will open; your screen should look similar to the previous screenshot.

To change the Dynamics GP desktop so that the area pages appear in Dexterity Test mode, copy the [cit]Background[/cit] folder from your Dynamics GP workstation installation into the folder containing your development dictionary.

Moving to test mode

The first thing you need to do is to make sure you can switch from tools mode into test mode, sometimes called **Debug** mode. To change over into test mode, use the following navigation:

Debug | **Test Mode** (you could also press *Ctrl* + *T* on your keyboard)

The Dynamics GP login window should be displayed. If, instead, you get the following error message, you need to do one more thing.

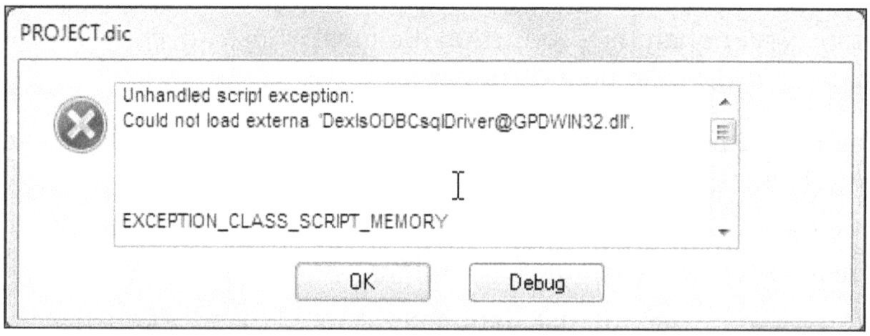

To resolve this error, simply copy the `GPDWIN32.dll` file from your Dynamics GP application folder into your Dexterity application folder.

Once you get the Dynamics GP login window, log in as the `sa` user. Select **Fabrikam** as the company. As soon as your login is complete, you will be greeted with the following dialog:

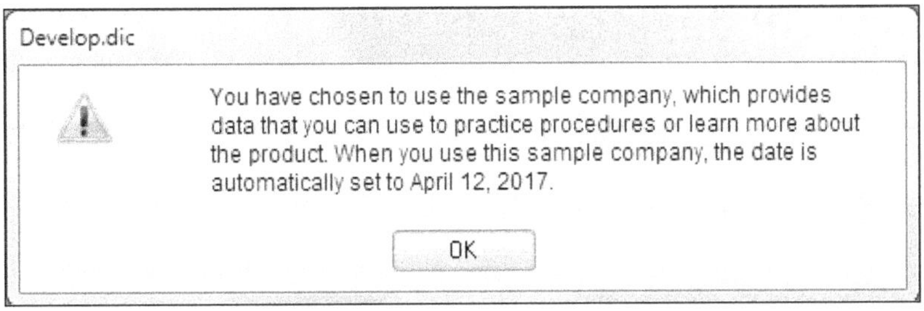

Select **OK** to dismiss the dialog. Dynamics GP will complete its launch and the following window will open:

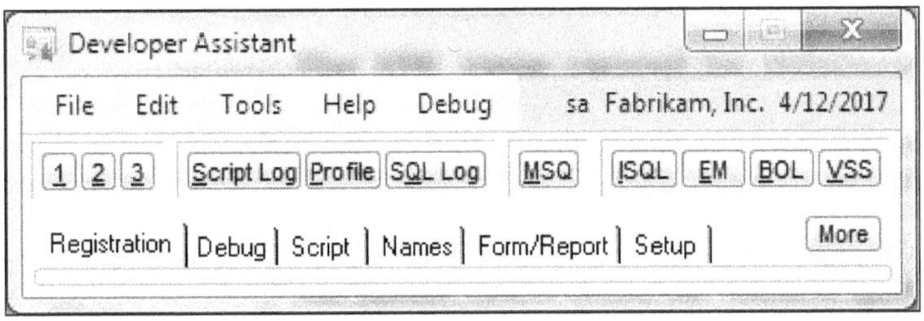

Close the **Developer Assistant** window. Having to close two windows each time you go into Test mode will get real old, real fast. To keep these windows from opening, you need to add two more entries to your `Dex.ini` file. Open the `Dex.ini` file in the **Microsoft Dexterity\Dex 11.0** folder and add the following switches at the bottom:

```
SAMPLEDATEMSG=FALSE
DevAssistHide=TRUE
```

The first switch prevents the sample company dialog from appearing. The second prevents the **Developer Assistant** window from automatically opening.

Dynamics GP desktop

When you switch into Test mode, the regular Dynamics GP homepage will not look the same as what you are accustomed to. The following screenshot should be similar to how your desktop will look:

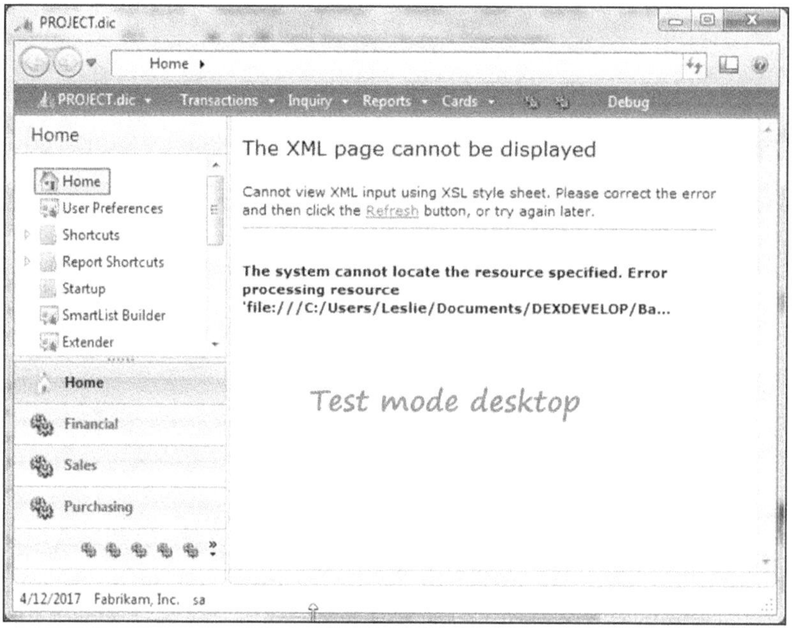

To change the Dynamics GP desktop so that the area pages appear in Dexterity Test mode, copy the `Background` folder from your Dynamics GP workstation installation into the folder containing your development dictionary.

Modifying user security

The **Lookup** windows in Dynamics GP are not stored in the `Dynamics.dic` dictionary; they are in the `SmartList` dictionary `EXP1493.dic`. When running in test mode, you will have access only to the resources in the `Dynamics.dic` dictionary. Because the Lookup windows do not exist in that dictionary, you must change the **Alternate/Modified Forms and Reports** settings to point the security to the original Lookup windows.

If you attempt to open a lookup window without changing the security settings, you will encounter the following error:

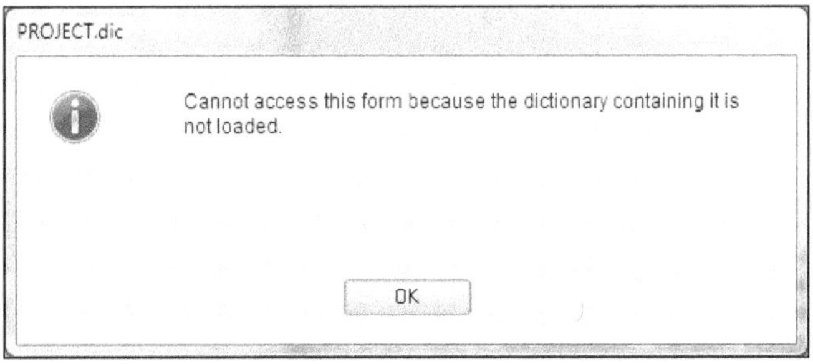

To resolve this error, press *Ctrl* + *T* to switch into test mode and then open the **Alternate/Modified Forms and Reports** window. To open this window, use the navigation **PROJECT.dic | Tools | Setup | System | Alternate/Modified Forms and Reports**, as shown in the following screenshot:

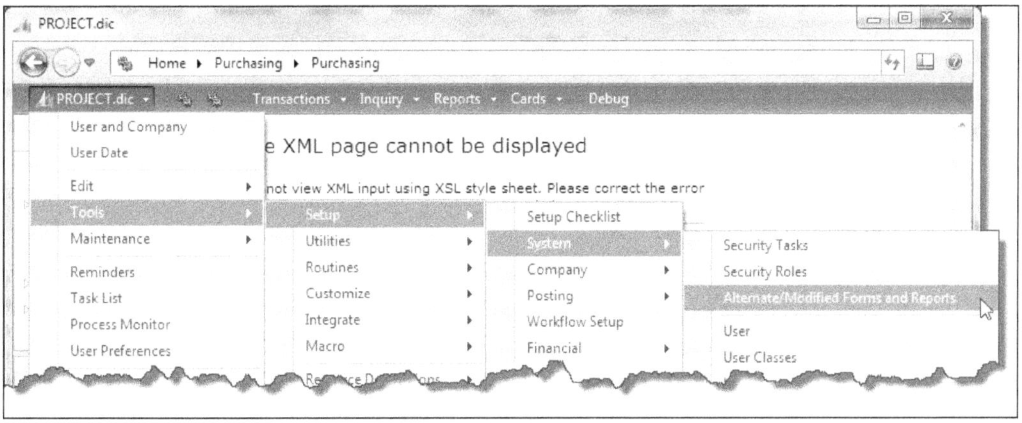

Instead of changing an existing ID, it's a good idea to create a new ID that will be used for testing and development. For now, let's just use the `DEFAULTUSER` ID.

You will be using the customer lookup window in your project, so you only need to change the security for the **Customers and Prospects** window. Set the field values for the top part of the **Alternate/Modified Forms and Reports** window as follows:

Field	Value
ID	**DEFAULTUSER**
Description	**Default User**
Product	**All Products**
Type	**Windows**

When you tab off the **Type** field, the bottom of the window will list each series that contains an **Alternate** or **Modified** window. It may take a couple of moments to load, so have patience. Expand the **Sales** series, navigate down to the **Customers and Prospects** node, expand it, and then mark the radio button next to **PROJECT.dic**; press **Save** to commit your changes. Your window should look similar to the following screenshot:

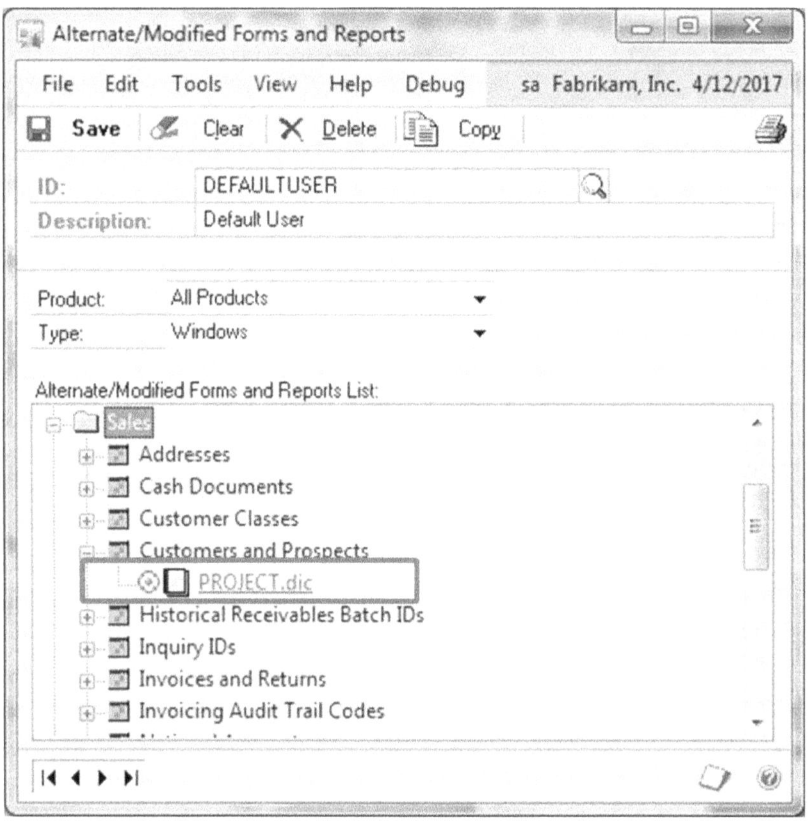

Now if you open the **Customer Maintenance** window from **Cards | Sales | Customer** and press the lookup button next to the **Customer ID** field, the following lookup window should open. This window is known affectionately as the old "green bar" lookup window:

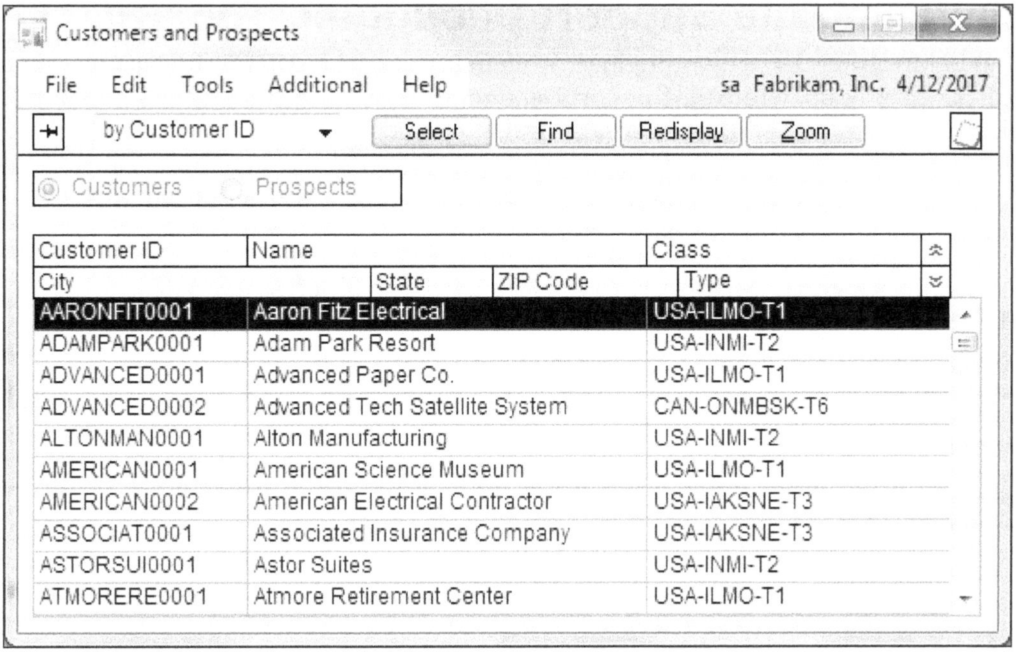

You'll notice that this old lookup window has far fewer features than the lookup window provided by the `SmartList` dictionary. One of the differences is that you cannot click on the column headings to sort the data. There are other differences, but that is the one that you will probably miss the most.

Installing DexSense

Download **DexSense**, an IntelliSense-like tool for Dexterity, from its creator Tim Gordon at Alpine Consulting `http://www.alpinelimited.com`.

This application makes coding considerably faster because it saves you a substantial amount of time that you would otherwise have spent in hunting down the names of your dictionary resources. It works much like the IntelliSense feature, which we all like so much, found in other Microsoft programming tools. DexSense is available as a free download for all Dynamics GP developers, and supports the Dexterity versions 8, 9, 10, and 2010.

Tim Gordon created this for the good of the community and he continues to enhance it and support it. After you see how terrific it is, please take the time to drop an e-mail to Tim at `tim@alpinelimited.com` and let him know how much you like it.

Installing the Support Debugging Tool (SDT)

David Musgrave, Escalation Engineer from the Asia Pacific Microsoft Dynamics GP Support Team, wrote the Support Debugging Tool. With this tool you have an unbelievably robust software package. David packed this tool with so many utilities and functions that it would take an entire book to explore it completely. The SDT is a necessary aid that will greatly assist you in debugging your application and identifying potential performance issues.

Currently, a login for Microsoft's PartnerSource portal is necessary to download the SDT. There is no charge for this software, but you may have to get it from your Dynamics GP partner if you do not have a PartnerSource login. For more information on the SDT, visit the Support Debugging Tool Portal at `http://aka.ms/SDT`

Blast off!

At last, your development environment is ready to go!

Overview of Dexterity

When developing integrated applications using Dexterity, you will be working with a copy of the actual Dynamics GP dictionary. You will have access to all of the preexisting resources, except the scripts, and can re-use them in your application.

While it is possible to change existing Dynamics GP forms and reports and use them in your application as alternate forms and reports, best practices dictate that you should avoid creating alternate forms and reports because of the tremendous amount of work you might be saddling yourself with to maintain them. For example, with each new release of Dynamics GP, you quite possibly will need to recreate these resources from scratch.

In addition, you will be limiting your users because they can only use one alternate of the same form at a time. If two developers modify the same form, somebody's functionality will lose.

Components of Dexterity

Several files are used together to create a Dexterity application. This section briefly describes certain files that are copied to your hard drive when you install Dexterity.

Filename	Description
`Dex.exe`	The Dexterity application.
`Dex.dic`	The dictionary that contains resources used by Dexterity, Dexterity Utilities, the Process Server, and the Runtime engine.
`Dex.chm`	The Dexterity help file.
`DexUtils.exe`	The Dexterity Utilities application.
`DexUtils.dic`	The dictionary that contains resources used by Dexterity Utilities.
`DexUtils.chm`	The Dexterity Utilities help file.
`Dynamics.exe or Runtime.exe`	The Dexterity runtime engine.
`Contain.exe`	The OLE container application.
`Dps.exe`	The Process Server application can be used to offload background processes to a different machine called a Process Server.
`Dpm.exe`	The Distributed Process Manager application performs load balancing among multiple Process Servers.
`IG.chm`	The help file describing how to create Dexterity applications that integrate with Dynamics GP. This is required reading! It is much easier to navigate through this information using the help file than the `IG.pdf` document that is in the `\Microsoft Dexterity\Dex 11.0\Manuals` folder.
`Import.chm`	The Import Utility help file.
`MiniDex.chm`	A help file describing the Customization Maintenance windows, Process Monitor windows, and Report Destination windows.
`ResDesc.chm`	The Resource Descriptions tool help file. This help file describes the Table, Window, and Field description windows that you can access from the **Tools** menu of nearly every Dynamics GP window.

Filename	Description
.dll files	Dynamics Link Libraries used to perform a variety of operations, such as calling 32-bit DLLs, .NET-managed shell assemblies, .NET eventing, access to c-tree Plus tables, and the data dictionary API (allows external applications to access Dexterity resources).
.tlb files	Type Libraries that allow COM support for Dexterity applications.

Resources and their relationships

Dexterity uses a unique method for defining resources. Each resource created in Dexterity stands alone and has its own resource ID. For instance, a field is a global resource that exists outside of a table. You can use that same field definition in any table regardless of the information stored in it. For example, if you create a 15-character string field, you can use that string field in any table without needing to redefine it.

Dexterity uses a bottom-up approach when creating resources. You create the lowest level of resource first, and then you build up to the highest level of resource.

The following diagram illustrates the bottom-up approach used by Dexterity:

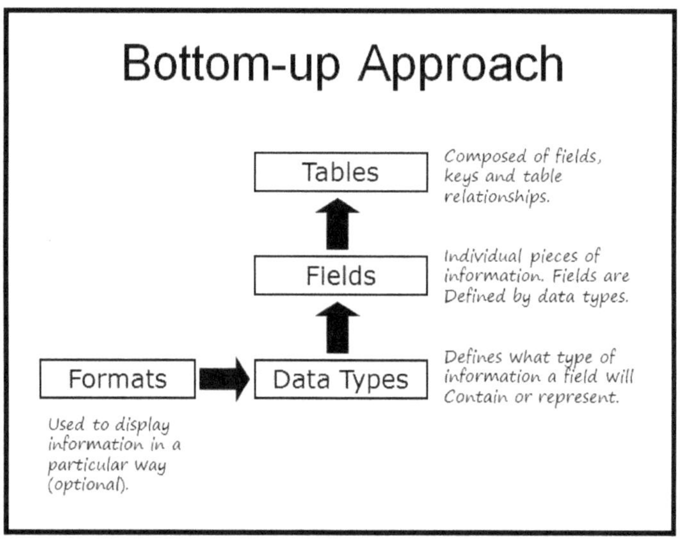

The following is a description of several types of dictionary resources in the order in which you would be likely to create them:

DataType

The DataType is the lowest level of resource, and you use it to define what type of information can be contained in a field and that field's characteristics. DataTypes control the following elements:

- Control Type
- Keyable length
- Storage size
- Number of decimal places
- Static values
- Format
- Composite definition

The Control Type is the main attribute of the DataType as it controls what *kind* of field you have. A field can be many things. It could be a string, picture, date, push button, or drop-down list, to name a few. Details of the different Control Types supported by Dexterity are in *Appendix A, Dexterity Control Types.*

Format

A **Format** controls how Dynamics GP displays the field. For example, you may want your ID field to be all capital letters; a format can convert all the letters typed into that field into capital letters. In the case of a phone number, you may want the area code in parentheses and a dash in its correct place. A format can accomplish this using placeholders. The capital X is the placeholder, and all other characters are displayed as they were typed. Any capital X is substituted with data from the field. For example:

Format	Data	Display
(XXX) XXX-XXXX	2145555555	(214) 555-5555

> Many DataTypes can use the same format. If you change the format, you change it for all of the DataTypes that use it.

Field

A **Field** is a distinct bit of information that is stored in a table or displayed on a window. Its attached DataType governs the kind of data you can store in the field. Like Formats and DataTypes, fields are global and you can use them on any resource in the dictionary.

For example, say field Phone1 uses a DataType we'll call **PHONE**. Multiple fields could be using the PHONE DataType. Let's say the Phonefmt format is attached to the PHONE DataType. If you modified the Phonefmt format by changing the format string, like changing the dash to a period, you would be changing the display of the field Phone1 wherever it appears in the application.

Like formats and DataTypes, fields are global resources and you can use them on any window or in any table.

Field properties

Each field has an assortment of properties that govern its behavior and appearance.

Object properties

Object properties control the function or behavior of a field. Some common object properties are discussed as follows:

- **AutoCopy**:The `AutoCopy` property determines whether a field is automatically written to the window or table when you execute a `copy to` or `copy from` sanScript statement
- **Cancel**: The `Cancel` property identifies the button whose script (Change script) will run when the user presses the *Esc* key on the keyboard. The `Cancel` property is only available for push button fields.
- **DataType**: The `DataType` property identifies the DataType attached to the field.
- **Default**: The `Default` property identifies the button whose script (Change script) will run when the user presses the *Enter* key on the keyboard. The `Default` property is only available for push button fields.

- **DefaultDblClick**: The `DefaultDblClick` property determines whether the Default push button's script (Change script) will run when you double-click on a line in the list. If this property were not set, nothing would happen when you double-clicked on the line. Typically, the **Select** button is marked as the Default push button.
- **Editable**: The `Editable` property determines whether or not the user can type into the field. If the `Editable` property is set to `False`, the user will not be able to change the value of the field.
- **Field**: The `Field` property sets the field's name. You use the field name when you refer to the field in your sanScript code.
- **Hyperspace**: The `Hyperspace` property is applied to push button fields only. If `Hyperspace` is set to `true`, the focus will not move away from the previous field when the push button is pressed. If the focus doesn't move, then only the script (Change script) for the push button runs and no scripts will run for the previous field. If the `Hyperspace` property is not set to `true`, any script written to identify a change in the previous field's value (Change script), or any script written to identify the event of leaving the previous field (Post script) would run in addition to the button's script (Change script) when the button is pushed. This property is commonly used on lookup buttons and clear buttons.
- **LinkedLookup**: The `LinkedLookup` property identifies the lookup button associated with the field. Typing *Ctrl* + *L* in the field will push the linked lookup button, thereby executing the script (Change script) attached to that button.
- **LinkedPrompt**: The `LinkedPrompt` property identifies the prompt associated with the field. By linking the prompt to the field, some of the prompt's visual properties will correspond to certain properties of the field.
- **Required**: The `Required` property determines whether a value is required in the field before a record can be saved. This property does not set the behavior; it sets a characteristic that you can check from the sanScript code.

Visual properties

Visual properties control the display characteristics of a field. Some common visual properties are discussed as follows:

- **AltLineColor**: The `AltLineColor` property determines whether alternate lines of a listbox appear in a different color. You can control the color using the `Field_SetAltLineColor()` function.

- **Appearance**: The `Appearance` property defines the type of border that is used around the field. The choices include 2D Border, 3D Border, and 3D Highlight. The 3D Highlight choice is the Dynamics GP standard; this is the setting that results in the underline appearing below the prompt. The borders must be visible for the line to show up.
- **Border**: The `Border` property controls whether or not the border is visible.
- **Position-Left**: The `Position-Left` property sets the position of the left side of the field in pixels.
- **Position-Top**: The `Position-Top` property sets the position of the top side of the field in pixels.
    - The intersection of the `Position-Left` property and the `Position-Top` property identifies the location of the upper left-hand corner of the field.
- **Size-Height**: The `Size-Height` property sets the height of the field in pixels.
- **Size-Width**: The `Size-Width` property sets the width of the field in pixels.
- **Visible**: The `Visible` property determines whether or not the field can be seen by the user.
- **WordWrap**: The `WordWrap` property determines if the text in a `Text` field will be wrapped when it hits the right edge of the field, or if it will continue as a single line until the *Enter* key is pressed. If the `WordWrap` property is set to `true`, the horizontal scroll bar is removed from the bottom of the field.
- **Zoom**: The `Zoom` property determines whether the push button or prompt will take on the visual characteristics of a zoom field. If the `Zoom` property is set to `True` on a prompt, the text will be blue and underlined, or it will be whatever is set in **User Preferences** for a zoom field. If the `Zoom` property is set to `True` on the push button, the mouse pointer will turn into the image of a little hand, pointing its index finger when the mouse passes over the button.

Linking prompts

Linking a field prompt marries the field to the prompt. On the face of it, doesn't seem important, but it is. Always, always link your field prompts. The prompt will match the field's behavior if you link them. Here are some situations where a linked prompt makes all the difference:

- Hiding a field: the prompt of a hidden field will also be hidden
- Disabling a field: the prompt of a disabled field will also be grayed out
- A required field: the prompt of a required field will be displayed according to the user's display preferences.

- When using VBA:
    - The prompt becomes the VBA field name
    - The **caption** property can be used to change the prompt
    - The **enabled** property will also act upon the prompt
    - The **required** property follows the user's display preferences
    - The **visible** property will also show or hide the prompt

To link a prompt, select **Tools | Link Prompt** from the menu bar or select *Ctrl + E* on the keyboard. Click on the field and drag the mouse to the prompt you want to link to. Once linked, both the field and the prompt will flash a black outline similar to the following screenshot:

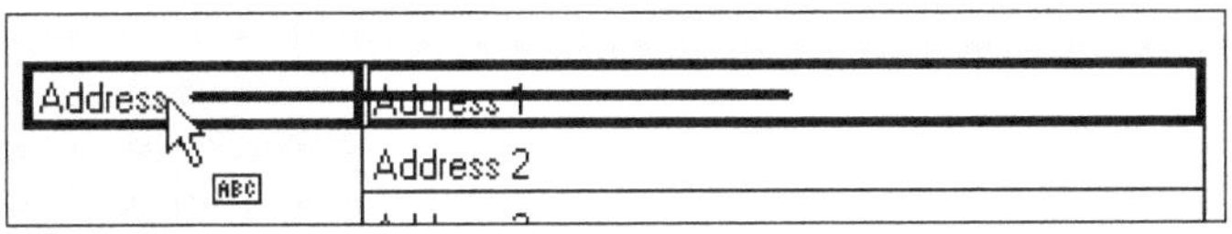

You can link multiple fields to a single prompt. However, you cannot link a field in a scrolling window to a prompt in the main or host window.

When you have finished linking the prompts, select **Tools | Link Prompt** again or select *Ctrl + E* on the keyboard to turn it off.

Linking lookup buttons

Similar to prompts, you need to link your lookup buttons to the field that is being looked up. Linking the lookup button provides the *Ctrl + L* functionality. For example, if you're on the **Customer ID** field and hit *Ctrl + L*, the **Customers and Prospects** lookup window will open. If you hadn't linked the lookup button to the **Customer ID** field, nothing would have happened when you hit *Ctrl + L*.

To link a lookup button, select **Tools | Link Lookup** from the menu bar. Click on the field you want to link the lookup with and drag your mouse to the lookup button. Once linked, both the field and the lookup button will flash a black outline similar to the following screenshot:

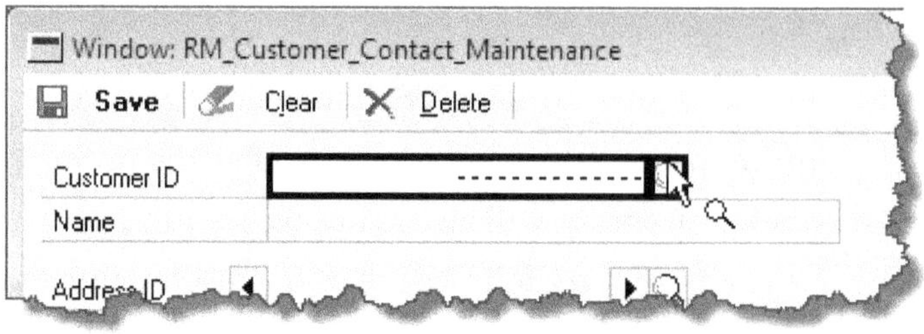

When you have finished linking the lookups, select **Tools | Link Lookup** again to turn it off.

Composite

A **Composite** field is a different animal; it is the combination of several fields acting together as a single field. The Account Number field is a popular example of a composite. Each segment of the Account Number is an independent field and appears as such in SQL. The composite resource creates a new field used by Dexterity, which is a grouping of all of the independent field segments. The segments become components of the composite, but you can still refer to them individually in sanScript.

Using composites to pass parameters in functions and procedures is a very handy technique. Instead of passing multiple fields, put all of the fields in a composite and simply pass the composite as a single parameter. The called script can evaluate the value of each segment for its use in the procedure or function. Each segment of a composite can be a different DataType.

There are several steps involved in creating a composite:

1. Creating segment DataTypes.
2. Creating segment fields.
3. Creating the composite resource (this is a separate base resource type).
4. Creating the DataType for the Composite.
5. Assigning a format to the composite DataType (a format is required for a Composite).
6. Creating the composite field.

This sounds like a lot of work, but it will go fine if you follow the steps and don't try to get ahead of yourself.

The following diagram is a graphical representation of the steps for creating a four-segment composite field; the numbers correspond to the number of the step that was set out earlier:

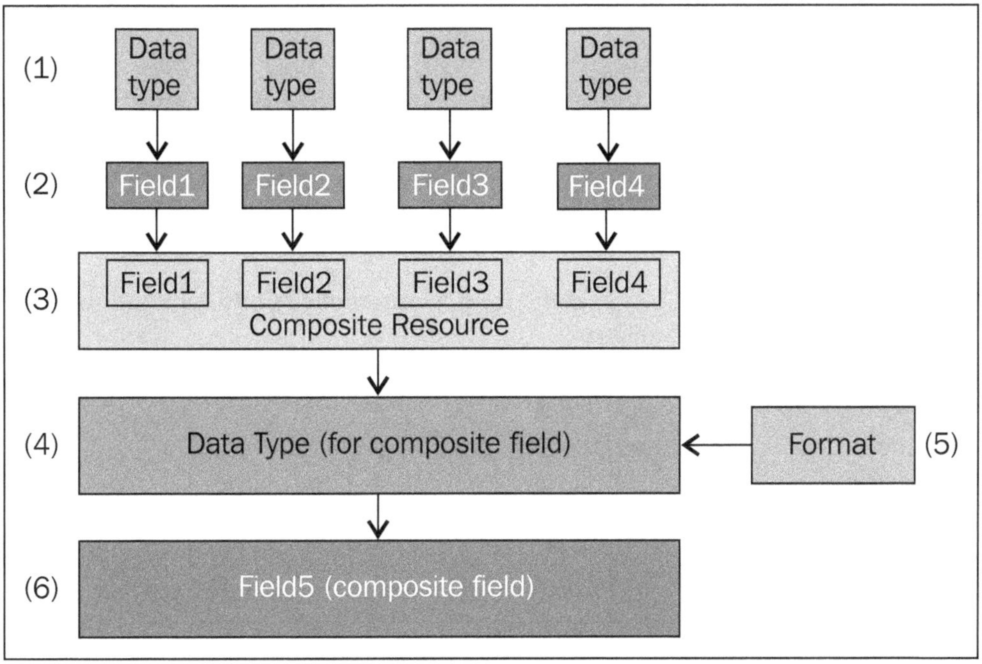

The previous explanation presumes that each of the segments is a new field. You can use existing fields in a composite as well. If you used existing fields, you would start with step three.

Composite fields have the same properties available to them as any other field.

Table

A **Table** is a single store of related information containing a group of fields. While a Dexterity table is no different from other database tables, they are unique in how they are constructed. Conventional databases create the table first and then define the fields unique to that table. With Dexterity, you can define the field once and then you use it in any table.

The table-naming convention for Dynamics GP tables was set out in *Chapter 1, Microsoft Dynamics GP Architecture*.

Form and window

A **Form** is a collection of related resources grouped together for a common purpose. A form can contain windows, tables, menus, commands, constants, and scripts. A window represents the user interface. Every screen in Dynamics is a window resource, and all window resources are contained in forms. You cannot create a window outside of a form. A form can contain many windows or no windows at all. While the windows are displayed and defined separately, they work together to perform a specific function.

For example, the **Item Maintenance** window contains basic information about a single item, such as the **Item Number** and **Description**. If you click on the **Options** button at the bottom of the window, the **Item Maintenance Options** window will open. This window contains more information, such as **Warranty Days** and **ABC Code**.

The **Item Maintenance** window and the **Item Maintenance Options** window are both contained in the same form.

Window properties

Each window has an assortment of properties that govern its behavior and appearance.

Object properties

Object properties control the function or behavior of the window. Some common object properties are discussed as follows:

- **AutoLinkTable**: The `AutoLinkTable` property identifies the table from which users can add additional fields to the window using the Modifier tool. While identifying an `AutoLinkTable` is optional, it's a very user-friendly element to include.
- **AutoOpen**: The `AutoOpen` property determines if the window will open automatically when the form opens. By default, the main window on the form will automatically open. The main window is the window that is listed first on the form. You can turn this behavior off, or cause one or more windows to automatically open using this property.

- **CloseBox**: The `CloseBox` property determines if a Close box will be active on the window. The Close box is identified in the following screenshot:

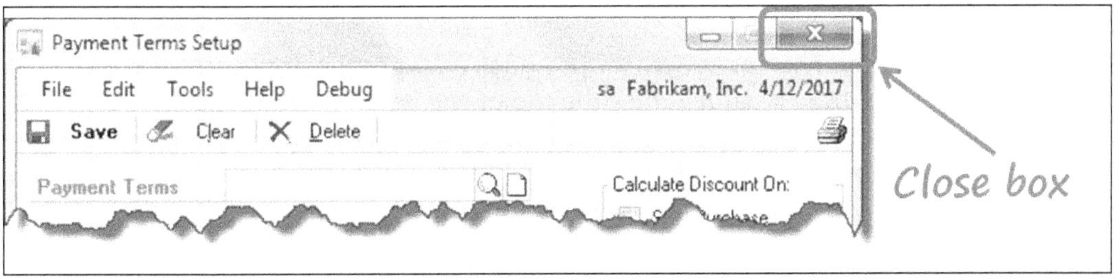

- **Name**: The `Name` property sets the name of the window object. You can use the window's name to refer to it in your sanScript code.
- **Title**: The `Title` property determines what will be displayed at the top of the window so it can be identified by the user. The window title has been pointed out in the following screenshot:

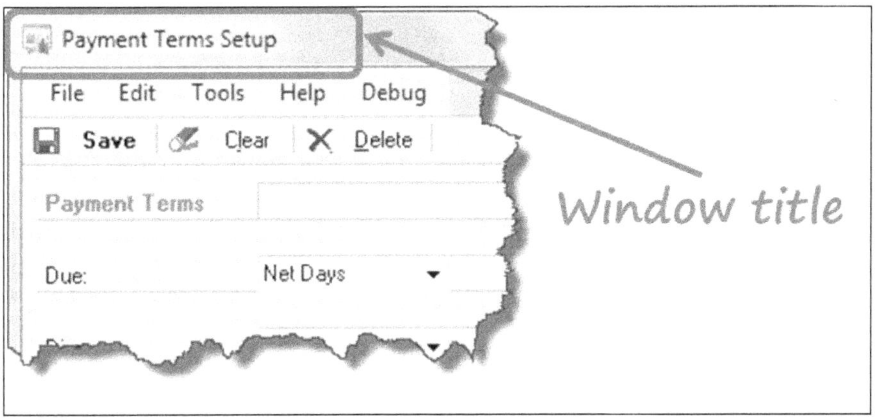

Visual properties

Visual properties control the display characteristics of the window. A common visual property is discussed as follows:

- **ControlArea**: The `ControlArea` property changes the appearance of the window to draw a band across the top of the window that is commonly used for the placement of action push buttons. The following screenshot identifies the control area of a window.

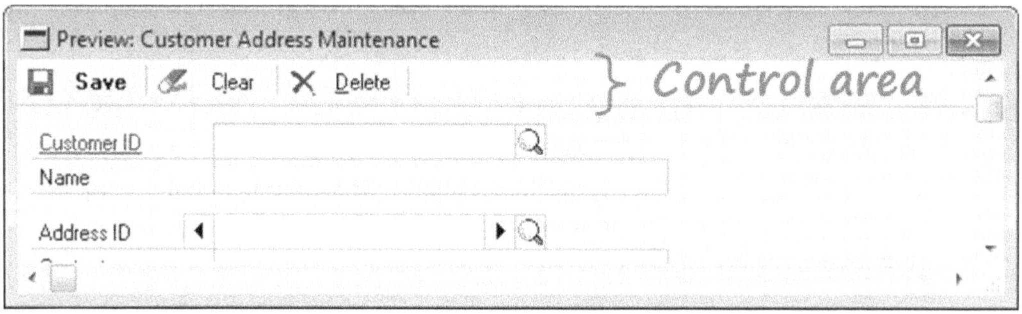

Scrolling window

While not its own separate resource, the scrolling window deserves its own section because of its unique characteristics. A scrolling window looks like a grid or table and presents data in a row and column format. It may look like a grid, but it behaves very differently from your classic Visual Basic or Visual Studio grid. You will learn more about the behavior of scrolling windows in *Chapter 5, Deploying a Dexterity Solution - sanScript - making it work* when you add your sanScript code to this application.

The first thing that is different about it is that you have to place it on another window. A scrolling window cannot exist on its own. Let's call the window on which you place the scrolling window the "host" window. You literally have to draw the scrolling window onto the host window. The next section will show you how.

Creating a scrolling window

To create a scrolling window, select the **Scrolling window** button in the Toolbox. Do not hold down your mouse button and do not try to drag this object on to the window. Just push the button and move your mouse away from the Toolbox. The mouse pointer will change to a little scrolling window pointer. The **Scrolling window** button and mouse pointer are identified in the following screenshot:

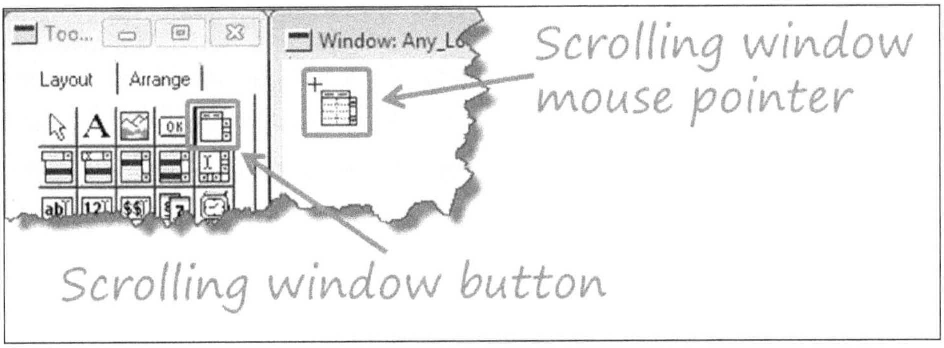

Hold down your left mouse button and draw a box on the window layout representing the size and placement of the scrolling window, as shown in the following screenshot:

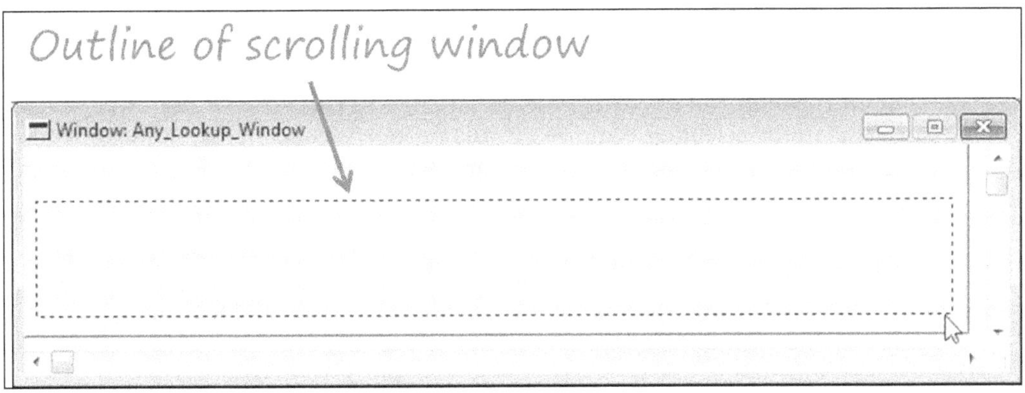

When you release the mouse button, a scrolling window will appear where your box was drawn and a dialog will open prompting you for information about the new scrolling window. The new scrolling window is shown in the following screenshot:

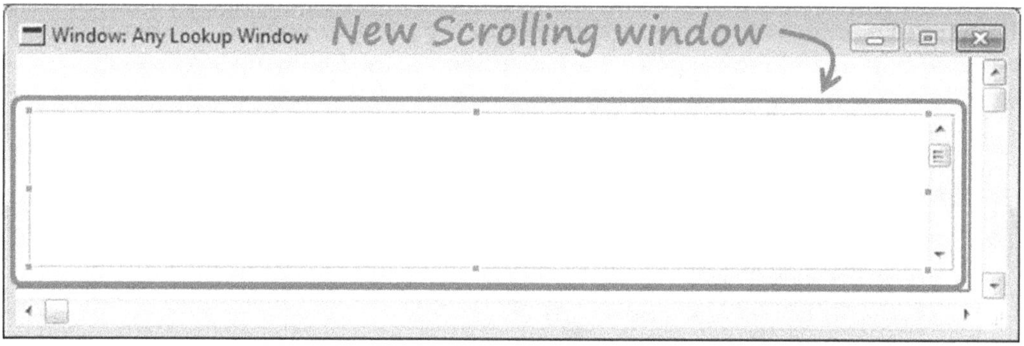

Scrolling window properties

Each scrolling window has an assortment of properties that govern its behavior and appearance.

Object properties

Object properties control the function or behavior of the scrolling window. Some common object properties are discussed as follows:

- **DefaultDblClick**: The `DefaultDblClick` property determines whether the Default push button's script (Change script) will run when you double-click on a line in the scrolling window. If this property was not set, nothing would happen when you double-clicked on the line. Typically, the **Select** button is marked as the Default push button.
- **LinkTable**: The `LinkTable` property sets the default table that will fill the scrolling window. Another feature of the Linked Table is that the user can use the Modifier tool to drag any of the Linked Table's fields into the layout area of the scrolling window.

- **LinkTableKey**: The `LinkTableKey` property sets which key of the Linked Table will be used to sort the data as it fills the scrolling window, if no other key is indicated.
- **Name**: The `Name` property identifies this scrolling window's name. You can use this name to refer to the scrolling window in your sanScript code.
- **WindowType**: The `WindowType` property sets the type of the scrolling window that you're working with. scrolling windows come in three types:
    - **BrowseOnly**: This type of scrolling window is a "look but don't touch" kind of scrolling window. Your **Customer_Contact_Lookup** window will include a browse-only scrolling window. You can view the data, but you cannot modify it, add to it, or delete it.
    - **Editable**: This type of scrolling window allows you to modify the current record values, but you can neither add new records nor delete existing records. A good example of this type of scrolling window is the **Audit Trail Codes Setup** window in **Posting Setup**.
    - **AddsAllowed**: This type of scrolling window allows you to add new rows to the window and delete existing rows from the window. The best examples of this type of scrolling window are the line item sections of the **Sales Transaction Entry** screen and the **Purchase Order Entry** screen. You will be using one of these types of scrolling windows to hold the phone numbers on your **Customer Contact Maintenance** window.

Visual properties

Visual properties control the display characteristics of the scrolling window. There are not many visual properties to choose from. The most common one is as follows:

- **AltLineColor:** The `AltLineColor` property determines whether the alternate lines of the scrolling window will appear in a different color. This property is normally set to `True`. You can control the color using the `Field_SetAltLineColor()` function.

Adding fields to a scrolling window

In order to put fields on a scrolling window, you must first open it to access its layout area. Fortunately, opening it is easy—just double-click on it. Once it is open, you can drag fields from the Toolbox to the scrolling window exactly as you do on the host window. The only trick with a scrolling window is that you have to put the fields in the part of the scrolling window that will be visible.

The scrolling window uses a dashed line to mark the bottom of the visible area. The scrolling window's layout area, as shown in the following screenshot, has only a single row visible:

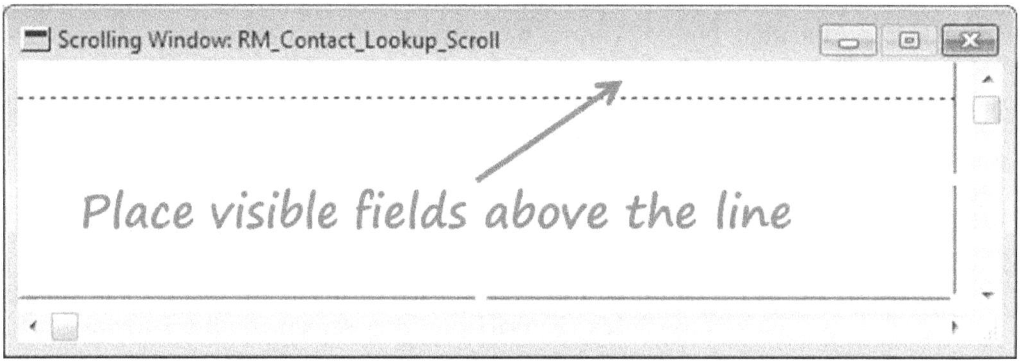

You cannot drag that line down to expand the visible area on the scrolling window; you have to change where the line falls using the Big and Small Line Item tools.

A scrolling window has two areas in which you can place data fields. The **Big Line Item** and the **Small Line Item** define those areas. All scrolling windows default to a single row, as was shown in the previous screenshot. The dashed line you see is the Small Line Item mark. You can create a Big Line Item, or expand the Small Line Item by opening the Scrolling window layout area and then selecting **Tools | Mark Small Line Item** or **Tools | Mark Big Line Item**.

Many people have difficulty marking the line item where they want it, but you'll know the secret. Mark the line *inside* (not below) the last row that you want in that section.

From the **Tools** menu select **Mark Big Line Item**. The mouse pointer will change to resemble a little scrolling window with a plus sign in the upper left-hand corner. Move the pointer to the last row of what will be the Big Line Item area and click on it. Refer to the following screenshot for guidance:

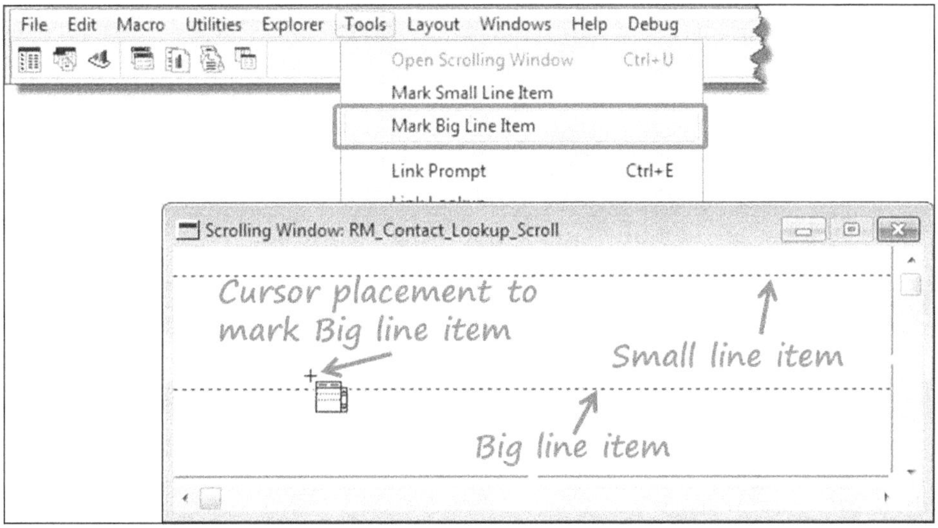

If you do it as mentioned previously, the border will be exactly where you want it. The following screenshot identifies the visual areas of a scrolling window with an expanded Small Line Item and the addition of a Big Line Item.

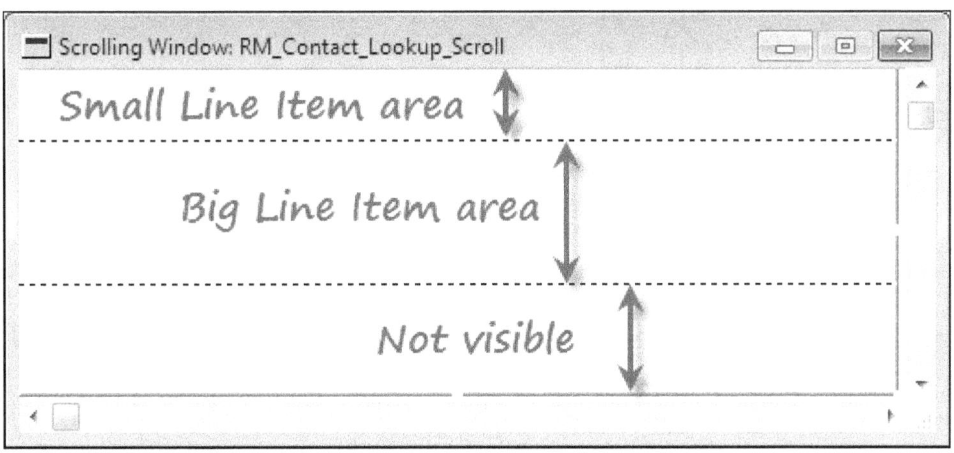

Navigating the Resource Explorer

The **Resource Explorer** is the first window you will see upon launching Dexterity. Starting from this window, you will complete virtually every task in your program. You will create and manage all of the resources in your dictionary from this window. Think of it as the landing page for Dexterity. There is a lot going on here, so let's take a look at some of the primary components.

The Resource Explorer is divided into four main areas:

- Resources Tree
- Resources List
- Toolbar
- Status area

Each of these areas are identified in the following screenshot:

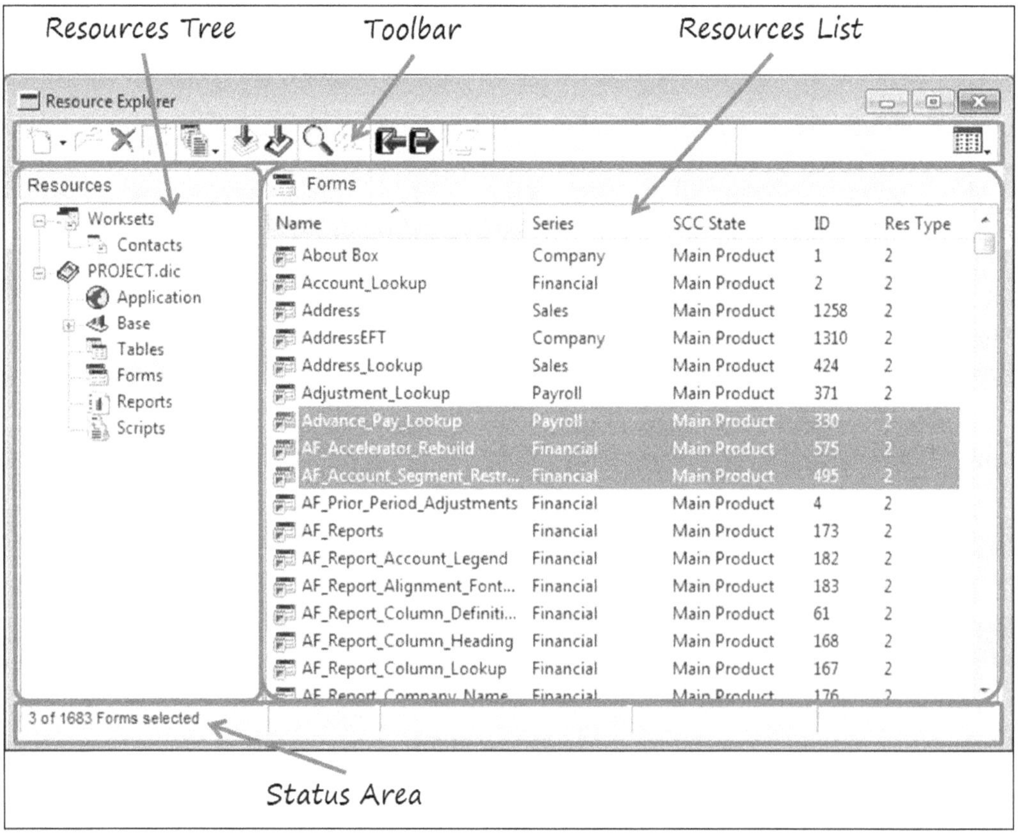

A brief description of each area in the **Resource Explorer** window is as follows:

- **Resources Tree**: The Resources Tree is the left pane of the **Resource Explorer** window. It displays a hierarchical structure that is used for navigating to the various resource types used to create a Dexterity application.
- **Resources List**: The Resources List is the right pane of the **Resource Explorer** window. It displays a detailed list of each resource of the type selected in the Resources Tree. In the previous screenshot, **Forms** is selected in the Resources Tree; each form in the **PROJECT.dic** is listed in the Resources List to the right.
- **Toolbar**: The **Resource Explorer** toolbar provides access to the actions that can be performed on resources, such as compiling scripts, importing or exporting resources, and copying resources.
- **Status Area**: At the bottom of the window is the Status Area, which displays information such as the total number of resources in the list and the quantity selected. Information about error messages also appears in the Status Area.

Worksets

A **Workset** is more of a tool than a traditional resource. Essentially, it is a group of shortcuts that you can use to quickly find the selected resources. You can include any combination of objects, such as forms, reports, fields, tables, and so forth in a Workset.

Using a Workset is a great way to isolate the select set of resources that you are working with. Without a Workset, you would have to scroll through the long list of items in the Resources List to locate the one you want.

You can create as many Worksets as you like, but you cannot nest them. Worksets are listed in the Resources Tree of the **Resource Explorer**. You will use a Workset to complete the Dexterity project in this book.

You can create Worksets using the **Workset** button in the **Resource Explorer** toolbar. The following screenshot identifies the **Workset** button:

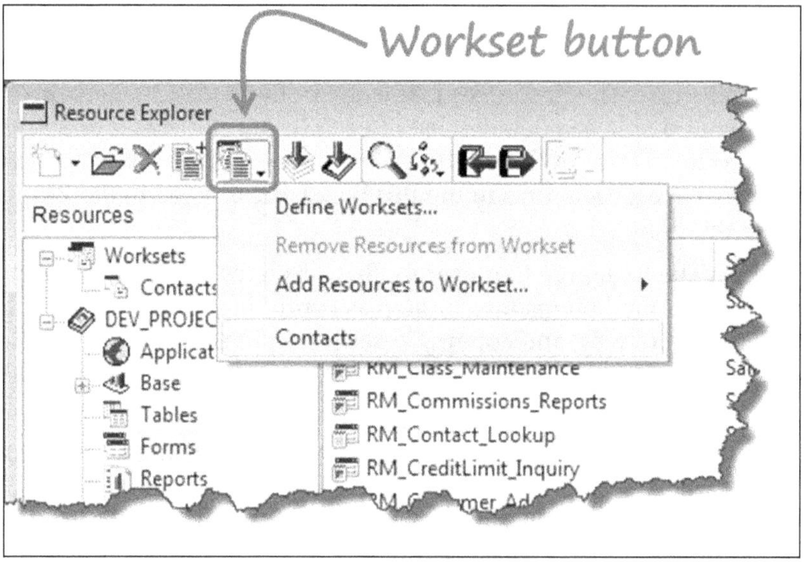

Summary

You covered a lot of ground in this chapter. Best of all, you are ready to start your Dexterity project. In this chapter, you gained knowledge on the development process using Dexterity, how to create your development environment, how to create your development dictionary, how Dexterity resources interrelate, and how to navigate the Dexterity Resource Explorer.

Your next step is to begin building your application by creating the user interface for your Dexterity project. Read on; we start that project in the next chapter.

4 Building the User Interface

In this chapter, you will create the user interface for your customer contacts integration. The customization will provide a means for users to store information about the contacts they have with their customers. In addition, they will be able to enter a limitless number of phone numbers for each contact.

As you develop this application, you will see the key concepts of Dexterity in action. While you can create a standalone application using Dexterity, our focus will be on creating an application that integrates with Dynamics GP.

Creating the user interface will take you through the following key topics:

- Creating base resources
- Creating tables and keys
- Creating forms and windows
- Creating scrolling windows
- Working with window fields
- Completing your windows

Because this is likely your first project, we'll venture through it like a tutorial so that you don't miss a thing.

Let's get started!

Overview

You will create a window named **Customer Contact Maintenance** in which you will enter the contact information, as well as create a new table and other resources needed to get the job done. You will also create the **Customer Contact Lookup** window that you will open from the **Contact ID** field. To get your new window to open, you will add a menu item to the native Dynamics GP **Customer Maintenance** window.

The completed **Customer Contact Maintenance** window will be similar to the following screenshot:

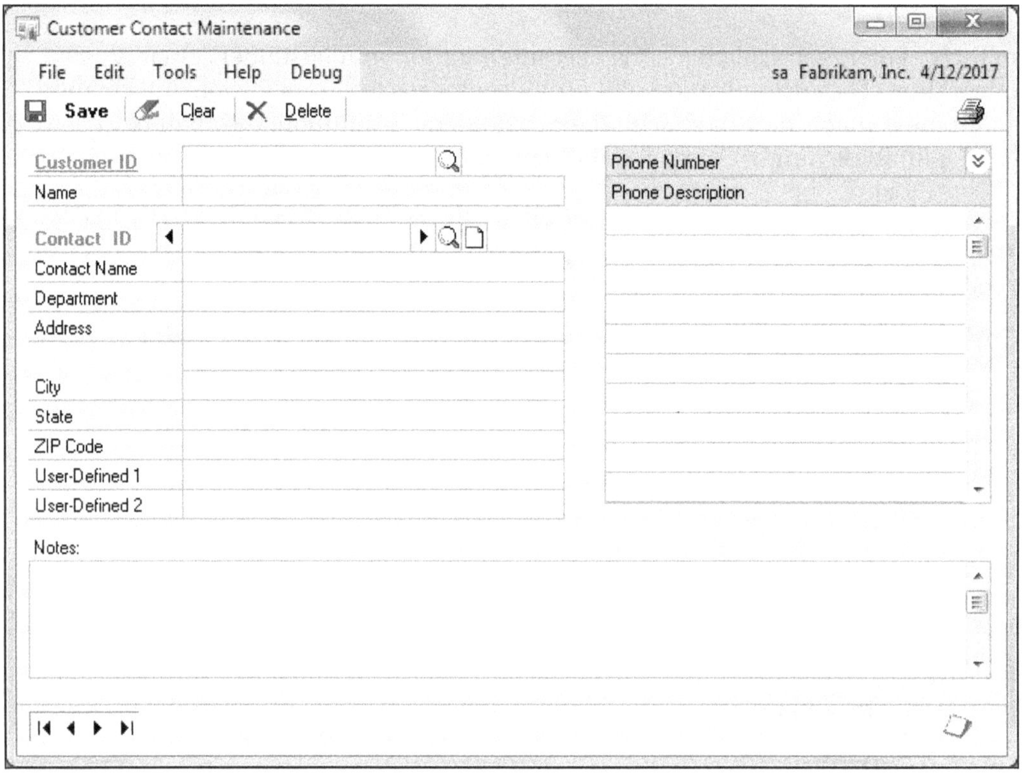

The completed **Customer Contacts** lookup window will be similar to the following screenshot:

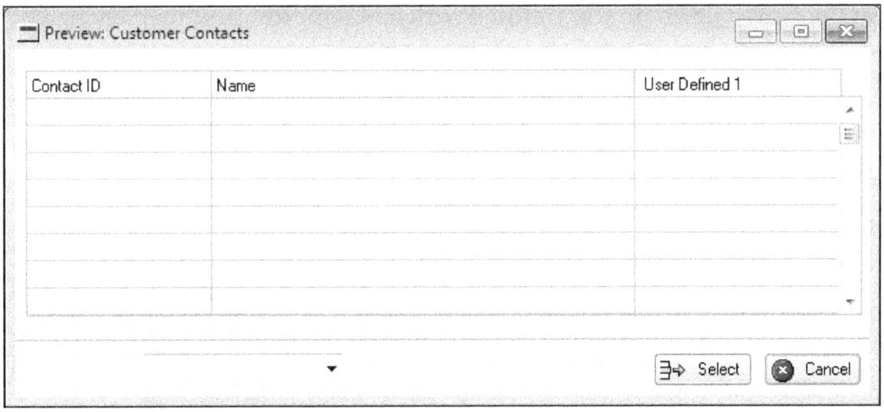

First, you will create a new **workset** and then create your base resources.

Workset

Recall that a workset is just a group of shortcuts to certain resources so that you might access them more quickly. Worksets really come in handy when navigating a large dictionary, and your development dictionary would certainly be considered large.

To create a workset, select the **Workset** button, identified in the following screenshot, from the **Resource Explorer** toolbar. Select **Define Worksets...** from the menu that opens, as shown in the following screenshot:

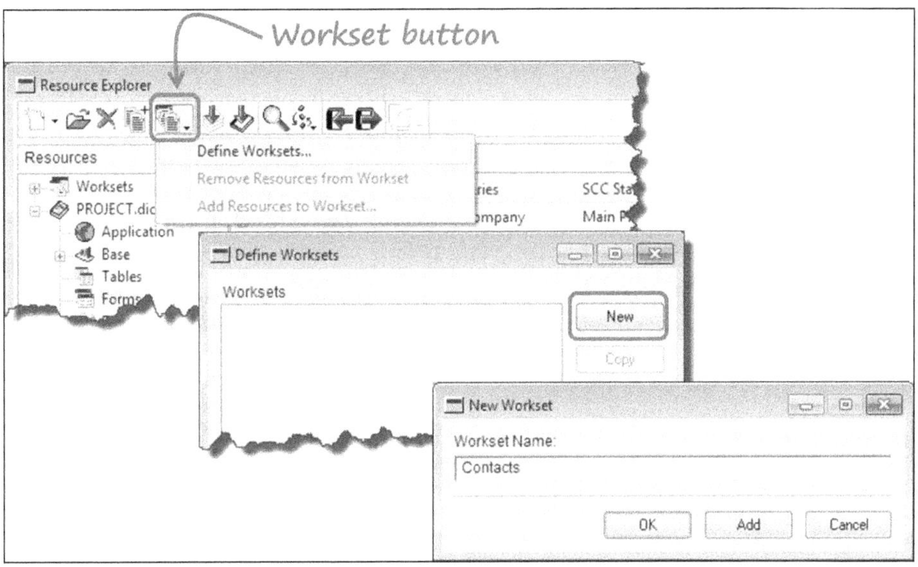

Click on the **New** button on the **Define Worksets** window, and then enter the **Workset Name** as **Contacts**. Select **OK** to complete creating the workset, and then press the **Close** button on the **Define Worksets** window.

Expand the **Worksets** resource in the **Resource Explorer** window and select your new **Contact** workset.

Your **Resource Explorer** window should look much like the window shown in the following screenshot:

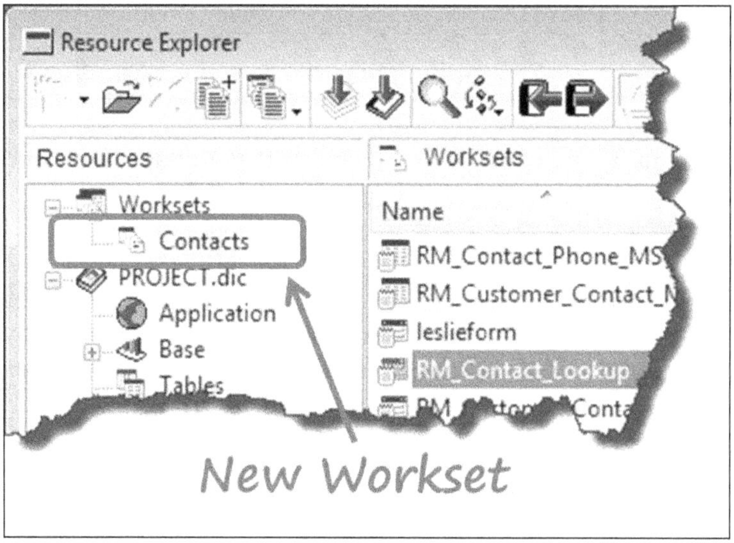

When you create resources with your workset selected, the resources will automatically be added to the selected workset. You can add resources to your workset at any time; it's just more convenient if it happens automatically. If you want to delete a resource, you cannot do that from the workset, you will need to go down to the regular resource list and delete it there.

For this entire project, if the value for the field is not specified, then accept whatever the default value is, even if it is blank. Also, if you are creating multiples of the same type of resource, push the **Add** button instead of the **OK** button. Your resource will be saved and the window will stay open, ready to accept your next record.

Base resources

Dexterity has two types of resources: **base resources** and **standard resources**. As you might expect, the base resources are the building blocks that you need to create first. A field is one of the base resources. A field needs a data type, which is another base resource. A data type can hold a format, which is another base resource. A window may have a picture on it, which is again another base resource. You get the idea. Each of the base resources was discussed in *Chapter 3, Getting Started with Dexterity*. In this chapter, you're going to actually create the base resources you need for your project.

Data types

With your workset selected, create a data type for the **Contact ID** field. In the **Resource Explorer** window, create a new data type resource by selecting the new resource drop-down arrow and then selecting **Data Type...**. Refer to the following screenshot for navigation:

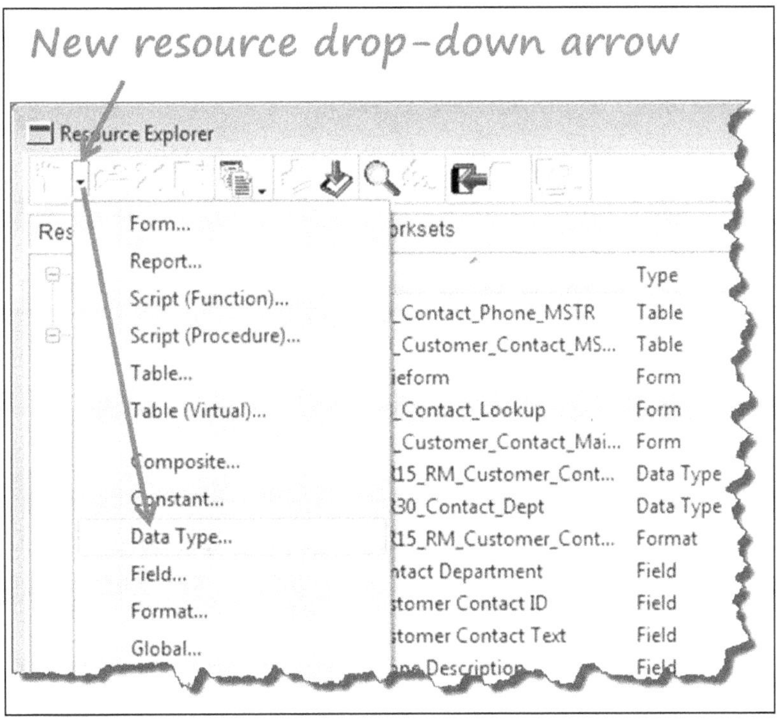

When you create the new data type, you are going to use an existing global format that came from the `dynamics.dic` dictionary that you turned in to your development dictionary. Right out of the chute you get to start capitalizing on existing resources. To enter the value for the **Format** requested (as shown in the following screenshot), you can just type it in, or push the ellipses button and select it from the list that opens.

In the **Data Type Definition** window, enter the following information:

Field	Value
Data Type Name	STR15_RM_Customer_Contact_ID
Control Type	String
Keyable Length	15
Format	STR_LBAU

Upon completion, the **Data Type Definition** window should look similar to the following screenshot:

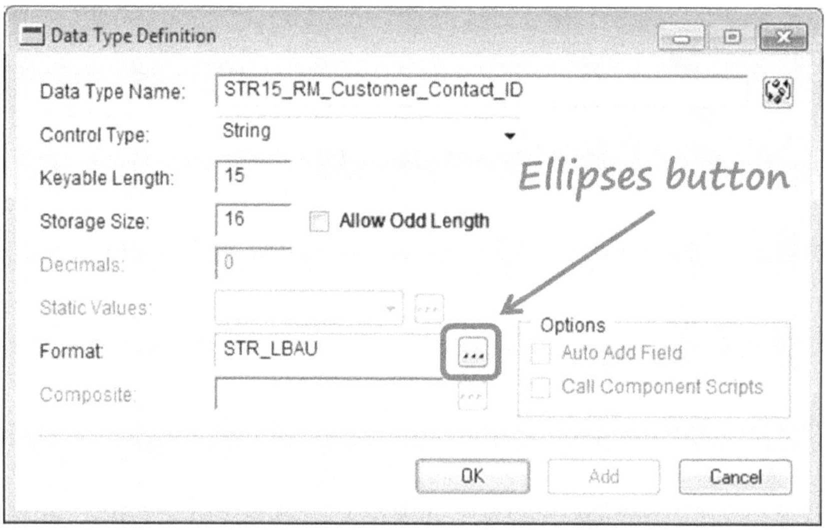

Create a second data type using the following information:

Data Type Name	`STR30_Contact_Dept`
Control Type	`String`
Keyable Length	`30`

Format

You will create one format for the **STR30_Contact_Dept** data type. In the **Resource Explorer** window, select the new resource drop-down arrow and then select **Format**.

In the **Format Definition** window, enter the following information:

Format Name	**STR15_RM_Customer_Contact_ID**
Format Type	**String**
Alignment	**Left**
Fill	**Asterisk**
String Options	**Alphanumeric Only**

Upon completion, your **Format Definition** window should look similar to the following screenshot:

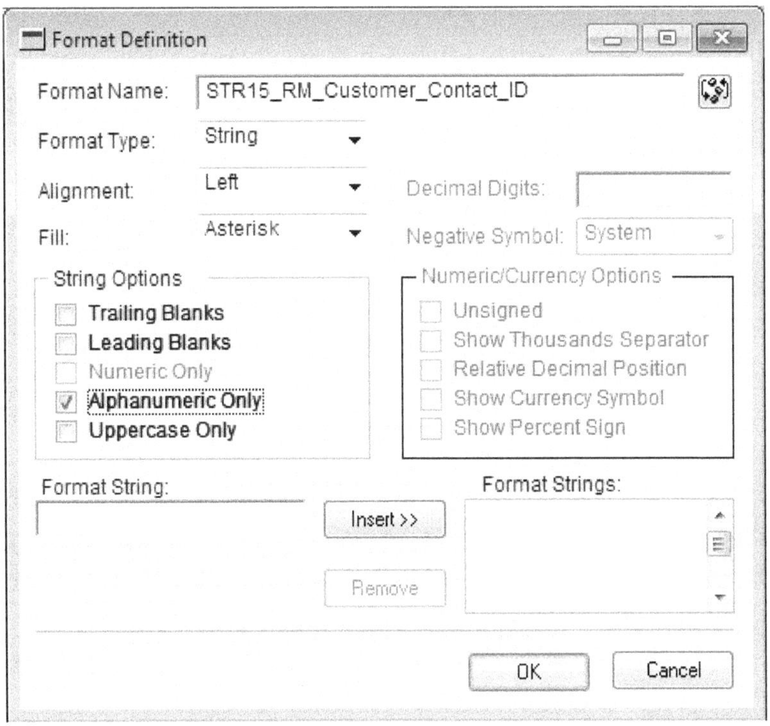

Select **OK** to save your new format and close the window.

Fields

You're going to create four fields. In the **Resource Explorer** window, select the new resource drop-down arrow and then select **Field**.

In the **Field Definition** window, enter the following information:

Field Name	**Contact Department**
Physical Name	**Contact_Department**
Data Type	**STR30_Contact_Dept**

Upon completion, your **Field Definition** window should look similar to the following screenshot:

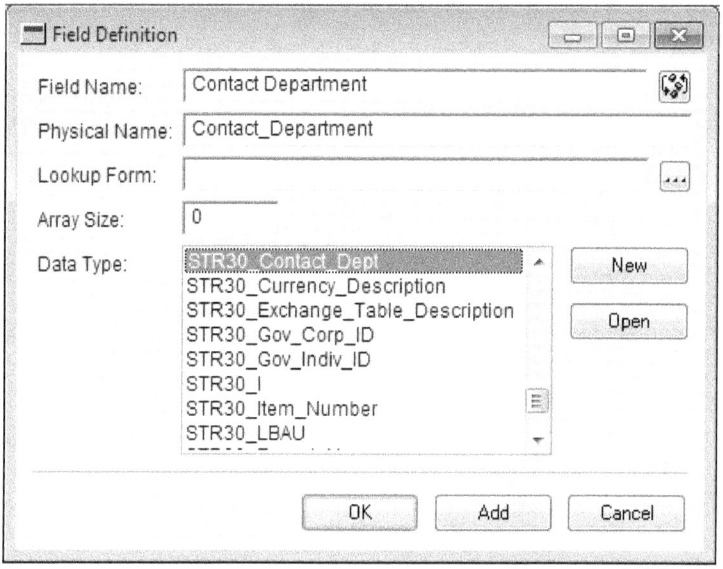

Select the **Add** button on the **Field Definition** window. The new field will be saved and the screen will clear so that you can continue adding new fields.

Create three additional fields using the following information:

Field Name	`Customer Contact ID`
Physical Name	`Customer_Contact_ID`
Data Type	`STR15_RM_Customer_Contact_ID`

Field Name	`Customer Contact Text`
Physical Name	`Customer_Contact_Text`
Data Type	`TX32000`

Field Name	`Phone Description`
Physical Name	`Phone_Description`
Data Type	`STR20`

Creating tables and keys

Your project requires two new tables: one in which to store the contacts and another in which to store their phone numbers. Because you would like to create tables with names longer than eight characters (Imagine!), you need to change an option in Dexterity.

Select **Edit | Options** from the menu bar. In the **Table Options** section of the **Options** window, mark the checkbox next to **Allow Long Physical Table Names**. With this option marked, you can create table names with up to 28 characters.

The 28-character limitation is imposed by Dexterity, not by SQL.

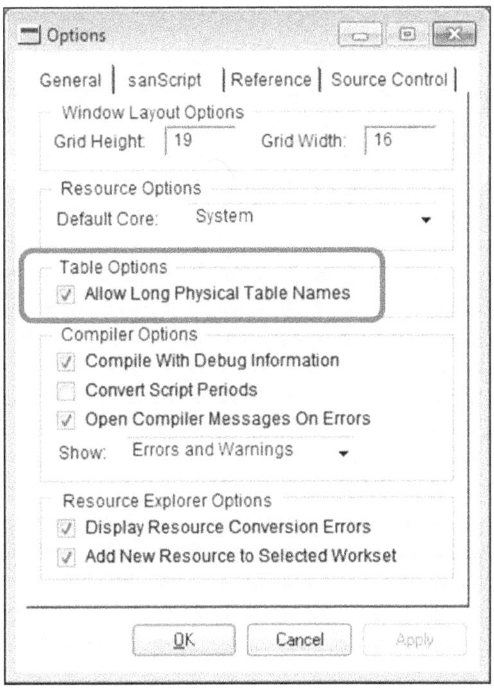

Tables

You will use the **Table Definition** window to create your tables. In the **Resource Explorer** window, select the new resource drop-down arrow and then select **Table**. The **Table Definition** window will open. A table definition contains the table's name, fields, relationships, and so on.

Customer Contact Master

First, you will create the table for storing the contact information. Leave any fields not mentioned in the following table at their default values, even if blank:

Table Name	`RM_Customer_Contact_MSTR`
Display Name	`Customer Contact Master`
Physical Name	`RM_Customer_Contact_MSTR`
Series	`Sales`
Database Type	`SQL`

Insert the following fields into the table by selecting each in the **Global Fields** column and then clicking on the **Insert** button. Unfortunately, you can only select one field at a time. Just to keep us in step, select your fields in the same order as they appear in the following table:

Field Name
`Customer Number`
`Customer Name`
`Customer Contact ID`
`Contact Person`
`Contact Department`
`Address 1`
`Address 2`
`Address 3`
`City`
`State`
`Zip Code`
`User Defined 1`
`User Defined 2`
`Customer Contact Text`
`Note Index`

You can't save the table yet! You don't have any table keys. So far, your **Table Definition** window should look similar to the following screenshot. Note the **Record Size** should be **510**:

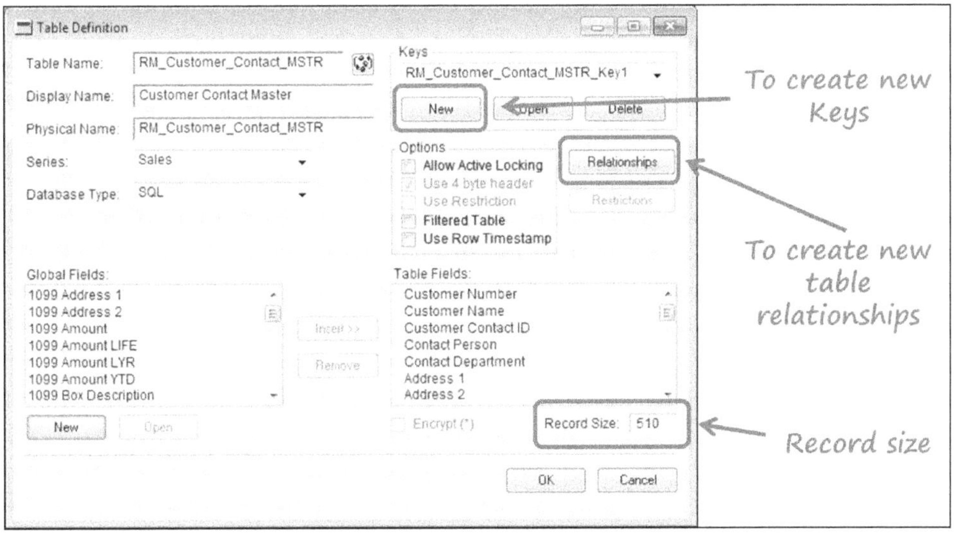

Next, you have to define some **keys** for this table. The table cannot be created until you define at least one key.

Table keys

You use keys to enforce unique records, create table relationships, determine sorting order, and so on. The records are sorted in the order of the key segments. As we said before, each table must have at least one key. A table needs a primary key to ensure duplicate records are not inserted into the table.

You will create two keys for the **Customer Contact Master** table:

- **Customer Contact Master Key1**

  To create the first key, open the **Key Definition** window by clicking on the **New** button in the upper right-hand corner of the **Table Definition** window. The **Key Definition** window will open. To insert a table field into the **Key Segments** area, select the field in the **Table Fields** area and then click on the **Insert** button.

  Use the following table to complete the key definition. It is important that your key segments appear in the same order as they appear in the table; therefore, insert the fields in the order listed. Leave any fields not mentioned at their default values.

Key Name	`RM_Customer_Contact_MSTR_Key1`
Key Segments	`Customer Number` `Customer Contact ID`
SQL Key Options	`Mark the Primary checkbox`

Click on **OK** to save the key and close the **Key Definition** window.

- **Customer Contact Master Key2**

  Now, you are going to create a second key. Select the **New** button on the **Table Definition** window once more. Use the following values for the second key:

Key Name	`RM_Customer_Contact_MSTR_Key2`
Key Segments	`Customer Contact ID`
	`Customer Number`

The second key is different from the first key in how it sorts the records. The records are sorted by the order of the key segments. Whenever you execute table operations, the records will be acted upon in the order of the selected key. If you do not specify a key, the first key is used.

For example, if you browse through a table using `get next table RM_Customer_Contact_MSTR`, the records will advance to the next customer number because the table defaults to the first key. If instead you use `get next table RM_Customer_Contact_MSTR by number 2`, then the second key will be used. The records will advance to the next **Customer Contact ID** instead of the next **Customer Number**.

You can specify the key using its number or its name. Therefore, the statements `get next table RM_Customer_Contact_MSTR by number 2` and `get next table RM_Customer_Contact_MSTR by RM_Customer_Contact_MSTR_Key2` yield the same results.

Your completed **Table Definition** window should include two keys and the **Record Size** in the lower right-hand corner should still be **510**.

Contact Phone Master

Create a second table to store the phone numbers.

Table Name	`RM_Contact_Phone_MSTR`
Display Name	`Contact Phone Master`
Physical Name	`RM_Contact_Phone_MSTR`
Series	`Sales`
Database Type	`SQL`

Insert the following fields in the table.

Field Name
`Customer Number`
`Customer Contact ID`
`Phone Number`
`Phone Description`

Table keys

You will create three keys for the **Contact Phone Master** table:

- **Contact Phone Master Key1**

  For the first key, open the **Key Definition** window by clicking on the **New** button in the upper right-hand corner of the **Table Definition** window. The **Key Definition** window will open.

  Use the following table to complete the key definition. It is important that your key segments appear in the same order that they appear in the table. Leave any fields not mentioned at their default values:

Key Name	`RM_Contact_Phone_MSTR_Key1`
Key Segments	`Customer Number`
	`Customer Contact ID`
	`Phone Number`
SQL Key Options	`Mark the Primary checkbox`

  Click on **OK** to save the key and close the **Key Definition** window.

- **Contact Phone Master Key2**

  Create the second key next. Select the **New** button on the **Table Definition** window again to open the **Key Definition** window. Use the following values to complete the **Key Definition** window:

Key Name	`RM_Contact_Phone_MSTR_Key2`
Key Segments	`Phone Number`
	`Customer Contact ID`
	`Customer Number`

  Click on **OK** to save the key and close the **Key Definition** window.

- **Contact Phone Master Key3**

  Finally, you are going to create the third key. Select **New** on the **Table Definition** window once more. Use the following values to complete the **Key Definition** window:

Key Name	`RM_Contact_Phone_MSTR_Key3`
Key Segments	`Customer Contact ID`
	`Phone Number`
	`Customer Number`

  Click on **OK** to save the key and close the **Key Definition** window.

  Click on **OK** to save the table and close the **Table Definition** window.

Table naming conventions

Deciding on a name for your new table should not be arbitrary. You learned in *Chapter 1, Microsoft Dynamics GP architecture* about the necessity for naming conventions and the current best practice. We have adhered to the naming conventions for the tables you just created:

Module abbreviation	Table contents	Table type
RM	Customer_Contact	MSTR
RM	Contact_Phone	MSTR

In addition, you have used a display name that is user-friendly. Too many developers get lazy and do not take the time to change the display name for the benefit of the user. Display names appear in the Report Writer, SmartList Builder, Extender, and others. You, however, will follow the naming conventions and make your applications just as friendly as the Dynamics GP core, because you're good!

Table options

There are five table options in the **Table Definition** window, but only a couple of them merit our attention right now. Those are **Allow Active Locking** and **Use Row Timestamp**.

Allow Active Locking permits you to lock an individual record so that no other user may change it. They may still read the record; they just cannot modify nor delete it. Very few tables in the Dynamics GP dictionary are set up for active locking. Dexterity uses **passive locking**. Passive locking allows multiple users to modify the same record at the same time; they just can't modify the same field concurrently. If they attempt it, Dexterity will display a warning dialog. Microsoft calls this locking scheme **Optimistic Concurrency Control** (OCC).

For example, let's say two users have the same customer card open. One of the users changes the **Name** field and saves it. When the other user changes the **Name** field and hits the **Save** button, the following error will display:

This customer record has been updated since you opened this window. Changes won't be saved.

On the other hand, if one user modifies the **Name** field and saves it, and the other user modifies the **City** field and saves it, both changes will save.

Not using active locking also allows high availability of the data and the system runs much faster. You should rarely mark a table to allow active locking.

> By using passive locking, multiple users can change or even delete a record at the same time.

Use Row Timestamp adds an additional field to the table named `DEX_ROW_TS`. When you select this option, the **UTC** (**Coordinated Universal Time**) and date of when the record was created or modified is automatically written to this field. Do not mistake this as an audit trail code, or as a means to provide any audit control on the data. The field is merely there to hold the date and time that the record was last modified.

Types of tables

You can create three different types of tables in Dexterity:

- **Physical tables**

  These are your standard, run-of-the mill tables. They are physically stored in the SQL database and their data is available to all for manipulation.

- **Temporary tables**

  These are tables with the physical name of `temp`. If the **Database Type** is `SQL` or `Default`, the table will be created in SQL's `tempdb system database`. If the **Database Type** is `c-tree` or `P.SQL`, a file will be created in the user's `Temp` folder with a name beginning with `TNT`. For Windows 7, look in `C:\Users\`*username*`\AppData\Local\Temp` for the files. Temporary tables and files are supposed to be deleted by Dexterity after they are used, but sometimes we have to help. Feel free to delete any files in the user's local `Temp` folder that begin with `TNT`. Please note that by default, the `C:\Users\`*username*`\AppData` folder is a hidden folder.

- **Virtual tables**

  These are not actually tables, they are **Views**. Virtual tables are not created in the **Table Definition** window. They are created in the **Virtual Table Definition** window. To access the **Virtual Table Definition** window, use **Resource Explorer**, select the new resource drop-down arrow, and then select **Table (Virtual)**.

  You add the tables and fields, define the table relationships, and establish the keys for your view in this window. When the virtual table is created, an actual SQL view is added to the database.

Creating forms and windows

Forms can contain windows, tables, menus, constants, scripts, and commands. Every screen you see in Dynamics GP belongs to a form. You need two forms for your customer contact application. Each form will contain a single window.

Maintenance form and window creation

The first form will hold the **Customer Contact Maintenance** window; the second form will hold the **Contact Lookup** window.

You're going to create the **Customer Contact Maintenance** form and window in a rather unconventional way. Instead of using the WYSIWYG graphical forms designer to create the form, you are going to copy an existing form, change its name, and then modify it.

The form you're going to use is the **RM_Customer_Address** form, because it most closely resembles the form you need. Find this form by selecting **Forms** in the **Resources** tree of the **Resource Explorer** window. Look in the **Resources** list and select the form named **RM_Customer_Address**. With the form selected (do *not* double-click, you do not want to open the **Form Definition** window), click on the **Export to Text File** button on the toolbar as shown in the following screenshot:

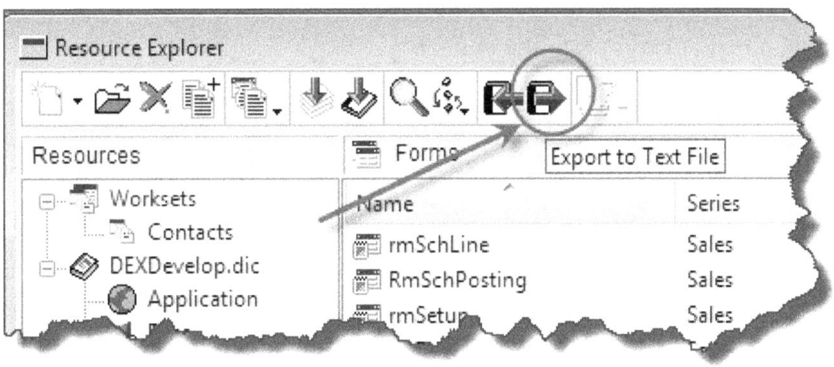

The **Export to Text File** window will open. Make note of the **Location for Export File** and the **File Name**; you will need it later:

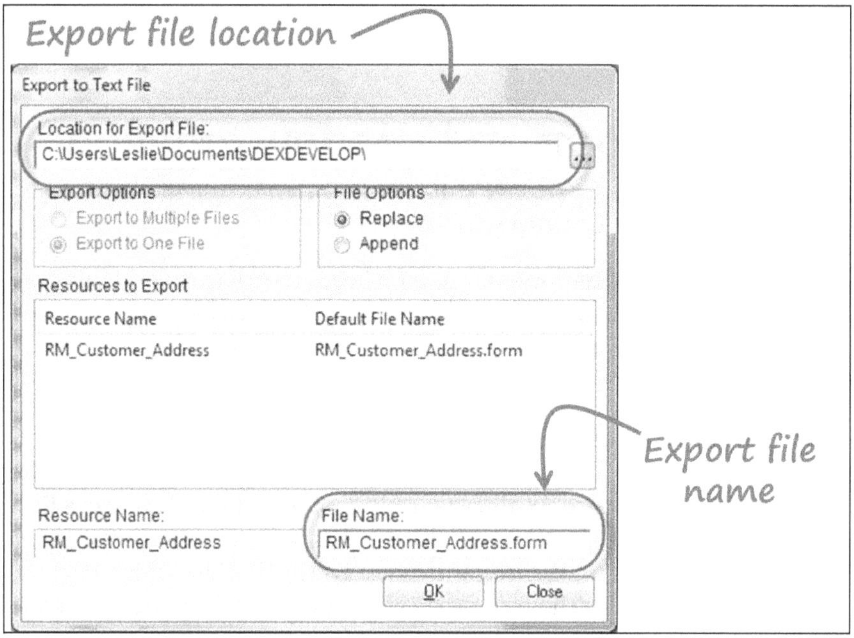

Select the **OK** button to complete the export.

Next, you are going to open the `RM_Customer_Address.form` file in Notepad and turn it into your **Customer Contact Maintenance** form. So, open the `RM_Customer_Address.form` file in Notepad. In order to see the filename in your browse menu, you may need to change the criteria to `All Files (*.*)`.

Execute a search and replace by pressing *Ctrl* + *H* on your keyboard. Fill in the fields with the data in the following table, and then select **Replace All**:

Find what:	`RM_Customer_Address`
Replace with:	`RM_Customer_Contact_Maintenance`

When you click on the **Replace All** button, the Notepad text on your computer screen will look like you executed an **Edit** | **Select All** command just for a second. First, all of the text will appear selected, and then it will return to normal. The search and replace is finished when that happens; you will not get a dialog on the screen telling you so.

Close the **Replace** window and then click on **Save** to save the `RM_Customer_Address.form` file. Next, you are going to import this file into your dictionary.

With your contacts workset selected, click on the **Import from Text File** button on the toolbar; see the following screenshot for the button's location:

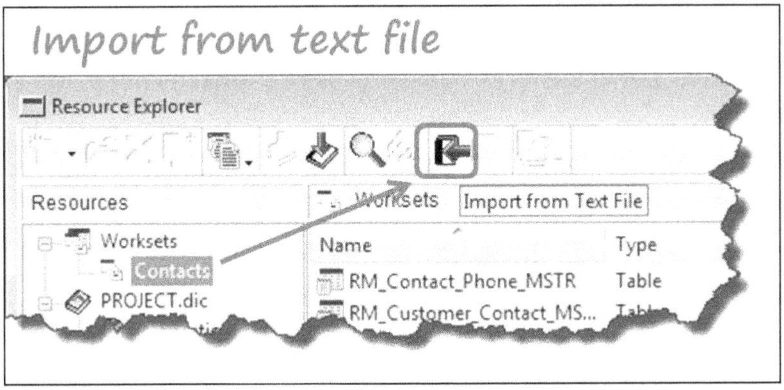

The **Import From Text File** window will open. Browse to the `RM_Customer_Address.form` file and click the **Import** button. You should have 77 conversion errors; they are of no consequence to us. Dismiss the dialog by pushing the **OK** button and then the **Close** button to close the **Import From Text File** window.

You should now have a form named **RM_Customer_Contact_Maintenance** in your forms resource list.

If it's not already included, add the form to your **Contacts** workset. With the new **RM_Customer_Contact_Maintenance** selected, click on the **Worksets** button on the **Resource Explorer** toolbar, then on **Add Resources to Workset**, and then on **Contacts**.

Open the **Form Definition** window by double-clicking on your new **RM_Customer_Contact_Maintenance** form.

Attaching tables

The first thing you need to do is attach the correct tables to your form and remove the ones you don't need. Select the **Tables** tab on the **Form Definition** window; select each table except the **RM_Customer_MSTR** table and push the **Detach** button. Sadly, you have to do these one at a time. Do not close the **Form Definition** window when you're done; you still have more work to do.

> Hitting the Space bar when the **Are you sure . . .** dialog opens will push the **Yes** button and dismiss the dialog.

Now you're going to add the tables you previously created to the **RM_Customer_Contact_Maintenance** form. From the **Tables** tab, click on the **Attach** button.

The **Table Lookup** window will open. Double-click on the **RM_Customer_Contact_MSTR** table to attach it to the form. Attach the **RM_Contact_Phone_MSTR** table in the same manner.

When you've finished, the **Form Definition** window should look similar to the following screenshot:

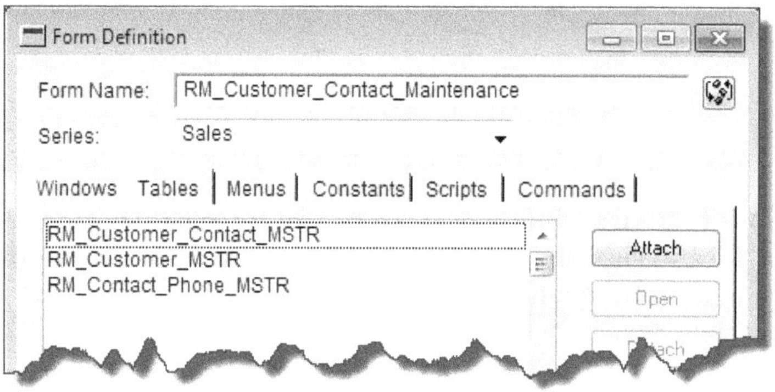

Setting window properties

Go back to the **Windows** tab and open the **RM_Customer_Contact_Maintenance** window. You can open the window by double-clicking on it or selecting the **Open** button on the right-hand side of the window. You should now be looking at the window in the WYSIWYG graphical design window. There is a lot going on here, but we do not need most of these fields so we are about to delete them. First, we need to change a couple of the window's properties.

On the right-hand side of the layout window, the **Properties** window should be displayed. If it is not present, push *Ctrl* + *M* to open it. As you can see by looking at the **Properties** window, this is where you can control visual characteristics as well as object features of your **RM_Customer_Contact_Maintenance** window. You're going to change two object features. Select the **Object** tab in the **Properties** window. Select **[Window]** from the object drop-down list at the top of the **Properties** window. Make the following changes to the window object properties:

Property	Initial Setting	Your Setting
AutoLinkTable	RM_Customer_MSTR_ADDR	RM_Customer_Contact_MSTR
Title	Customer Address Maintenance	Customer Contact Maintenance

Refer to the following screenshot, from the *Before* and *After* view of the **Properties** window:

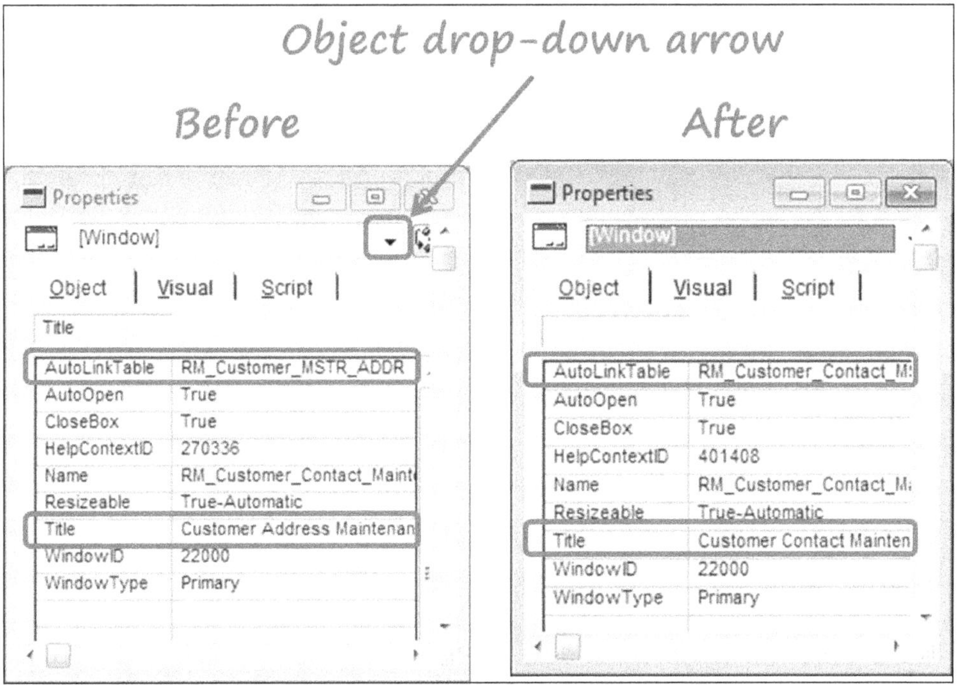

Removing window fields

With that bit of housekeeping done, let's take care of all of those extra fields on the window. Expand the layout window so that you can see all of the fields below the displayed window. If you cannot see any fields below the window, then you need to turn on the **Show Invisible Fields** setting. You also need to make sure the **Show Field Names** option is not selected.

To do that, select **Layout** from the menu bar. Make sure there is a checkmark next to the **Show Invisible Fields** menu item, and no checkmark next to the **Show Field Names** menu item. Clicking on the menu item will toggle between it being checked or unchecked.

Delete fields from the window according to the following screenshot. You should delete the circled fields. You can select and delete them one at a time, or you can *lasso* the fields by holding down the left mouse button and drawing a box around them. Only the fields completely within the box will be selected.

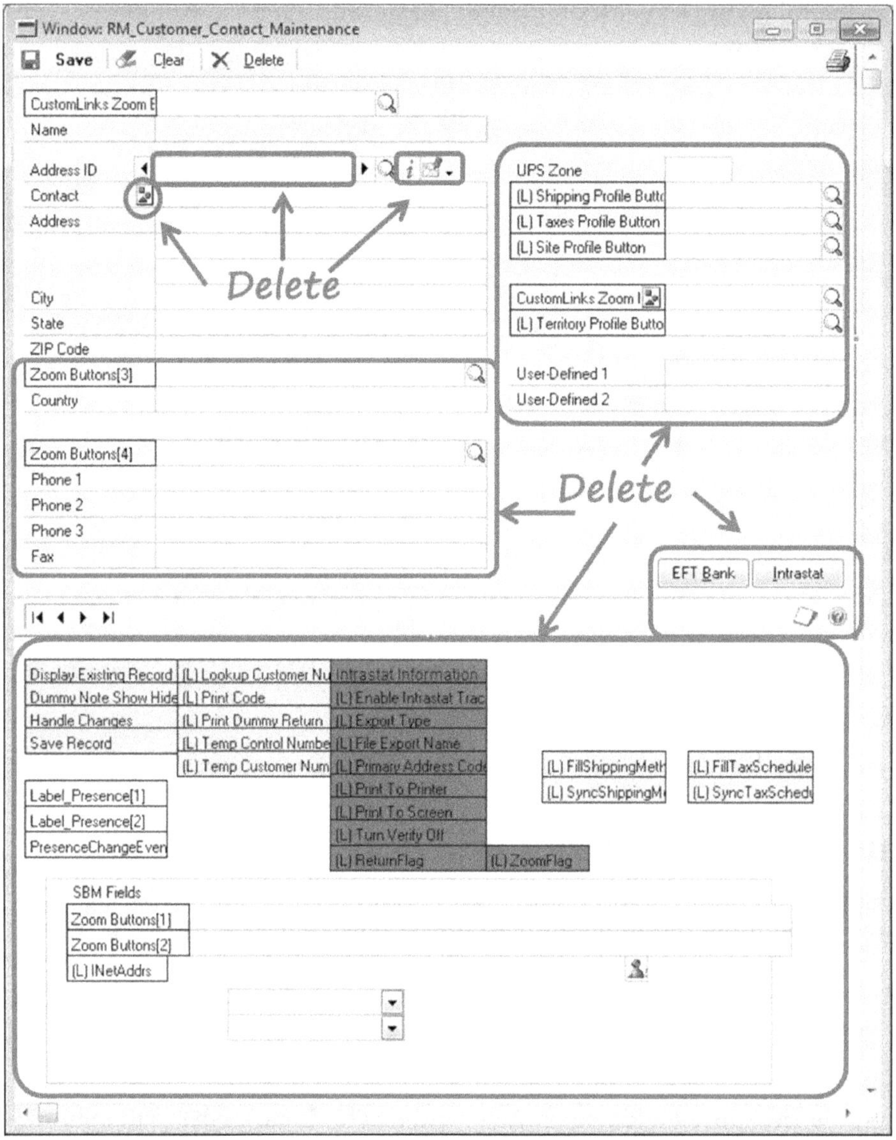

When you are finished removing the extra fields, your window should look similar to the following screenshot:

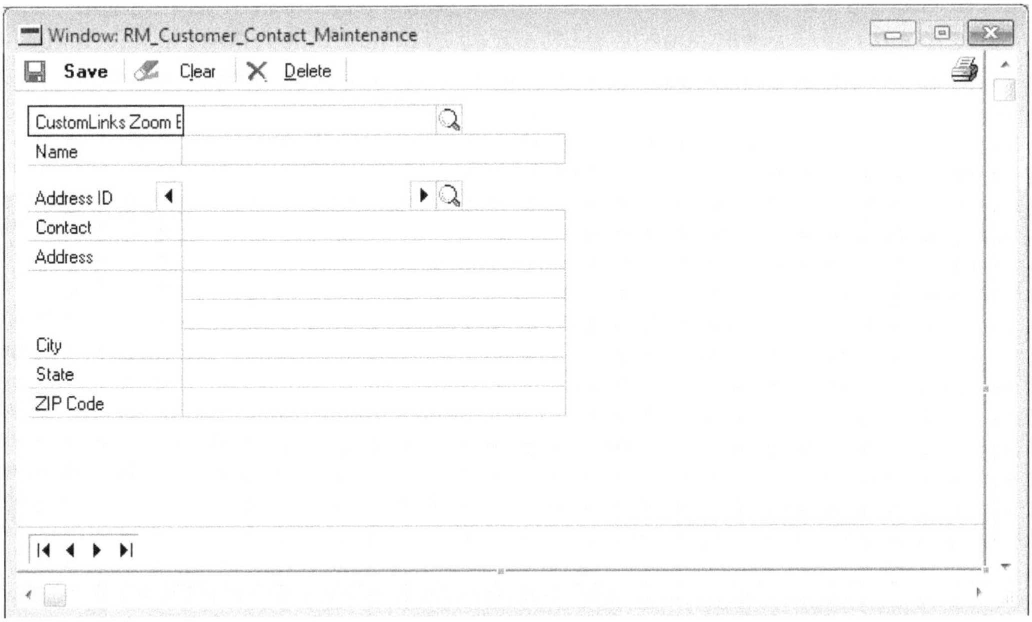

Close and **Save** the window. Click on the **OK** button on the **Form Definition** window to close it.

> Using the import/export technique can save you hours of development time.

Adding fields to the window

Now that you've deleted all of the unwanted fields from your window, it's time to add some new fields. Before starting this adventure, turn the **Show Field Names** option back on from the **Layout** menu. Next, in order to make room for the **Contact Department** field, you need to move all of the fields below the **Contact Person** field down one line. If you use the field properties set out in the following section, you'll get them into the correct position.

Global fields

To add a global field to the window, drag it from the Toolbox and place it on the layout area. As you drag it onto the window, an outline of the field will become visible to aid you in positioning it on the window. Refer to the following screenshot to get an idea of how the cursor will look, and where the fields are positioned:

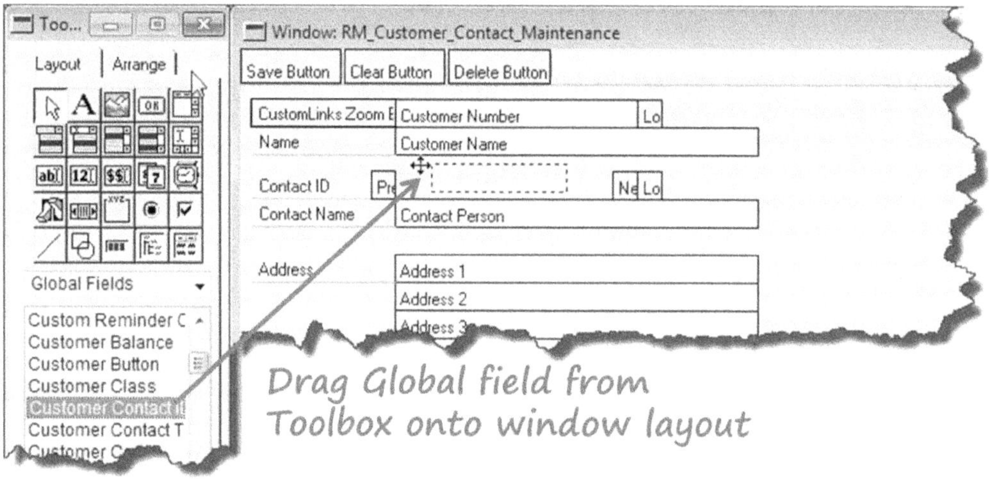

Add the following global fields to the **Customer Contact Maintenance** window. Don't concern yourself with placement right now; you'll set this when you set your properties.

Global Field Name
`Customer Contact ID`
`Contact Department`
`User Defined 1`
`User Defined 2`
`Customer Contact Text`
`Show Detail Button`
`Show Summary Button`

Eventually, `Show Detail Button` and `Show Summary Button` will be stacked on top of one another next to the yet-to-be-created scrolling window. For now, leave them separate so that you can select each of them for attaching your code. Set properties for the text prompts and global fields as indicated in the following tables:

Window text properties

[Text] Address

Tab	Property	Value
Visual	Position-Left	8
Visual	Position-Top	140
Visual	Size-Width	103

[Text] City

Tab	Property	Value
Visual	Position-Left	8
Visual	Position-Top	197
Visual	Size-Width	103

[Text] State

Tab	Property	Value
Visual	Position-Left	8
Visual	Position-Top	216
Visual	Size-Width	103

[Text] Zip Code

Tab	Property	Value
Visual	Position-Left	8
Visual	Position-Top	235
Visual	Size-Width	103

Global field properties

Address 1

Tab	Property	Value
Visual	Position-Left	110
Visual	Position-Top	140
Visual	Size-Width	256

`Address 2`

Tab	Property	Value
Visual	`Position-Left`	`110`
Visual	`Position-Top`	`159`
Visual	`Size-Width`	`256`

`Address 3`

Tab	Property	Value
Visual	`Position-Left`	`110`
Visual	`Position-Top`	`178`
Visual	`Size-Width`	`256`

`City`

Tab	Property	Value
Visual	`Position-Left`	`110`
Visual	`Position-Top`	`197`
Visual	`Size-Width`	`256`

`State`

Tab	Property	Value
Visual	`Position-Left`	`110`
Visual	`Position-Top`	`216`
Visual	`Size-Width`	`256`

`Customer Contact ID`

Tab	Property	Value
Visual	`Position-Left`	`110`
Visual	`Position-Top`	`83`
Visual	`Size-Width`	`154`

Contact Department

Tab	Property	Value
Visual	Position-Left	110
Visual	Position-Top	121
Visual	Size-Width	256

User-Defined 1

Tab	Property	Value
Visual	Position-Left	110
Visual	Position-Top	254
Visual	Size-Width	256

User-Defined 2

Tab	Property	Value
Visual	Position-Left	110
Visual	Position-Top	273
Visual	Size-Width	256

Customer Contact Text

Tab	Property	Value
Visual	Position-Left	8
Visual	Position-Top	310
Visual	Size-Height	95
Visual	Size-Width	624
Object	WordWrap	True

Show Detail Button

Tab	Property	Value
Visual	Position-Left	614
Visual	Position-Top	34
Visual	Appearance	2D Border
Visual	Border	True

`Show Summary Button`

Tab	Property	Value
Visual	`Position-Left`	`598`
Visual	`Position-Top`	`34`
Visual	`Appearance`	`2D Border`
Visual	`Border`	`True`

Local fields

In addition to the global fields you previously added, you are going to add a single local field below the visible area of the window. A local field is not stored in a table, but is instead used either to hold an interim value on the window or to house a piece of code. Local fields are prefixed with an `(L)` and can be used on any window that is part of the form on which it was created. Even though two forms can contain a local field with the same name, they are not the same field.

You need to create a local field named `Temp Control Number` on your **RM_Customer_Contact_Maintenance** window. To create this field, select the drop-down arrow on the Toolbox and select the **Local Fields** list item. Refer to the following screenshot for the location of these entries:

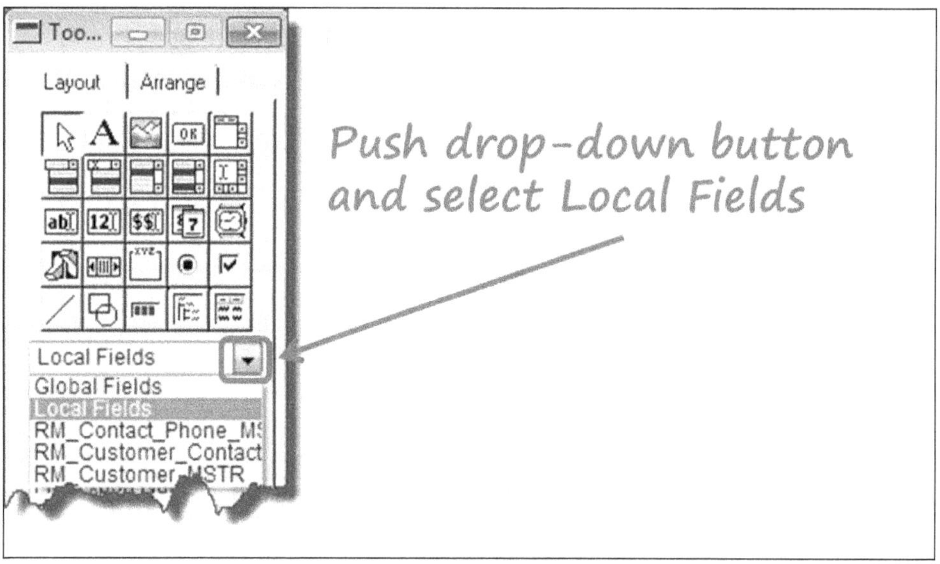

The **Local Field Definition** window will open. This local field will hold the value your user has selected, so that your application can check to make sure there are no unsaved changes that haven't been handled before switching to the new selection. Complete the local field definition according to the following screenshot:

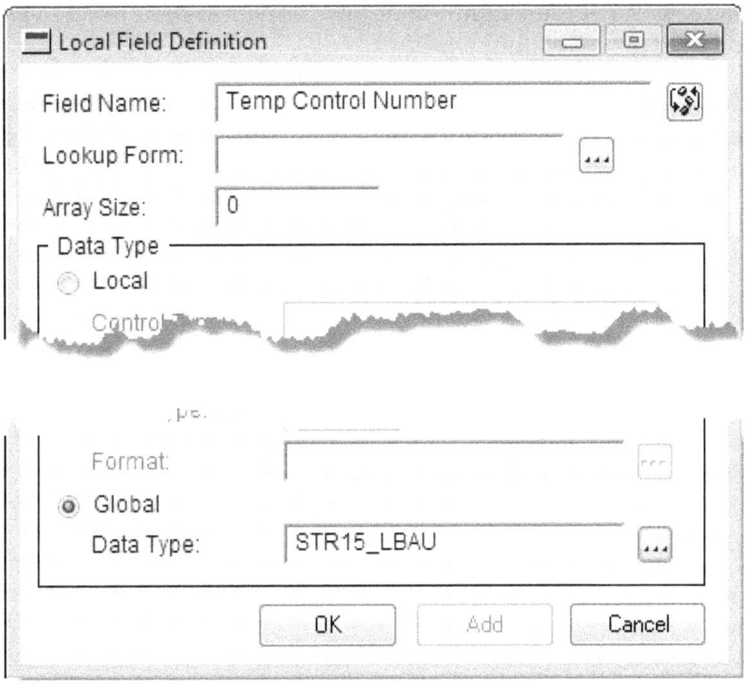

Push the **OK** button to save the temporary field and then drag the field from the **Local Fields** list to the space below the visible area of the **RM_Customer_Contact_Maintenance** window. Note that the `(L)` prefix is automatically added to the field by Dexterity.

Set the field properties for this local field as follows:

```
(L) Temp Control Number
```

Tab	Property	Value
Visual	`BackColor`	`Yellow`
Object	`AutoCopy`	`False`
Object	`Editable`	`False`

New scrolling window

According to the specifications for your application, you need a spot to enter a limitless number of phone numbers for each contact. To satisfy that requirement, you must add a scrolling window to the **Customer Contact Maintenance** window. Refer to the screenshot at the beginning of this chapter to determine the approximate placement for your new scrolling window.

Draw the scrolling window on your **Customer Contact Maintenance** window and complete the **Scrolling Window Options** dialog to match the following screenshot:

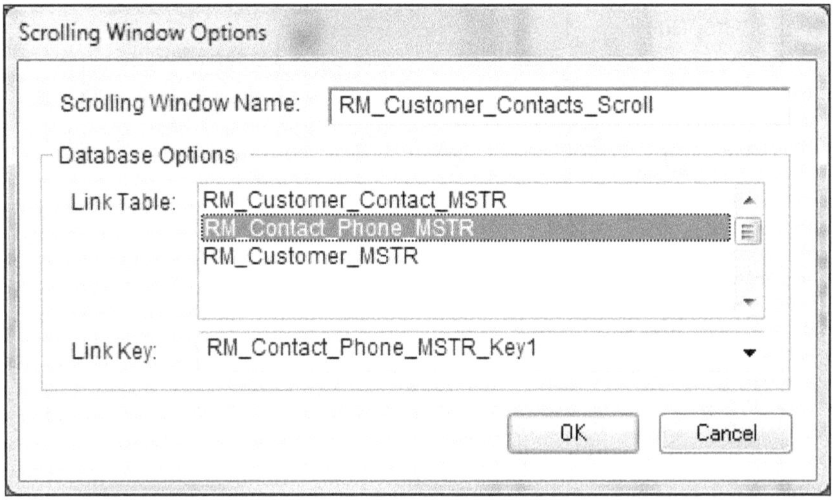

Set the visual properties for the scrolling window according to the following table:

`RM_Customer_Contacts_Scroll`

Tab	Property	Value
Visual	`Position-Left`	`424`
Visual	`Position-Top`	`101`
Visual	`Size-Height`	`192`
Visual	`Size-Width`	`192`

Open the scrolling window by double-clicking on it. In order to display two rows of information for each phone number, you need to add a Big Line item to the scrolling window.

Add the following fields to the scrolling window and set their properties according to the following tables:

`Phone Number`

Tab	Property	Value
Visual	Border	False
Visual	Position-Left	0
Visual	Position-Top	0
Visual	Size-Height	20
Visual	Size-Width	192

`Phone Description`

Tab	Property	Value
Visual	Border	False
Visual	Position-Left	32
Visual	Position-Top	19
Visual	Size-Height	20
Visual	Size-Width	192

In accordance with Dynamics GP standards, be sure to use the Line tool to draw a line between the two fields.

Lookup form and window creation

The second form holds the **Contact Lookup** window. This window will open allowing you to pick a contact from a list, thereby populating the **Customer Contact Maintenance** window. You'll create this form and window using the conventional method - from scratch. Look in the *Overview* section of this chapter to see what your completed lookup window will look like.

In the **Resource Explorer** window, select the new resource drop-down arrow and then select **Form**. The **Form Definition** window will open. Enter the following data:

Form Name	RM_Contact_Lookup
Series	Sales

Select the **Tables** tab and then press the **Attach** button. From the **Table Lookup** window, select the tables listed as follows, one at a time; push the **OK** button after each selection to complete the process:

- `RM_Customer_MSTR`
- `RM_Customer_Contact_MSTR`

Select the **Windows** tab and then click on the **New** button to create a new window. Close the window layout to save the changes. You have not actually created the window until you save it. You'll want to save it before you spend any time designing it. Open the window backup by double-clicking on it in the **Form Definition** window.

Window fields

Next, you're going to add a couple of push buttons and a field to the window. Most importantly, you're going to add a scrolling window that will become the actual lookup area.

Before you get started, let's make sure your desktop matches mine. The toolbox should be on the left; if it isn't, select *Ctrl + B* on the keyboard (or **Layout** | **Toolbox** from the menu bar). The **Properties** window should be to the right of your layout window; if it isn't, select *Ctrl + M* on the keyboard (or **Layout** | **Properties** from the menu bar).

As you go through the layout and design instructions, always accept the default value for any settings not specifically mentioned in the instructions. Also, it's a good idea to periodically save your work. There is no **Save** button on the layout window, so you'll need to close and then re-open the window to save it.

First, you're going to set the window properties. Click on a blank section of the window; the **Properties** window should show **[Window]** in the **Object** drop-down box. If it doesn't, you can select **[Window]** from the **Object** drop-down box. The **[Window]** object will be the last item on the list. On the **Properties** window, set the properties as indicated in the following table:

`Window`

Tab	Property	Value
Object	`AutoLinkTable`	`RM_Customer_Contact_MSTR`
Object	`Name`	`Customer_Contact_Lookup`
Object	`Title`	`Customer Contacts`
Object	`WindowType`	`Modeless Dialog`

Tab	Property	Value
Visual	`Size-Height`	`320`
Visual	`Size-Width`	`620`

> The 320 x 620 size is the Dynamics GP default for a lookup window.

Next, you're going to add some fields to your window.

Drag the **Sort by**, **Cancel Button**, and **Select Button** global fields from **Toolbox** onto your window. Don't concern yourself with exact placement at this time, you'll set placement using your properties settings.

Like you did with the **RM_Customer_Contact_Maintenance** window, add the global fields to your window by dragging them from the list of global fields in **Toolbox** and drop them onto the window.

One of the great things about working in the development dictionary is the rich assortment of fields you don't have to create yourself. Even the pictures are already on the buttons!

Next, you need to draw a separator line near the bottom of the window, above the buttons. To draw the line, push the Line tool button in the toolbox, do not hold down your mouse button and do not try and drag this object onto the window. Move the mouse away from the toolbox and you'll see it has changed into a **+/** symbol, as shown in the following screenshot:

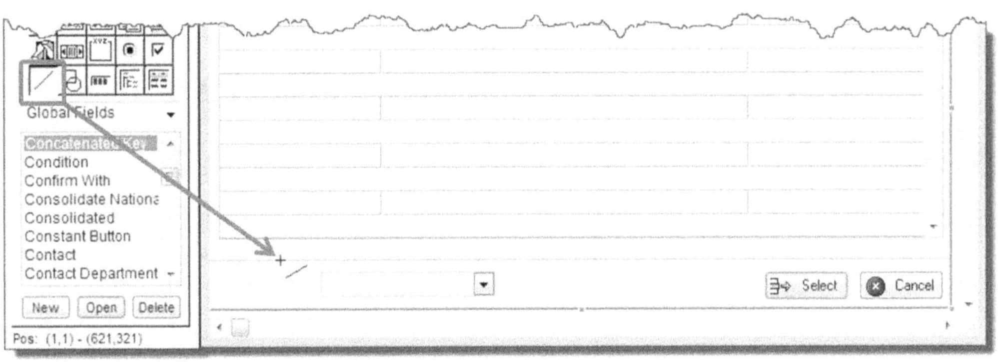

Hold down your left mouse button and draw the line. Don't worry about positioning or length, you will set that in the properties window.

Set the properties of your line as follows:

`Separator Line`

Tab	Property	Value
Visual	`Position-Left`	`0`
Visual	`Position-Top`	`280`
Visual	`Size-Width`	`620`

Set the properties of your fields as follows:

`Sort By`

Tab	Property	Value
Visual	`Position-Left`	`93`
Visual	`Position-Top`	`290`
Visual	`Size-Width`	`145`

`Cancel Button`

Tab	Property	Value
Visual	`Position-Left`	`542`
Visual	`Position-Top`	`290`
Object	`Cancel`	`True`

`Select Button`

Tab	Property	Value
Visual	`Position-Left`	`462`
Visual	`Position-Top`	`290`
Object	`Default`	`True`

Scrolling windows

It's time for the **scrolling window**.

Draw a scrolling window on the face of the **Customer_Contact_Lookup** window. Make it big enough to cover nearly the entire window.

Complete the visual properties of the **Scrolling Window Options** window as shown in the following screenshot:

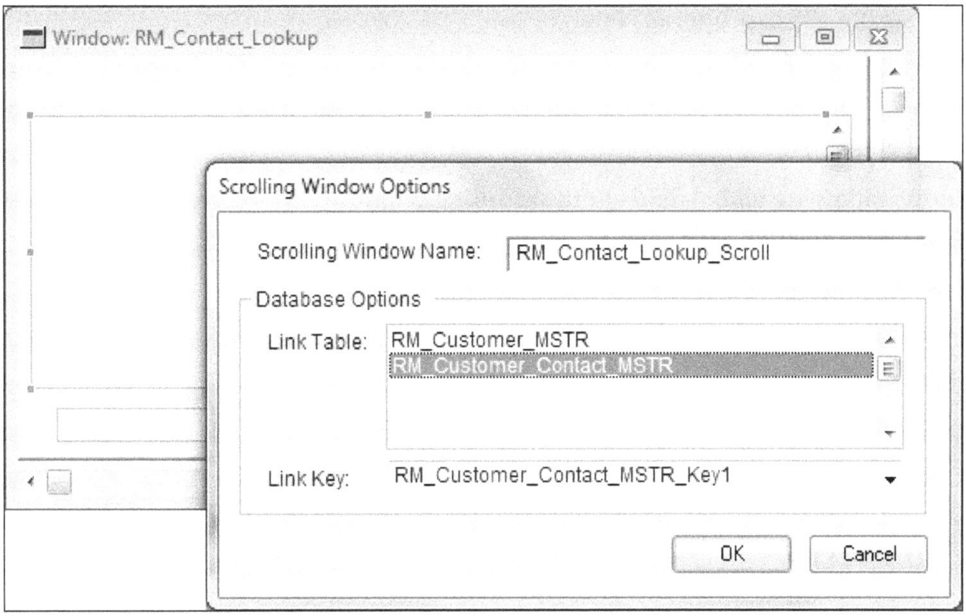

Setting properties for a scrolling window is a little different from other fields. The visual properties for the scrolling window are set from the host window. In our case, the host window is the **Customer_Contact_Lookup** window.

Select the scrolling window and you'll see you have only the **Visual** tab in the **Properties** window. In order to access the object properties, you need to double-click on the scrolling window to open it. For now, just set the visual properties according to the following table:

`Scrolling Window`

Tab	Property	Value
Visual	`Position-Left`	`8`
Visual	`Position-Top`	`15`
Visual	`Size-Height`	`230`
Visual	`Size-Width`	`590`

You'll set the object properties later.

Before you add the fields to the scrolling window, change the layout settings so that you can see the field names. Select **Layout** from the menu bar and then select **Show Field Names**.

Add the following global fields to the scrolling window and set the field properties according to the following tables:

- `Customer Contact ID`
- `Contact Person`
- `User Defined 1`

`Customer Contact ID`

Tab	Property	Value
Visual	`Border`	`False`
Visual	`Position-Left`	`3`
Visual	`Position-Top`	`0`
Visual	`Size-Height`	`20`
Visual	`Size-Width`	`132`

`Contact Person`

Tab	Property	Value
Visual	`Border`	`False`
Visual	`Position-Left`	`137`
Visual	`Position-Top`	`0`
Visual	`Size-Height`	`20`
Visual	`Size-Width`	`301`

`User Defined 1`

Tab	Property	Value
Visual	`Border`	`False`
Visual	`Position-Left`	`442`
Visual	`Position-Top`	`0`
Visual	`Size-Height`	`20`
Visual	`Size-Width`	`144`

After you put the data fields in place, your completed scrolling window will be similar to the following screenshot:

Next, you need to dress up the scrolling window a bit. Since each fields in the scrolling window is going to have its `Border` property turned off, why not set them all at once, instead of one at a time?

Hold down the *Shift* or *Ctrl* key and click on each one of the fields.

Set the property according to the following table:

Scrolling window fields

Tab	Property	Value
Visual	`Border`	`False`

Now that you've turned off the borders, you need to use the Line tool to visually separate the fields and the row. Draw a line the width of the scrolling window beneath the fields. After that place a line between each field. This is the Dynamics GP standard on how a scrolling window is supposed to look. Close to save the scrolling window as well as the **Customer_Contact_Lookup** window when you've finished.

Open the **Customer_Contact_Lookup** window, and then your scrolling window (remember, you just have to double-click on it), so that you can access its object properties.

Set the object properties according to the following table:

Scrolling window object properties

Tab	Property	Value
Object	`DefaultDblClick`	`True`
Object	`WindowType`	`Browse Only`

Working with window fields

In order to finish the layout of your two windows you need to know how to put static text on the window and how to link prompts and lookups.

Adding static text

Static text includes any words you see on the window that are not data fields. The value never changes, hence the word *static*. The most obvious static text is a field prompt. Some static text is identified on the following screenshot:

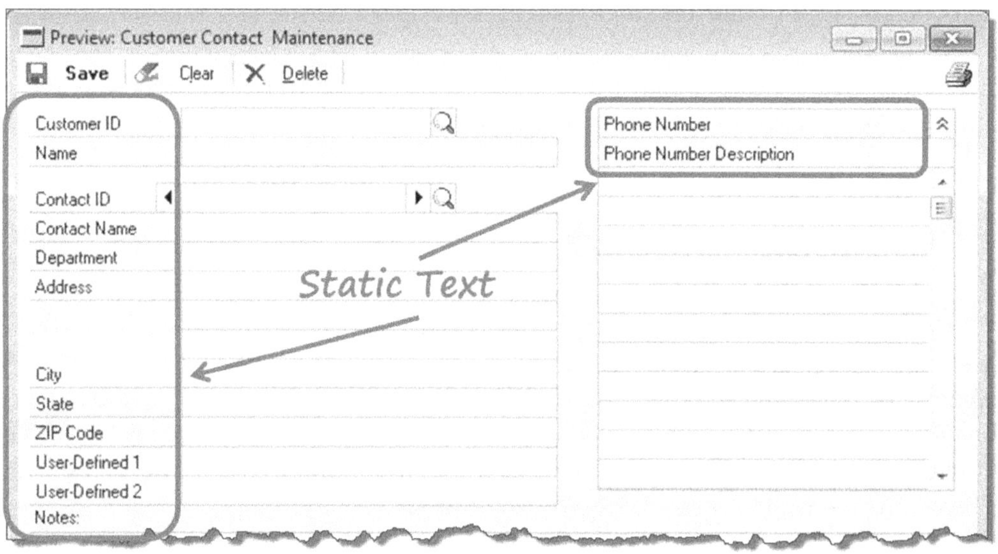

To type text on a window, use the toolbox button with the letter **A** on it.

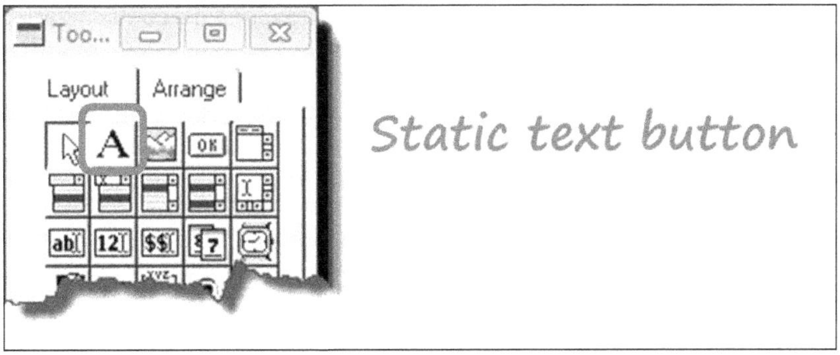

Push the button and then click on the window where you want to add or modify text. When the cursor changes to a flashing vertical line and the mouse pointer is shaped as an **I** bar, you are ready to type.

Column headings

To complete your lookup window, you need column headings. The column headings are actually static text on the host window, not the scrolling window. In order to know where to place the headings, you need to uncheck the **Show Field Names** option. Open the layout menu to access the **Show Field Names** item.

With the **Show Field Names** option off, you should be able to place your headings by following the lines you drew on the scrolling window. Your lookup window will look like the following screenshot after you add the appropriate column headings to your lookup window:

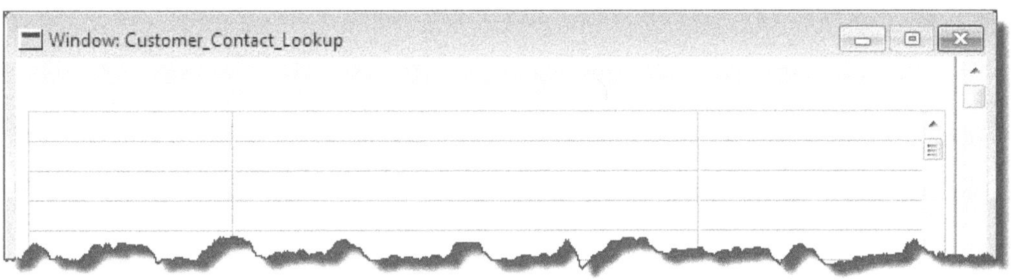

Summary

In this chapter, you've created the entire user interface for you customization, including the scrolling windows.

Don't forget about the import/export trick you used to bypass the tedious job of building the **Customer Contact Master** window from scratch.

Your next step is to code your application and bring it to life. In the next chapter, you'll dive into using sanScript, the life's blood behind Dexterity.

5

sanScript – Making It Work

You will give life to your application in this chapter using sanScript! Soon, a looking glass button on your **Customer Contact Maintenance** window will open a lookup window. You will scroll through the data using browse buttons, and zoom around the windows with ease. Working with an assortment of features will give you a good idea of what it is like to develop using Dexterity.

Table operations come first: creating, retrieving, updating, and deleting records. Table operations will lead you to ranges. You'll use ranges to fill scrolling windows and filter data. The introduction to triggers will provide a means for you to interact with third-party applications and to provide rich functionality without the need for alternate forms.

Welcome to Dexterity!

Key topics in this chapter include:

- Introduction to sanScript
- Scrolling windows
- Triggers

It's time to start making your application function.

Introduction to sanScript

sanScript is the engine of Dynamics GP. Like any other programming language, you will never really appreciate what it can do until you see it in action. This chapter will provide that action. sanScript uses a natural language syntax making the scripts very easy to read and easy to learn. It comes with a rich function library, a graphical forms designer, an embedded report writer, a source-level script debugger, and supports COM and .NET technologies. A well-written application in Dexterity provides a seamless integration with Dynamics GP. Your users will not be able to detect when they have switched from the native Dynamics GP code, to yours.

You can get more information about the history and structure of sanScript in the first three chapters of this book.

Scripts

Before you get started writing scripts, you need to understand a few basics about sanScript. There are not that many rules, but still, if you don't follow them your script will never compile. sanScript does not support Microsoft's Intellisense. A fully qualified field name can get quite long, so some help in looking these names up can be a lifesaver. For example, the fully qualified field name of the `Customer Number` field on the `Customer Contact Maintenance` window looks as follows:

```
'Customer Number' of window RM_Customer_Contact_Maintenance of form
RM_Customer_Contact_Maintenance
```

To assist you with the long field names, you have two tools. The first, native to Dexterity, is the **Names** window. To get the field name from the **Names** window, select the **Names** button from the Script Editor. When the **Names** window opens, select the **Forms** tab and choose the correct form name. With the correct form selected, click on the **Windows** button and highlight the desired window. Click on the **Fields** button and then double-click on the field you want to reference. The **Names** window will close and you will be returned to the Script Editor. Since Dexterity does not support the right-click mouse action, you need to use the *Ctrl + V* keyboard shortcut to paste the field name into the Script Editor. The following screenshot shows the components of the **Names** window:

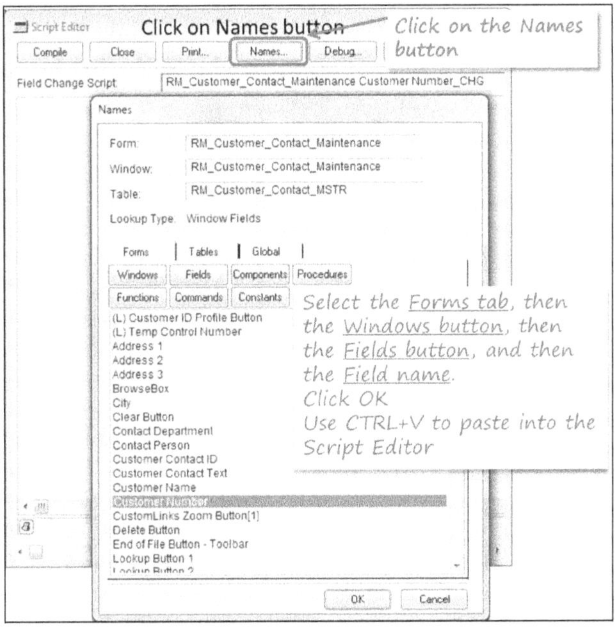

The second tool is the DexSense application written by *Tim Gordon*. In one of the steps of preparing your development environment, you installed DexSense. DexSense works from within the Script Editor. As you type, it monitors your keystrokes waiting for the sequence of keys indicating that you need to select an object name. At that point, a small window opens listing the available object names. When you select the desired object and press the *Enter* key, the object name will paste into the Script Editor. The DexSense window is shown in the following screenshot:

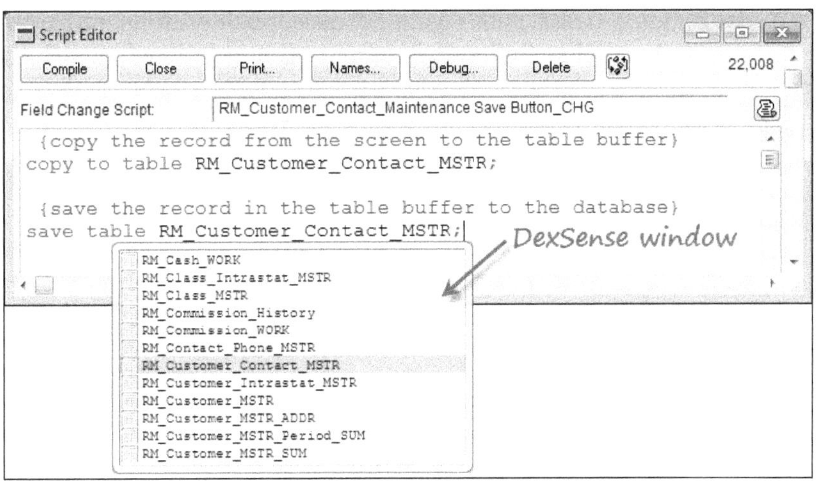

Syntax rules

Every language has standards and conventions. The following is a quick list of some of the basic rules of sanScript:

- All statements end with a semi-colon
- Statements may span multiple lines without a continuation character
- sanScript is case sensitive
- Reserved words are all lowercase.
- Constants are all uppercase
- Object names containing spaces must be set off by 'single quotes'
- Comments are bounded by French braces {} and can go anywhere
- Comments can span multiple lines
- Scripts can be 24,000 characters long including spaces and comments

Script flow

You attach scripts to events using the Script Editor. When the event occurs, the script assigned to that event executes. Determining the correct event to place your scripts in is of the utmost importance.

Generally speaking, each object has several states or events. We will be focusing on three of them: **Pre**, **Change**, and **Post**:

- The **Pre** event occurs just as you're opening a form or window, or entering a field.
- The **Change** event occurs when you change the value of a field and exit the field. For toggling fields, such as checkboxes, push buttons, radio groups, list boxes, and visual switches, the change event occurs when you click on the field.
- The **Post** event occurs just as you're closing a form or window, or exiting a field.

Forms and windows do not have a Change event.

The Change script is the default script. When you issue a `run script` statement, you execute the Change script of the object. You need to qualify any other scripts.

Scripts run in the following order:

Script Flow Structure

- Pre Form
  - Pre Window *
    - Pre Field
    - Change Field
    - Post Field
  - Post Window
- Post Form

*Pre Window also runs with **`restart form`** and **`restart window`**

Script naming conventions

Script naming conventions are very straightforward. What is even better, Dexterity automatically suggests a script name that follows the convention! Script names are made up from three parts: `[window name] [field name]_[type]`. The `[type]` segment refers to the type of script; it is abbreviated to `PRE`, `CHG`, `POST`, and so on.

For example, the name of the Change script for the **Save** button on the **Customer Contact Maintenance** window is `RM_Customer_Contact_Maintenance Save Button_CHG`, where `RM_Customer_Contact_Maintenance` is the window name, `Save Button` is the field name, and `CHG` is the type.

Procedures and functions do not suggest decent names, so you will always want to rename them. There are no hard-and-fast rules for naming functions and procedures. Generally, you would want to name them according to the action being performed: names such as `INITIALIZE` and `HANDLE CHANGES` are popular.

Table operations

To know any database language is to have an intimate knowledge of table operations. Therefore, building a foundation in sanScript would naturally begin with learning a lot of **CRUD**. To know your CRUD, you need to be able to do the following:

- **C**reate a record
- **R**etrieve a record
- **U**pdate a record
- **D**elete a record

Once you have mastered these actions, the rest is just syntax.

Understanding the interaction of data between the tables and the windows is paramount to understanding Dexterity and sanScript. Data displayed in the windows gets there from either you pounding it in or you retrieving it from the database via the **table buffer**. A table buffer is a temporary storage area in memory and acts as an intermediary between you and the data. Only one record at a time can occupy a table buffer, and it will stay there until you tell the table buffer to put it back or to clear it out. In computer-speak, the table buffer contains values from the last table operation.

If you want to display the table buffer values on a window, you need to get a *copy* of those values from the table buffer and copy them up to the window. Now you have three versions of the data. The one stored in the physical table, the copy in the table buffer, and the copy on the window.

When a user makes a change to the values on the window, nothing changes in the table buffer nor the physical table. Changing values on a window is not a table operation. The windows are not bound to the tables; the changes are stored in memory until you decide what to do with them. If you decide you would like to update the values in the physical table, you need to copy the record from the window to the table buffer, and then from the table buffer to the physical table. Nothing happens automatically, you will need to code each of these actions deliberately.

That's the gist of it. Just keep in mind that the table buffer's copy of the record doesn't change until you tell it to change. As with any programming language, there are several exceptions to this dictate, but we'll keep it straightforward in this writing.

It is critical that you understand this concept of window, table buffer, and table before you can go forward with Dexterity. Recite every morning:

> "*The table buffer can only hold one record at a time.*"

As you tackle more advanced concepts in your future programming endeavors, you'll discover more about managing records in the table buffer, but for now, let's go with the *one record* concept.

Where do table buffers come from? I am so glad you asked! As you learned in the last chapter, you can attach tables to forms. Unless you have instructed the form not to, when you open a form, it creates a table buffer for each of its attached tables. Each table buffer is loaded into memory ready to receive the values of a single record. It stays in memory until the form is closed.

You can attach the same table to multiple forms. If multiple forms open with the same table attached, you get a table buffer for each form. The table buffers are independent of each other. Nevertheless, they are the property of the form they came with. The form that opened it owns the table buffer. If you open more than one form, you get more than one table buffer. When you close the form, the table buffer closes as well.

To facilitate these table operations, you will use the following sanScript statements:

- `get`
- `change`
- `remove`
- `save table`
- `release table`
- `copy to table`
- `copy from table`

The word *table* refers to the table buffer in some statements, and the physical table in others. As you might imagine, picking the right statement is sometimes confusing until you get accustomed to it. The quote "a picture is worth a thousand words" comes to mind whenever the table buffer discussion begins.

Perhaps the following diagram will help clarify the concept:

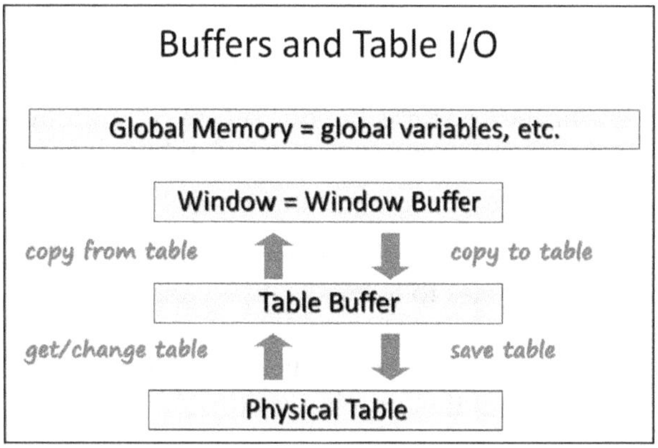

Keeping the previous diagram in mind, the following is a brief description of each of the statements mentioned:

get

- Use `get` to retrieve a record from the database to the table buffer
- No lock is placed on the record
- You cannot change the values
- Inquiry windows and lookup windows use `get` to retrieve records
- Retrieving a record from the **RM_Customer_MSTR** table looks like this:

```
get table RM_Customer_MSTR;
```

change

- Use `change` to retrieve a record from the database to the table buffer
- A passive lock is placed on the record
- You *can* change the values
- Maintenance windows and transaction windows use `change` to retrieve records
- Retrieving a record from the **RM_Customer_MSTR** table looks like this:

```
change table RM_Customer_MSTR;
```

remove

- Use `remove` to delete the record in the table buffer from the physical table. You may also use `remove range` to delete a range of records. If you use the range modifier, and have not defined a range, you delete all of the records, ouch!
- `remove` does not clear the table buffer.
- Removing a record from the **RM_Customer_MSTR** table looks like this:

```
remove table RM_Customer_MSTR;
```

> Always use `Table_IsRangeSet(`*tablename*`)` to check for the existence of a range before using the `remove range` statement!

save table

- Use `save table` to save the contents of the table buffer to the physical table
- It releases any lock on the record whether or not the save operation is successful
- `save table` does not clear the table buffer
- Saving a record to the **RM_Customer_MSTR** table looks like this:

```
save table RM_Customer_MSTR;
```

release table

- Use `release table` to release the lock placed on the record by the `change table` statement. You must release the lock before you can read another record into the table buffer.
- `release table` does not clear the table buffer.
- It has no impact on the physical table.

copy to table

- Use `copy to table` to copy the values of all *auto-copy* fields from the selected form to the table buffer.
- It copies values from the current form to the table buffer.

- Notice we said form and not window. Whenever we say form, we are including all of the windows on the form.
- The window does not have to be open for the field values to copy. The values are copied to the table buffer in the order in which the windows are listed on the form definition window.

> **auto-copy** is a field property that defaults to TRUE. If a field is not an auto-copy field, the **copy** statements will skip that field.

copy from table

- Use `copy from table` to copy the value of all auto-copy fields from the table buffer to the windows on the selected form
- It does not release the lock on the row
- It does not clear the table buffer

Creating a record

You will begin coding table operations by creating a record.

Because everything has to go through the table buffer, creating a record is a two-step process. First, you copy the record from the window to the table buffer, and then you save the record from the table buffer to the physical database table.

Turning focus to your project, let's write the code for the **Save** button.

Launch Dexterity and open the `Project.dic` dictionary. Launch **DexSense** from **Start Menu** | **All Programs** | **Dexsense** | **DexSense** and wait for it to connect. **DexSense** runs as a tray icon in your system tray. Once it connects, the tray icon should turn to blue; if it is red, review the DexSense manual for troubleshooting steps.

After **DexSense** is up and running, open the **RM_Customer_Contact_Maintenance** form and then open the **RM_Customer_Contact_Maintenance** window. If the window is displaying the field names, you may want to turn them off. To hide the field names, select **Layout** from the toolbar and then uncheck the **Show Field Names** menu item.

Double-click on the **Save** button to open the **Script Editor** window. Your script will copy the values on the form into the table buffer, and then save the values from the table buffer to the database table. If the record already exists in the database table, this script will update it.

Enter the script as shown in the following screenshot into the Script Editor. Close the **Script Editor** window when finished and compile the script when prompted:

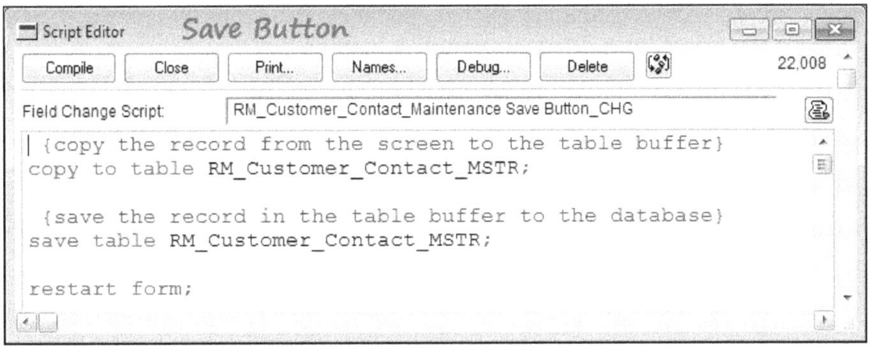

Congratulations! You have satisfied the first and third requirements of CRUD; you can now create and update a record.

As you type in the Script Editor, your DexSense application should open a small window and present you with a list of object names. All you need to do is select a name from the list and DexSense will copy the text into the Script Editor. DexSense is a tool that can save you hours of time looking for and typing in long descriptions. However, keep in mind that it does not validate your sanScript code, nor does it display sanScript functions or statements.

Since we are so close to the **Clear** button, let's add a script to it. When you push the **Clear** button, you want your script to clear all field data on the window and return the screen to its default state. Double-click on the **Clear** button to open **Script Editor**. The following script will do the trick:

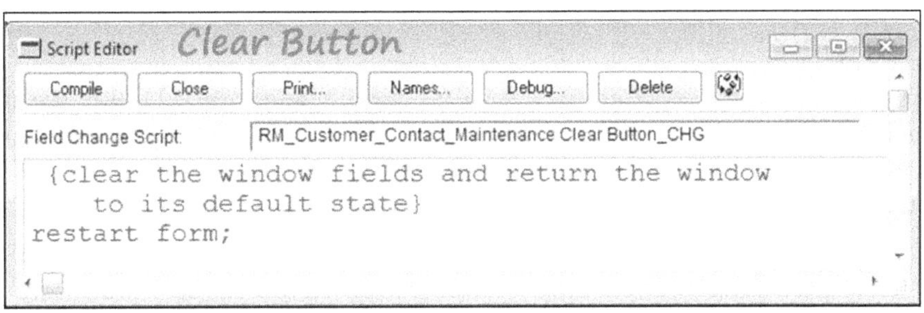

The Clear Button script looks simple, but it accomplishes many things. The **restart form** statement sets your form up to accept another record. Specifically, the **restart form** statement completes the following:

- Clears the window fields (the window buffer)
- Clears the change flag on the form
- Sets focus to the first field in the tab order of the main window
- Runs the window `PRE` script of the main window on the form
- Runs the field `PRE` script of the first field in the tab order on the main window (because focus is moved to that field)
- Releases any lock on the record in the table buffer
- Does *NOT* clear the table buffer
- Does *NOT* run the form `PRE` script.

Retrieving a record

Like saving a record, retrieving a record is also a two-step process. First, you set the table buffer with the key field values, and then you tell the system to go looking for the matching record in the database. If it finds the record, the table buffer will fill with the remaining field values from the auto-copy fields in the table. You specify whether you want to change the record, or just read the record, by the statement you've used to request it.

Using the `Change` statement will allow you to modify the record values.

Using the `Get` statement retrieves the record as read-only.

Customer

For your project, if the user changes the value of the **Customer ID** field, you'll presume they want to retrieve that record. Therefore, you need to put your code on the `change` script of the **Customer ID** field. If the **Customer ID** doesn't exist, then you are going to tell the user to try again. Double-clicking on a field in the form layout window will open the `change` script for that field. Since you want to attach a script to the **Customer ID** field, double-click on the **Customer Number** field to open its Change script. Enter the code from the following screenshot into **Script Editor**. Any text bounded by French braces `{ }` is part of a comment and will not affect the code if not included:

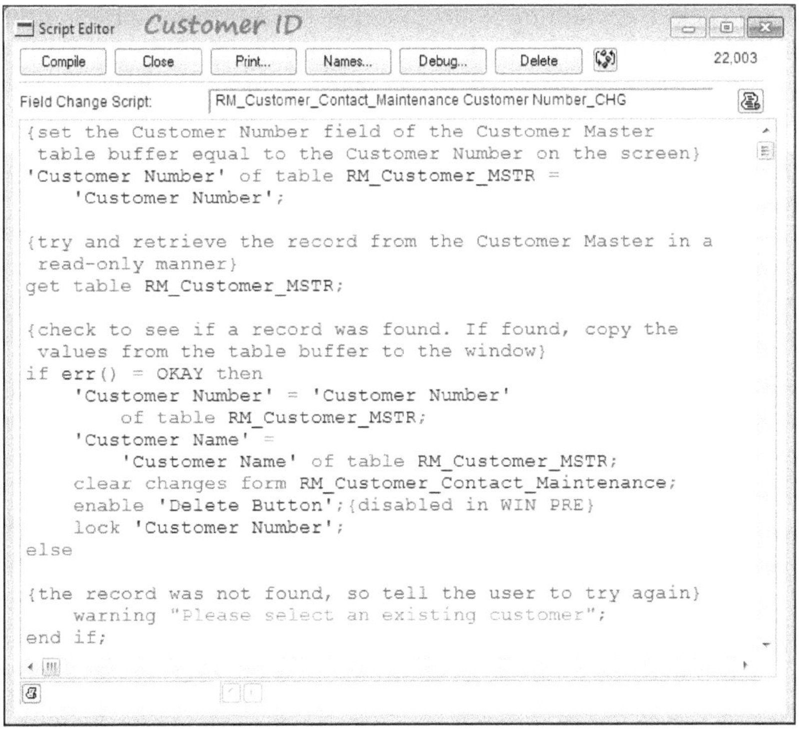


```
{set the Customer Number field of the Customer Master
 table buffer equal to the Customer Number on the screen}
'Customer Number' of table RM_Customer_MSTR =
    'Customer Number';

{try and retrieve the record from the Customer Master in a
 read-only manner}
get table RM_Customer_MSTR;

{check to see if a record was found. If found, copy the
 values from the table buffer to the window}
if err() = OKAY then
    'Customer Number' = 'Customer Number'
        of table RM_Customer_MSTR;
    'Customer Name' =
        'Customer Name' of table RM_Customer_MSTR;
    clear changes form RM_Customer_Contact_Maintenance;
    enable 'Delete Button';{disabled in WIN PRE}
    lock 'Customer Number';
else

{the record was not found, so tell the user to try again}
    warning "Please select an existing customer";
end if;
```

Script analysis

This script does not use the `copy from table` statement because you only want to copy **Customer Number** and **Customer Name** from **RM_Customer_MSTR**. If you use the `copy from table` statement, any field on the window that matches the field name in the table would be overwritten. You don't want that because you do not want to populate the address fields on the **Customer Contact Maintenance** window with the customer's address. You are going to be populating those address fields with the contact's address, not the customer's address.

Customer zoom

Dynamics GP windows often contain window elements called **zoom fields**. A zoom field allows the user to *drill down* on that field for additional information. Two things are in play to visually alert the user of a zoom: the color of the prompt is blue and underlined, and the mouse changes to a pointing finger when it passes over the prompt.

When the user clicks the pointing finger, another window opens. Depending on what kind of field it is, the additional information returned by the zoom varies. An ID field should bring up the related set-up, or maintenance window. A summary field should bring up the related detail.

Users expect these window elements to be present; therefore, to deliver a truly seamless customization, your windows must include them too.

The **Customer ID** field on your **Customer Contact Maintenance** window should be a zoom. Because you copied an existing Dynamics GP form, the window already contains all of the necessary elements for a zoom.

First, you need to delete **CustomLinks Zoom Button**; you won't be using it. Simply highlight it and delete it. Refer to the following screenshot to locate the button field:

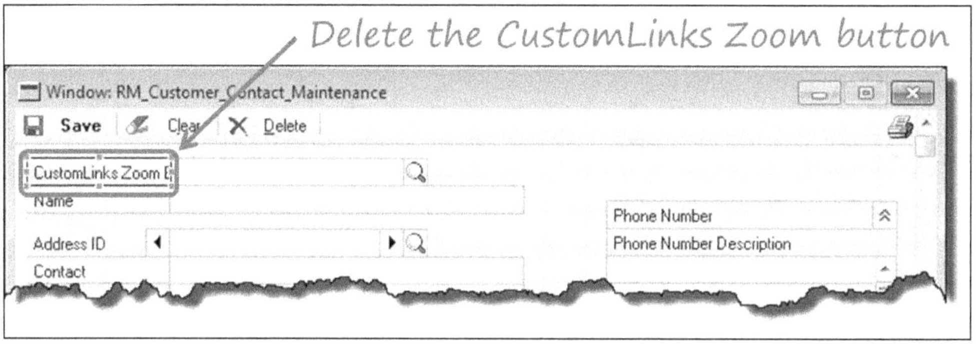

Move down the **Customer ID** profile button that is on top of the **Customer ID** field prompt and you will see that the characters of the field prompt are blue and underlined; refer to the following screenshot:

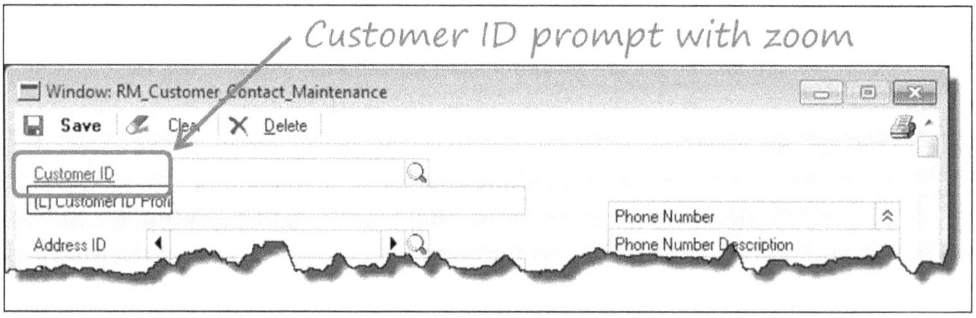

This behavior (the blue color and underline) is controlled by the `Zoom` property of the field prompt; it has nothing to do with the linked field itself.

Setting any prompt's `Zoom` property to `True` will cause it to be blue and underlined. (The **User Preferences** settings actually control the color, but the default is blue and underlined).

The pointing finger is achieved by setting the `Zoom` property to `True` of the invisible push button that is positioned on top of the prompt. The **Customer ID** profile button that you moved down to uncover the field prompt is the invisible button in this case.

When your user clicks on the **Customer ID** field prompt, the **Customer Maintenance** window should open. Since you cannot attach code to a bit of static text, you use the invisible push button to do your bidding. You attach your code to the invisible push button; when the user clicks on the prompt, they are really pushing your button!

The following screenshot shows the **Properties** window of the **Customer ID** prompt and the invisible push button:

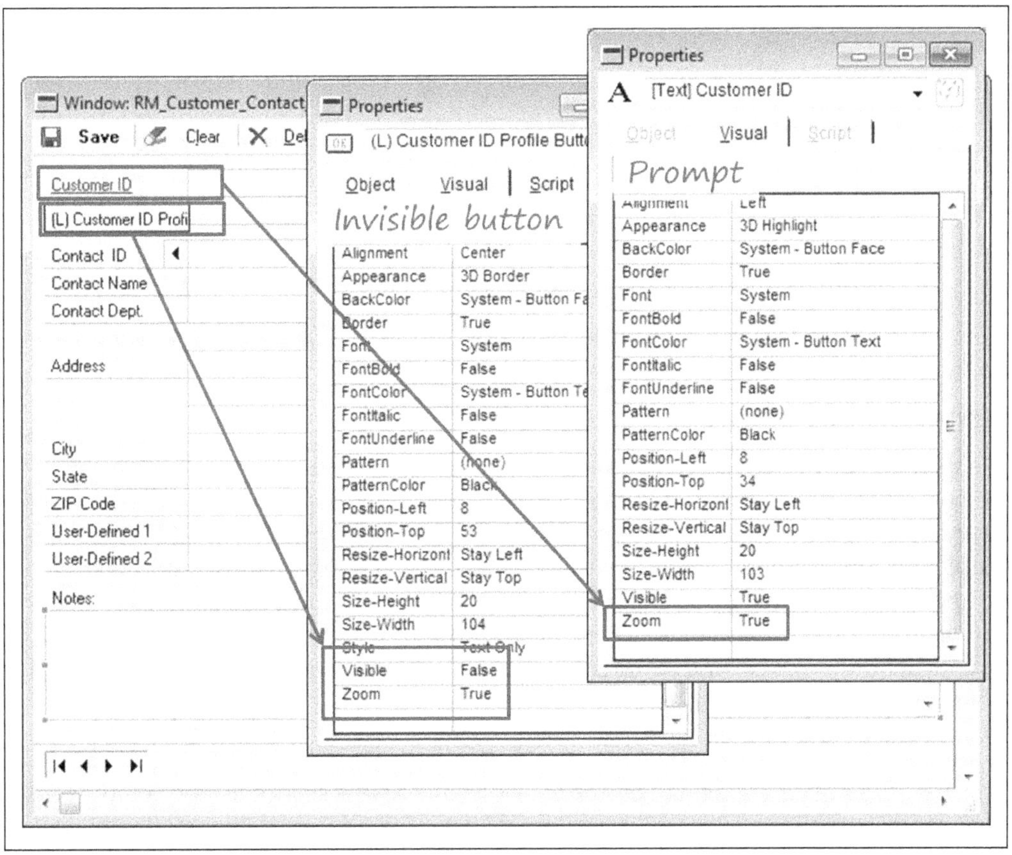

Put the invisible button back on top of the **Customer ID** prompt. Double-click on the invisible push button (**(L) Customer ID Profile Button**) to open its Change script in **Script Editor**. Copy the script below to **Script Editor**. When the user clicks the zoom, this script will execute and open the **Customer Maintenance** window displaying the customer from the **Customer Contact Maintenance** window:

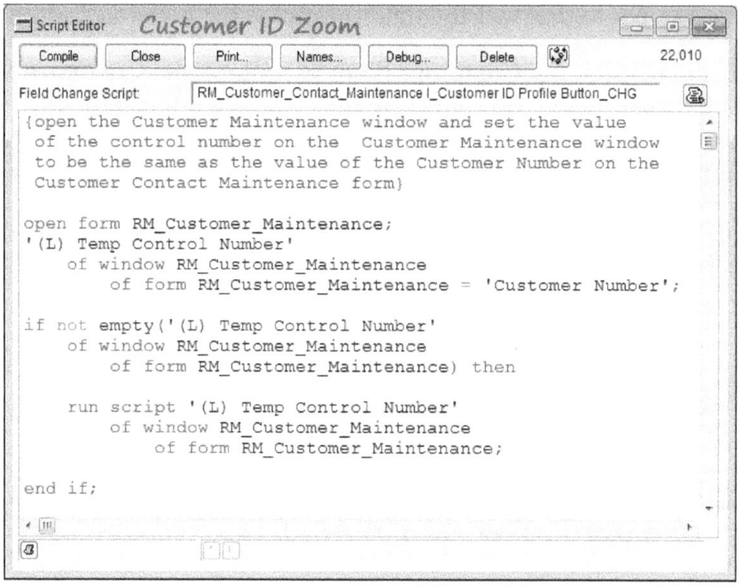

Move **(L) Customer ID Profile Button** back on top of the **Customer ID prompt**.

Browse buttons

The next bits of functionality you need to take care of are the browse buttons at the bottom of the **Customer Contact Maintenance** window. They are supposed to browse through the customer records in the **RM_Customer_MSTR** table. This one is simple: set the value of the table buffer to whatever is on the window and just read up and down the table in **Customer Number** order. We could use a different order, but we do not have a sort by field on the window to indicate a different order. With no instruction to the contrary, the first table key is used. The only segment in the first table key is **Customer Number**.

Before coding your browse buttons, you first need to drag the global field named `Display Existing Record` to the area below the visible section of the window.

Double-click on each browse button to open the **Script Editor** window and then type in the relevant code from the following screenshots.

The left-most button retrieves the first record in the table. If you are on the first record in the table, your script causes a beep sound to play:

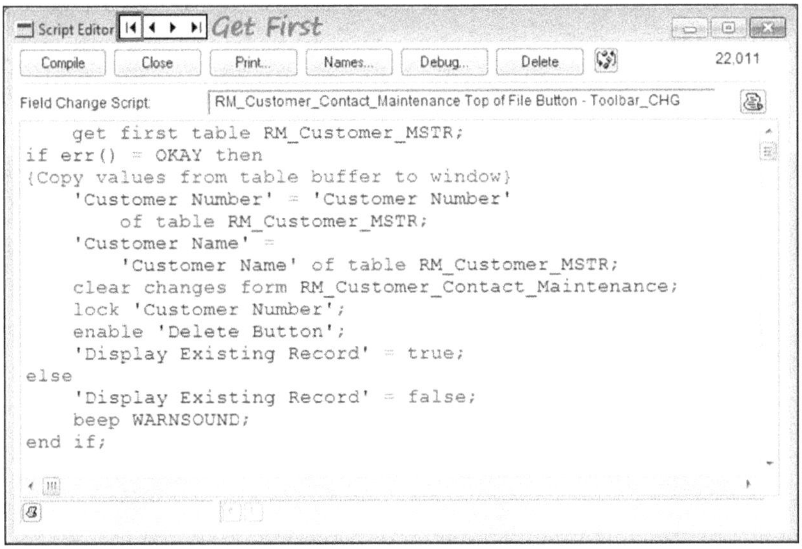

The get previous button retrieves the record just before the displayed record. If the displayed record is the first record in the table, your script causes a beep sound to play.

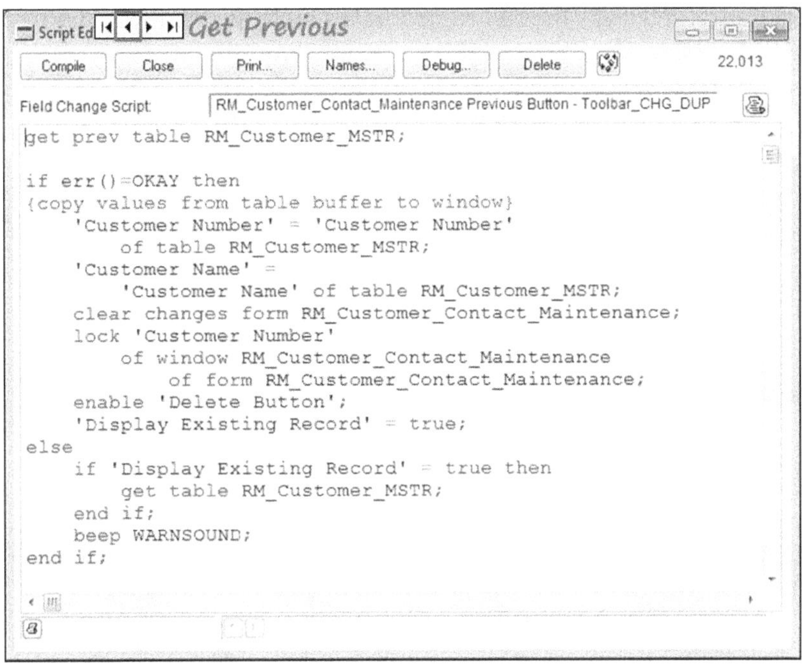

The get next button retrieves the record after the displayed record. If the displayed record is the last record in the table, your script causes a beep sound to play.

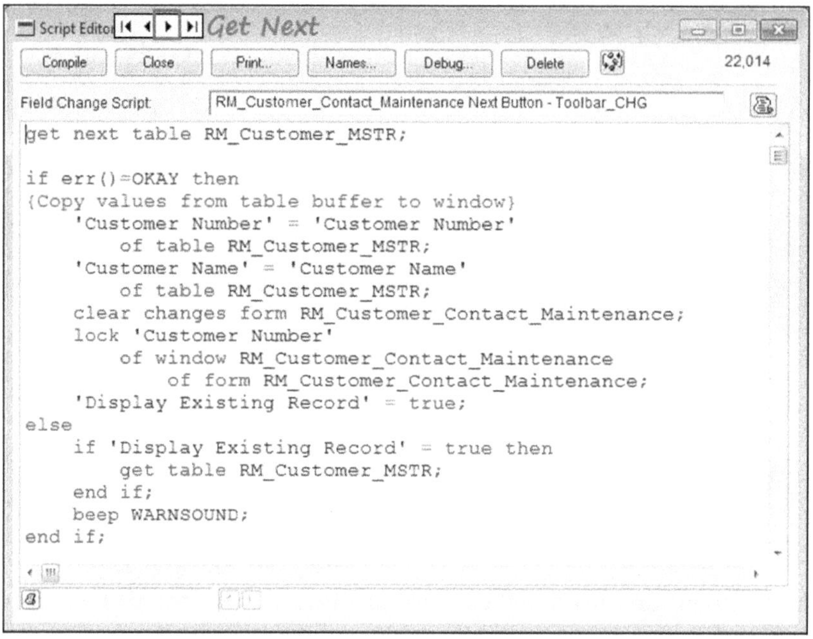

The get last button retrieves the last record in the table. If the displayed record is the last record in the table, your script causes a beep sound to play.

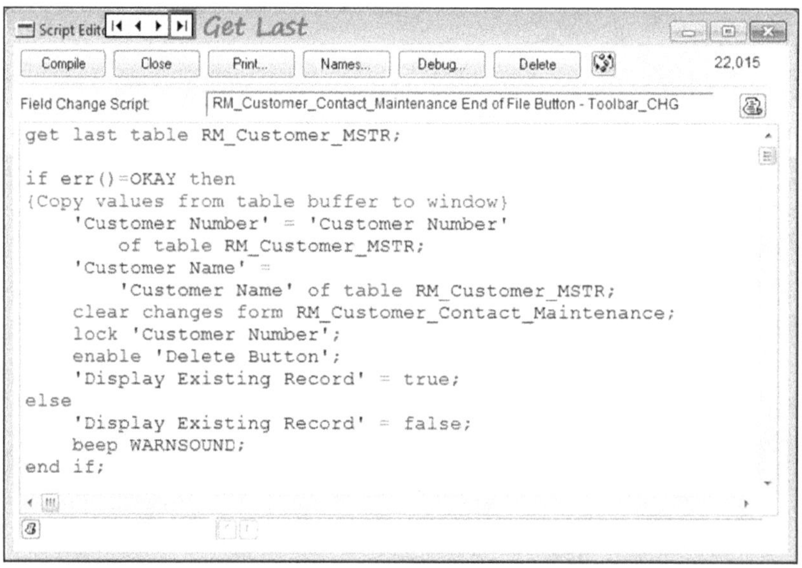

Updating a record

At this point, with your **Save** button active, you can retrieve records by typing in the customer ID or browsing to it using the browse buttons. Updating a record involves simply bringing it up to the screen, changing it, and then pushing the **Save** button. Looks like you have Create, Retrieve, and Update in the bag; Delete is just moments away.

Deleting a record

The following sanScript statement provides the last component of learning the table operations of a new language. To delete a record, you copy the record into the table buffer and then use the `remove table` statement to delete it.

Double-click on the **Delete** button to open the **Script Editor** window and then type in the code from the following screenshot:

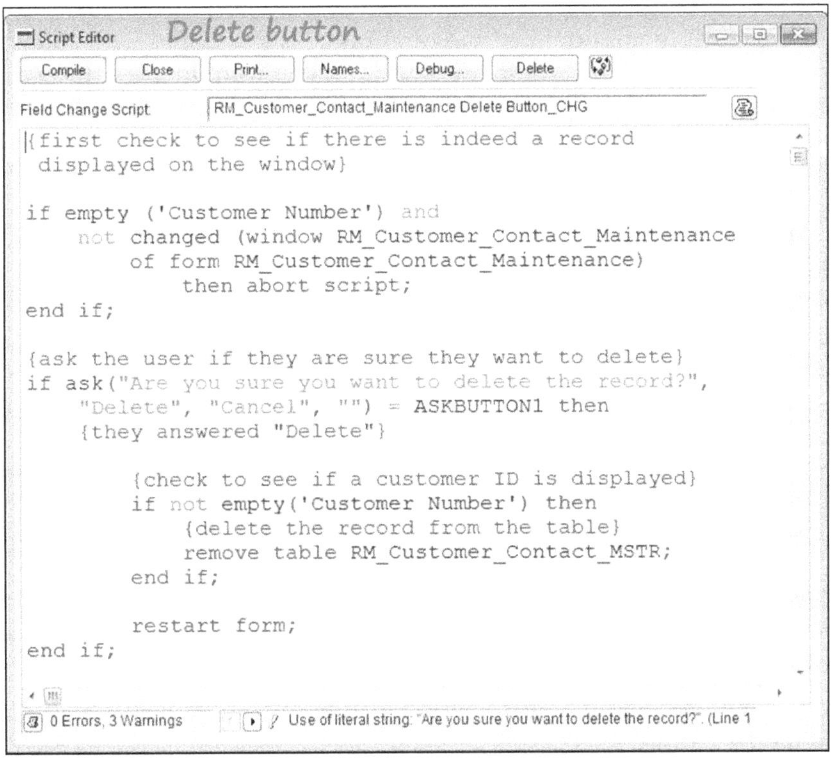


```
{first check to see if there is indeed a record
 displayed on the window}

if empty ('Customer Number') and
    not changed (window RM_Customer_Contact_Maintenance
        of form RM_Customer_Contact_Maintenance)
            then abort script;
end if;

{ask the user if they are sure they want to delete}
if ask("Are you sure you want to delete the record?",
    "Delete", "Cancel", "") = ASKBUTTON1 then
    {they answered "Delete"}

        {check to see if a customer ID is displayed}
        if not empty('Customer Number') then
            {delete the record from the table}
            remove table RM_Customer_Contact_MSTR;
        end if;

        restart form;
end if;
```

Congratulations! You are now CRUD qualified.

Ranges

A **range** is a subset of data. Nearly every script you write will be acting against a range of records rather than the entire table. If you are adept at defining ranges, you have come a long way toward mastering your task. Using sanScript, you describe the beginning of the range, and then issue a `range start` statement. You describe the end of the range, and then issue a `range end` statement. The range is then set and until you remove it any table operation on this form will act against this range as if the records in the range comprised the entire table.

In your application, contacts are grouped by a customer, so you need a way to filter your data so that you only display contacts for one customer at a time. Setting a range allows you to bring up a customer and only see the contacts associated with that particular customer.

After establishing a range, a `get first` statement will retrieve the first record in the range, and a `get last` statement will retrieve the last record in the range. The table name becomes the range name once the range is set; it's as if the only records in the table are the records in your defined range. A range is associated with a key; if you do not specify a key, the table's first key is used. If you want to use a different key, you must specify that key with every table operation statement. You can refer to the key by name or by number.

Setting a range

Next, you're going to write the code to retrieve a list of phone numbers for a distinct customer contact. Since you only want to retrieve phone numbers for a single contact of a particular customer, you are going to use **Customer ID** and **Contact ID** as a filter on the **RM_Contact_Phone_MSTR**. Whether you browse through existing contacts or use the lookup window, you only want to see phone numbers associated with the selected customer contact.

After the user selects a **Customer ID** field, there are three ways for the user to retrieve a **Contact ID** value:

- Type the contact ID into the **Contact ID** field
- Scroll to the contact ID using the left and right browse buttons on either side of the **Contact ID** field
- Open the **Contact ID lookup** window and pick the contact from the lookup window

Let's open the lookup window and pick the contact off the list. Double-click the lookup button next to the **Contact ID** field and type the sanScript mentioned in the following screenshot into the **Script Editor** window:

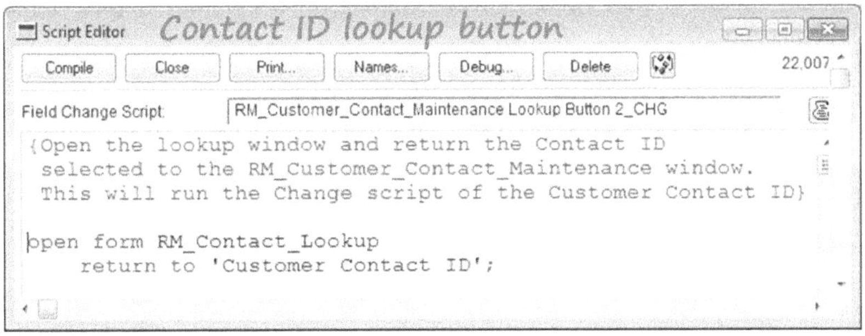

It looks simple enough, and it is, so long as you set a range on the **RM_Customer_Contact_MSTR** table.

You're going to use a range so that you only have to deal with those contacts that are attached to the selected customer. There are a lot of opinions about how to set a range in Dexterity. You're going to learn how to set a range that is easy to do, easy to read, easy to troubleshoot (read debug), and works *every time*. In Dexterity circles, it is known as a **well-behaved range**.

In the beginning, a well-behaved range was necessary because of some subtle, and not so subtle, differences in the supported databases. It had to do with whether the range was exclusive or inclusive. Today, Dynamics GP uses only SQL; therefore, we do not have so many problems with ranges. Nevertheless, ranges are the single most important thing you should be adept at in order to be a superior Dexterity programmer. If you learn nothing else, learn ranges. Let's begin.

To establish a range, you must give the system a beginning point and an end point. To establish a well-behaved range, you need to provide that beginning and end point for each segment in the table key you are using. Here are the basics to setting a range:

1. Clear existing ranges.
2. Set the beginning key values.
3. Start the range.
4. Set the ending key values.
5. End the range.

The secret is to set a value for each segment of the key. It may seem repetitive to you, but believe me, it's the way to go. Let's say you have a table with a four-segment key. The table's **Key Definition** would look similar to the following screenshot:

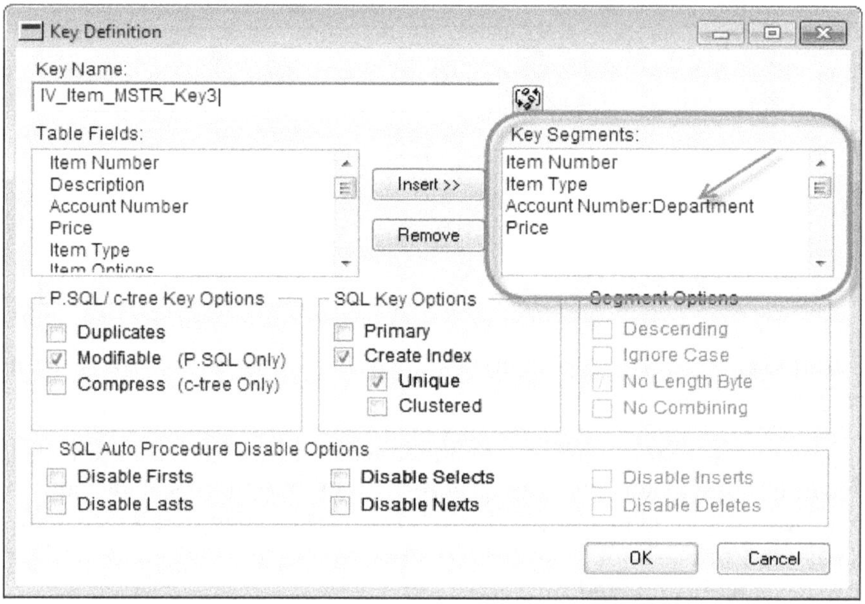

The item identified by the arrow, is a segment in a composite field. You have some additional flexibility when you work with composites. Instead of dealing with the entire composite, you can use each segment of a composite just as you would an independent field. If you were creating an inquiry, and needed to sort the records according to **IV_Item_MSTR_Key3** above, you would set a range for each field in the key.

For instance, let's suppose you needed a report that included only items with the type of `Warranty`, sold by departments 500 through 900, at a price of $10,000 or above. Dexterity ranges can handle this sort of a request quite nimbly. For those segments where there is no absolute low or high end, you use the statements of `clear` for the low end and `fill` for the high end.

`clear` is the lowest possible value accepted by the field, and `fill` is the highest possible value. You need to consider all item numbers in your report, so your begin value for the item number is `clear` and your end value is `fill`.

The easiest way to portray the beginning and ending points for a range is to fashion a table like the following one that includes each segment in the table's key that you are using. Here's what the table would look like for your inventory report:

Segment Number	Segment Name	Begin Range	End Range
1	Item Number	Clear	Fill
2	Item Type	Warranty	Warranty
3	Account Number: Department	500	900
4	Price	10,000	Fill

You don't need to type this script anywhere as part of this project; it is merely an example.

If the code for a warranty is equal to `4`, the sanScript code for setting the four-segment range described above would look like the following screenshot:

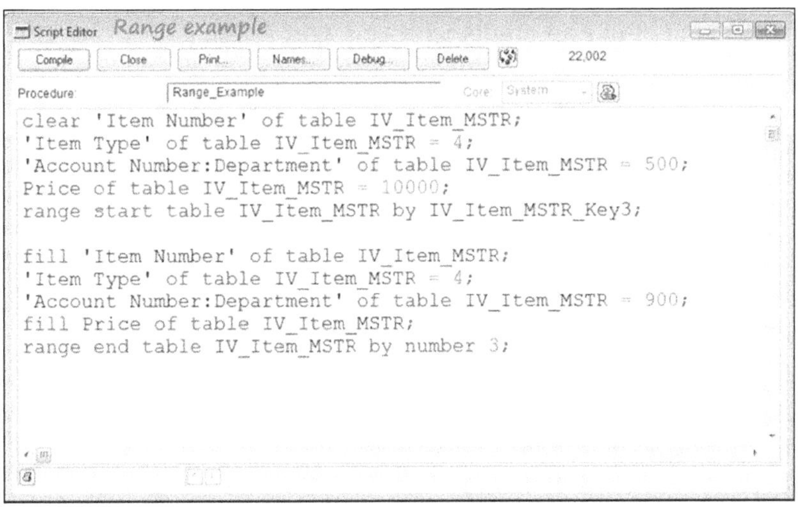

You see how each segment of the key is included in both the `range start` section and in the `range end` section? Also, you can refer to the key either by the key number or the name of the key. In the previous script, the `range start` statement uses the key's name, whereas the `range end` statement uses the key's number.

If you deal deliberately with every segment of your key, you will never go wrong.

Creating a virtual key

If you do not have the key you need predefined in the table, you can create a **virtual key**.

Let's say you need to display contacts of a specific customer that live in a specific state. Does that sound like a range to you? Of course it does! The first thing you need is a table key that includes all of the elements required for the restriction. You need a key containing the following segments:

- `State of table RM_Customer_Contact_MSTR`
- `"Customer Number" of table RM_Customer_Contact_MSTR`
- `"Customer Contact ID" of table RM_Customer_Contact_MSTR`

A quick look at your table definition will reveal that you do not have a key with the previously mentioned segments. Rather than modifying your table definition, Dexterity has a means to alleviate situations like this with the introduction of the virtual key.

A virtual key is kind of like an on-demand key. You define it for a one-time use, and then it goes away. First, let's define the virtual key.

To see how this works, you need to add a few more fields to your window. Add a local string field with the following properties:

`(L) RestrictState` (local field)

Tab	Property	Value
Object	`DataType`	`RestrictState`
Object	`Field`	`RestrictState`
Object	`LinkedLookup`	`Lookup Button 3`
Object	`SetChangeFlag`	`False`
Object	`Tooltip`	`Enter state restriction`
Visual	`Position-Left`	`318`
Visual	`Position-Top`	`83`
Visual	`Size-Width`	`48`

Your window will look similar to the following screenshot with the extra fields added:

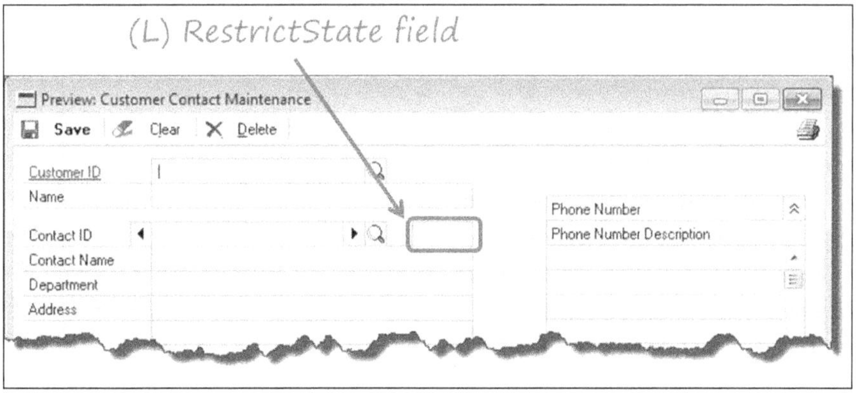

How this will work is that if you enter a state restriction in the local field, a virtual key will be used. By using this key, you will limit the contacts retrieved to just those that live in the state indicated. If the **(L) RestrictState** field is empty, then all of the contacts will be retrieved. You'll need to add your code to the `PRE` script of the **Customer_Contact_Lookup** window.

Using **Resource Explorer**, double-click on the **RM_Contact_Lookup** form, and then open the **Customer_Contact_Lookup** window. In the **Properties** window, click on the **Scripts** tab and then double-click on the **Pre** script. Your **Properties** window will look similar to the following screenshot:

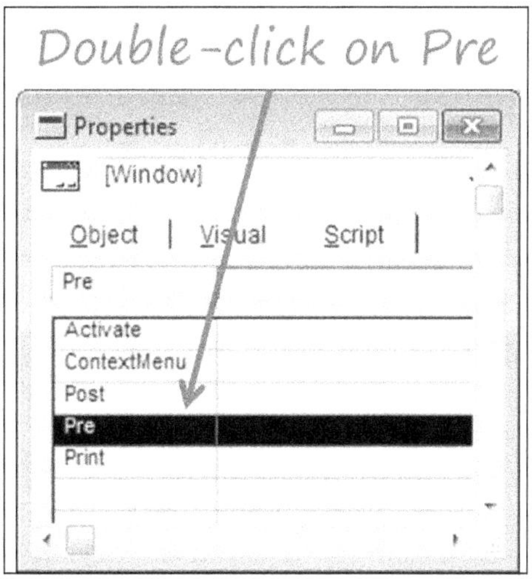

Next, you need to add the script that creates your virtual key, sets a range using the new key, and then populates the scrolling window of phone numbers according to the range.

First, create the virtual key and assign it to the table by typing the following script:

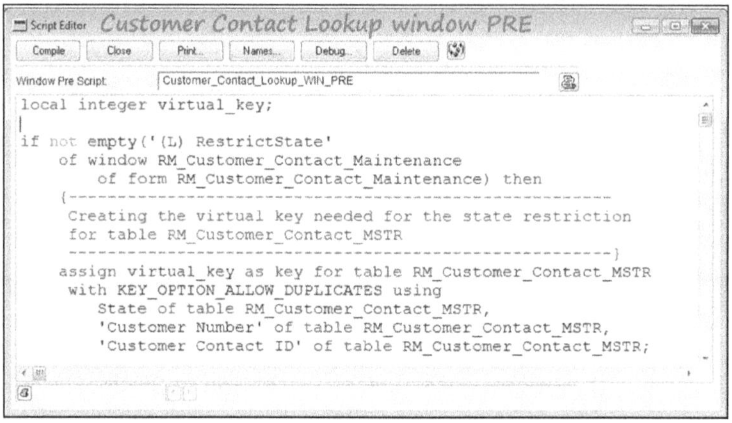

Next, you need to set the range using each segment of your virtual key. The `range start` and `range end` clauses are highlighted in the following code depicted in the screenshot so you can more easily pick them out. Notice that in the `range start`, `range end`, and `fill window` statements the `by` qualifier is used to direct Dexterity to use your new virtual key:

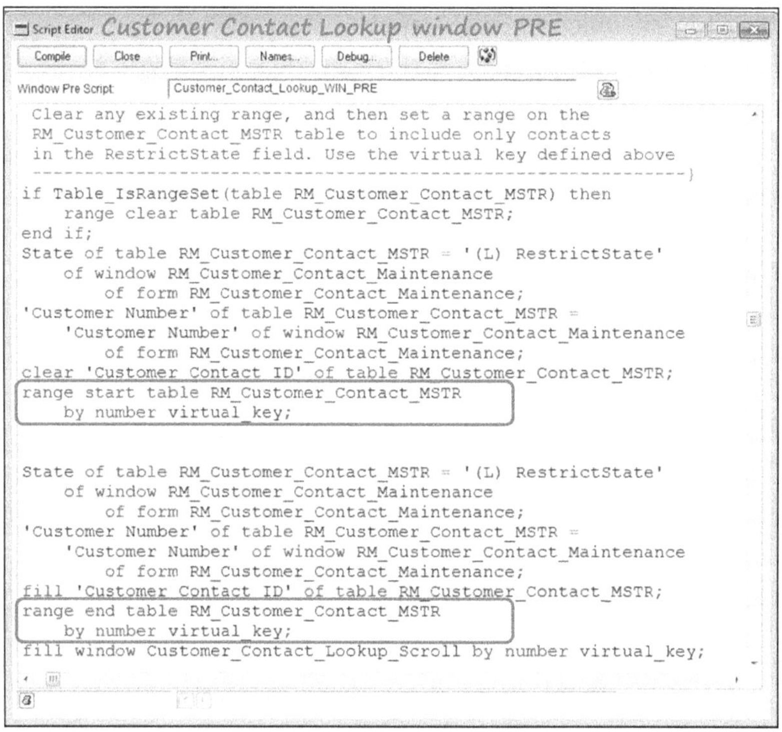

If no state restriction has been entered on the window, set your range using the table's first key using the following script. The `range start` and `range end` statements have been highlighted:

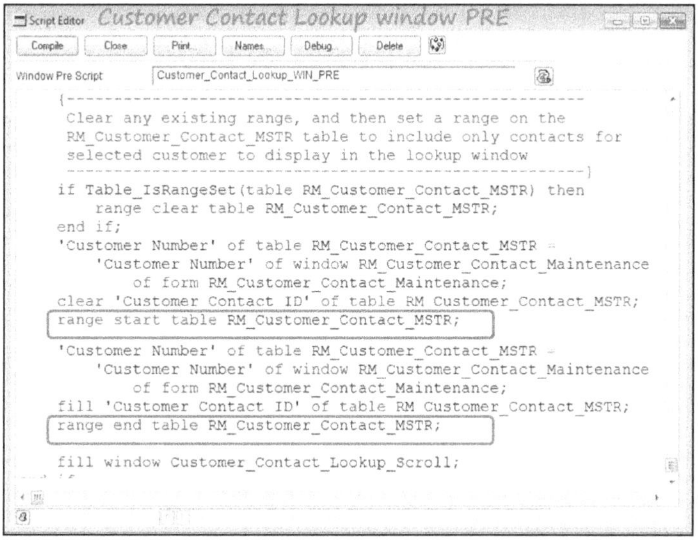

Now that you have your contact selected, you need to retrieve that contact's phone numbers. You will need to set another range to accomplish this filtering. This code will be attached to the **Change** script of the **Contact ID** field. Double-click on the **Contact ID** field to open Script Editor. Type the following script to set the range on the `RM_Contact_Phone_MSTR` table:

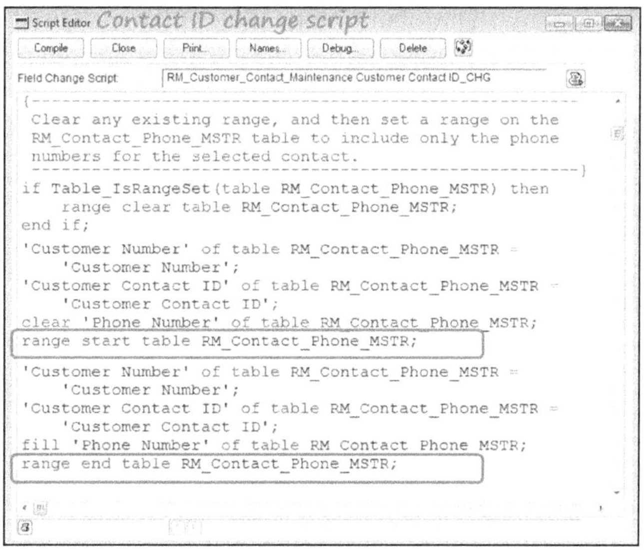

To complete your contact selection, you need to fill the scrolling window with the selected contact's phone numbers and the **Customer Contact Maintenance** window with the rest of the information about your selected contact. Complete the **Contact ID** Change script with the following code:

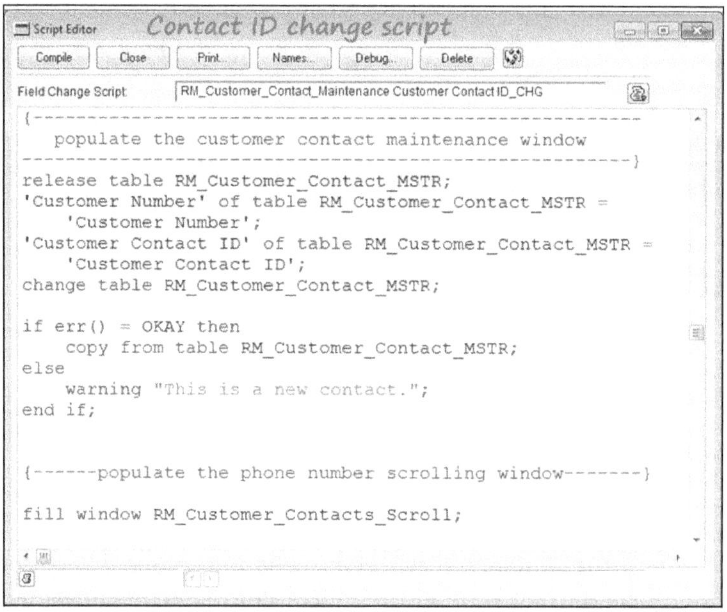

range where

An alternative to using a virtual key is the `range where` statement. Using `range where`, you can pass a `where` clause to SQL to put additional restrictions on the data returned. You would be writing *pass-through SQL* if you used this approach. For example, if you wanted to limit the customers that appear in a lookup window to just those customers that lived in IL, your code would look something like the following code:

```
range clear table RM_Customer_MSTR;

range table RM_Customer_MSTR where
physicalname('State' of table RM_Customer_MSTR) + "'= IL'";
fill window Customer_Lookup_Scroll;

range clear table RM_Customer_MSTR;
```

When writing any kind of pass-through code, whether it be pass-through sanScript or pass-through SQL, you need to be mindful of the string terminator used by the recipient language. For instance, the `where` clause previously used, if written in T-SQL (Transact SQL), would look like the following code:

```
WHERE STATE = 'IL'
```

Looks straightforward enough, but one challenge is turning the Dexterity table and field names into the equivalent physical names found in SQL. The second challenge is getting all of the quotes and apostrophes and such to come out right.

Luckily, sanScript has a function that will translate the *technical names* of our objects into the *physical names* found in SQL. It is the `physicalname()` function you see being used. Using this function, you put the full technical name of the object between the parentheses and `physicalname('State' of table PM_Vendor_MSTR)` translates into `STATE`.

Using the `physicalname()` function, you can get the physical name of a global field, the physical name of a table, or the column name of a field in a SQL table.

Next we have to fashion a sanScript statement that will resolve to `= 'IL'`. It looks easy on the face of it, but these statements can get quite involved. The complication
comes with the **string terminators**. The basics are that Dexterity uses the quotation sign as a string terminator and T-SQL uses the single quote. You could get some code that was very hard to read if you tried using the keyboard characters of `""` and `''`. Imagine if your source data included quotes and apostrophes! It would be easy for your code to fail.

The old stand by for testing your pass-through SQL is to use a warning dialog or debug dialog for short things and a text field for longer things. You would display your code in these fields and work through it until you got it right. Another popular idea is to use constants or the `char()` function instead of the actual keyboard character.

You have thousands of predefined constants in the development dictionary. Among them are a set of constants that return punctuation and special characters. The following tables list those constants that can take the place of special characters.

Of course, you can always create your own. Notice that the double quote is missing; that's odd. Use `char(34)`.

Constant	Value	Constant	Value
CH_ACCENT	`	CH_LEFTPAREN	(
CH_AMPERSAND	&	CH_LESSTHAN	<
CH_ASTERISK	*	CH_MASK	X
CH_ATSIGN	@	CH_PERCENT	%
CH_BACKSLASH	\\	CH_PERIOD	.
CH_BRACKET_LEFT	[	CH_PLUS	+
CH_BRACKET_RIGHT	]	CH_POUND	#
CH_CARAT	^	CH_QUESTIONMARK	?
CH_COLON	:	CH_RIGHTPAREN	)
CH_COMMA	,	CH_SEMICOLON	;
CH_DASH	-	CH_SINGLEQUOTE	'
CH_DOLLARSIGN	$	CH_SLASH	/
CH_EQUAL	=	CH_SPACE	\ \""
CH_GREATERTHAN	>	CH_TILDE	~
CH_LEFTPAREN	(	CH_UNDERSCORE	_

Using the previous constants, our `where` clause would look like this:

```
physicalname('State' of table PM_Vendor_MSTR) + char(34) + CH_
SINGLEQUOTE + CH_EQUAL + IL + CH_SINGLEQUOTE + char(34);
```

Scrolling windows

Scrolling windows are everywhere! You find them on lookup windows, line items on invoices, line items on purchase orders, distribution screens, kit components, accounting periods, units of measure, they're everywhere! Very often, when you are working with a scrolling window, you are working with a range. If you are making a list of those things to really master when working with Dexterity, ranges are number one, scrolling windows are number two.

First, what is a scrolling window? Simply put, it is the grid on a window. It usually presents itself as rows and columns, but it has much more ability than that. We have been told that a Dexterity scrolling window is like nothing else. You cannot compare it with VB, Java, COBOL, nothing.

Scrolling windows come in three flavors:

- `BrowseOnly`
- `Editable`
- `AddsAllowed`

Before we delve into the different scrolling window sections, let's go over some characteristics common to them all. The following screenshot shows a typical scrolling window:

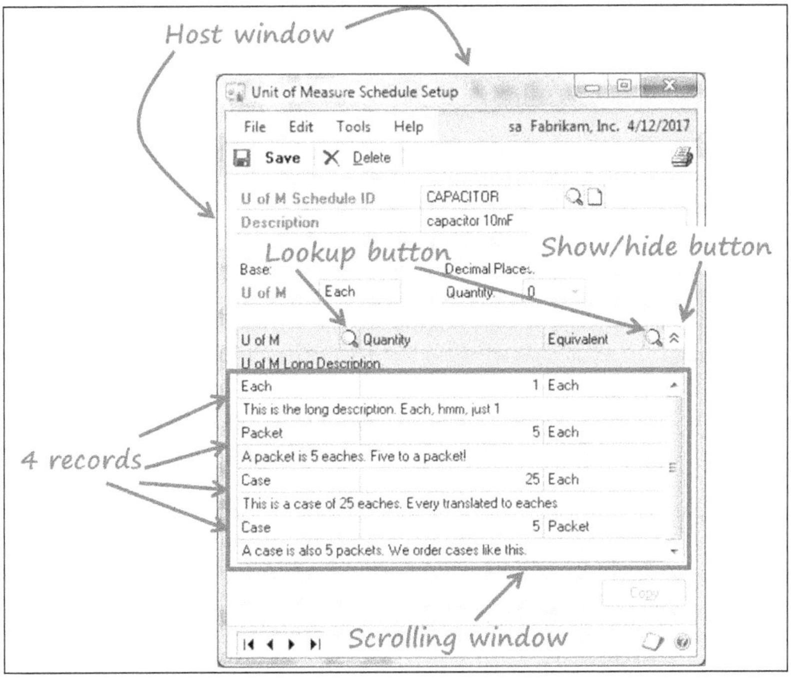

A **scrolling window** is an object on another window, much like a push button or a field. The window that houses the scrolling window, we will refer to as the *Host window*. That is not an official name, but it makes it easier to talk about. The scrolling window itself is the box just below the headers. Make note that the headers reside on the host window, *not* on the scrolling window.

The looking glass buttons used to look up **U of M** and **Equivalent** are likewise on the host window.

Big and Small Line item

The button with the chevron on it changes the display of a record in the scrolling window from one line to two, and back. We labeled it as the **Show/hide** button in the previous screenshot. The following screenshot shows the comparison between a scrolling window's `Show` state and its `Hide` state:

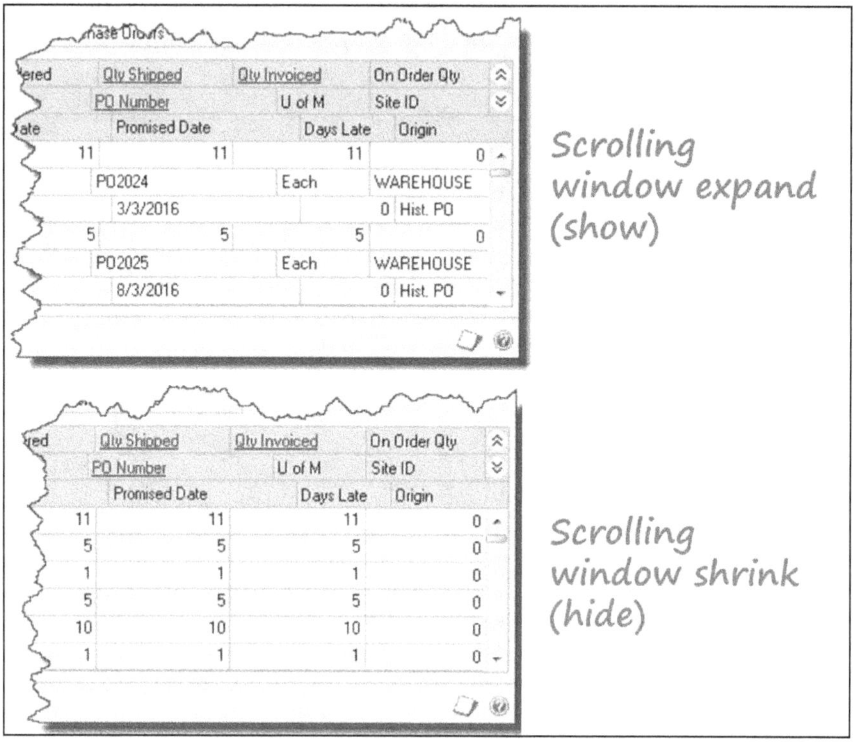

To achieve the two views, you separate the scrolling window into two partitions. The partition at the top is called the **small line item**, and the partition at the bottom is called the **big line item**. The big line item is optional. The following screenshot shows the layout view of the item number lookup window with the big and small line item markings revealed:

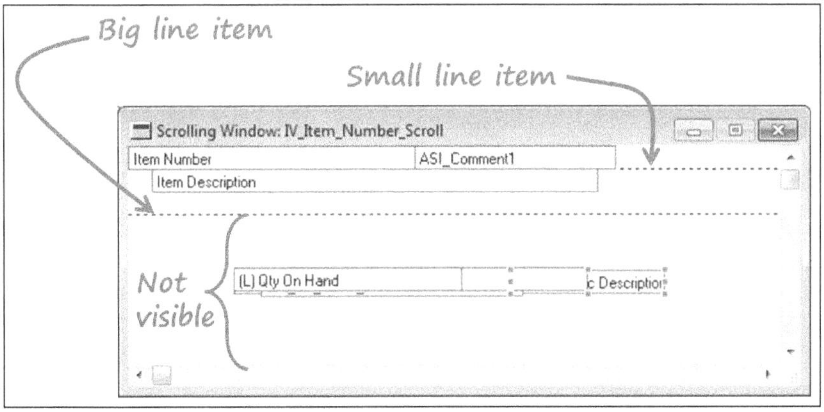

By marking the small line item so that it is at the bottom of the scrolling window, you can achieve a one-row window. You can use this as a window embedded in a window that you can show or hide at the press of a button. You do not have to make the two line items the same height. You can put checkboxes, drop-down lists, push buttons, and the whole shebang on your scrolling window. However, the *shebang* cannot include another scrolling window. You are not allowed to nest scrolling windows. As you can see from the markings in the previous screenshot, items placed below the big line item are not visible at runtime.

To mark the big or small line item, navigate to the **Tools** menu and select either **Mark Small Line Item** or **Mark Big Line Item**. Your mouse pointer will change to a scrolling window icon as shown in the following screenshot. The biggest trick to marking these partitions is to know where to position the cursor to make the mark. You make the mark by simply clicking on the left mouse button. The following screenshot shows you exactly how to position the pointer. So many people put the mouse pointer under the area where they want the mark. They end up getting the mark too low. All you need to do is position your cursor in the last row that you want included in the partition. Do not put it below the last row, but *in* the last row. The line marking the bottom of the section will appear:

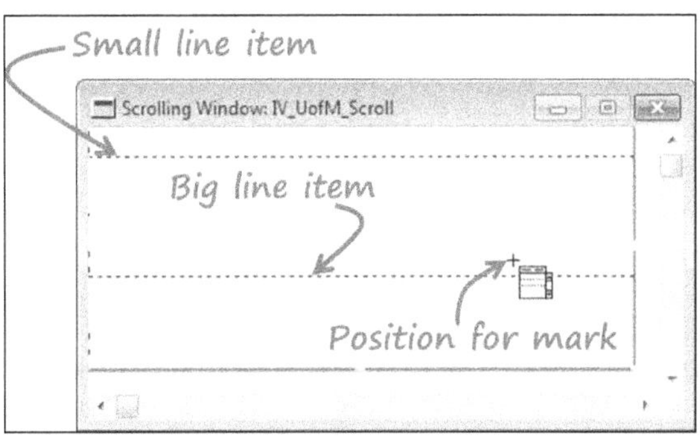

As we go through the different types of scrolling windows, know that most of the features of one are available for the others. You're going to learn the high points today, so let's get after it.

BrowseOnly windows

A **BrowseOnly** scrolling window is a *look but don't touch* window. It prevents you from making any changes to the data, no additions and no deletions; the window is for scanning only. Lookup and inquiry windows are commonly `BrowseOnly` windows. You tell Dexterity what type of window you want to create by setting the **WindowType** property. The following screenshot shows both the **Visual** properties and the **Object** properties:

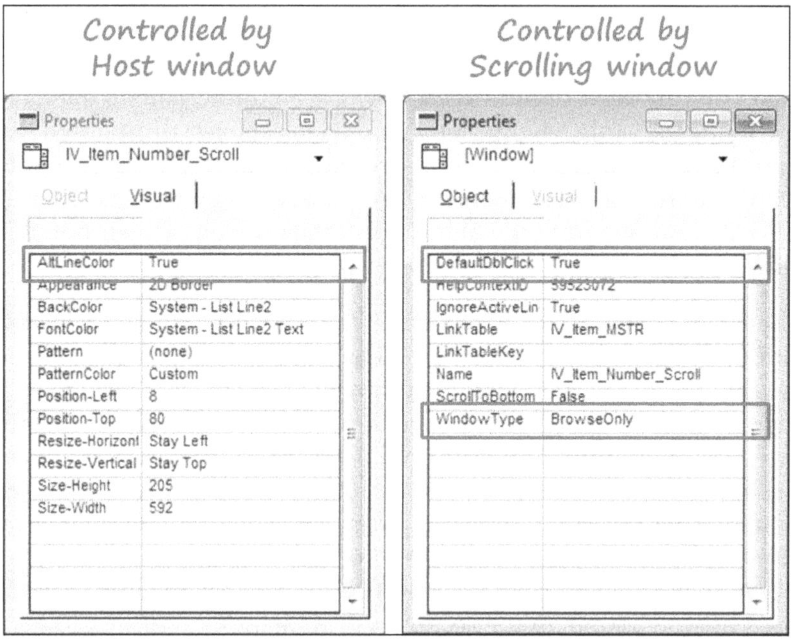

You modify the visual properties of the scrolling window from the host window. If you think of the scrolling window as just another object on the window, then it makes sense for the visual properties to be available. You adjust the size of the scrolling window from the host window too. Size and position are considered visual properties. One visual property that is notable is the **AltLineColor** property. The default setting is **True**, but you can change it here.

You tweak the object properties of the scrolling window with the scrolling window open. We will cover the object properties in the section, *Lookup windows*.

Lookup windows

We have a `BrowseOnly` scrolling window in our project that is on the **RM_Contact_Lookup** window. A lookup window is loaded with functionality. When you build your lookup windows, they need to behave just like a Dynamics GP lookup window. A closer look at one of the lookup windows will inspire you to copy one already finished. You can build the window from scratch, but it would be very tedious.

We've highlighted several features of a Dynamics GP lookup window in the following screenshot:

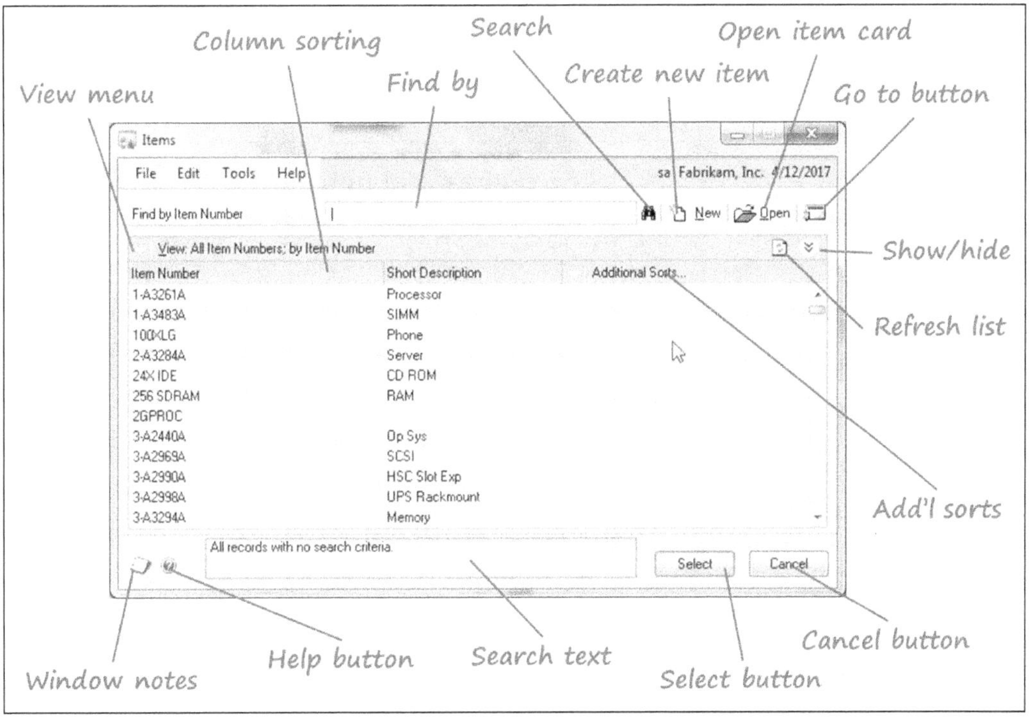

Lookup windows are in fact what you will use most often to find data. SmartList gets a lot of press, but lookup windows are what we rely on. Everyone knows how to use a lookup window: click on the looking glass button and a scrolling window opens. You don't even have to click; you can use the keyboard shortcut *Ctrl* + *L*. After keyboarding or clicking, the grid fills with your records. You can double-click on one of the records, and the selected record pops onto your entry screen, slick.

For *Ctrl + L* to work, you must link the lookup button to the lookup field.

Calling the lookup form

As part of the SDK, you have instructions on how to call procedures to open over 100 Dynamics GP lookup windows. To find these procedures, look on the opening page of the SDK under **Procedures and functions** as shown in the following screenshot:

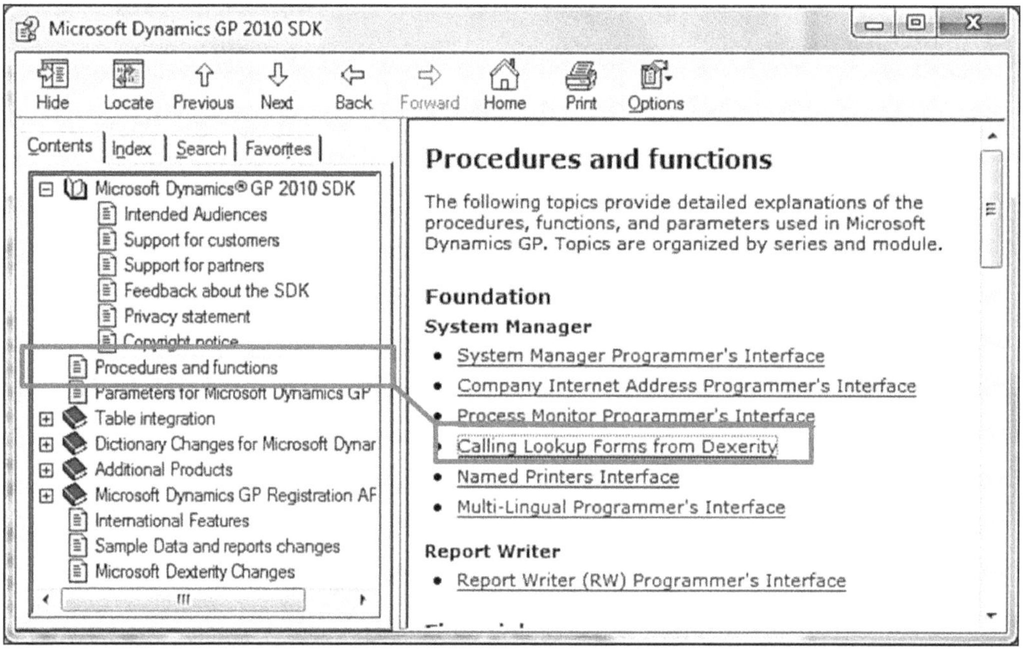

You're going to put some code in your project to open the Dynamics GP **Customers and Prospects** lookup window.

As stated earlier, you need to be able to do the following:

1. Open the lookup window.
2. Return the value of the selected item.
3. Close the lookup window.

Sounds straightforward enough; let's do it.

Normally you would use two sanScript statements to accomplish your lookup. The first one goes on the calling form, and the other goes on the called form. The form where you want the value returned is the calling form. The form where the value comes from is the called form. Because we are using an existing Dynamics GP procedure for the **Customers and Prospects** lookup window, you do not need any script on the called form; it's already there.

Attach the script mentioned in the following screenshot to the change event of **Lookup Button 1**. This lookup button is next to the **Customer Number** field on the **RM_Customer_Contact_Maintenance** window. Double-click on the lookup button to open the **Script Editor** window:

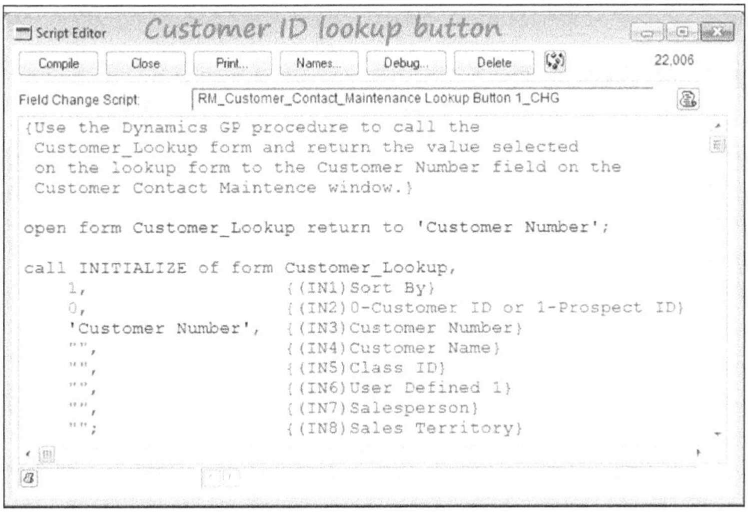

You can return any value you want from the lookup window to any field you choose in the calling window. Here, you are returning the value to the **Customer Number** field, but you could just as easily return it to the **Address 1** field.

The **INITIALIZE** procedure is an existing resource in the `Dynamics.dic` dictionary. Instructions in the SDK explain the eight parameters, and give you the correct syntax to pass them. You cannot look at the code for this procedure, but you can utilize it with the previously mentioned call.

> The **INITIALIZE** procedure contains over 100 lines of sanScript source code.

You already entered the script for the lookup button next to the **Customer Contact ID** field back in the *Setting a range* section.

From the **Debug** menu, select **Test Mode** and give your customer lookup button a try. You can't double-click and return a value because you don't have any scripts for that.

Defaulting the double-click

To get the *double-click to select a value* action working, you need the `DefaultDblClick` object property of the scrolling window set to `True`, and the `Default` object property of the **Select** button set to `True`. How this works is that by setting the window as a `DefaultDblClick` object, the `change` script of the **Default Button** will automatically run. Check the properties of the objects on your windows and set them appropriately.

Editable windows

You can change the field values in an *editable* scrolling window. However, you cannot add new records nor delete any existing records. The **Audit Trail Codes Setup** window shown is a good example of an editable window. You can change the values in the **Reprint**, **Origin**, **Next Number**, and **Source Document** columns, but you cannot add any new ones, nor delete the ones that are there. You will not find too many of these windows in the system:

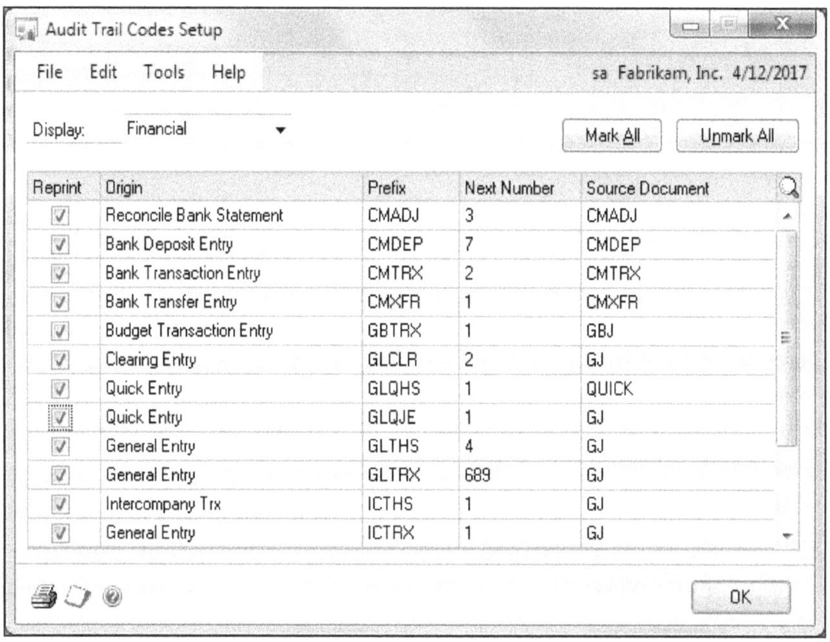

No matter what kind of scrolling window you are creating, you must link the scrolling window to one of the tables attached to the form. The attached table normally contains the fields you want to add to the scrolling window. Additionally, you will be required to select a table key. Unless you specify otherwise, the scrolling window will fill according to the linked key.

Since you can change the values on this window, you must have the ability to save those changes. There is no **Save** button on the window, so what gives? Scrolling windows have a set of events attached to the row itself. You still have all of the events for each of the fields on the scrolling window, but you have a new set.

Line events

You can attach sanScript code to the **Line Events** of a scrolling window. Moving from line to line is indeed an event.

LineFill

The **LineFill** event occurs each time a new line displays in the scrolling window. When the scrolling window first fills, the `LineFill` event runs repeatedly until the scrolling window is full. The `LineFill` event happens before the `LinePre` event.

LinePre

The **LinePre** event occurs just as the line gains focus. You can use this event to bring in data from another table or window and display it with the selected record. The `LinePre` event occurs after the `LineFill` event.

LineChange

The **LineChange** event occurs just as the line loses focus and one of the fields on the line has changed. You will find this event most useful for saving records.

LinePost

The **LinePost** event occurs just as the line loses focus. You would use this event to clear any records brought into the line by the `LinePre` event.

LineInsert

The **LineInsert** event occurs when you select **Insert Row** from the **Edit** menu. **Insert Row** is only active in the **Edit** menu if you have attached a script to the `LineInsert` event. You could use this event if you need to place records in a certain order.

LineDelete

The **LineDelete** event occurs when you select **Delete Row** from the **Edit** menu. **Delete Row** is only active on the **Edit** menu if you have attached a script to the `LineDelete` event.

AddsAllowed windows

The final type of scrolling window is the `AddsAllowed` type. You can add, edit, update, or delete records displayed in an `AddsAllowed` scrolling window. This is the type of window used by sales transaction entry, purchase order entry, item transaction entry, and so on.

`AddsAllowed` is the only type of scrolling window that can use the `LineInsert` and `LineDelete` Events.

Triggers

Triggers watch for events and then respond to those events. This section will introduce you to the five different types of object triggers supported by Dexterity. You will learn what they do, how to register them, and some considerations for using them. At the end of this section, you will register a form trigger on the **RM_Customer_Maintenance** form to open your **Customer Contact Maintenance** window.

Triggers can help you do the following:

- Add functionality where source code already exists
- Change functionality where source code already exists
- Make changes to the same form another developer has modified
- Avoid creating alternate forms so that you do not have to recreate them with each new release

Dexterity supports the following five types of object triggers:

- Form
- Focus
- Database
- Procedure
- Function

You can attach a trigger anywhere you can attach sanScript code. You can monitor events such as opening a form, deleting a record, or calling a procedure.

Before you can use a trigger, you have to register it with the runtime engine. When you register a trigger, you tell the system what kind of trigger it is and what to do when it fires. Each trigger has a specific procedure to describe itself and pass along instructions. You can register triggers from any procedure in your application, but if you want to follow best practices (which you do), you will use a procedure named **startup**.

The **startup** procedure runs prior to the form's `Pre` event of the main menu. Nearly every third-party application uses triggers and registers those triggers in a procedure named `startup`. Not all of these startup procedures can run at the same time, so they run in the order their products are listed in the `Dynamics.set` file. Similarly, whenever two developers have triggers responding to the same event, the triggers will fire in the order they are listed in the `Dynamics.set` file.

Triggers are client-side only, they do not respond to SQL events. Do not confuse Dexterity triggers with SQL triggers. Just as SQL is blind to Dexterity, Dexterity is unaware of SQL. If Dexterity did not complete the action, the trigger event will never fire. For instance, if you used Dexterity to call a SQL stored procedure that deleted a record, the Dexterity database trigger will not fire.

Using triggers can significantly change the business logic of Dynamics GP. Combining the changes you are making with the changes other developers are making could lead to unanticipated results! Rigorous testing in a multi-dictionary environment is critical for you to use triggers successfully.

Form trigger

A **Form trigger** fires whenever the form opens. It adds an item to the **Additional** form menu. If the form does not already have an **Additional** form menu, this trigger will create one. While you can have a form trigger perform any action you like, typically you would use it to provide navigation for your custom application.

The following screenshot shows the results of a properly registered form trigger:

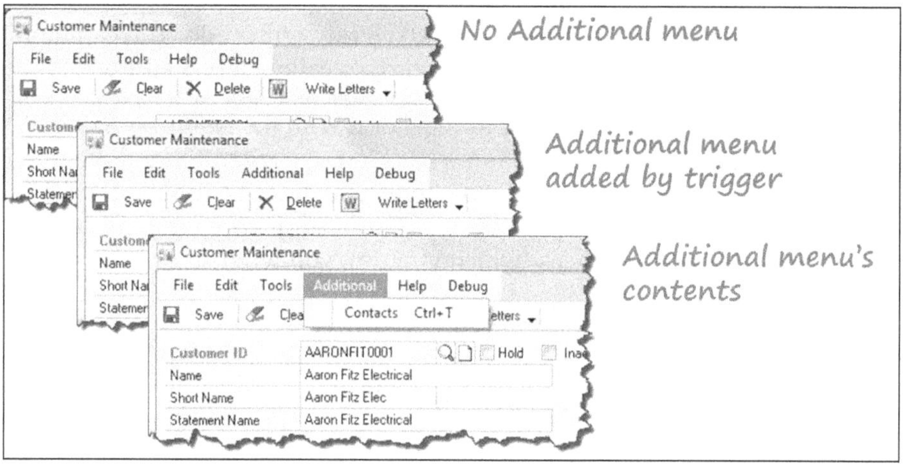

Form trigger registration

The registration procedure bears the following syntax:

```
Trigger_RegisterForm(form form_name, menu_item_name, accelerator_key, script processing_procedure {, tag})
```

Let's take this procedure apart so you can understand better what each parameter is telling the runtime engine, as given in the following list:

- `Trigger_RegisterForm`: This alerts the application that it's about to get a form trigger definition.
- `form` *form_name*: This states the technical name of the form that will fire this trigger whenever it opens.
- *menu_item_name*: This states the name of your menu item that will come under the **Additional** form menu. In the previous screenshot of the **Customer Maintenance** window, the *menu_item_name* is **Contacts**.
- *accelerator_key*: This states which key to use with the *Alt* + any key combination to select the menu item. In the previous screenshot, the *T* key is the accelerator key.

- `script` *processing_procedure*: This states the name of the global procedure to run whenever this menu item is selected.
- *tag*: This states the name of the variable you would like to use to store a number assigned by Dexterity to uniquely identify this trigger.

Form trigger considerations

Keep the following things in mind when you use form triggers:

- You should use a message resource instead of a literal string to name your menu. If you're not the only developer who registers a trigger against this form, it's possible for the **Additional** menu item to be the same word. The end user can use **Modifier** to change the message resource if conflicts in naming occur. There would be no functional conflicts with having the same name, but it could be confusing. As a bonus, by using a message resource you can more easily translate your code into another language!
- Like the menu name, conflicts in the accelerator key could be created because of multiple developers. For the same reasons as the menu, you should use a message resource to define the shortcut key.
- A form trigger will fire even if the window has been modified, is an alternate window, or is a modified alternate window.

Cross-dictionary considerations

You can register a form trigger against any dictionary in the application using the function `Trigger_RegisterFormByName()`. This function adds the product ID of the dictionary containing the form as its first parameter. The full syntax is shown as follows:

```
Trigger_RegisterFormByName(product_ID, form_name, menu_item_name, accelerator_
key, script processing_procedure {, tag})
```

Focus trigger

A **Focus trigger** responds to **focus** events such as moving in or out of a field, opening or closing a form, or pushing a button. Focus events are often called **user interface events**. You can register a focus trigger against any action where you can attach sanScript code.

Focus trigger registration

The registration procedure follows the following syntax:

```
Trigger_RegisterFocus(anonymous (qualified_resource), focus_type, attach_type,
script processing_procedure {, tag})
```

Let's take this procedure apart so you can understand better what each parameter is telling the runtime engine, as given in the following list:

- `Trigger_RegisterFocus`: This alerts the application that it's about to get a focus trigger definition.
- `anonymous(`*qualified_resource*`)`: This states the fully qualified technical name of the resource whose focus event is being monitored. A fully qualified name includes the entire string of objects you need to identify the specific resource.

  Use the `anonymous()` function so that Dexterity will not try to open the resource. Because you register the trigger in the startup script, Dexterity does not yet recognize it.

  Fully qualified resource names look as mentioned in the following list:

  a. **For a menu**: `menu` *menu_name* of `form` *form_name*
  b. **For a command**: `command` *command_name* of `form` *form_name*
  c. **For a field**: *field_name* of `window` *window_name* of `form` *form_name*
  d. **For a window**: *window_name* of `form` *form_name*
  e. **For a scrolling window**: *scrolling_window_name* of `form` *form_name*
  f. **For a form**: *form_name*

- *focus_type*: This identifies the focus event the trigger is monitoring. You identify the event using a constant or its integer value. The constants and their corresponding integer values are listed in the following table:

Focus event constants	Integer value
TRIGGER_FOCUS_PRE	0
TRIGGER_FOCUS_CHANGE	1
TRIGGER_FOCUS_POST	2
TRIGGER_FOCUS_PRINT	3
TRIGGER_FOCUS_FILL	4
TRIGGER_FOCUS_INSERT	5
TRIGGER_FOCUS_DELETE	6
TRIGGER_FOCUS_CONTEXT_MENU	7

- *attach_type*: This states whether the trigger's processing procedure should run before or after any sanScript code attached to the event.
- `script` *processing_procedure*: This states the name of the global procedure to run whenever the trigger fires.
- `{,` *tag*`}`: This states the name of the variable you would like to use to store a number assigned by Dexterity to uniquely identify this trigger.

Focus trigger considerations

Keep the following things in mind when you use focus triggers:

- The focus event will fire the trigger whether or not sanScript is attached to the event. Therefore, if no sanScript is attached, the processing procedure will run immediately.
- Another developer may issue a reject script that will preclude your processing procedure from running. Remember, the triggers fire in the order they are listed in the launch file. You can manipulate the firing order by moving your application up to a higher position in the launch file.

Cross-dictionary considerations

You can register a focus trigger against any dictionary in the application using the function `Trigger_RegisterFocusByName()`. This function adds the product ID of the dictionary containing the focus object as its first parameter. The full syntax looks as follows:

```
Trigger_RegisterFocusByName(product_ID, focus_type, attach_type,_name, script
processing_procedure {, tag})
```

Database trigger

A **database trigger** responds to *successful* table operations. You can monitor multiple table events in the same trigger.

Database trigger registration

The registration procedure uses the following syntax:

```
Trigger_RegisterDatabase(anonymous(table table_name), form form_name,
table_operations, script processing_procedure {, tag})
```

Let's take this procedure apart so you can better understand what each parameter is telling the runtime engine as given in the following list:

- `Trigger_RegisterDatabase`: This alerts the application that it's about to get a database trigger definition.
- `anonymous(table` *table_name*`)`: This states the name of the table this trigger is monitoring.

  Use the `anonymous()` function so that Dexterity will not try to open the resource. Because you register the trigger in the startup script, Dexterity does not yet recognize any table names.

- `form` *form_name*: If a form name is provided, then only database operations that originate from the given form's table buffer will fire the trigger. You might use a form restriction, for instance, if you wanted to monitor any address changes from a specific screen. Use a zero if no form restriction is used.
- *table_operations*: This identifies the table operations that the trigger is monitoring. You can use an integer, or one or more constants to specify the operation(s). To monitor more than one table operation, add the values of the constants together. For example, to monitor a table add and a table update, the integer value would be 4 + 8 = 12:

Database event constants	Integer value
`TRIGGER_ON_DB_READ`	1
`TRIGGER_ON_DB_READ_LOCK`	2
`TRIGGER_ON_DB_ADD`	4
`TRIGGER_ON_DB_UPDATE`	8
`TRIGGER_ON_DB_DELETE`	16

- `script` *processing_procedure*: This states the name of the global procedure to run whenever the trigger fires.
- `{,` *tag*`}`: This states the name of the variable you would like to use to store a number assigned by Dexterity to uniquely identify this trigger.

Database trigger considerations

Keep the following things in mind when you use database triggers:

- The table operation must be *successful*.
- Database triggers are *NOT SQL* triggers and do not respond to operations completed outside of Dexterity.
- Table operations not caused by a Dexterity command will not fire the trigger. For example, calling a SQL stored procedure using Dexterity, will not fire the trigger.
- The table buffer passed to the processing procedure is available to that procedure even if the table operation fails.
- If you are using a range, the trigger will fire for each record in the range.
- You can restrict the trigger to monitor operations that were initiated by a certain form.

Cross-dictionary considerations

You can register a database trigger against any dictionary in the application using the function `Trigger_RegisterDatabaseName()`. This function adds the product ID of the dictionary containing the table as its first parameter. The full syntax looks as follows:

```
Trigger_RegisterDatabaseByName(product_id, table_name, form_name, table_
operations, script processing_procedure {, tag})
```

Procedure trigger

A **procedure trigger** fires when the specified procedure is called using the `call` statement. You will use procedure triggers to monitor posting procedures, aging procedures, closing procedures, and so on.

Procedure trigger registration

The registration procedure uses the following syntax:

```
Trigger_RegisterProcedure(script (procedure_name) {of form form_name},
attach_type, script processing_procedure {, tag})
```

Let's take this procedure apart so you can better understand what each parameter is telling the runtime engine as given in the following list:

- `Trigger_RegisterProcedure`: This alerts the application that it's about to get a procedure trigger definition.
- `script` (*procedure_name*) `{ of form` *form_name*`}`: This identifies the fully qualified technical name of the procedure being monitored. Form procedures can be monitored so long as the qualified name is used. The qualified name includes the name of the form.
- *attach_type*: This determines whether the trigger's processing procedure should run before or after the procedure being monitored.
- `script` *processing_procedure*: This states the name of the global procedure to run whenever the trigger fires.
- `{,` *tag*`}`: This passes the name of the variable you would like to use to store a number assigned by Dexterity to uniquely identify this trigger.

Procedure trigger considerations

Keep the following things in mind when you use procedure triggers:

- Ignoring the parameters from the original procedure is considered a best practice.
- Modifying the values of parameters is not suggested because of unexpected results. If you do use parameters, use the same ones in the same order. This is especially true if the parameter is an `out` parameter. For `out` parameters, it is mandatory that your parameters exactly match the monitored procedure's parameters, as the next process after that procedure would be expecting this `out` parameter's value. It could prove disastrous, especially if it is a table buffer which is passed as an `out` parameter. Conceptually, your trigger would be an intermediate process which has interfered with the execution flow, and therefore should not break it by ignoring the `out` parameters.
- Pay particular attention to background calls; the parameters are passed to the processing procedure and the background procedure at the same time.

Cross-dictionary considerations

You can register a procedure trigger against any dictionary in the application using the function `Trigger_RegisterProcedureByName()`. This function adds the product ID of the dictionary containing the procedure as its first parameter. The full syntax looks as follows:

```
Trigger_RegisterProcedureByName(product_ID, procedure_name {of form form_
name}, attach_type, script processing_procedure {, tag})
```

Function trigger

A **function trigger** fires when the specified function runs. Like procedure triggers, you will use a function trigger to monitor things such as advancing document numbers or monitoring transaction types, and so on.

Function trigger registration

The registration procedure uses the following syntax:

```
Trigger_RegisterFunction (function (function_name) {of form form_name},
attach_type, [function processing_function | script processing_procedure] {, tag})
```

Let's take this procedure apart so you can better understand what each parameter is telling the runtime engine as given in the following list:

- `Trigger_RegisterFunction`: This alerts the application that it's about to get a function trigger definition.
- `function` *(function_name)* `{ of form` *form_name*`}`: This states the fully qualified technical name of the function being monitored. Form functions can be monitored so long as the qualified name is used. The qualified name includes the name of the form.
- *attach_type*: This states whether the trigger's processing procedure or function should run before or after the function being monitored.
- `function` *processing_function*: This states the name of the global function to run whenever the trigger fires. This is the only type of trigger that can respond by running a function instead of a procedure.
- `{,` *tag*`}`: This states the name of the variable you would like to use to store a number assigned by Dexterity to uniquely identify this trigger.

Function trigger considerations

Keep the following things in mind when you use function triggers:

- If a procedure (versus a function) runs when the trigger fires, that procedure cannot have any parameters. Consequently, the procedure does not have access to any of the parameters of the monitored function.
- If a function responds when the trigger fires, that function *must* have parameters. Those parameters must match exactly the parameters of the monitored function. Accordingly, the processing function *does* have access to all of the parameters of the monitored function.

Cross-dictionary considerations

You can register a function trigger against any dictionary in the application using the function `Trigger_RegisterFunctionByName()`. This function adds the product ID of the dictionary containing the monitored function as its first parameter. The full syntax looks as follows:

`Trigger_RegisterFunctionByName(`*product_ID*`,` *function_name* `{of form` *form_name*`},` *attach_type*`, [function` *processing_function* `| script` *processing_procedure*`] {,` *tag*`})`

Create your form trigger!

To give one of these triggers a try, let's register a form trigger that will create a menu item to open your **Customer Contact** window.

Processing procedure

First, create the trigger processing procedure. This must be a global procedure, so you will create it directly in the **Resource Explorer** window. In the **Resource Explorer** window, create a new script resource by selecting the new resource drop-down arrow from the toolbar and then selecting **Script (Procedure)**.

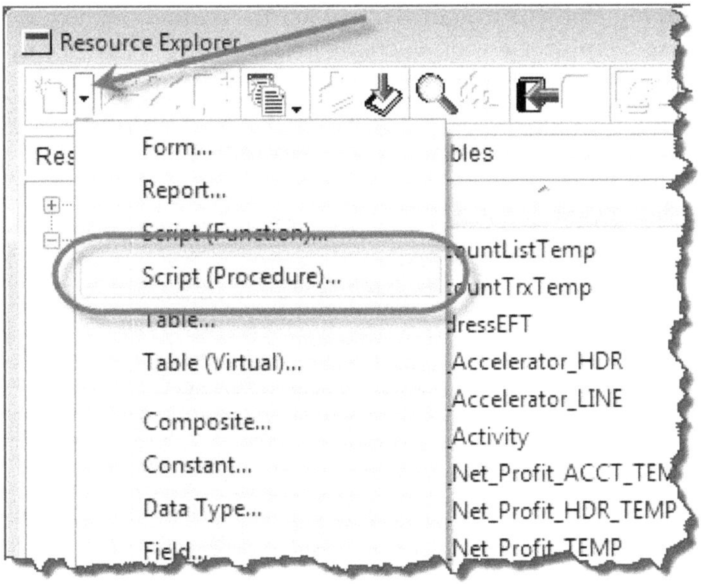

Name the procedure `OpenCustomerContacts` and enter the script mentioned in the following screenshot. This script will open the **Customer Contact Maintenance** window displaying the record from the **Customer Maintenance** window:

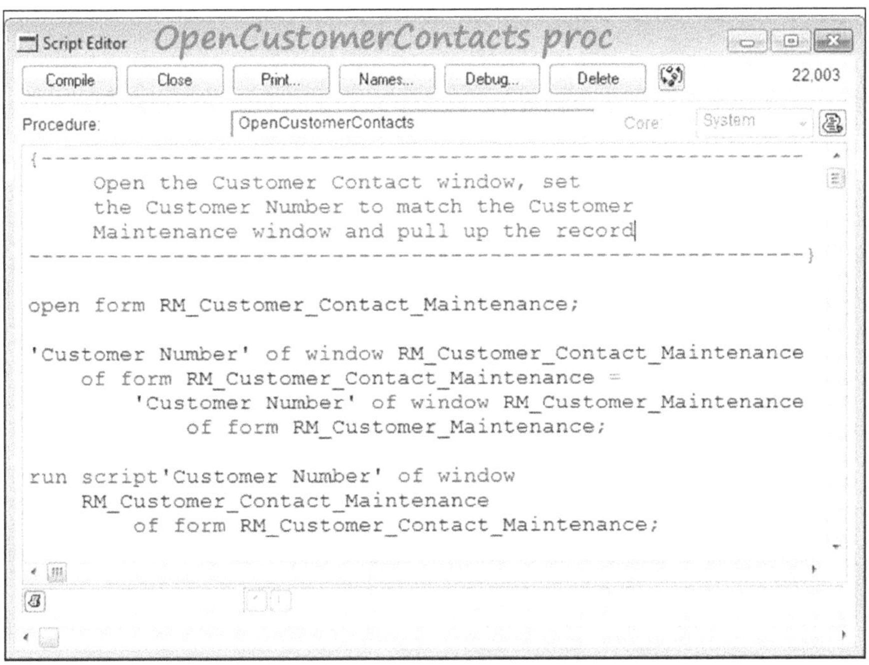

Next, we have to register this trigger. To register it, create the `Startup` procedure. From the **Resource Explorer** window, create another script resource by selecting the new resource drop-down arrow from the toolbar and then selecting **Script (Procedure)**. Name the new procedure `Startup` and enter the script mentioned in the following screenshot to register the new form trigger:

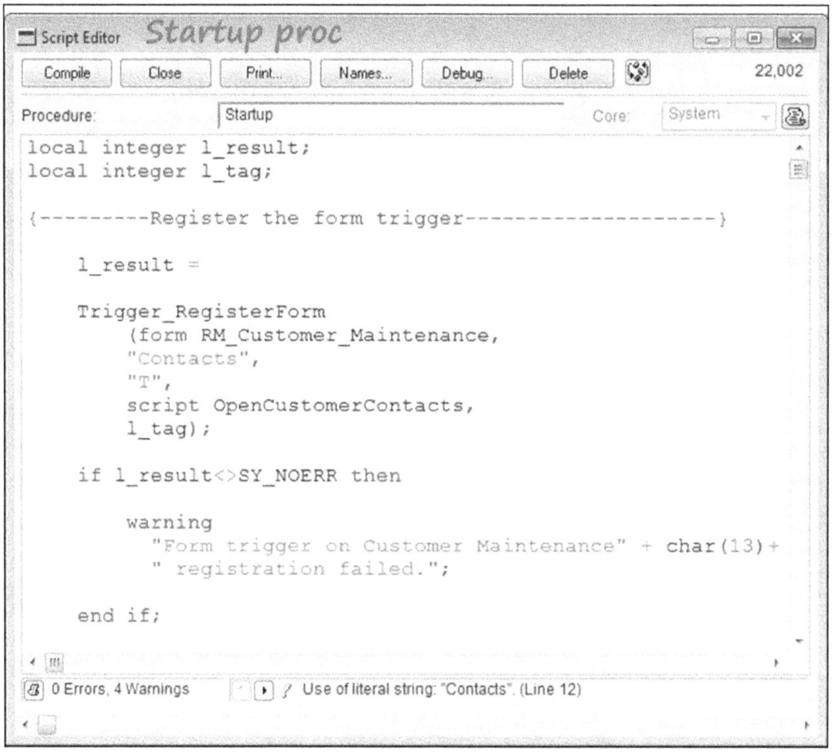

Go into the test mode and try out your new navigation! Your new **Additional** menu will be on the **Customer Maintenance** window.

Summary

You covered a lot of ground in this chapter! Starting with the basic rules of sanScript, you started writing procedures right away. The new concept of table buffers is a good foundation for understanding table operations. Moving on, you worked with the most important concept in database programming - ranges. Ranges moved into scrolling windows. Where you not only created a range, but also created a range based on another range! You tried out the new technique of setting up a temporary index using a virtual key. Using the `range where` statement gave you a glimpse at pass-through SQL. You closed the chapter with the most powerful tool in the Dexterity toolset: **object triggers**.

In the next chapter, we will take this customization to its ultimate goal - deployment.

6
Deploying a Dexterity Solution

This chapter will guide you through the steps necessary to deploy your Dexterity solution. All of that debugging has finally gotten you to the point where you're ready to go public! You think, anyway.

Now it's time to get acquainted with the Dexterity Utilities program and set yourself up so that your application is easily upgradable. It's also time to make sure your application plays nice with other third-party programs. Remember, Dexterity's test mode only plays with one dictionary. A minimal Dynamics GP installation includes more like 20 dictionaries! In addition, most installations include non-Microsoft applications, such as yours.

It would probably be a good idea to install a few of the more popular third-party applications such as Rockton Software's SmartFill and eOne Solutions' SmartConnect or SmartView into your test environment. The more, the merrier.

You need to make sure your triggers fire under as many scenarios as it's reasonable to test. You might also consider writing some code into your installation routine that moves your application to the top of the launch file. It's always better to be the first application on the list (under Dynamics) so that nobody else can issue a *reject script* statement on you.

It's time for you to make some decisions about whether you are going to create a nice Windows Installer package, or if your users are going to have to do all of the work themselves.

Don't forget that your users may not have the kind of equipment that you do. You need to figure out what it takes to run your application so that everyone can surely meet those requirements before they try to install it.

It's up to you to make sure your users can deploy your application with ease and confidence.

Key topics in this chapter include the following:

- System requirements
- Versions and builds
- Table creation routines
- Completing the application
- Creating the chunk file
- Testing in a multi-dictionary environment
- Distributing the completed application

System requirements

Sometimes we overlook one of the most basic components of our new application. Does the client's system meet the requirements for Dynamics GP? Fortunately, this is well documented.

General requirements

Check the following link for the current list of general requirements for Dynamics GP 2010:

`http://www.microsoft.com/en-us/dynamics/using/gp-system-requirements.aspx`

Check the following link for requirements if you need any of the Dynamics GP 2010 web applications:

`http://www.microsoft.com/en-us/dynamics/using/gp-web-apps.aspx`

We've seen clients running unsupported versions of SQL Server, and even outdated versions of Windows on the client machines. You can't assume everyone keeps up to date like you do.

The following table lists the basic minimums for Dynamics GP release 2010:

Client	Server
Not applicable to client	SQL Server 2008 R2
	SQL Server 2008 with SP1 or later
	SQL Server 2005 with SP3 or later
Windows 7 (Professional or Ultimate Editions), or Windows Vista SP2 or later (Business or Ultimate Editions), or Windows XP with SP3 or later (Professional Edition)	Microsoft Windows Server 2008 with SP2 or later, or Windows Small Business Server 2008 with SP2 or later, or Windows Server 2003 R2, or Windows Server 2003 with SP2, or later. Windows Small Business Server 2003 R2
2 GB of available RAM	2 GB of available RAM (more is better)
2010 Microsoft Office system, or 2007 Microsoft Office system	
Windows Internet Explorer (7, 8, or 9)	
Adobe (8, 9, or 10)	

Dynamics GP supports both 32-bit and 64-bit environments; it can also be installed in a virtual environment. Review the details about the supported virtual environments at `http://support.microsoft.com/kb/937629`. You can install the rich client on the desktop, or on a terminal server. You need to be sure your application will run in each of these environments, or be very clear as to its limitations.

In addition to the general requirements previously listed, some features of Dynamics GP have other requisites.

Feature-specific requirements

If your enhancement leverages any of the additional features of Dynamics GP, be sure to disclose any extra software your application needs. You don't want any surprises when your end user tries to use your application and then finds out they need several thousand dollars' worth of additional software. Requirements of several features are provided as follows:

- **Word form documents**: The **Word Form** feature provides a means to use Microsoft Word to create forms and reports from XML files produced by Report Writer. The software required for the Word Form feature includes the following:
    - Microsoft Office 2007 32 bit or Microsoft Office 2007 64 bit
    - Microsoft Office 2010 32 bit or Microsoft Office 2010 64 bit
    - Open XML SDK 2.0 for Microsoft Office: The SDK is installed automatically with Microsoft Dynamics GP 2010
    - Microsoft Dynamics GP add-in for Microsoft Word: This add-in is installed from the Dynamics GP 2010 DVD media under the **Additional Products** section, as shown in the following screenshot:

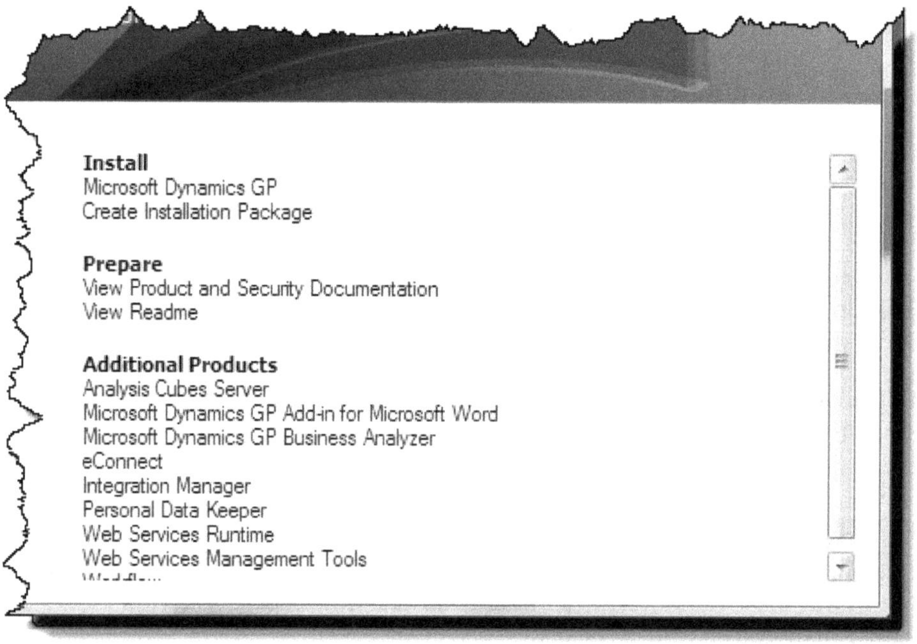

- **E-mail functionality**: The e-mail functionality feature allows you to mail documents such as invoices and purchase orders to customers and vendors. You can send the documents singularly or by batch. To use this feature, the following is the requirement:

    - Microsoft Office 2007 32 bit or Microsoft Office 2010 32 bit
    - Extended MAPI compliant e-mail application

> E-mail functionality is not yet compatible with 64-bit versions of Microsoft Office.

- **Charts and KPIs (Key Performance Indicators)**: Microsoft SQL Server Reporting Services 2008 or 2008 R2 32 bit or 64 bit
- **Unified communications**: Microsoft Office Communicator 2007 R2 or later
- **Word form template generator**: Microsoft Office 2007 or 2010 32 bit

> Word form template generator is currently only available to partners and is not yet compatible with 64-bit versions of Microsoft Office.

- **Business analyzer**: Microsoft SQL Server Reporting Services 2008 or 2008 R2 32-bit or 64-bit
- **Business Portal 5.0**: These are for 32-bit environments only
    - IIS (Internet Information Services) 6.0 or IIS 7.0
    - Windows SharePoint Services 3.0 32-bit, or Microsoft Office SharePoint Server 2007 Enterprise Edition 32-bit SP 2 or later
    - Web Services for Microsoft Dynamics GP 2010

> Web Services for Microsoft Dynamics GP 2010 is required only if you use Microsoft Office SharePoint Server 2007 Enterprise Edition 32-bit SP 2 or later

    - An instance of Microsoft SQL Server

> To view charts and KPIs on the Executive Center page, you need Microsoft SQL Server Reporting Services 2008 or later.

    - Internet Explorer 7, 8, or 9

- **Business Portal 5.1**: This is for 64-bit environments only
    - IIS 7.0
    - Microsoft SharePoint Foundation Server 2010, Microsoft SharePoint Server 2010 Enterprise Edition, or Microsoft SharePoint Server 2010 Standard Edition (for Business Portal 5.1 R2 or later)
    - Internet Explorer 7, 8 or 9
    - Web Services for Microsoft Dynamics GP 2010

> NOTE: To view Charts & KPI's on the Executive Center Page, you need Microsoft SQL Server Reporting Services 2008 or later.

- **Workflow in 32-bit environments**
    - IIS 6.0 or IIS 7.0
    - Windows SharePoint Services 3.0, or Microsoft Office SharePoint Server 2007 Enterprise Edition 32-bit SP 2 or later
    - Internet Explorer 7, 8, or 9
    - Microsoft Office 2007 32-bit or 64 bit, or Microsoft Office 2010 32 bit or 64 bit
    - Web Services for Microsoft Dynamics GP 2010
- **Workflow in 64-bit environments**
    - IIS 7.0
    - Microsoft SharePoint Foundation Server 2010, Microsoft SharePoint Server 2010 Enterprise Edition, or Microsoft SharePoint Server 2010 Standard Edition
    - Internet Explorer 7, 8, or 9
    - Microsoft Office 2007 32 bit or 64 bit, or Microsoft Office 2010 32 bit or 64 bit
    - Web Services for Microsoft Dynamics GP 2010

Each of the products previously listed require that you install a specific version of the software. This is true of Dynamics GP and your application as well. In the next section, we will discuss why versions and builds matter.

Minding versions and builds

When you create your chunk file for deployment, you will be asked to assign it a major version, minor version, and build number. The runtime engine reads the version and build numbers and compares them to any existing dictionary for that product ID. If the version numbers match, the build numbers are examined. If the build number in the chunk is equal to or greater than the build number of the existing dictionary, a new dictionary will be created by your chunk.

If the version numbers do not match, or the build number is lower than the existing dictionary, the file will remain a `.cnk` file and a message similar to the following message will be written to the `installerrors.txt` file:

```
Version 11.0.17 of Contacts.cnk is not compatible with version 11.1.15
of CONTACTS.DIC
```

This keeps you from accidentally installing an older version of a dictionary. On the downside, these rules could keep your application from updating if it is on a different major version than Dynamics. The fix is often as simple as deleting the old dictionary and reinstalling the new one. If you lose track of your version and build numbers, you can retrieve them using the `Runtime_GetModuleInfo()` function. You should create an **About** window that will display this information as part of your application.

Another version and build guardian is the `DU000020` table in the `Dynamics` database. This table is the `duCompanyVersions` table and it keeps track of the version and build numbers for each product expected to be installed on the workstation. When you attempt to log in, Dynamics GP checks the dictionaries installed locally to the versions in the `DU000020` table. If the version on your workstation does not match, you will be presented with the following error:

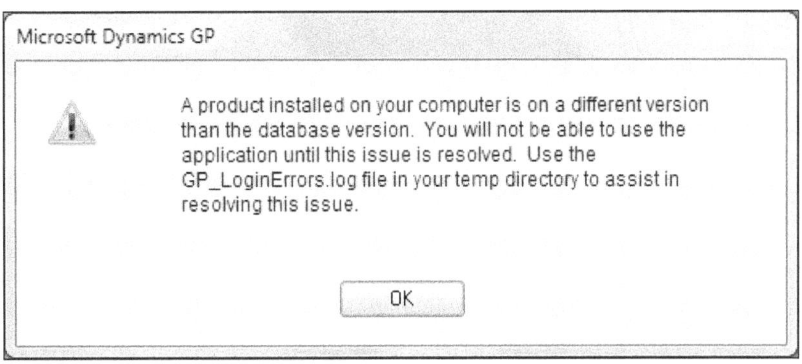

The **GP_LoginErrors.log** is stored in the `Temp` folder of your user profile on the workstation that generates the error. For Windows 7, you can find the log file here:

```
C:\Users\user_name\AppData\Local\Temp\GP_LoginErrors.log
```

One problem with the previously mentioned location is that the `AppData` folder is hidden in a typical windows installation.

As an example, let's say a record in your `DU000020` table in the `DYNAMICS` database held the following values:

CompanyID	PRODID	versionMajor	versionMinor	versionBuild
-1	1493	9	0	1734

Assuming you do not actually have Version 9 installed on your workstation, you would be prevented from logging in until the situation was corrected. If you could get to your `GP_LoginErrors.log` file, it would hold the following message:

```
Product Name: SmartList     Database Version: 901734 Client Version:
1101734
```

It's not terribly informative, but you get the gist of what's going on. Fortunately, changing the database version from 9 to 11 makes everything right with the `DU000020` table and you would be allowed to log in. Warning! You would be overriding a Dynamics GP setting, so be very certain the data in the table is wrong.

Each service pack, feature pack, and so on, comes with a new build of the `dynamics.dic`. If you created your product using a different build of `dynamics.dic`, you need to confirm that the changes made to the new `dynamics.dic` do not conflict with your dictionary. Always use the same version and build number to develop your product that your product will be deployed in.

We'll talk about how you designate the version and build for your application in this chapter's section, *Creating the chunk file*.

Table creation routines

The tables used by your application are not automatically created. Back in the Pervasive PSQL 2000 and FairCom ctree Plus days, the table was created automatically the first time you accessed it. SQL doesn't work that way. If you try to use the application without creating your tables, you will get an error message similar to the one shown in the following screenshot. The message will not be displayed until the system tries to access one of your tables:

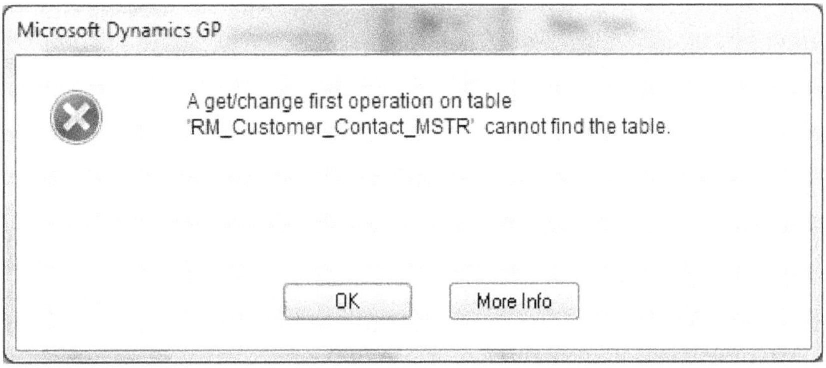

Likewise, if the table exists but its structure does not match the table definition in your application, you will receive the following error:

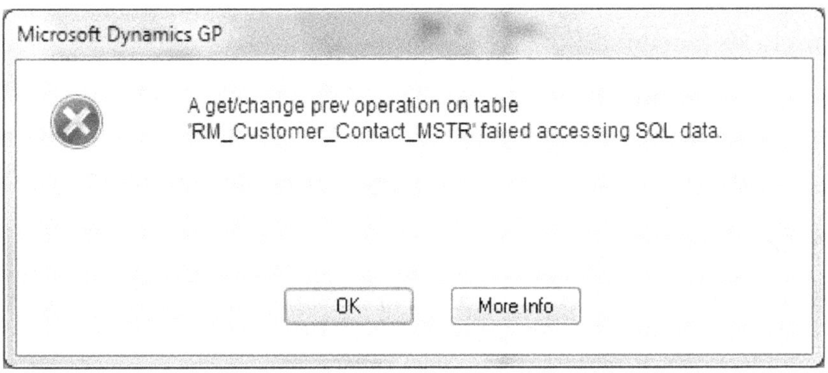

The **More Info** button provides the reason your application could not access the SQL data; namely, there was a column in the Dexterity table definition that did not exist in the physical table:

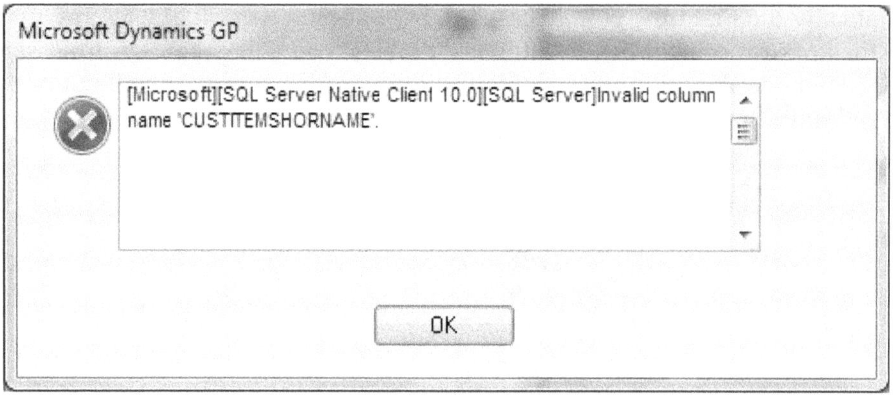

Your job now is to provide an easy way for your user to create the SQL `Tables` and `*zdp` procedures required by your application, to grant permissions to those tables, and to bind the default values to those tables.

The method you choose to build the tables depends in part on how you deploy your application. If it's deployed in-house, the method will probably be much easier to craft.

Two things to keep in mind:

- If you are not already logged in to the SQL Server, you will be presented with a standard SQL login dialog. You want to avoid that because it's messy.
- The person creating the tables must have sufficient SQL authority to do so. Typically you will satisfy this requirement by only allowing `sa` or `DYNSA` to create the tables.

This section is limited to the initial creation of the tables; upgrading an existing table is covered in *Chapter 11, Upgrading Customizations*.

You'll learn about the following methods in this section:

- Using the **SQL Maintenance** window
- Building a separate utilities window
- Automatically creating the tables upon launch

Using the SQL Maintenance window

You can launch the **SQL Maintenance** window from within the user interface of Dynamics GP. The navigation is shown as follows:

Microsoft Dynamics GP | Maintenance | SQL

After the **SQL Maintenance** window opens, change **Database** to the name of the database in which you want to create the tables. This is typically your company database. Select the name of your application from the drop-down box for **Product**. The products are listed in the order they appear in the `Dynamics.set` file. The **Company** will default to the company you are logged in to, and cannot be changed.

Any table resource defined in your application will populate the scrolling window. Our **Customer Contacts** example has two tables as shown in the following screenshot:

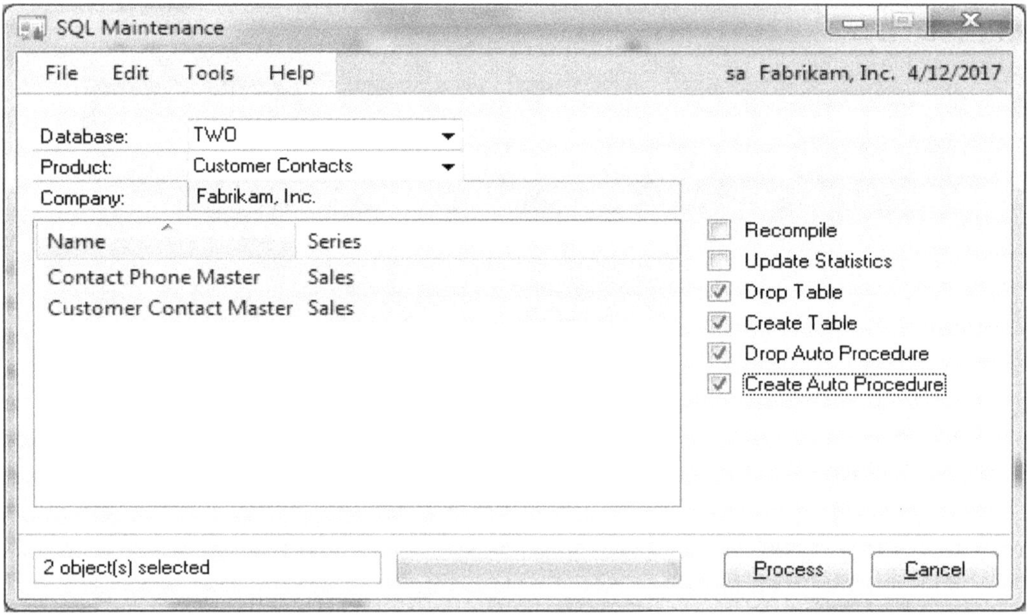

The **SQL Maintenance** window has several checkboxes on the right-hand side. To create your tables and **auto procedures,** check the lower four boxes.

The system will create the two tables according to the Dexterity table definitions and grant the appropriate permissions to the `DYNGRP` database role. Additionally, the system will create the `*.zdp` stored procedures to handle basic table operations in SQL.

That's all there is to it. You have the two tables created, permissions granted, defaults bound, and the `.zdp` procedures in place. It's short and simple. But wait! Can you see any risk involved in using this procedure? The user could accidentally drop any table in any Dynamics GP database, and even the Dynamics database itself! Not fun.

The next method will take you a little longer, but it gets the job done in a more controlled environment.

Building a utilities window

Building a separate utilities window used to be a very popular method. New procedures came on the scene a few years ago that will hopefully make this technique obsolete. You want your application to be installed without risking accidental disaster. For quick and dirty table creation without the risk of betting the farm, this is a good way to go.

Using this method, you can build a separate window that contains local push buttons. The script behind the buttons will add the tables, create the procedures, grant permissions, and bind defaults.

You will be using pass-through SQL to create a stored procedure named `am_AutoGrant`. The `am_AutoGrant` procedure grants SELECT, INSERT, UPDATE, and DELETE permissions to the `DYNGRP` database role. It also grants the `DYNGRP` role the right to execute the `am_AutoGrant` procedure.

Getting the syntax exactly right for pass-through SQL usually takes a few runs. Using a text field as a preview is very helpful in making sure your commas and apostrophes end up in the right place. We will be using this technique for your window.

The utilities window method does require your user to find this window and push the buttons. For this reason, you need to make this window really easy to find. The best answer is to cause the window to open automatically based on criteria about the state or existence of the tables.

For practice, we are going to create the window shown in the following screenshot, followed by the instructions after the screenshot. The text in the preview window is an example of what a pass-through SQL statement preview would look like:

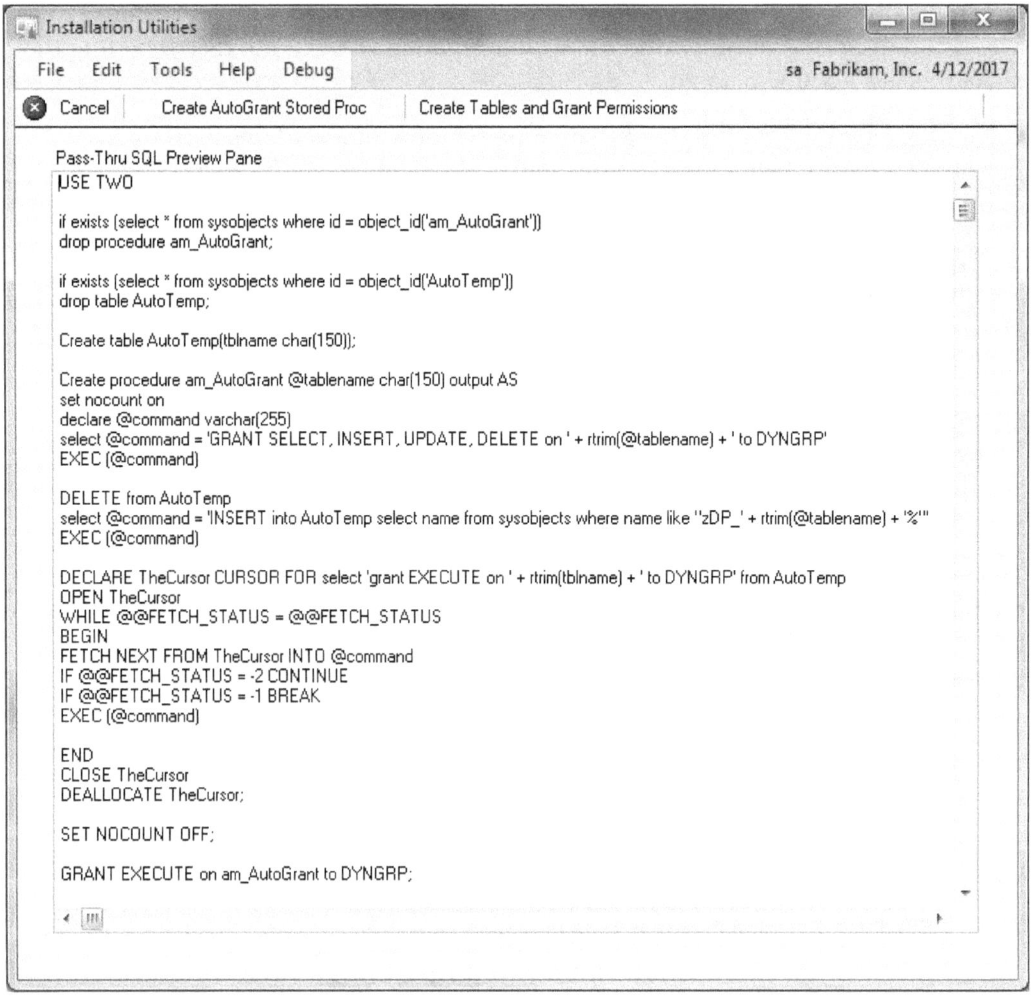

Create the `Installation_Utils` form using the following information:

Form Name	`Installation_Utils`
Series	`Company`

Create the `Installation_Utils` window using the following information:

Name	`Installation_Utils`
Title	`Installation Utilities`
Back Color	`True`
Control Area	`True`
Size-Height	`580`
Size-Width	`530`

Create a new global field to display your SQL statement using the following information:

Field Name	`SQL_Statements`
Physical Name	`SQL_Statements`
Data Type	`TX32000`

You're going to use this global text field to preview your pass-through SQL statement as you build it. Simply drag the global field over to the window and size it using the **Properties** window.

Global Field	`SQL_Statements`
Position-Left	`15`
Position-Top	`66`
Size-Height	`494`
Size-Width	`496`

You're going to use a couple of local push buttons to fire off your code. Place the controls on the window according to the previous screenshot. Create the new local resources with the following information:

Global Field	`Cancel Button`

Local Field Name	`Create Sproc`
Control Type	`Push Button`
Static Values	`Create AutoGrant Stored Proc`

Local Field Name	`Create Tables`
Control Type	`Push Button`
Static Values	`Create Tables and Grant Permissions`

You will add the words that appear on the button in the **Button Items** window as shown in the following screenshot:

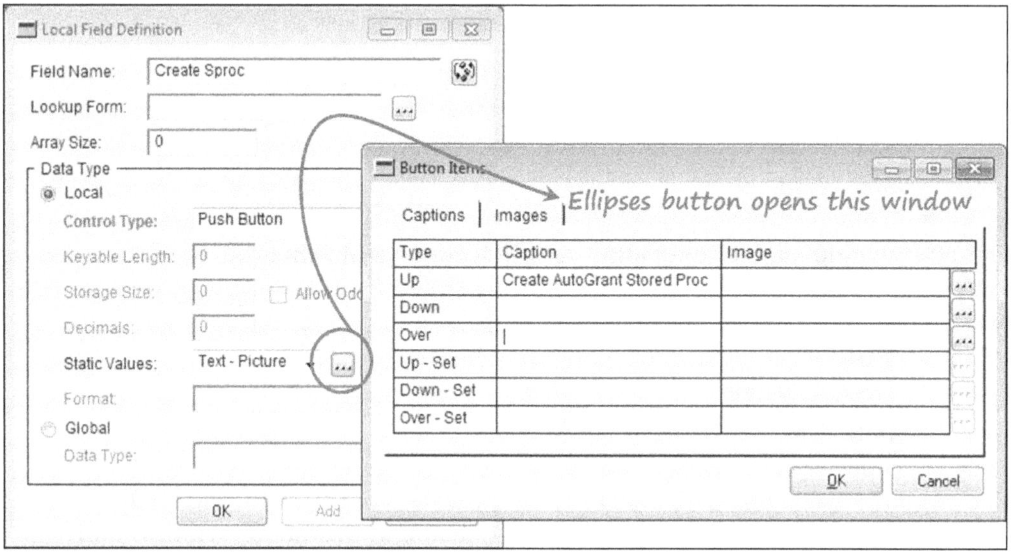

Next, add Change scripts to each of the buttons.

The **Cancel** button will simply clear the form. Create the following Change script:

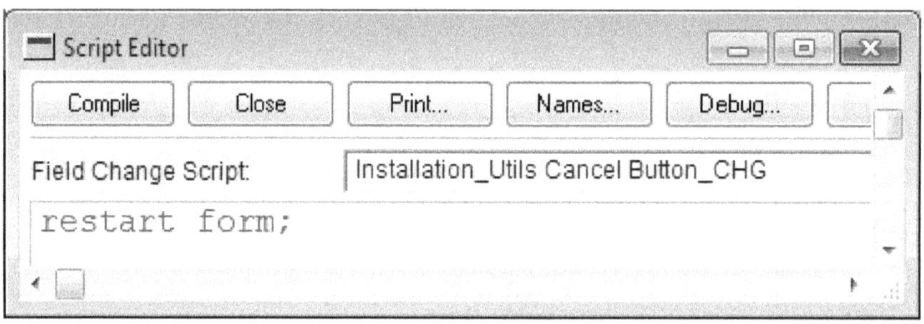

The **Create AutoGrant Stored Proc** button's, Change script accomplishes the following:

- Uses pass-through SQL to create the SQL stored procedure named `am_AutoGrant`. The `am_AutoGrant` procedure uses SQL statements to grant SELECT, INSERT, UPDATE, and DELETE permissions to the `DYNGRP` database role.
- Uses pass-through SQL to give the `DYNGRP` role GRANT and EXECUTE permissions to the `am_AutoGrant` SQL stored procedure.

It's necessary to give the DYNGRP role access to the stored procedure so that you can access the procedure from Dexterity. If you don't give access to DYNGRP, you will need to use SQL Server permissions to grant access.

Create the Change script as shown in the following screenshot for this local button; it will not compile. Save the uncompiled script by selecting **Script** from the menu bar, and then **Save Source** (or *Ctrl* + *S*). Finally, click on the **Close** button on the **Script Editor** window:

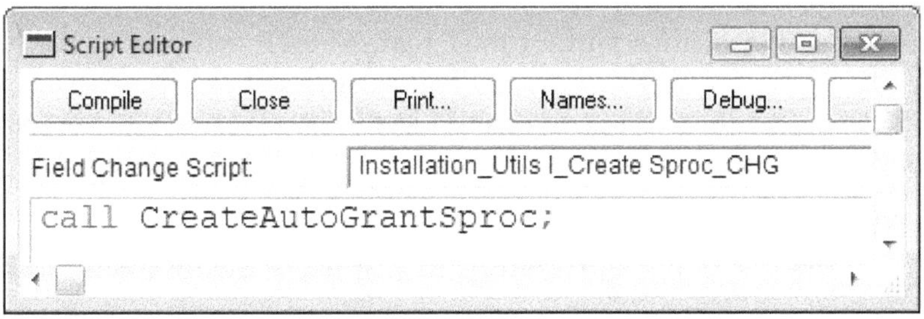

Now, let's create the CreateAutoGrantSproc global procedure. To create the new global procedure, select the new resource drop-down arrow and then select **Script (Procedure)**.

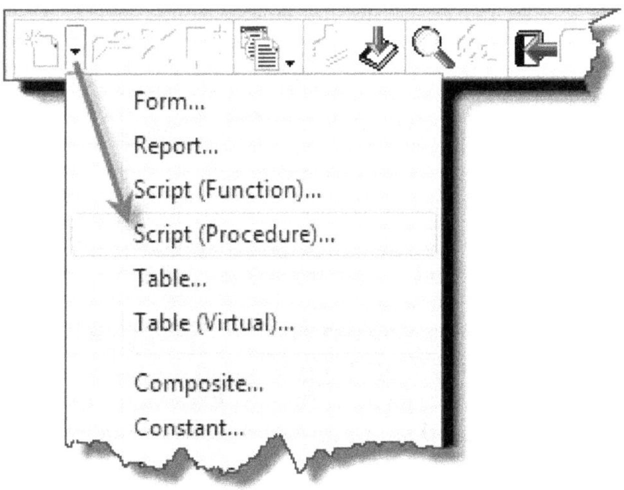

Name the procedure **CreateAutoGrantSproc**:

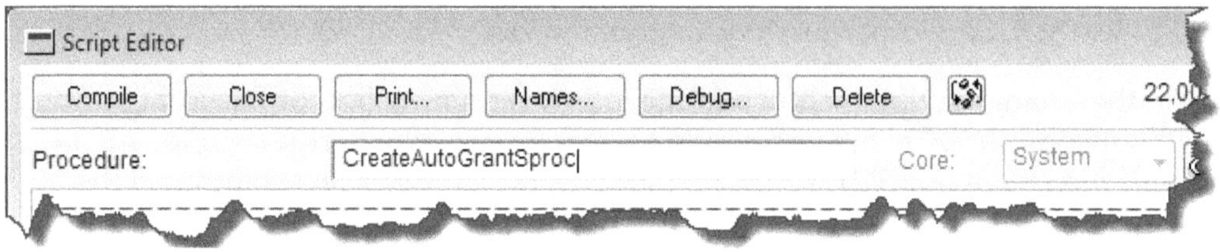

You can download the entire script for the **CreateAutoGrantSproc** procedure at `http://www.packtpub.com/support`. This script is taken largely from the *Integration Guide* manual that comes with Dexterity and various Microsoft Knowledge Base articles.

The **Create Tables and Grant Permissions** button will call the `Create_Tables` procedure. The `Create_Tables` procedure accomplishes the following tasks:

- Uses the Dexterity `Table_SetCreateMode` function to create the SQL `Tables` and `zdp` procedures
- Uses the `GrantAccess` stored procedure to give permission to the `DYNGRP` role for the `Tables` and `zdp` procedures

Create the following Change script for this local button:

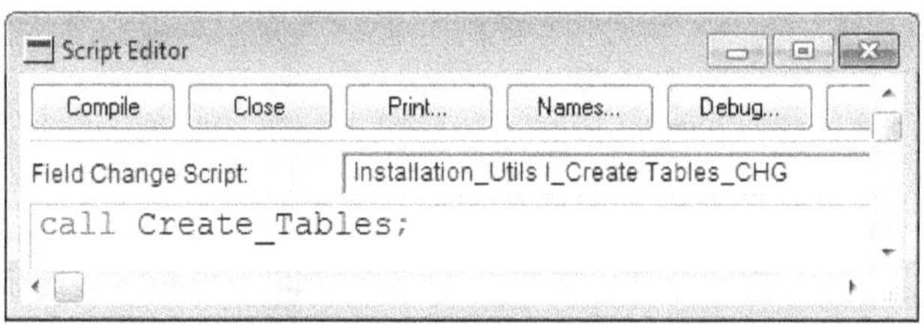

Now let's design the Create_Tables global procedure. Select the new resource drop-down arrow and then select **Script (Procedure)**. Name the global procedure **Create_Tables**. This procedure will use Dexterity sanScript to create your third-party tables (rather than pass-through SQL) and set permissions to those tables.

You can download the entire script for the `Create_Tables` procedure at `http://www.packtpub.com/support`.

OK, now we have two tables created, permissions granted, defaults bound, and auto- stored procedures in place. Just a note on the defaults; this is largely so that you can access the tables from outside of Dexterity without creating null values. Dexterity tables do not allow nulls, so you need to fill them with default values when the field would otherwise be blank.

Automatically creating the tables upon launch

Instead of requiring the user to launch a utilities window, you can trigger off the `Add_Successful_Login_Record` procedure. Triggering off this procedure guarantees the user is already logged in to the SQL Server, thereby avoiding the second login window.

When the procedure executes, it will create your `Tables` and `*zdp` procedures automatically if they do not already exist. Your trigger processing procedure is called `Create_Tables`. If the tables already exist, no action will be taken by Dexterity.

Include the following code in a global procedure named `Startup`. When the trigger is activated, the `Create_Tables` procedure that you previously created will execute:

```
{---------Register the Procedure trigger----------------}

  l_result =
  Trigger_RegisterProcedure(
    scriptAdd_Successful_Login_Record,
    TRIGGER_AFTER_ORIGINAL,
    scriptCreate_Tables,
    l_tag);
  ifl_result<> SY_NOERR then
    warning "The Add_Successful_Login_Record " +char(13)+
          " trigger registration failed.";
  end if;
```

Completing the application

To complete an application is to stand back and make sure it's tight. To make sure you have followed the user interface design standards and to really look at the application from the user's point of view. This section will help you begin to develop your own completion checklist.

Forms and windows

Consider each of your windows and be sure you have met the following standards:

Linking your prompts

If you don't link your prompts, you won't get a visual notice of the required fields, hidden prompts when the field is hidden, and *grayed-out* prompts when they are disabled. This is easy to forget; don't let it happen to you. Refer to the *Linked prompts* section to aid you in this quest.

To link prompts, select **Tools** from the menu bar, and then **Link Prompt**. The **Tools** menu is shown in the following screenshot:

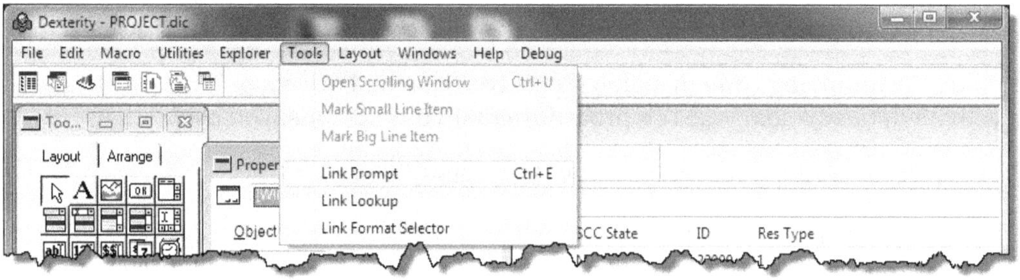

Linking your lookups

If you don't link your lookups, *Ctrl* + *L* will not open the lookup window as it is supposed to. If you have moved your buttons around, this is especially critical. You don't want your user hitting *Ctrl* + *L* and have the lookup window of another field open.

To link your lookups, select **Tools** from the menu bar, and then **Link Lookup**. The previous screenshot displays the **Tools** menu.

Adding tool tips

Tool tips can really help your application look polished. While *context-sensitive* help (read F1) may be above the call of duty, your users deserve tool tips. At a minimum, make sure every lookup button has *Ctrl* + *L* as a tool tip.

You can create tool tips in the **Properties** window. The following screenshot highlights where you can find the **Tooltip** object property:

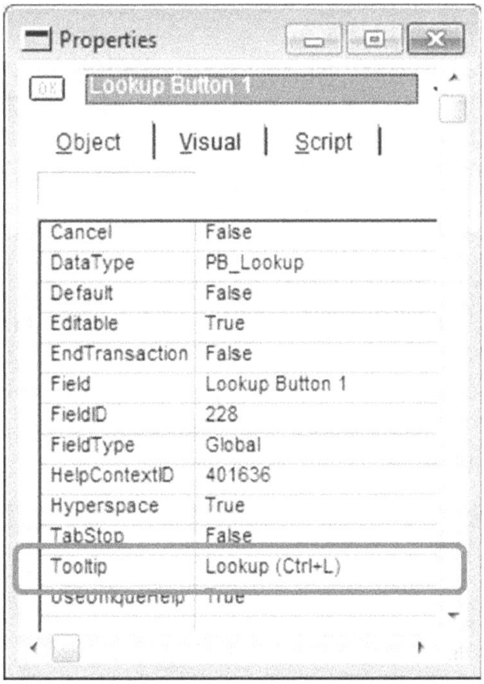

Hyperspacing your lookup buttons

The **Hyperspace** property applies only to push buttons. It is an object property that prevents the Change script or Post script from running on the field that has focus when the button is pushed. Imagine having the Change script and Post script running when you push the **Lookup** button. It will keep asking you if you want to add a new record instead of opening the lookup window and navigating to the closest match. Check your **Clear** buttons and your lookup buttons to make sure this property is set.

Linking your formats

Maybe a bigger plea would be to *use* linked formats. The format of a string or currency field can vary based on the value of another field. Think about the decimal places currency field that is part of the item card. How the quantity field displays for an item is based on the selection in the decimal places currency field.

For strings, the position of the format in the **Format Definition** window determines the index number of the format. Take a look around your application and make sure the field's format changes according to the appropriate selection.

Setting your tab order

As you develop a window, you move, change, and add window fields at different times. While you may have set the tab order during your original work, circling back around is a good idea to make sure the tab order still makes sense. Be consistent, window-to-window, and follow the method used by Dynamics GP when you can.

Setting the windows' opening positions and sizes

Get that window out of the upper left-hand corner! Consider the user sitting in front of the monitor. Where should your window open? Should it open relative to another window? Take the table names window (**Microsoft Dynamics GP | Tools | Resource Descriptions | Tables**). In my opinion, it is one of the worst-designed windows in the application. Love the information it provides, but I want it to look like the following screenshot:

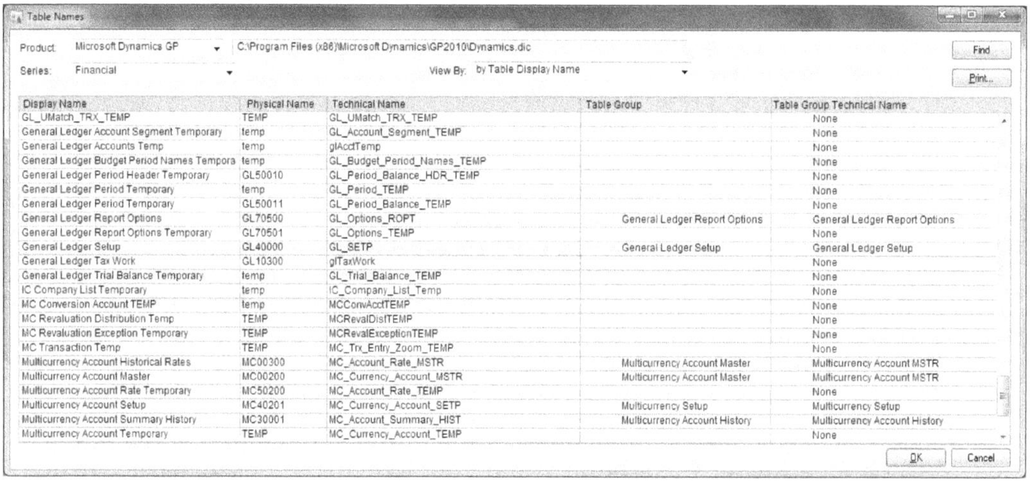

Take a look at your own windows and consider resizing any that may be awkward or limiting.

Complying with user interface standards

This is especially important if you want your application to achieve the *Certified for Microsoft Dynamics GP* accreditation. Check your windows for some of the most common errors:

- The toolbar often has several problems. Don't get caught in this; it really looks bad (and sloppy) if not done properly. The following list will help you out:
    - The buttons should be 72 pixels wide and 24 pixels tall
    - The buttons should have a 16 pixel vertical line between them
    - There should be a 16 pixel vertical line after the last toolbar button
    - There should be a horizontal line at the lower border of the toolbar/Control area
- The prompts are not underlined due to improper border properties.
- The fields and prompts are not lined up or do not align to the grid.
- Browse buttons are missing, in the wrong position, or not formatted properly.

Tables

Tables have far fewer things to check, but look for:

- Each master file table should have a note index field
- Make sure your tables have friendly display names
- Follow the standard naming conventions when naming your tables

Reports

There are several reports available that you may even use during the development process. However, before you deem the application complete, run these as part of the completion process.

Referential diagnostics

This report will alert you to any resource that references something that doesn't exist. Your testing could easily miss this kind of problem.

You run the **Referential Diagnostics** report in Dexterity Utilities. Open your dictionary as a source dictionary. Select the **Reports** button-drop and then select **Referential Diagnostics** from the list:

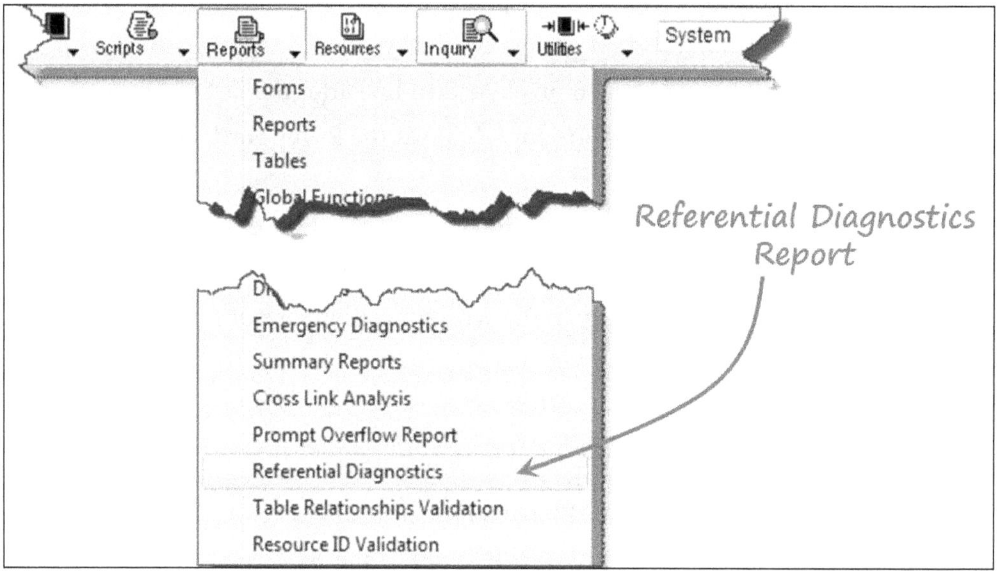

The report will list very specifically those items that require attention. An **invalid direct reference** occurs when one resource refers to another resource and the second resource does not exist. For example, if you delete a table that you used in the `Installation_Utils` form, you get the form error that is listed in the following report excerpt.

An **invalid transitive reference** occurs when one resource is referred to by another resource and the first resource does not exist. For example, if you remove a field from the `RM_Customer_Contact_MSTR` table that is used in a relationship or a key, you get the table error listed in the following report excerpt.

The following is an excerpt from a **Referential Diagnostics Report** file:

```
Invalid Direct References

Field  Customer Contact ID references a nonexistent Data Type .
Form  Installation_Utils references a nonexistent Table .
Form  Installation_Utils references a nonexistent Table .


Invalid Transitive References

Table RM_Customer_Contact_MSTR:  Key  RM_Customer_Contact_MSTR_Key1 references a field not in this table.
Table RM_Customer_Contact_MSTR:  Key  RM_Customer_Contact_MSTR_Key2 references a field not in this table.
Table RM_Customer_Contact_MSTR:  Relation  RM_Contact_Phone_MSTR references a field not in this table.
```

You need this kind of report to alert you of changes you may have made after you compiled your code.

Linked prompts

This report will alert you to any field on the selected form that does not have a linked prompt. Making sure your prompts are linked is also mentioned in the **Forms and windows** section covered earlier. You can run the linked prompts report in Dexterity Utilities. Open your development dictionary as a source dictionary, select the **Reports** drop-down button and then select **Linked Prompt** from the list.

On the **Linked Prompt Report** window (shown in the following screenshot), all of the dictionary's form resources are listed on the left. Select just those forms you want to analyze and insert them into the list on the right. Select the **Print** button and name the text file that will be the report:

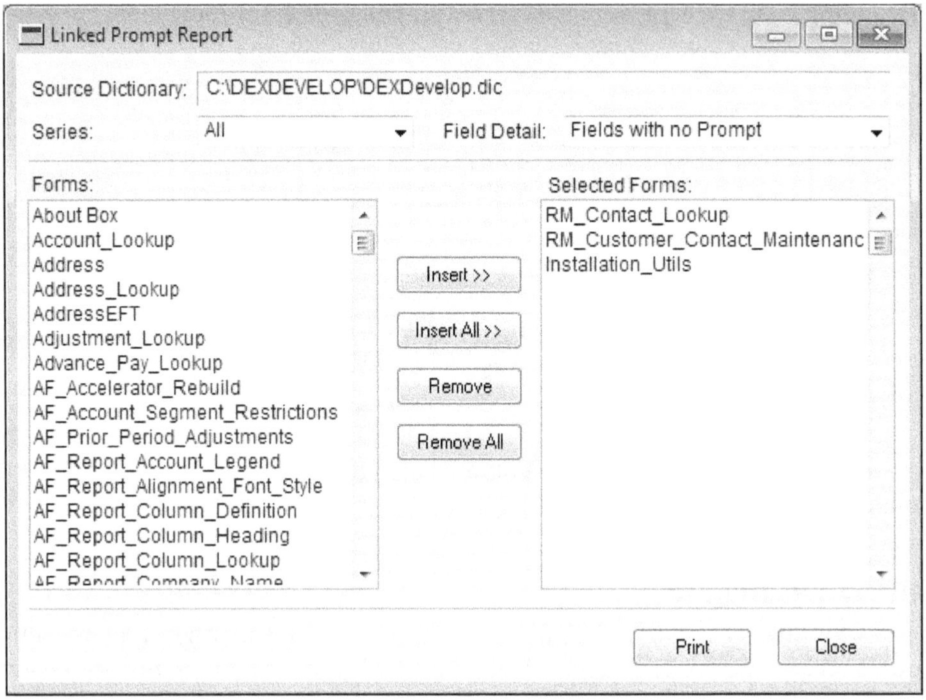

This is a very handy report. As you can see from the report excerpt in the following screenshot, some fields don't need a prompt; those fields are easy to identify. However, we can also see that the **Contact Department** field needs a linked prompt as shown in the following screenshot:

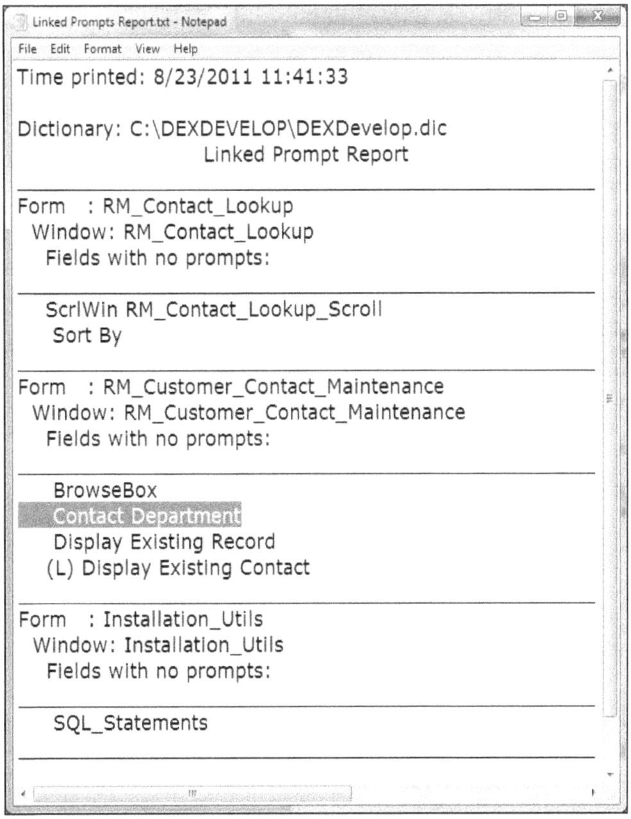

You need to use the development dictionary as the source dictionary; otherwise, the report will show **[Not Found]** next to any field that uses a resource from `Dynamics.dic`.

Table relationships validation

This report will alert you to relationships that you have defined incorrectly. For instance, you might have a data type mismatch, such as an integer to a string. Again, we make a lot of changes during the development process, this report should be on your checklist as a *must do* item.

An excerpt of the report follows, showing several relationship errors:

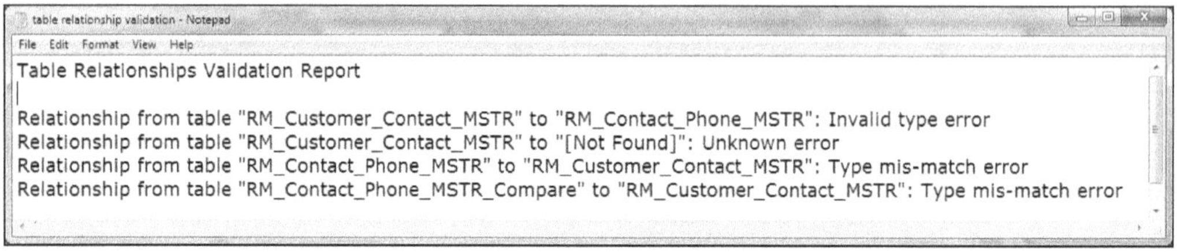

Table Relationships Validation Report

Relationship from table "RM_Customer_Contact_MSTR" to "RM_Contact_Phone_MSTR": Invalid type error
Relationship from table "RM_Customer_Contact_MSTR" to "[Not Found]": Unknown error
Relationship from table "RM_Contact_Phone_MSTR" to "RM_Customer_Contact_MSTR": Type mis-match error
Relationship from table "RM_Contact_Phone_MSTR_Compare" to "RM_Customer_Contact_MSTR": Type mis-match error

Other reports are available in Dexterity Utilities that will help you create documentation for your application. But now, let's create our chunk file.

Creating the chunk file

Just like a child leaving home for the first time, your application needs to survive on its own. Moving from the comfort of the development dictionary to a standalone dictionary is a process known as **chunking**.

A chunk has the extension `.cnk` and stands literally for a *chunk* of code. The chunk file is a self-extracting file that contains all of the Dexterity resources of your customization. Once extracted, a chunk file becomes a dictionary file ready for execution by the `Dynamics.exe` runtime. The following KB article will provide additional information on this topic:

- *How to create a chunk file in Dexterity in Microsoft Dynamics GP KB 894700*

  `http://support.microsoft.com/kb/894700`

The process of isolating your resources from the development dictionary is achieved by the Dexterity Utilities application. The short process extracts all of the resources with an ID of 22,000 or above. You individually transfer any alternate forms or reports, add the application information, and hit the **Go** button. Voilà, you've created a `.cnk` file.

Take the produced chunk file, drop it in the `GP2010` folder, and then launch Dynamics GP. The chunk will turn into a dictionary and be added to the `Dynamics.set` file. Your Dexterity application will now be included with the other applications and the functionality of your product will be available to all.

Let's create the chunk file for our **Customer Contacts** customization.

The following is the checklist for creating a chunk file:

1. Extract all resources numbered 22,000 and above into a new dictionary
2. Transfer any alternate forms reports
3. Enter dictionary details
4. Enter product information
5. Create the .cnk file

Extracting resources

Close Dexterity and then open the Dexterity Utilities application. In the Dexterity Utilities application, select **File | Open Source Dictionary** and open the Project.dic file. The file will open, but no dialog will appear informing you that the file has been loaded.

Select the **Utilities** drop-down list on the toolbar, and then **Extract** from the list, as shown in the following screenshot:

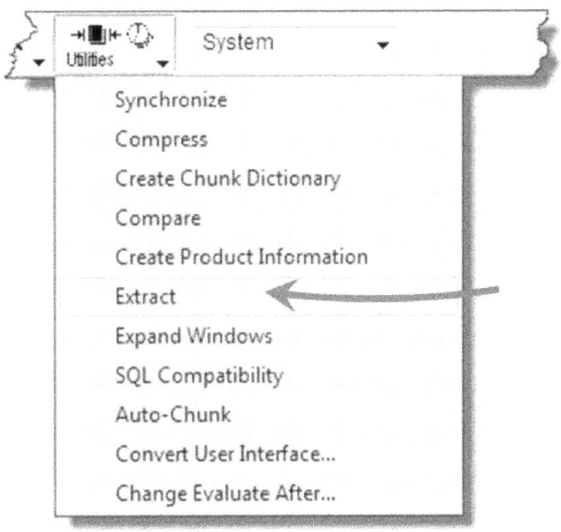

The **Extract** window will open identifying the source dictionary name and its location:

Selecting **OK** on the **Extract** window will open the **Extracted Dictionary Name** window. In this window, you type in the name of the dictionary that will be created to hold the extracted resources. This time we will accept the default of `Extract.dic` as shown in the following screenshot:

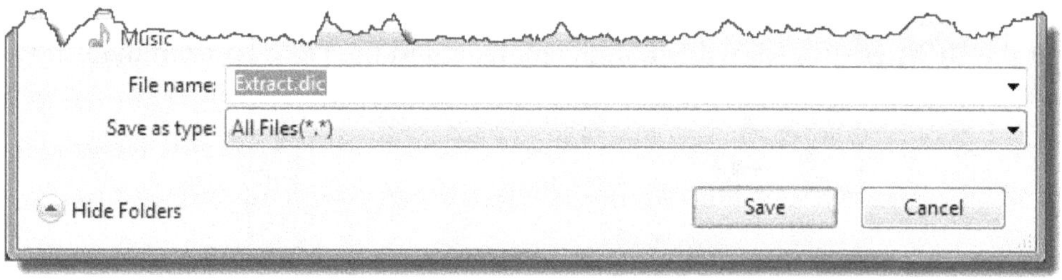

Select **Save** on this window. A small progress window will display as the third-party resources are extracted. Only the resources you created are included in the `Extract.dic` file.

Next, we need to transfer any alternate forms or reports.

Transfer dictionary module

If you have created any alternate forms or reports, you need to use the **Transfer | Dictionary Module** tool. You create an alternate form or report when you modify the original. Because you modified the original, its resource ID is less than 22,000. Your extracted dictionary only includes resources numbered 22,000 or above; therefore, you need to select the modified resource and transfer it individually to the extracted dictionary. Select the **Transfer** drop-down list on the toolbar, and then **Dictionary Module** from the list. The navigation is shown in the following screenshot:

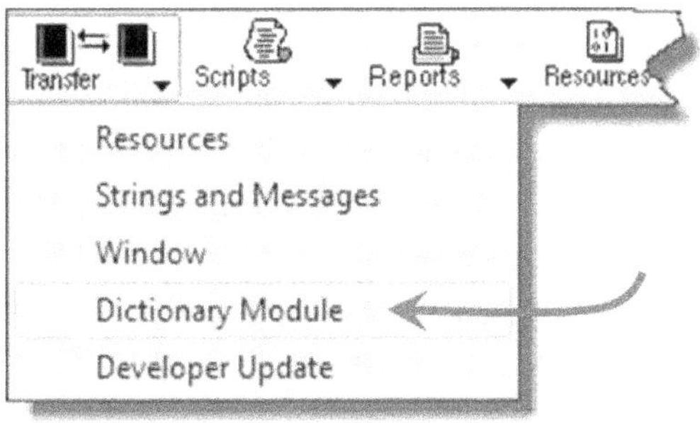

If you had modified the **Customer Maintenance** window to include a button that would open your **Contacts Maintenance** window, you would need to transfer the modified **Customer Maintenance** form to the extracted dictionary. To accomplish this transfer, select **Sales** as the **Series** and **Forms** as the **Dictionary Module Type**. In the **Source Dictionary Modules** list, select **[Form] RM_Customer_Maintenance** and then click on the **Transfer** button. No dialog will appear, but the **Total Size** field in the lower right-hand corner will increase by the size of the selected modules:

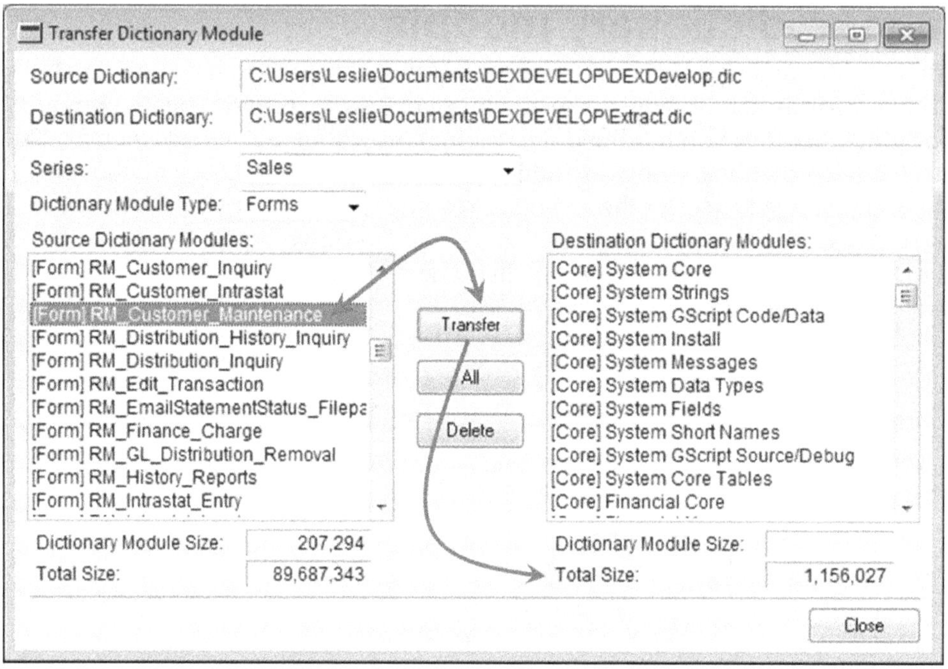

Close the source and destination dictionaries. Open the **Extract** dictionary as an editable dictionary. Select **Auto-Chunk** from the **Utilities** drop-down list:

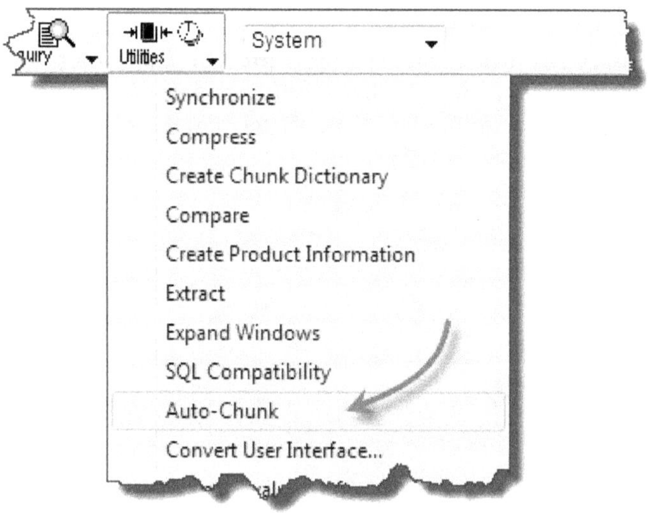

Use the **Auto-Chunk** window to define the names of the resulting chunk file and the ultimate merged dictionary. You can specify installation and version information. Let's explore this window and discuss each of the relevant fields:

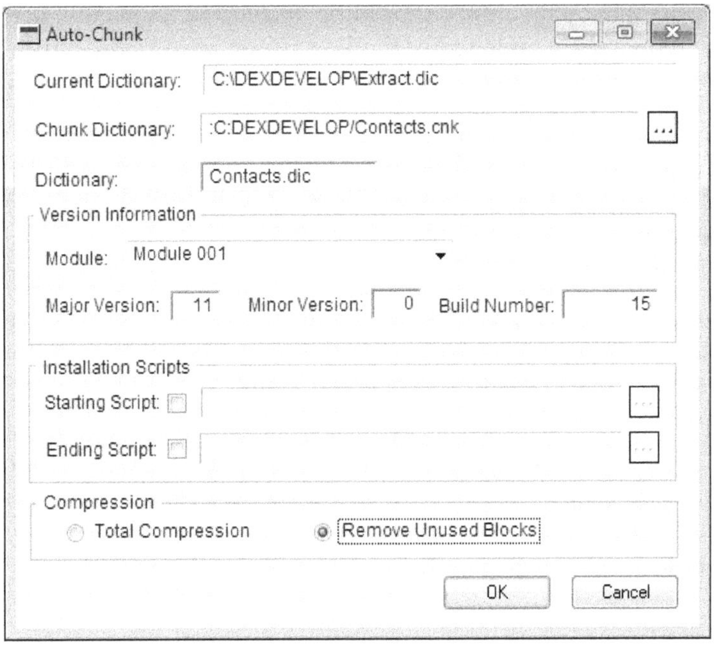

The following list describes the various fields present in the **Auto-Chunk** window:

- **Current Dictionary**: This is the name of the editable dictionary whose resources will be transferred to the chunk dictionary.
- **Chunk Dictionary**: This is the name of the chunk dictionary created by the **Auto-Chunk** process. To name the chunk dictionary, click on the ellipses button and type the desired name for the `.cnk` dictionary in the **File name** field.

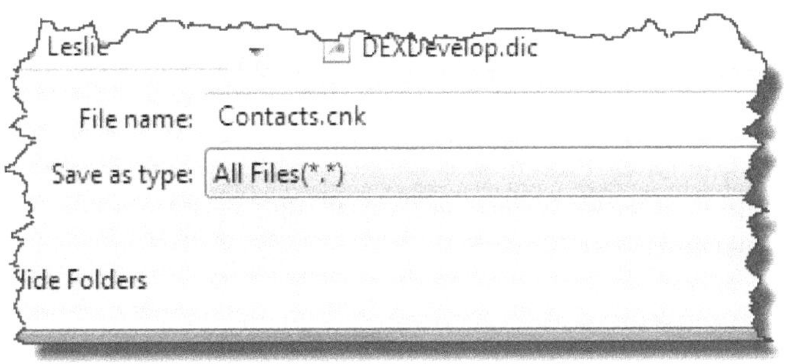

The chunk dictionary's name does not have to be the same as the merged dictionary; in practice, they often are the same.

- **Version Information**: Version information is used by the runtime engine to determine whether a chunk should be *unchunked*. Typically, the major and minor versions are matched to the Dynamics GP version and then your build number is evaluated to make sure it isn't an older version of your product.
- **Module**: Today the **Module** field is largely unused. In the past, words were used in the names of modules. It was converted to a list of numbers and can be used in code if you want. Leaving this at **Module 001** will be fine.
- **Major Version**: Normally this is set to the same number as the application you are integrating with. For example, Dynamics GP 2010 is version 11.
- **Minor Version**: In the past, when there was a major version that had a substantial change, this number was incremented. For instance version 3.0 versus 3.15. Today, use this to match the numbers of Dynamics GP.
- **Build Number**: Use this field to number each release of your software for the same major version. How often you do this is up to you, but you should definitely increment it each time you deploy a different chunk of your product. This number should start at 0001 for each major version and is then incremented from there.

> Enter the following in your Internet search engine to find more information about version and build numbers:
>
> `MSDN Blogs>Developing for Dynamics GP>Quick Tip: Best Practice for Dexterity Version and Build Numbers.`
>
> This article is accessible at `http://goo.gl/ovn71`.

- **Installation Scripts**: These are global procedures you write that can run after the *unchunking* process. You could use an installation script to read or write to a text file, for instance. It is no longer relevant to create both a before and after script because both scripts run at essentially the same time.

> You cannot check values of system variables using an installation script because these variables are set after your installation scripts have run.

- **Compression**: This determines whether your source code will be available in your completed dictionary. **Total Compression** removes the source code; **Remove Unused Blocks** leaves the source code in the dictionary.

  Whether or not you leave the code in the dictionary is largely a matter of ownership. If you strip the code out, don't lose the development dictionary, or at least the `Extract.dic` file. For chunk files created to test multi-dictionary environments, leave the code intact. With the code present, you can debug your application during multi-dictionary operation.

Click on the **OK** button at the bottom of the window. A dialog box will open asking you if you want to add product information. See the following screenshot:

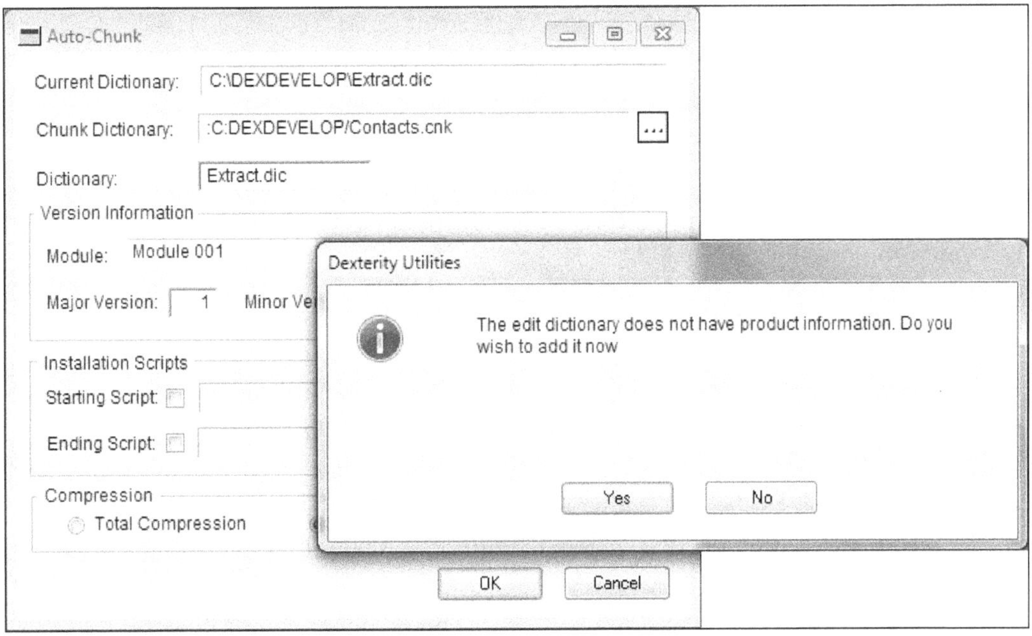

Click on **Yes** and the **Product Information** window will open (see the following screenshot). Let's review each field on the window so that you know how your answers here impact your installed dictionary:

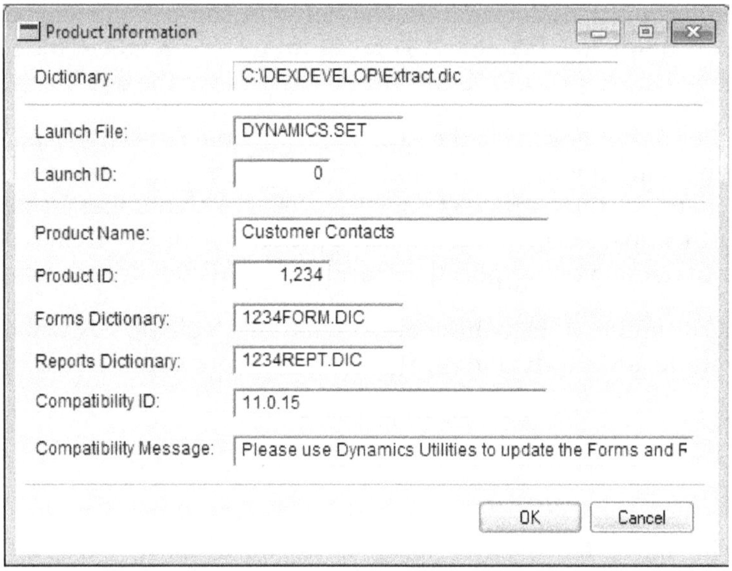

The following list will guide you through all the fields in the **Product Information** window:

- **Dictionary**: This bears the name of the dictionary created by the **Extract** utility.
- **Launch File**: This bears the name of the file you will be using with the main application to unchunk your application. Be careful to spell this correctly, otherwise your file will not unchunk.
- **Launch ID**: This value is always `0` for Dynamics GP.
- **Product Name**: This is what your product will be called in the `Dynamics.set` file.
- **Product ID**: This is the unique number you received from Microsoft that is assigned to your application. Don't make up this number; it is essential that the number be unique. Microsoft will not charge you to give you a product number, so do it.
- **Forms Dictionary**: This is the name of the dictionary that will be created to hold any user-modified forms. This is not created until the *user launches the* `Modifier` *module*.

- **Reports Dictionary**: This is the name of the dictionary that will be created to hold any user-modified reports. This is not created until the *user launches the* `Report Writer` *module*.
- **Compatibility ID**: Whatever you put here will be checked against what's in this position on the new chunk. If they don't match, the compatibility message will be displayed.
- **Compatibility Message**: The message displayed to the user if the Compatibility IDs do not match.

Click on the **OK** button once you have filled out this window and you will be returned to the **Auto-Chunk** window. Click on the **OK** button on this window and you will create your chunk file.

> You will be performing the chunking process many times during development. Record a macro so that you don't have to re-type all of the previously mentioned information every time. Make sure that you change the version in your macro file before you run it to create the final chunk. Otherwise, the chunk will have the same version as it was when the macro was recorded. You need to update your build number each time you create a new chunk file for distribution.

Now it's time to test your application with all of the other dictionaries loaded.

Testing in a multi-dictionary environment

First, let's install your application. Copy the chunk file into the Dynamics GP application folder. Out of the box, the location of the application folder is as follows:

```
C:\Program Files\Microsoft Dynamics\GP2010
```

Chunk doesn't unchunk

Sometimes the testing process runs into problems before it even gets off the ground. Let's imagine that your chunk file won't unchunk.

Various things can be creating this problem. Some reasons are obvious, some not so obvious. Let's go through the list of possible culprits. Don't feel bad if you find yourself here. We've all been there; how do you think we got the list?

The following are the things to check if your chunk file doesn't unchunk:

1. Check the `GP_LoginErrors.log` file in `C:\Users\username\AppData\Local\Temp\`
2. Make sure you put your chunk file in the correct folder. It goes in the same folder where the `Dynamics.exe` lives.
3. Make sure your chunk file is not marked `read-only`.
4. Make sure you have adequate permissions at the operating-system level.
5. Make sure the launch file is named `DYNAMICS.SET` and the launch file you entered on the **Product Information** window was `DYNAMICS.SET`.
6. Check for a previous installation using a different extracted dictionary name.
7. The chunk file should be 12 characters or less. Dex Utilities will not let you put in a longer name, but someone could have changed it.
8. Start Dynamics using the `run as administrator` selection.

If you find another problem, be sure to add it to the list and share it with the Dynamics GP community!

Now that you've got your product installed, let's test it.

Testing tools and techniques

Download a copy of the testing framework for Dynamics GP 2010 R2! This kit will make testing (almost) fun and easy. The testing framework is a collection of scripts, sample data files, examples, and utilities you can use for testing.

Several samples are included that show you how to load data, run a macro, execute a report, save results to a file, and then compare those results to a known baseline. Get your copy at:

`http://archive.msdn.microsoft.com/MDGPTestFramework`

Regardless of the testing tool you use, keep the following points in mind:

- Make sure you can return to your starting point. Without a repeatable baseline, it will be impossible to adequately evaluate the changes you make.
- Your test environment should mimic the production environment as closely as possible.
- Never, ever, test in production.
- Include both a server and a workstation. So many other developers forget to test the application on anything but the server.
- Test both on Fabrikam and a newly created company.
- Log in as a user other than `sa` or `DYNSA`.
- Make one change at a time. This should be in all caps; it's that important.
- Document everything.

If your trigger processing procedures are not running, put your application in different positions in the `Dynamics.set` file. Many things are controlled by the application's position in the launch file. For one, triggers fire in the order the applications are listed. Test how your application reacts if a different product's processing procedure prevents your procedure from running.

Additional resources available

- **The macro language**

  Recording test cases using the macro language is a great way for carrying out the same set of actions repeatedly. A white paper discussing how to use macros when testing an application is available on the web at the following address:

  `www.microsoft.com/download/en/details.aspx?displaylang=en&id=16233`

- **Dexterity documentation**

  Read the Dexterity documentation on testing in *Chapter 34* of *Volume 2* of the *Dexterity Programmers Guide for Dynamics GP 2010*:

  `http://www.microsoft.com/download/en/details.aspx?displaylang=en&id=8651`

- **The Support Debugging Tool**

  The Support Debugging Tool is a brilliant set of tools and utilities that you can use when developing, supporting, and testing Dynamics GP. At this time, this tool is only available to a Dynamics GP partner, so you will need to ask your reseller for a copy. There is no charge for the tool. Partners should go to the following web address to download the Support Debugging Tool:

  `http://blogs.msdn.com/developingfordynamicsgp/pages/support-debugging-tool.aspx`

- **Date-related issues**

  An article that discusses some resolutions to issues related to dates that you might encounter while testing your application is available at the following site:

  `https://community.dynamics.com/product/gp/gptechnical/b/developingforgp/archive/2009/12/18/testing-date-driven-features-in-gp.aspx`

Now that your application has been tested and debugged, it's time to distribute it to your end user.

Distributing the completed application

You have two ways to deliver your application. They are as follows:

- As a chunk file.
- As a Windows Installation file.

Sending the chunk

Dexterity customizations are one of the easiest applications to deliver. A vanilla integration, as is the **Customer Contacts** application, requires only that you copy the `.cnk` file to the installation folder of Dynamics GP. It's as simple as that. Fire up Dynamics GP and your application is installed.

Windows Installer services

You can use Windows Installer services to create an installer program for your Dynamics GP integration. Any tool that will create an installation file will work fine, but a template is provided for you to use with the WiX (Windows Installer XML) toolset. WiX is a free download from `http://wix.codeplex.com/`.

Chapter 44 of the *Dynamics GP 2010 Integration Guide* contains step-by-step instructions on how you can use the sample template to create your own installer. An installation file looks so much more professional than asking the user to copy a chunk file into the client folder. If you need to include several files with your installation, such as a manual or help file, creating the installation file is the preferred method.

Summary

This chapter took you through the process of completing and deploying your Dexterity application. Starting with the system requirements, you explored the importance of versions and builds, and how they are created. You looked at three different methods for creating the SQL tables that you defined for your application.

With your table creation procedures in place, it was time to extract your application from the development dictionary. With chunk in hand, you started down the long road of testing and debugging. To your delight, you learned about the new testing framework available to help you test in a controlled environment!

We talked about distributing your completed application using just the chunk file, or creating an installer for it. You have new tools in your toolkit to help you deploy your application, such as a completion checklist and some white paper articles. We'll be leaving Dexterity now and switching over to using Modifier with VBA to create customizations.

7
Creating Customizations with Modifier

In this chapter, you'll learn how to build a customization using the **Modifier** tool. The full name of the module is **Modifier with VBA**. In this chapter, however, we will cover only the Modifier component; in *Chapter 8, Creating Customizations with VBA*, we will add the VBA component.

In this chapter, you will learn about the following topics:

- How the Modifier tool fits into the development environment
- How to modify windows and window properties
- How to change global resources

Using Modifier, you can customize forms from any dictionary in the application. You can make global changes to pictures, strings, and formats that affect the entire application, including third-party products. The things you can do with the Modifier provide a method to create new fields, remove existing fields, rearrange fields, and change the tab order of fields on a window.

You cannot create any actions in your customization with Modifier alone. *Chapter 8* will explore how to fuel your Modifier additions using VBA.

Overview of Modifier

Modifier is an end-user tool that allows you to make changes to the user interface without using a full-blown application development environment. Rather than creating a new window, you can seize an existing window and change nearly everything about it.

Changes to windows made with theModifier are stored in a separate dictionary known as a **Forms** dictionary; each application has its own Forms dictionary. For example, the Forms dictionary for the core application is named `Forms.dic`; the Forms dictionary for Fixed Assets is named `F309.dic`. You can find the name and location of the Forms dictionary for each Dexterity application in the `Dynamics.set` file.

Two tools in one!

As the name implies, the Modifier with VBA comprises of two components. The first, Modifier, has been around as long as Dynamics GP itself. It's a scaled-down version of the Dexterity window design tool.

Essentially, Modifier allows you to make visual changes to the windows. You can also make certain object changes such as making a field required, or changing its format. You have access to a fixed list of properties and methods that you can manipulate with the Modifier.

You can add new fields and objects, such as push buttons and pictures, to a window, but they are for display only. Your changes are static; they cannot "do" anything. You can add a push button to a window, but when you push it, nothing happens. You cannot attach code to any object using the Modifier alone. Enter VBA.

VBA is a subset of Microsoft Visual Basic 6.0, which is an event-driven structured programming language. VBA comes with its own **IDE (Integrated Development Environment)** and is built into most Microsoft Office applications. Using VBA, you can give life to that push button you added with Modifier, you can populate those new fields with your own data, and you can alter the business logic of Dynamics GP. You'll learn more on VBA in *Chapter 8*.

Let's start exploring the Modifier.

Modifying windows and window fields

Hands-on learning is probably the best way to experience a new program. With that in mind, we're going to dive right in and start manipulating windows and window fields. Windows and window fields have characteristics known as **Properties**. When you open a window in the Modifier, the **Properties** window is typically visible in the area to the right of the window's layout. You will also get the chance to modify certain window and field properties while working through our ensuing sample.

Launching the Modifier

The Modifier is not a separate application that runs independent from Dynamics GP. It's actually a part of Dynamics GP itself—kind of like Report Writer.

Let's launch the Modifier and take a tour. We are going to open the general ledger's **Transaction Entry** window in the Modifier.

Open the **Financial** home page, go to the **Transactions** content pane, and then select **Journal Entry**. See the following screenshot for the navigation:

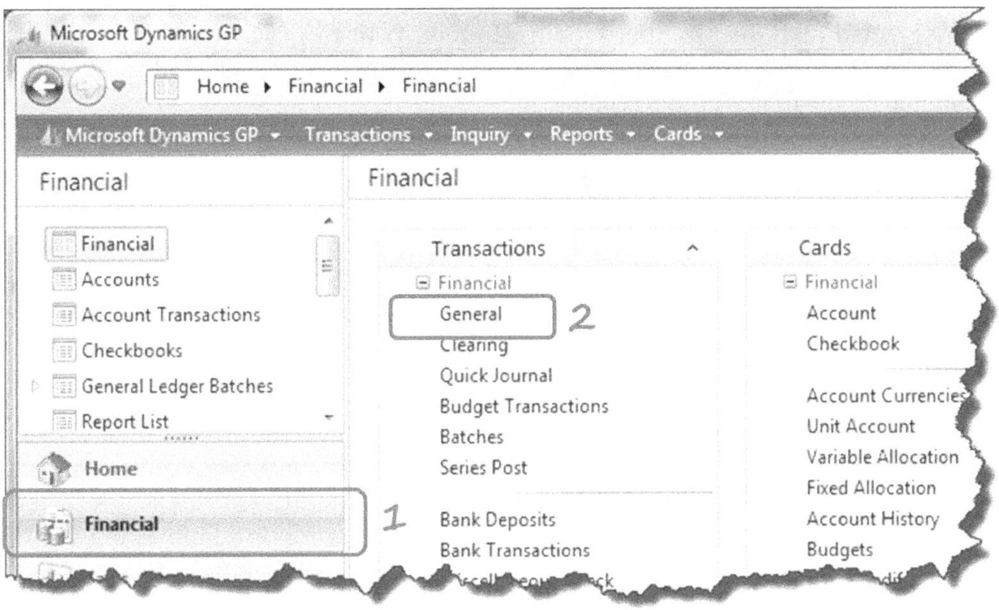

While you have the **Transaction Entry** window open, open the **Transaction Entry** window in the Modifier through **Tools | Customize | Modify Current Window**:

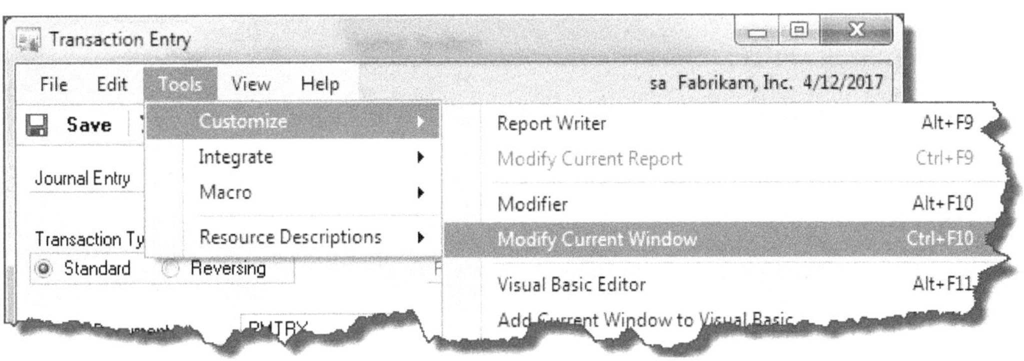

> The **Tools** menu is on the window, not the desktop.

Alternatively, you can press *Ctrl* + *F10* on the keyboard.

> If you see the options grayed out, make sure you have a license for the Modifier, and that it is enabled from the **Registration** window.

You are greeted with the Modifier's main window. Your form's layout window along with the **Toolbox** and **Properties** windows will be displayed. Your window should look similar to the following screenshot:

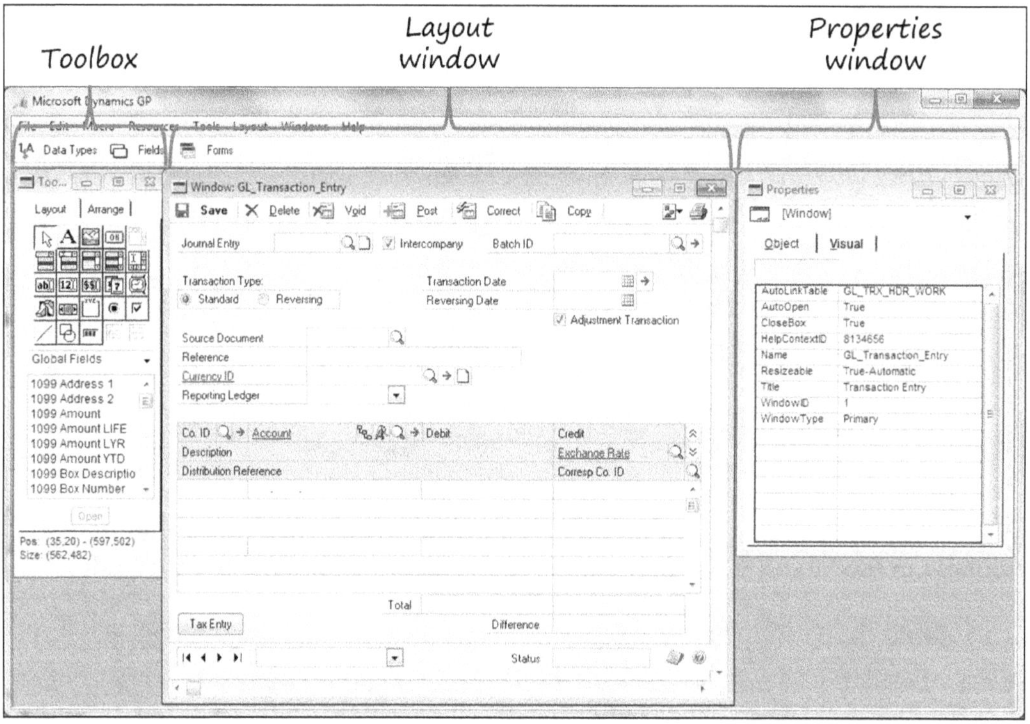

If this appears familiar to you, it should. It looks just like the Dexterity WYSIWYG Form Designer window.

The window properties

The **Properties** window controls two kinds of properties: **Object** and **Visual**. You can change any of the window's visual properties, but only a few of the object properties.

The following table provides a brief description of certain window object properties. An asterisk (*) after the name of the property means you can change it using the Modifier.

Property	Description
AutoLinkTable	If a window has an **AutoLinkTable** property, then you can easily add any of that table's fields to your window. All you need to do is select it from the list of fields and drag it out onto the window layout. Sadly, not all windows have the **AutoLinkTable** property; the **Sales Transaction Entry** window, for example, does not have this property. You cannot add or change the **AutoLinkTable** property.
AutoOpen	If this is set to `True`, which it normally is, the window will open automatically when the form opens. Secondary windows such as the **Date Entry** window on the **GL_Transaction_Entry** form are normally set to `False`. Secondary windows open using navigation on the main window, like a push button or an expansion arrow.
CloseBox*	If this is set to `True`, which it normally is, the window will contain a box in the upper right-hand corner with an **X** on it. When you click on that box, the window closes. If it doesn't have a close box, the window will need to have an **Exit** menu, or some other closing mechanism. The screenshot following this table identifies the close box.
Name	This is the technical name of the window. You use this name if you are a Dexterity programmer. Your users will not see this name in the ordinary course of business.
Title*	This appears at the top of the window and it is what your users see. The title is also known as the window's *display name*. The following image identifies the window's title.

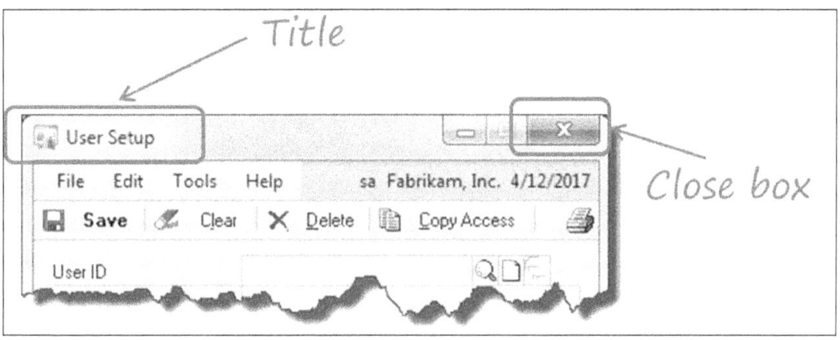

Size

You can set the size of a window in the following two different ways:

- Change its **Size-Height** and/or **Size-Width** visual property
- Use the mouse to size it as displayed

To use the **Properties** window, you need to select the object by either clicking on it or selecting it from the drop-down list. For the window, click on a gray area wherein no other objects exist. You then need to select the property you want to change and type the new value in the settings box. The following screenshot identifies each of these areas:

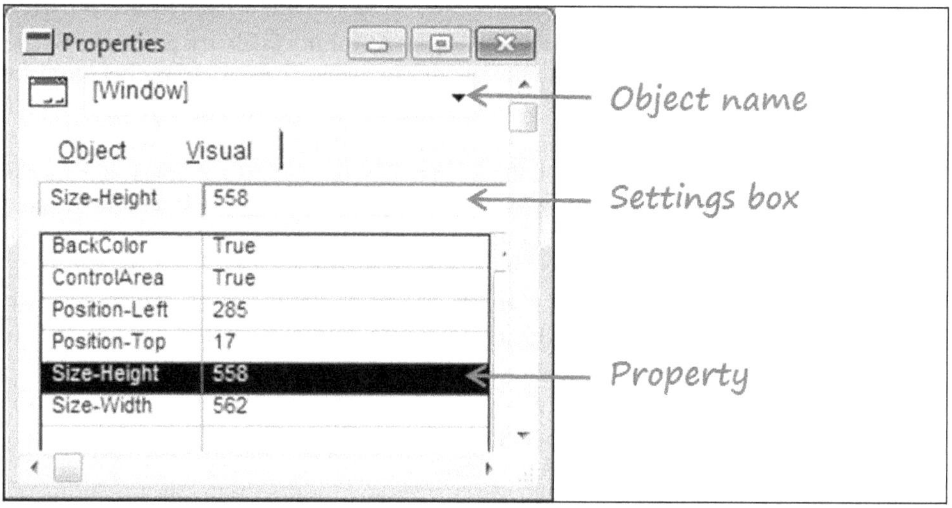

Opening position

If you don't like where the window opens, you can change it. Using the **Properties** window, simply change the **Position-Left** or **Position-Top** setting. Alternatively, you can just move the window to the place you want, and set it there.

If you are in Modifier, switch back to Dynamics GP (**File** | **Microsoft Dynamics GP**). We'll go through an example of how to change a window's opening position using the **Buyer Maintenance** window.

Open the **Buyer Maintenance** window from the **Cards** section of the **Purchasing** home page.

Note the opening position of the window. It should open in the upper left-hand corner of the desktop, as in the following screenshot:

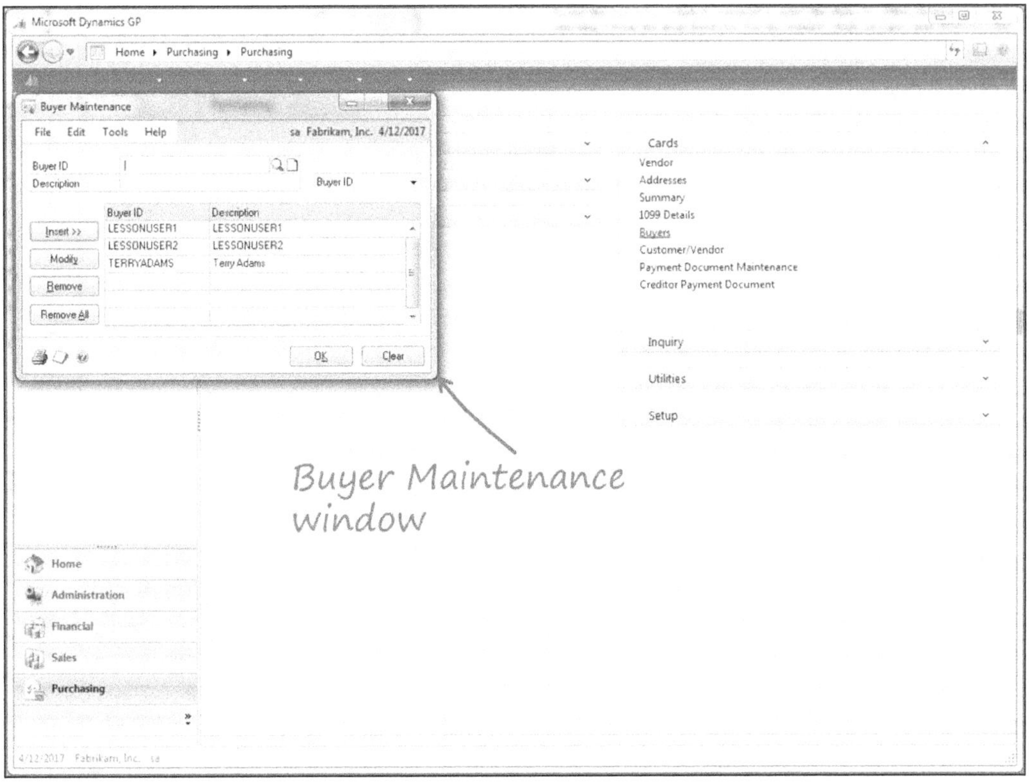

Now, press *Ctrl + F10* to open the window in Modifier. Once in Modifier, go to the application's menu bar and select **Layout | Position Window**.

The **Position Window** window will display in the same location that the **Buyer Maintenance** window opens (see the following screenshot), showing you its exact location on the desktop:

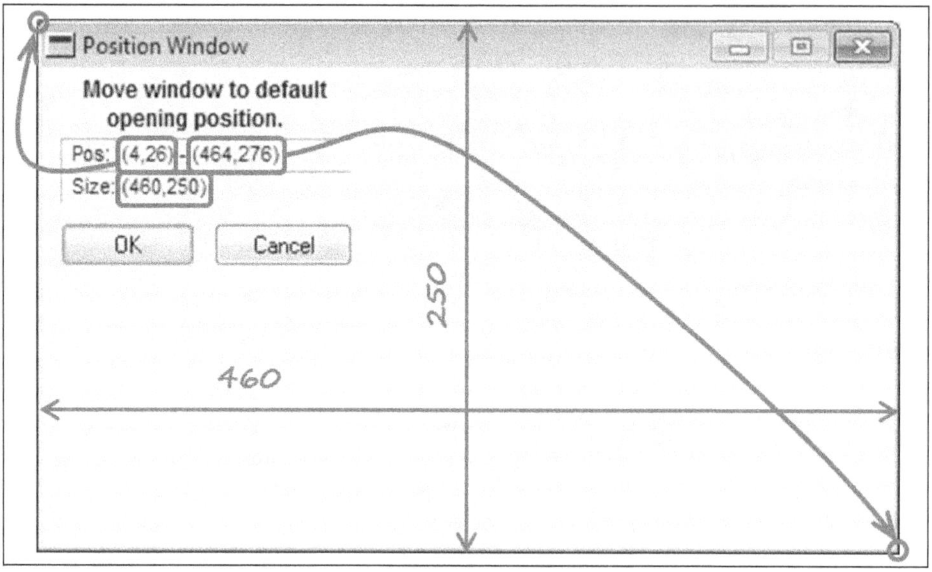

Referencing this screenshot, look at the numbers next to **Pos:**. These numbers indicate the location of the upper left-hand corner of the window and the lower right-hand corner of the window. All measurements on the **Position Window** are in pixels.

The first set of numbers (**4, 26**) means the upper left-hand corner is four pixels from the left side of the desktop and 26 pixels down from the top. The second pair of numbers (**464, 276**) means the lower right-hand corner is 464 pixels from the left side of the desktop and 276 pixels down from the top. These numbers are highlighted in the following screenshot:

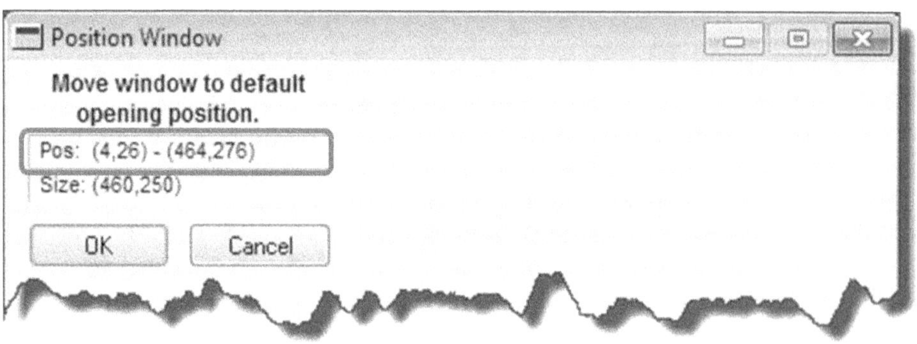

Beneath the position numbers, are numbers indicating the size of the window. The pair of numbers to the right of **Size:** tell you the width and height of the window. In this example, our window is 460 pixels wide and 250 pixels tall. These numbers are highlighted in the following screenshot:

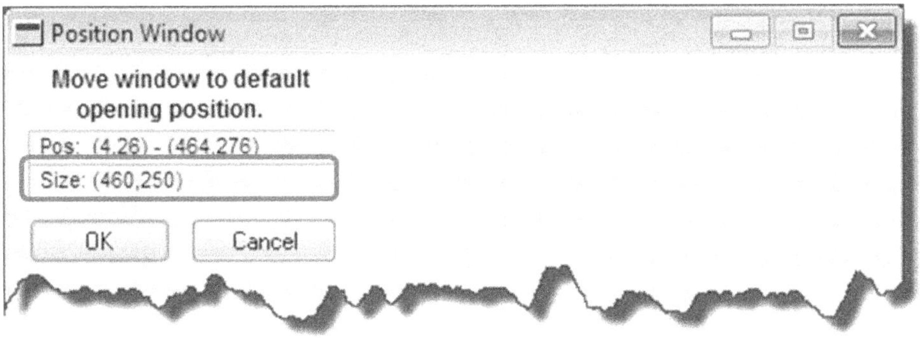

To adjust the opening position of the **Buyer Maintenance** window, click the left mouse button and hold in the **Position Window**'s title. While still holding down the mouse button, drag the **Position Window** object to where you want the **Buyer Maintenance** window to open. Once you are satisfied with its position, click the **OK** button.

> Be sure you are moving the **Position Window** object and not the **Buyer Maintenance** window itself.

To check your work, press *Ctrl* + *5* to open the **Preview** window. The **Preview** window will open in the same position that the **Buyer Maintenance** window will open in Dynamics GP.

The tab sequence

The tab sequence refers to the order in which the cursor advances through the window fields when you press the *Tab* button on the keyboard. If you add new fields, or rearrange existing fields, you will most likely need to change the tab order.

To set the tab sequence of a window, open the particular window, select **Layout** from the menu bar, and then **Set Tab Sequence**:

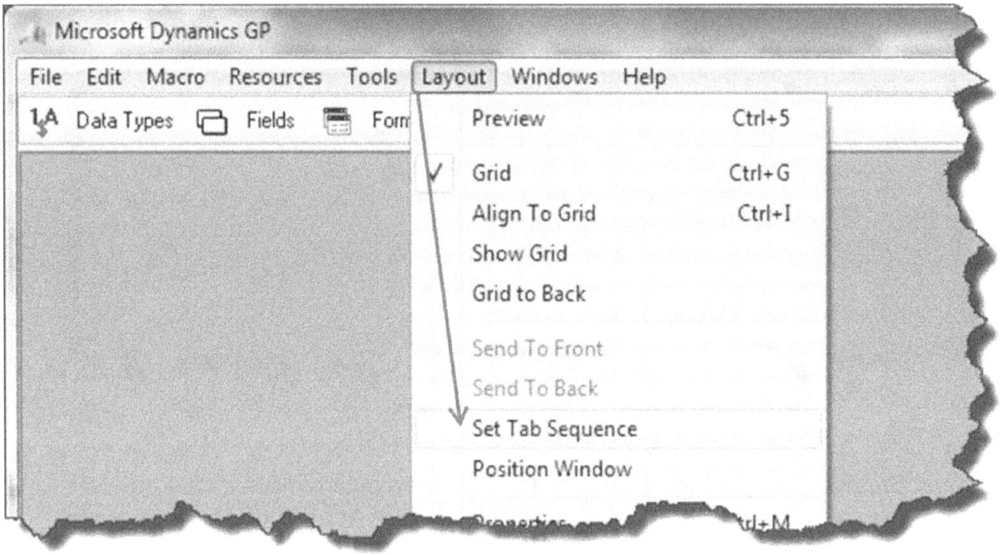

Double-click on the first field you want in the tab sequence and then press the *Tab* key. Even if the cursor lands on the first field, double-click on it and press *Tab*. Double-click on the second field in the tab sequence and press *Tab* again. Double-click on the third field in the tab sequence and press *Tab* again. Continue this until you finish the entire tab sequence.

In order to test your tab settings, press *Ctrl + 5* to open the **Preview** window. The cursor should be sitting in the first field of your tab order. Tab around the window and see if it behaves the way you want.

It is often difficult to get the hang of this in the beginning. The key is to remember to double-click on the field and to ALWAYS press the *Tab* key even if it advances to the field you want. You are probably wondering if you can set the tab sequence by putting an index in the **Properties** window—you cannot. You can only say whether or not the field is included in the tab sequence. You can see the **TabStop** property in the following screenshot; **TabStop** is one of the few **Object** properties you can change in the Modifier:

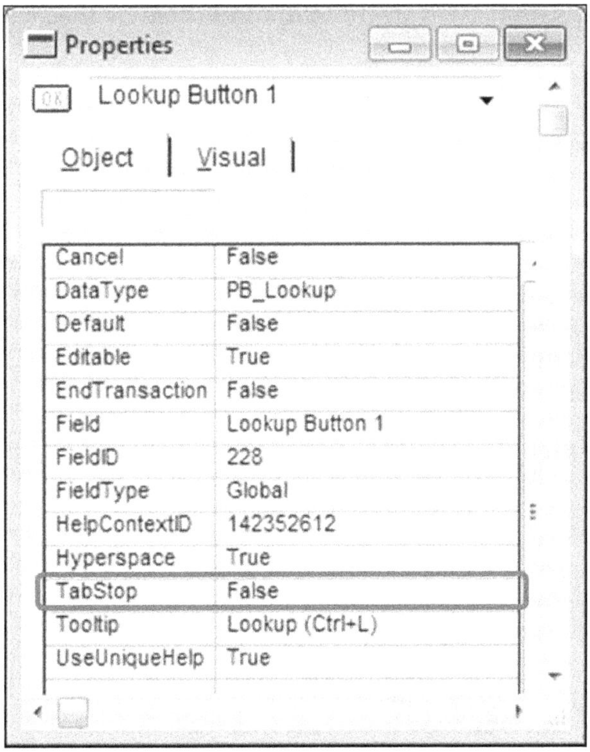

Give it a try! Set the tab sequence on the **Buyer Maintenance** window as indicated on the following screenshot:

The following table summarizes the tab stops indicated by the screenshot:

Tab Stop	Field
1	**Description**
2	**Sort by** (drop-down list)
3	The **Remove** button
4	The **OK** button
5	**Buyer ID**

This is a crazy tab sequence, but it will make it easy for you to see if your changes were properly executed.

When you finish with the tab order exercise, close the Buyer Maintenance window, and return to Dynamics GP using **File** | **Microsoft Dynamics GP**.

> Do not save the changes you made to the **Buyer Maintenance** window.

The window layout

A window layout refers to a window that you have open in the Modifier. Heretofore, you have used the *Ctrl* + *F10* keyboard shortcut, or the **Modify Current Window** menu item to open the Modifier. This time, you are going to launch Modifier through the main menu system: **Microsoft Dynamics GP** | **Tools** | **Customize** | **Modifier**.

The first thing that comes up is a dialog box asking you which product you want to work with. If you expand the drop-down list, you will see each of the products listed in the `Dynamics.set` file (the launch file).

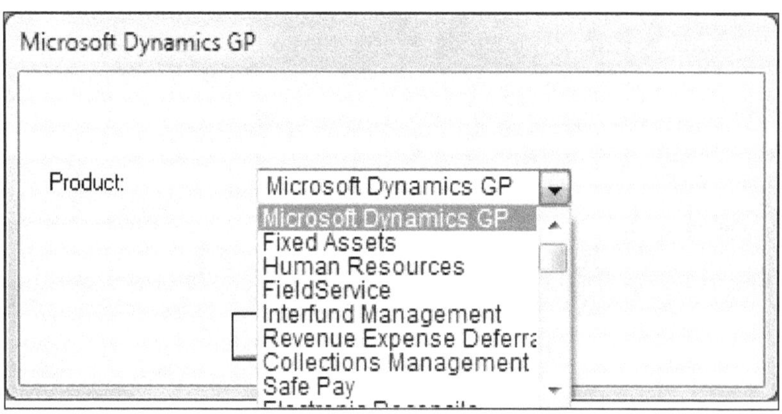

If you cannot make any sense out of the order in which the products are listed, do not feel alone. The products are actually listed in the order in which they appear in the `Dynamics.set` file. The descriptions come from the `Dynamics.set` file as well.

Select **Microsoft Dynamics GP**, and click on the **OK** button.

If this is the first time anyone has opened the Modifier, there will be a slight lag while the system creates the dictionary to hold the modified reports. In the case of Dynamics GP, that dictionary is named `FORMS.DIC`. You set the name of your application's modified forms dictionary when you create your Chunk (`.cnk`) file.

When the Modifier opens, you will see a gray desktop with a single window named **Modified Forms**. As you can see from the following screenshot, the Modifier tool can be a lonely place the first time you go into it:

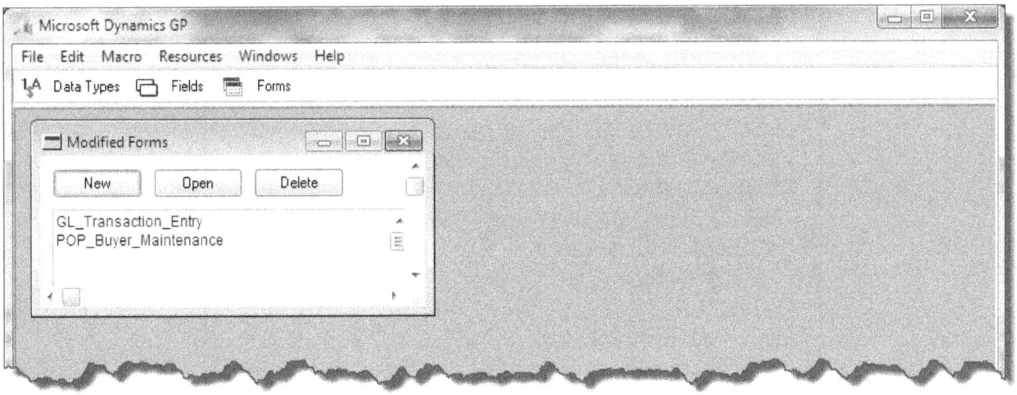

If you opened the **GL Transaction Entry** window and the **Buyer Maintenance** window earlier in this section, your Modifier window will look similar to the preceding screenshot. If you did not open these two windows in Modifier, the **Modified Forms** window will be empty.

When you use the *Ctrl + F10* option to open the Modifier, the product and window are selected automatically, and the target window is opened, revealing its layout. If you open Modifier from the menu as we did above, and you want to modify a window that is not listed, you need to click on the **New** button on the **Modified Forms** window, and then pick the form yourself.

The window that opens when you click on the previously mentioned **New** button, is the **Select a form to modify** window. Looking at the following screenshot, you can see a list of names that may be unfamiliar to you:

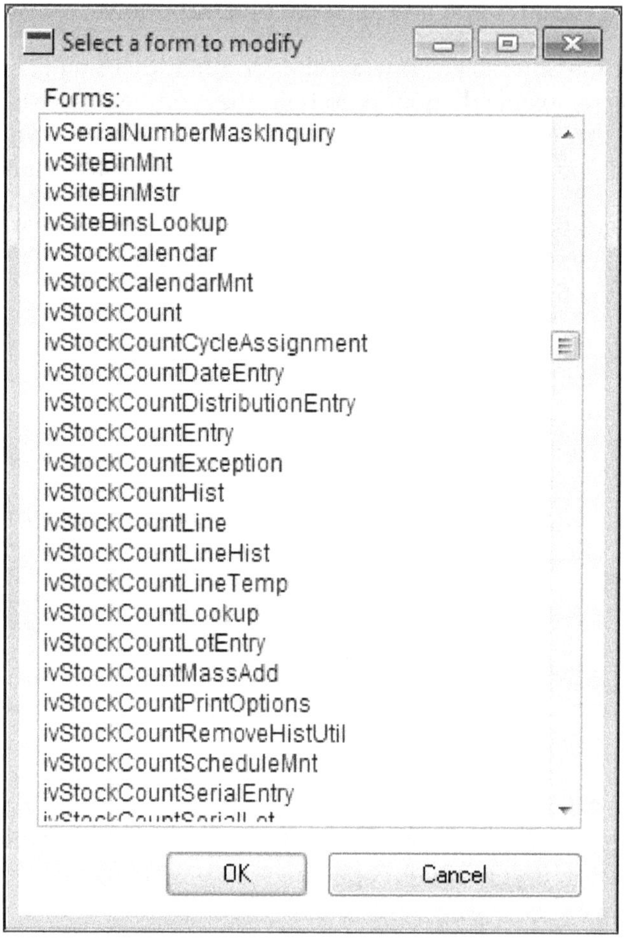

They are unfamiliar because the list contains all the forms included in the `DYNAMICS.DIC` file. There are nearly 1,700 of them. It's easy to appreciate the *Ctrl+F10* shortcut after you spend time searching through a seemingly endless list of forms.

You may have noticed that we were referring to the objects as **windows** earlier, but here everything says **forms**. What's the difference? You may remember from our discussion in *Chapter 3, Getting Started with Dexterity,* that forms contain windows. The window is not an object on its own; all windows live on forms. For instance, if you want to modify the **Customer Maintenance** window, you need to modify the `RM_Customer_Maintenance` form.

For our project, you want to modify the **Transaction Entry** window. Therefore, you need to open the `GL_Transaction_Entry` form. If your **Modified Forms** window lists the `GL_Transaction_Entry` form, select it and click the **Open** button. You can also double-click on it to open it.

If your **Modified Forms** window does not list the `GL_Transaction_Entry` form, click the **New** button and pick the `GL_Transaction_Entry` form from the list.

At this point, the **Form Definition** window should be open. The form definition window listing the **GL_Transaction_Entry** form is shown in the following screenshot:

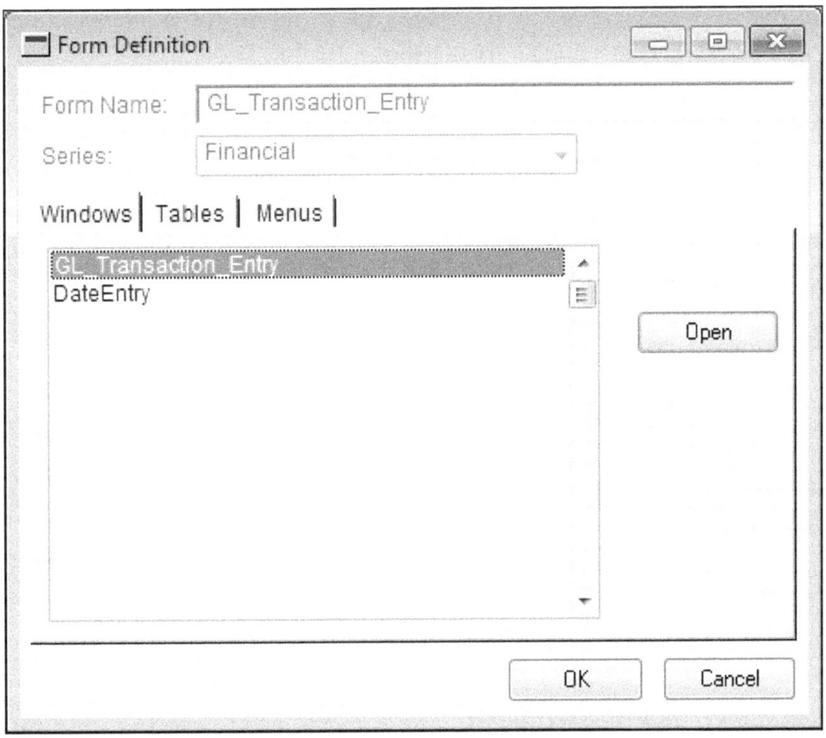

Select the **GL_Transaction_Entry** window, and click the **Open** button; the window layout will open. Your Modifier desktop will look similar to the following screenshot:

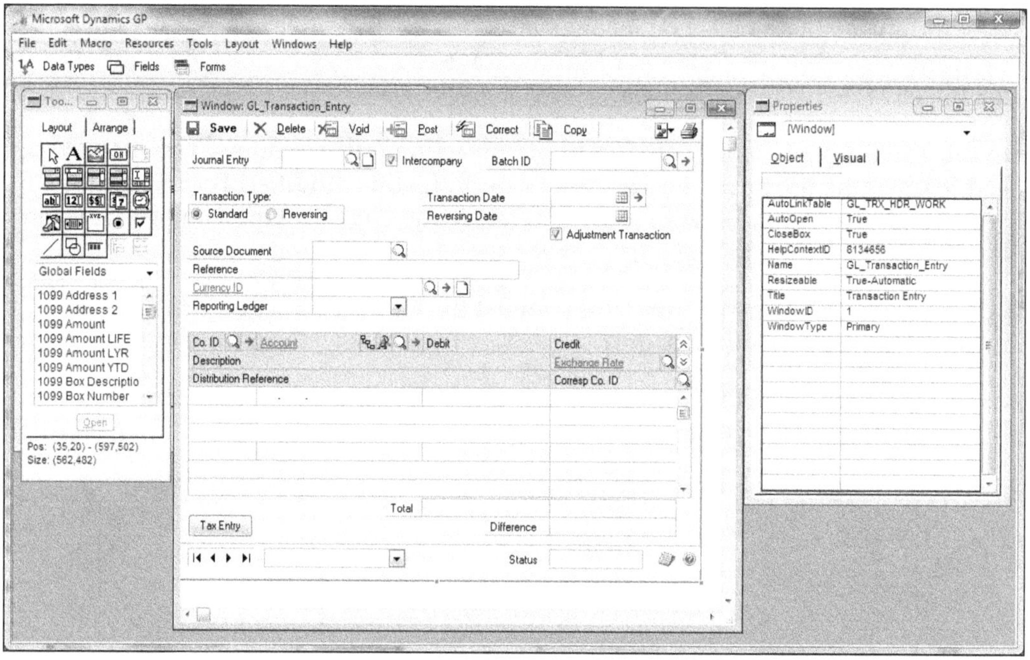

A little housekeeping before we get started—let's take a quick tour of the **Layout** menu, shown in the following screenshot:

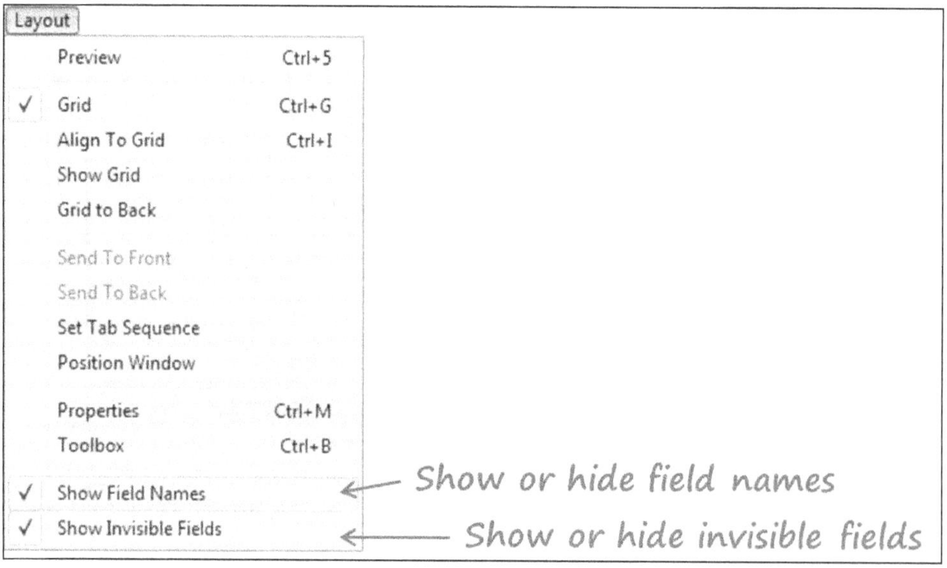

You will use the bottom two choices, **Show Field Names** and **Show Invisible Fields**, frequently. Toggle each of them on and off, and look at how the layout window changes. With both of these menus selected, as they are in this screenshot, your layout window will look similar to the following screenshot:

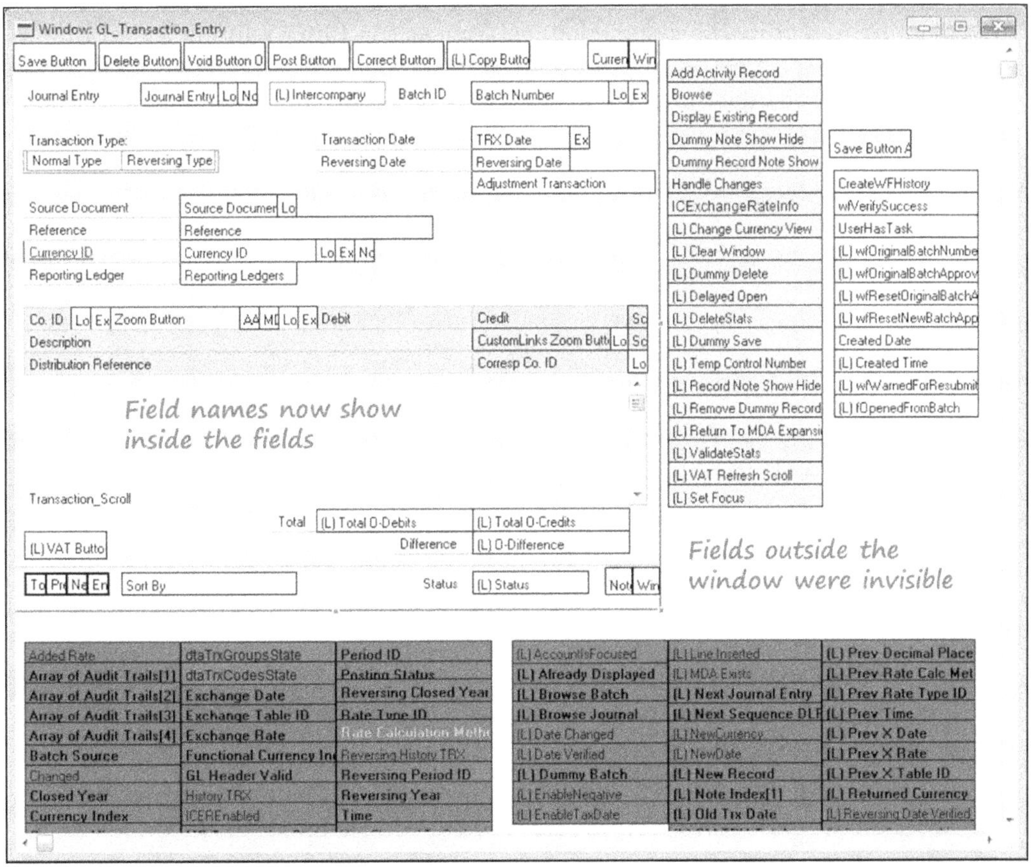

Some of the fields are in different colors, depending on the standards set by the Dexterity programmers who created this window. The color standards for invisible fields are as follows:

- **Yellow background**: Code is attached to this field
- **Pink background**: There is no code attached, but the field may be used to store an interim value
- **Teal background**: The field is a composite field used as an object
- **Bright green background**: The field is a reference field

Leave the **Show Invisible Fields** option checked, and uncheck the **Show Field Names** option. Now you are ready to modify your window's layout.

Modifying the General Entry window

You are going to modify the general ledger **Transaction Entry** window by adding fields and their associated prompts to both the main and the scrolling window. In addition, you will change the tab sequence of the window and adjust the location of certain fields.

When you finish your modifications, your window should look similar to the following screenshot:

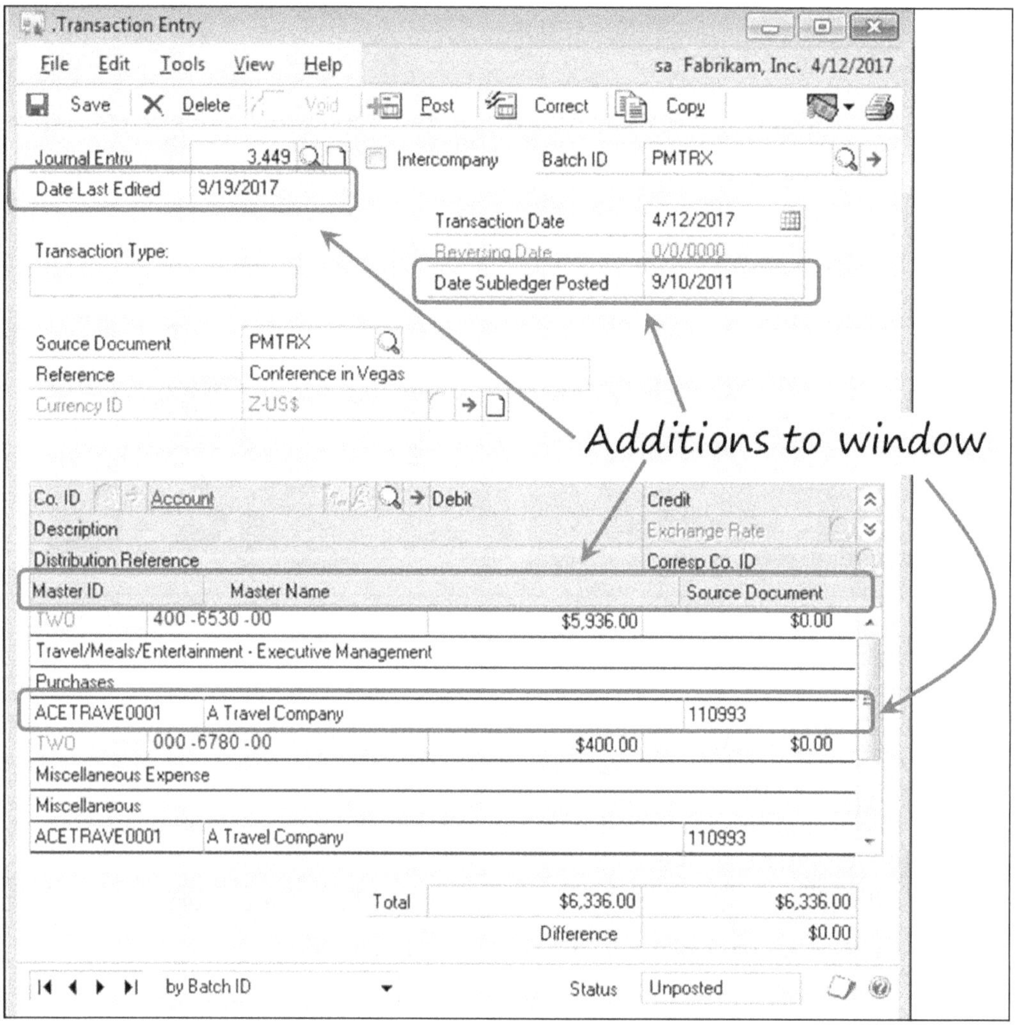

For perspective, in the following screenshot, you can see the original **Transaction Entry** window on the left, and the modified **Transaction Entry** window on the right:

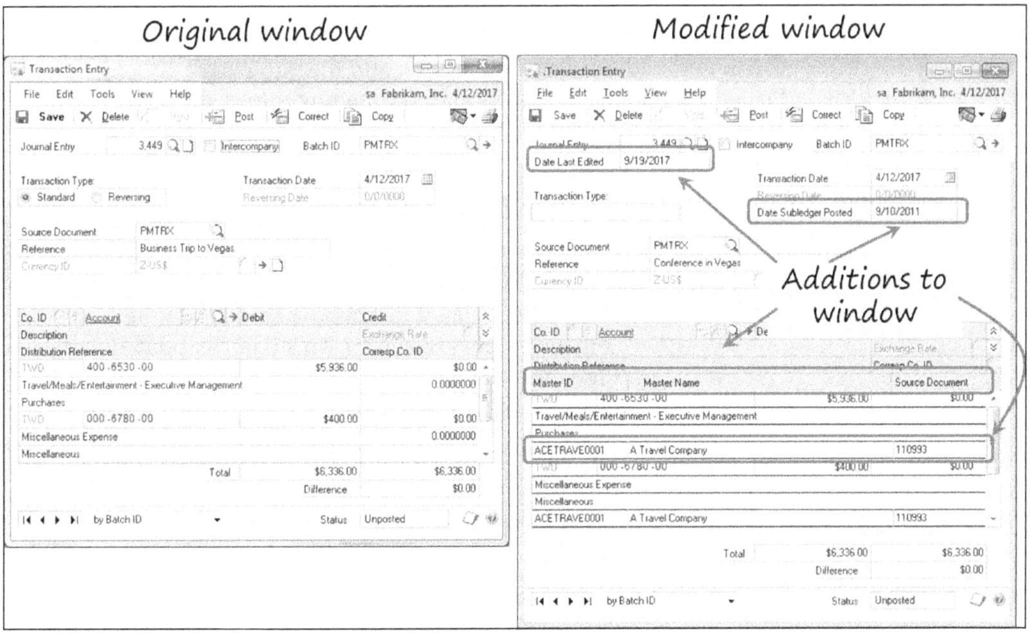

While the details are difficult to distinguish, you can see that your modified window will be a little longer than the original. It is longer because your modified scrolling window will have four rows per line item instead of three. In addition, you are adding a new field below the **Journal Entry** field.

Let's begin modifying the **Transaction Entry** window, and learn some key concepts about Modifier along the way.

You will be adding the following objects to the window in the specified locations:

Window Location	Field	Field Prompt
Transaction Entry main window	Last Date Edited	**Date Last Edited**
Transaction Entry main window	Originating Posted Date	**Sub ledger Posted Date**
Scrolling window	Originating Master ID	**Column heading: Master ID**
Scrolling window	Originating Master Name	**Column heading: Master Name**
Scrolling window	Originating Document Number	**Column heading: Source Document**

The column headings describing the fields added to the scrolling window are on the Transaction Entry screen above the scrolling window. They are not located on the scrolling window itself.

> When you add the Date Last Edited and the Sub ledger Posted Date to the Transaction Entry main window, don't forget to link your prompts.

The added fields are described in the following table:

Field Name	Field Description
Last Date Edited	The system date of the computer the last time someone edited the entry. This date is updated each time the entry is edited.
Originating Posted Date	The date the subsidiary ledger transaction was posted
Originating Document Number	The document number of the subsidiary ledger transaction
Originating Master ID	The Vendor ID, Customer ID, Checkbook ID, and so on, from the subsidiary ledger transaction
Originating Master Name	The name associated with the Originating Master ID described previously

A **subsidiary ledger** is a module that sends transactions to the general ledger. Examples of subsidiary ledgers include:

- Payables Management
- Receivables Management
- Bank Reconciliation
- Payroll
- Fixed Assets

Adding and modifying window fields

First, you are going to add the date on which the entry was posted to the sub-ledger. Put it underneath the **Reversing Date** field. To see this field, switch the **Show Field Names** menu item back on by selecting **Layout | Show Field Names**. You'll need to move the **Adjustment Transaction** field out of the way. Move it to underneath the **(L)Intercompany** field. Use the **Visual** tab of the Properties window to make changes to the field's location:

Adjustment Transaction Field	
Position-Left	223
Position-Top	53

To add the field, drag the **Originating Posted Date** field from the **Toolbox** on the left to the layout window, as shown in the following screenshot. Place it below the **Reversing Date** field:

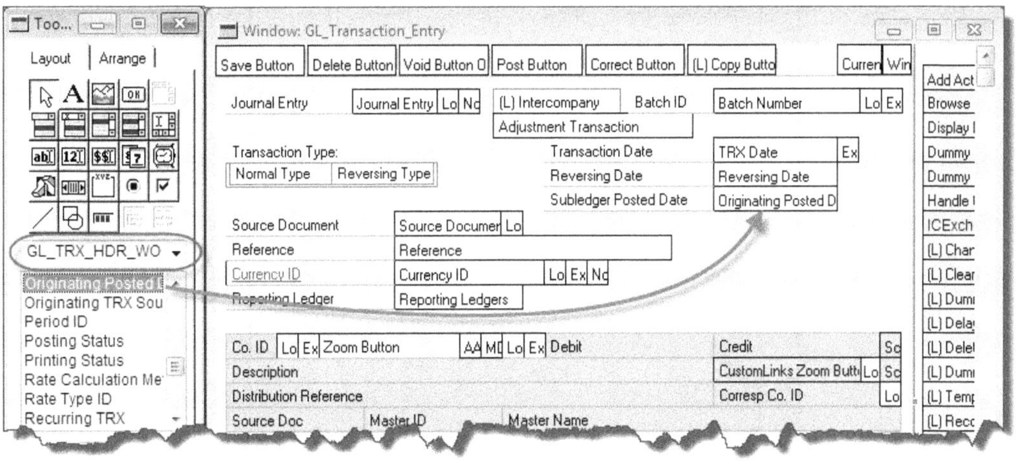

Set the properties of the **Originating Posted Date** field to the following values:

Originating Posted Date	
Position-Left	399
Position-Top	110
Size-Height	20
Size-Width	102

Add the **Sub ledger Posted Date** field prompt, and set the properties according to the following screenshot:

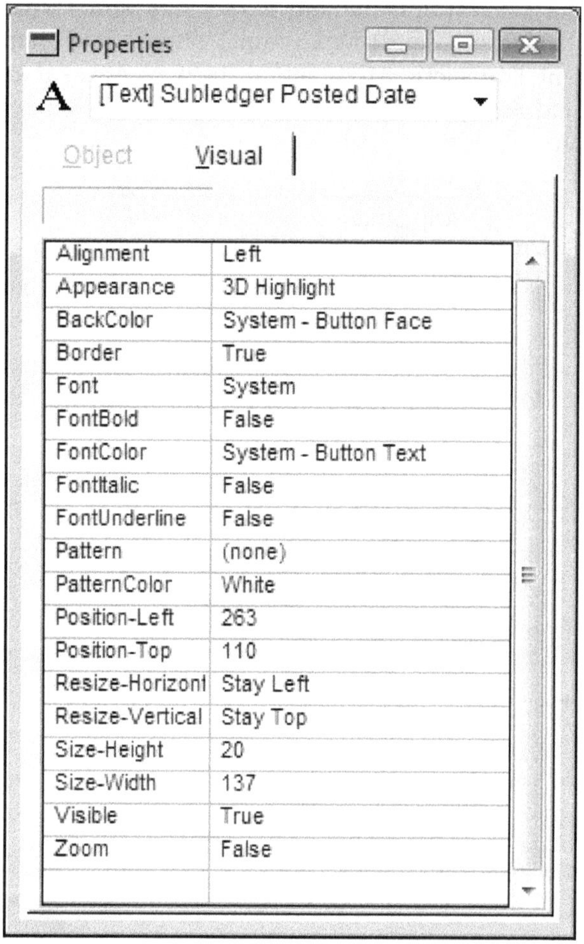

Continue by adding the **Last Date Edited** field to the window.

Go back to your **Originating Posted Dat**e field, and check its **Object** and **Visual** properties against the **Properties** windows, as show in the following screenshot:

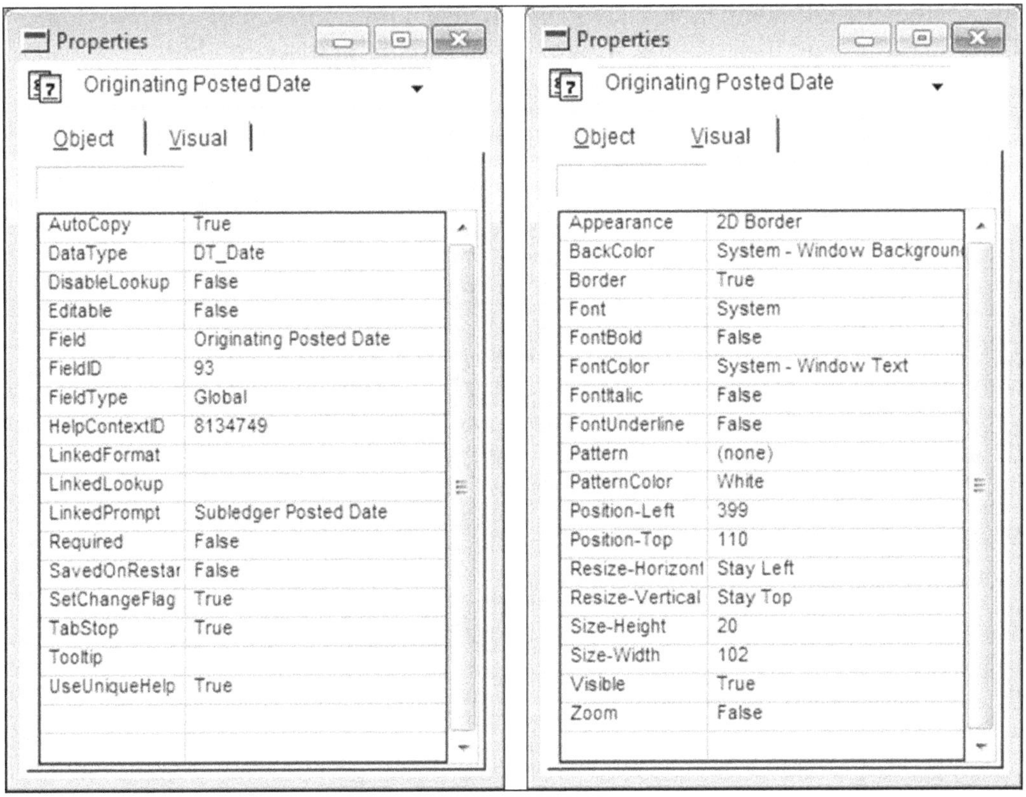

Add your field prompts and link them to the appropriate fields. We covered linking prompts in *Chapter 3, Getting Started with Dexterity*. Finally, set your tab order. We also covered setting the tab order in *Chapter 3*.

Adding fields to the scrolling window

Next, you are going to add three fields to the scrolling window. You need to make the **Transaction Entry** window bigger, and make the scrolling window longer to accommodate the extra fields. Move the other fields around so they fit underneath the larger scrolling window.

As a guide, the following table shows certain visual properties of the **Transaction Entry** window, and the scrolling window object on the **Transaction Entry** window:

Transaction Entry window	
Position-Left	`285`
Position-Top	`17`
Size-Height	`558`
Size-Width	`562`

Scrolling window object	
Position-Left	`8`
Position-Top	`300`
Size-Height	`154`
Size-Width	`528`

Double-click on the scrolling window to open it. We need a fourth line in the scrolling window, so we need to move the **Big Line Item** down one row. The trick to getting the big line item in the right place is to click *inside* the last row (marked by a dotted line). So many people try to click the big line item under the last row, or try to click right on the line, and that just doesn't work.

You cannot drag the Big Line Item into the right position; you must use the **Tools** menu and the mouse.

The following screenshot shows where to position the cursor to get the big line item to move down one row:

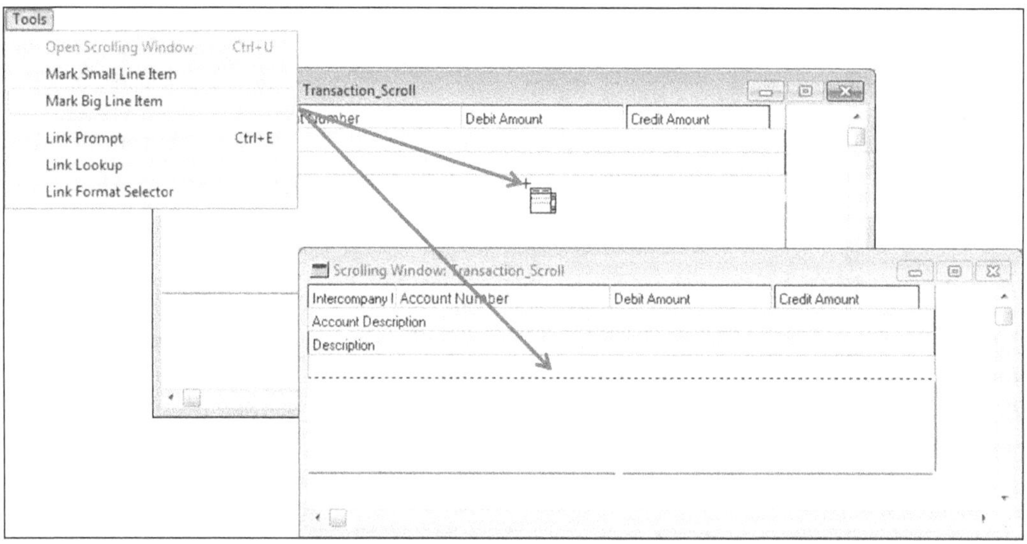

From the **Toolbox** window, drag the three fields from the **GL_TRX_LINE_WORK** table on to the scrolling window as shown in the following screenshot:

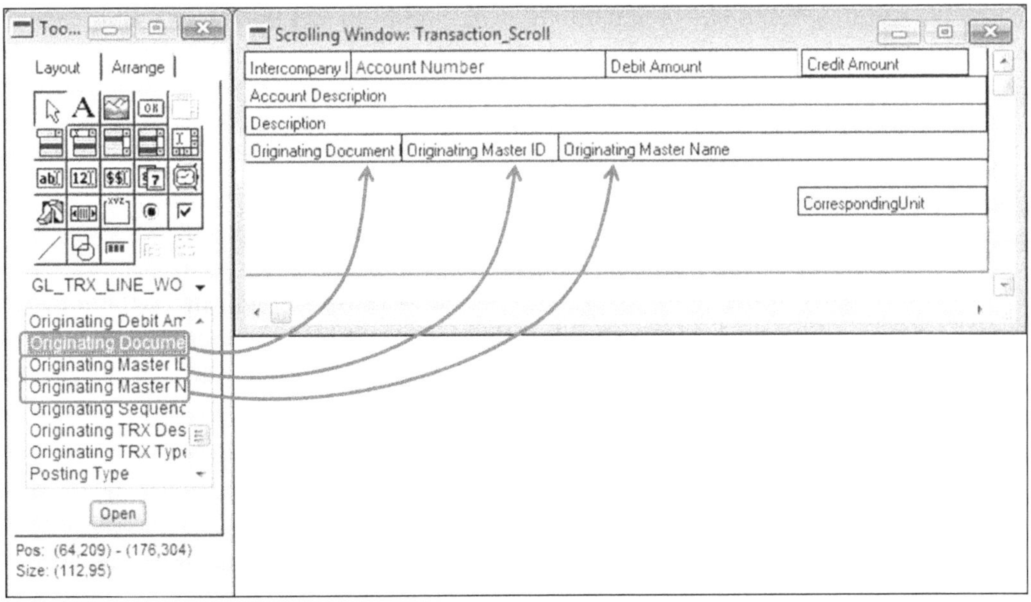

Set the properties for each of the fields to the values specified in the following table:

Originating Document Number	
BackColor	`Yellow`
Border	`False`
Pattern	`75% Shading`

Originating Master ID	
BackColor	`Yellow`
Border	`False`
Pattern	`75% Shading`

Originating Master Name	
BackColor	`Yellow`
Border	`False`
Pattern	`75% Shading`

An important part of completing a scrolling window is to make sure you have created a visual grid by using the line tool in the **Toolbox** window. Your fields will not show borders, so you need to provide the lines. Also, be sure you set the tab order of the fields inside the scrolling window.

Close the scrolling window, save your changes, and return to the **Transaction Entry** window.

The final step is to add the column headings for the new fields you added to the scrolling window. If you are showing field names in the layout, it's nearly impossible to get the headings properly lined up with the columns. Change your layout so that field names are not being displayed. If you did your grid right, you should be able to see the lines in the scrolling window. Now, you can easily line up your column headings.

Finish your modifications and return to GP. Go to **Microsoft Dynamics GP menu** | **Tools** | **Setup** | **System** | **Alternate/Modified Forms and Reports** and change the **Alternate/Modified Forms and Reports** settings to point to the modified window as shown in the following screenshot:

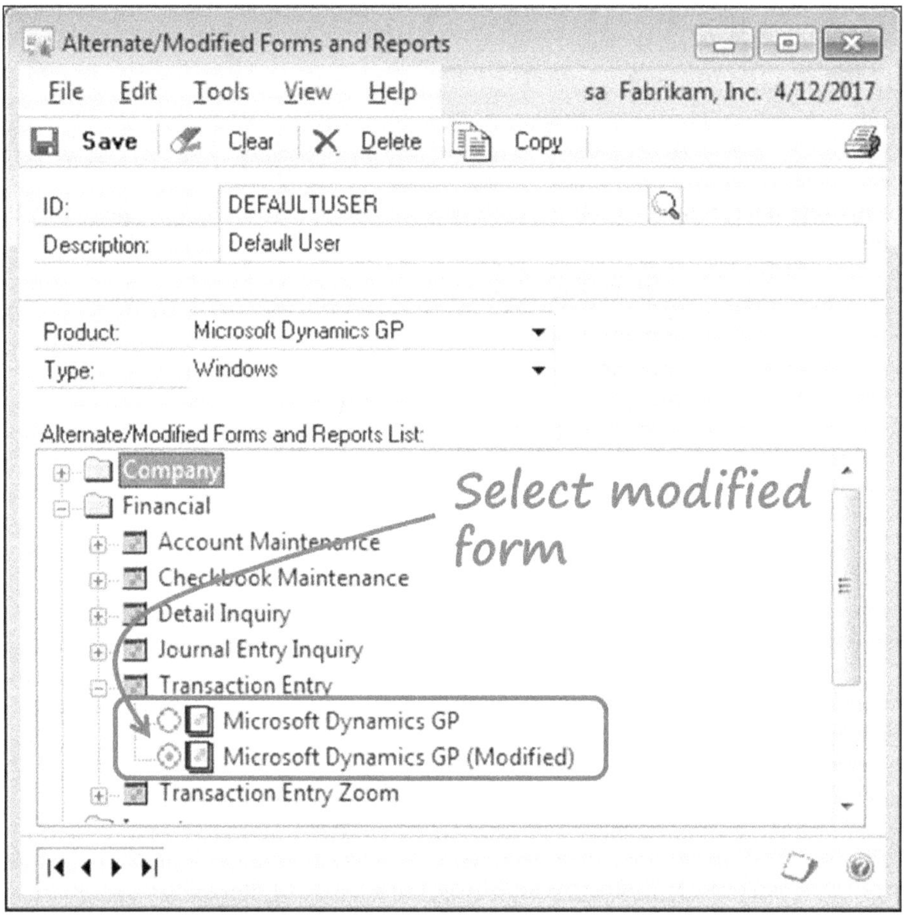

Open the **General Entry** window and check out its new look!

Modifying static text

A window prompt is a piece of static text that normally describes something; often that thing is a field. If it were in front of a field, you would call it a field prompt. You may want to modify these prompts to reflect terminology more appropriate for your business, or to clarify ambiguous terms. You also may want to modify text such as window titles, words on push buttons, and items in drop-down lists.

As an example of something often changed, let's use the **Paid Sales Transaction Removal** window in the **Receivables Management** module. This window has often been the source of confusion—What does "remove" mean anyway? Remove it to where?

In order to open the **Paid Sales Transaction Removal** window, navigate to **Sales** | **Routines** | **Paid Transaction Removal**. The following window will open:

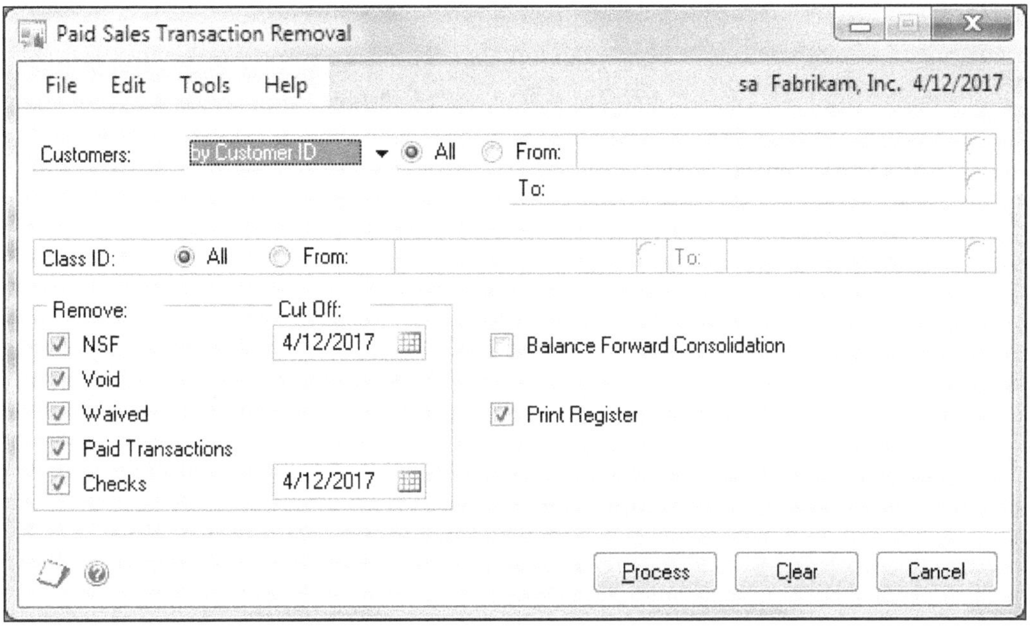

It's a nice compact window, but not terribly self-evident as to its function. Let's think of this as the *before* window. You can make it better!

In Dynamics GP, you would use this window to transfer closed, completed, or finished receivables transactions to the history file. Although this should be a part of the monthly closing, your users may not run it for fear that they might remove the transactions forever.

You're going to transform this window to make it friendlier and more understandable. The following screenshot shows your new *after* window:

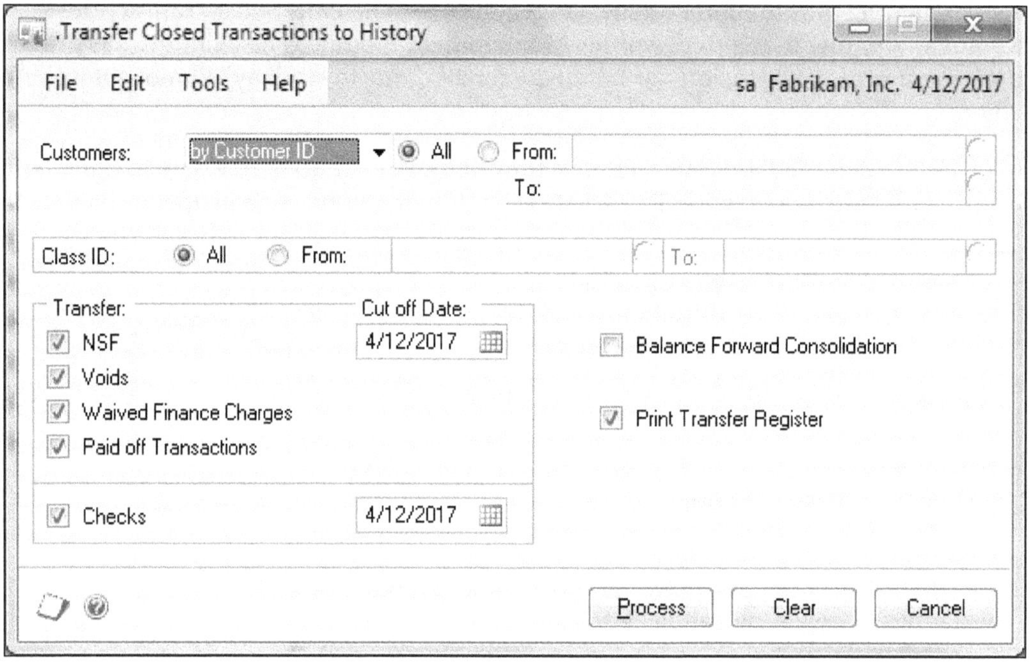

The window elements you are going to change are listed below, followed by instructions on how you can change them:

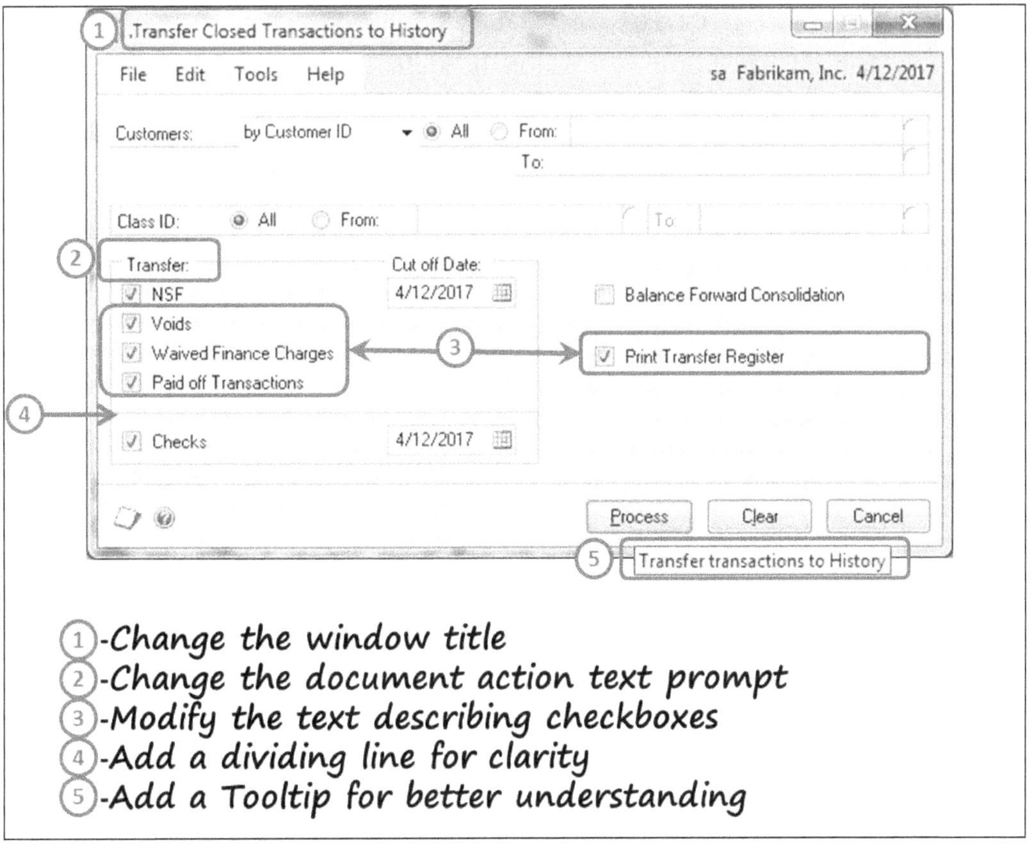

[Press *Ctrl* + *F10* to open this window in Modifier.]

Change the window title

The window title is what your users see in the top band of the window. You change the title in the **Object** tab of the **Properties** window.

Change the text prompt

In order to change the text prompt, click on the **Static Text** tool in the **Toolbox** window, as shown in the following screenshot:

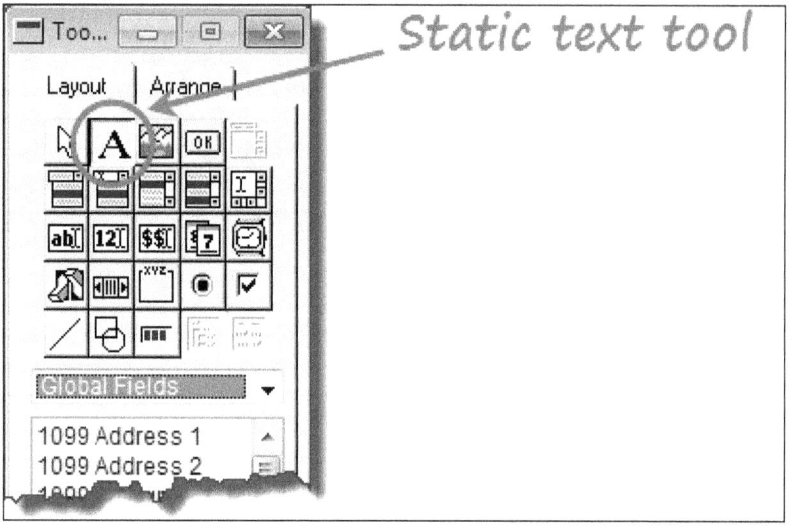

The mouse pointer will turn into a text pointer as in the following screenshot. Highlight the text you want to change, and type in the changes:

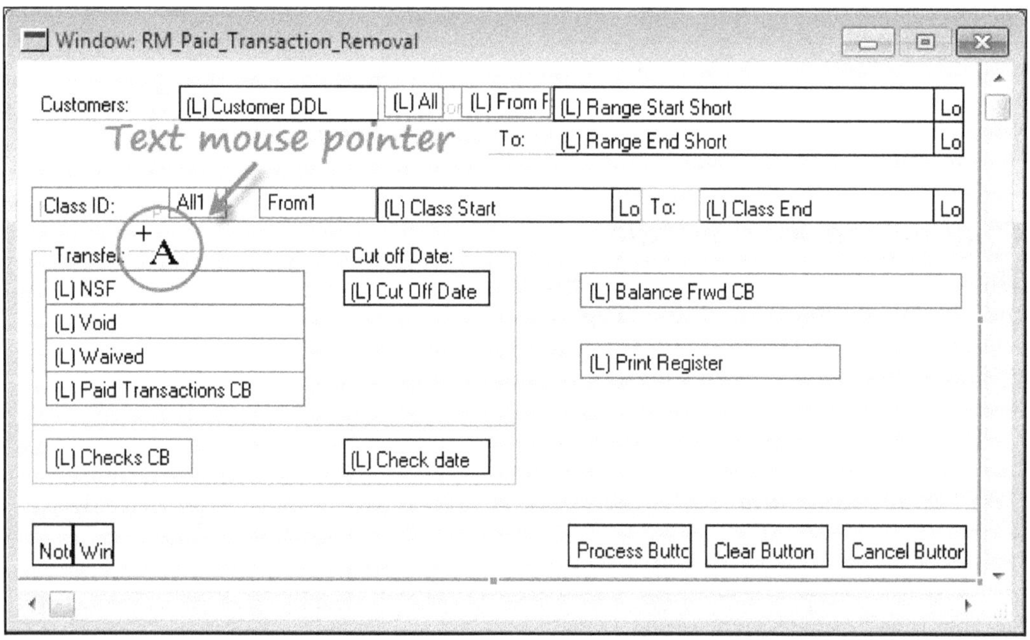

Change the static text of a checkbox

The text that appears to the right of the checkbox is a part of the checkbox field itself rather than a word written on the screen. Similarly, the words and images appearing on push buttons, the items on drop-down lists, the images on visual switches, and the labels for radio buttons are all static values. You can change these values in Modifier by changing the **DataType** attached to the field object.

In this section, you'll change the text for a checkbox on the **Paid Transaction Removal** window to something more desirable.

Follow the instructions as numbered in the following screenshot to change the text of the **(L)Waived** field checkbox:

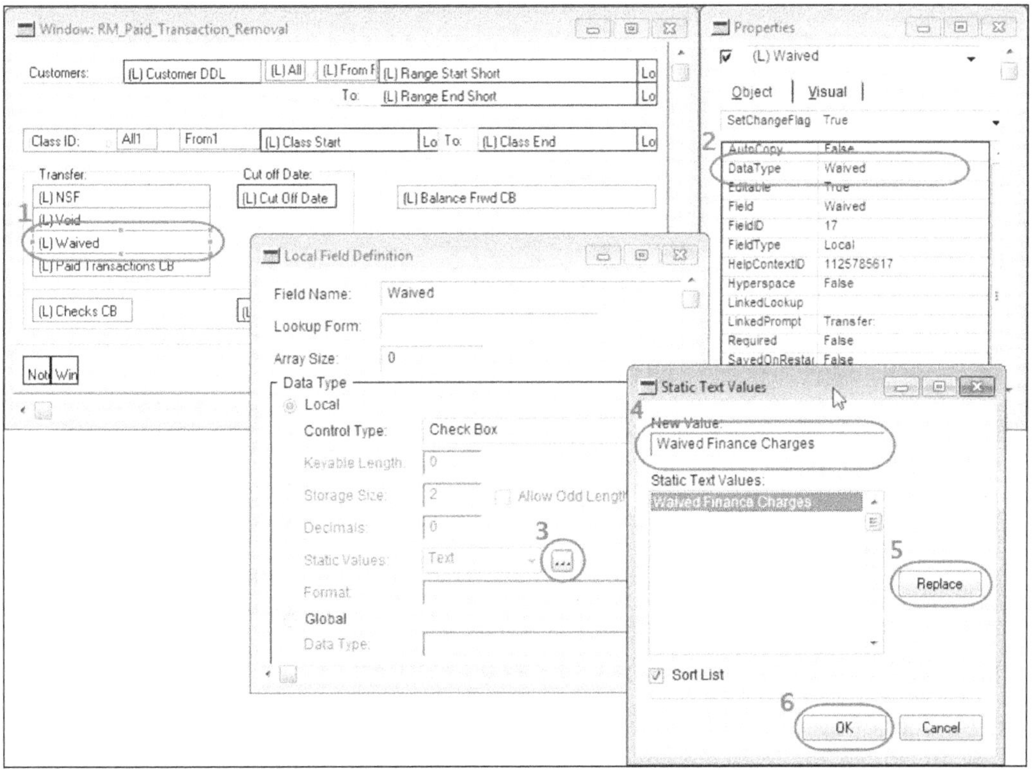

1. Select the **(L)Waived** field on the layout window.
2. Double-click on the **DataType** property in the **Properties** window.
3. Click on the ellipses button on the **Local Field Definition** window.
4. Click on the static text value in the **Static Text Values:** window, and change the value to `Waived Finance Charges` in the **New Value:** field.

5. Click on the **Replace** button.
6. Click on the **OK** button.
7. Press *Ctrl* + *5* to preview your changes.

Add a dividing line

The section of the screen containing the checkboxes is confusing, in that you cannot easily tell which dates are associated with which checkboxes. In order to more clearly identify the bottom date field as pertaining to the **Checks** item, use the line tool to draw a horizontal line separating the checkboxes.

The line tool is located in the **Toolbox** window.When you select the line tool, the mouse pointer changes to the image identified in the following screenshot:

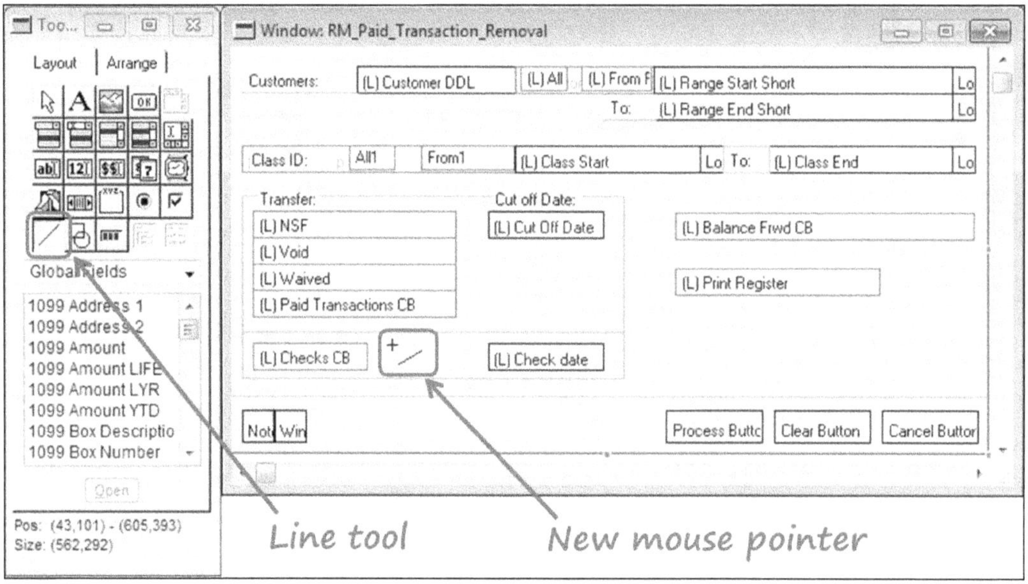

Simply draw the line where you want to set off the area.

To the right of the line tool you'll find the shape tool. You can use the shape tool to draw an ellipse (circle), rectangle, and rounded rectangle on a window.

Adding a tool tip

Adding a tool tip is easy, and fun! It's fun because you can surprise your users with your own clever message when they hover over a field. This text doesn't wander into those treacherous waters, so you will add the ho-hum tool tip `Transfer transactions to history` to the **Process** button. To add the tool tip, follow the instructions following the screenshot below:

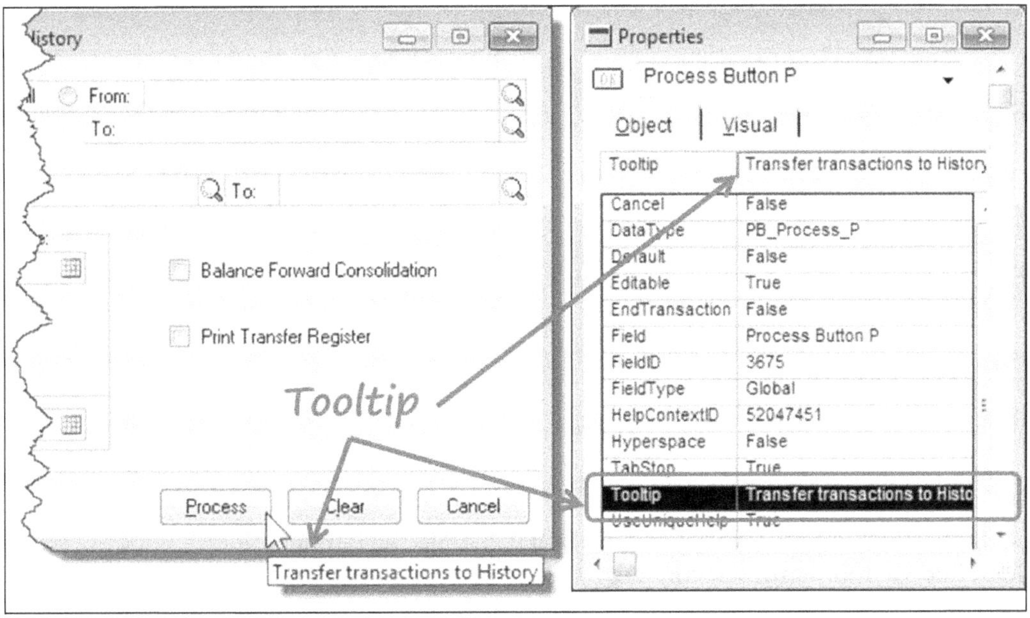

1. Select the **Process** button.
2. Select **Tooltip** from the **Object** tab of the **Properties** window.
3. In the **Settings** box, type the text for your tool tip.

Check your work by pressing the *Ctrl + 5* shortcut key.

Adding or changing graphic elements

Graphic elements is a fancy way of saying pictures. Let's say you want to add a picture, such as a company logo, to one of your windows (or all of your windows!). Maybe you've already added a picture, and now the picture has changed. If you have that same picture on thirty windows, changing the picture by bringing up each window will get real old, real fast.

In this section, you are first going to add a banner to the **Master Posting** window. After that, you are going to change the banner by changing the picture in the picture library. When you open the **Master Posting** window, you will see that the original banner has changed to the new picture. Basic sounding, but very powerful when you are trying to design a theme for your windows.

The first banner you add to the window will look like the one in the following screenshot:

Without touching the window again, you'll change the banner and the window will look like the following screenshot:

Adding a new picture

Start out by creating a simple WordArt object. We created the object above using a `48` point, `Calibri` font.

Open the **Master Posting** window from **Administration** | **Routines** | **Master Posting**.

Press *Ctrl* + *F10* to open the window in Modifier. Make the window longer to make room for the banner, and move all the fields on the window towards the bottom, leaving some empty real estate at the top. We set the **Visual** properties of the window to the following:

Property	Setting
Size-Height	`482`
Size-Width	`647`

We moved the fields down so that the **Series** field was `148` pixels from the top. Do whatever is pleasing to you, but these measurements will work for this example.

Once you have your window stretched out and your fields moved, follow these steps to add the banner:

1. Select and copy your WordArt object to your clipboard, and return to Modifier.
2. Select the picture tool from the **Toolbox** window; the mouse pointer will change into an outline of the picture tool as shown in the following screenshot:

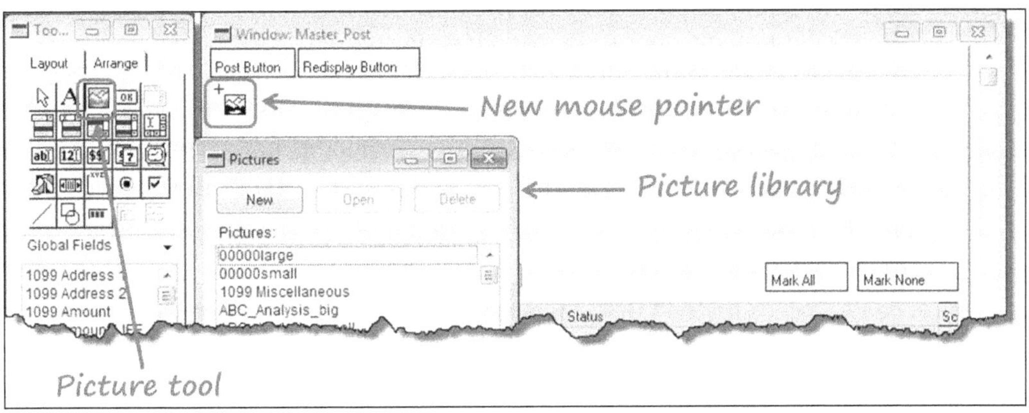

3. Click on the upper left-hand corner of the **Master Posting** window in the position where you would like the picture to begin. The **Pictures** window will open. This window displays a list of items—this is the **picture library**.
4. Select the **New** button on the **Pictures** window, and the **Picture Definition** window will open.
5. Type `Master Posting Banner` in the **Picture Name:** field, and then select the picture area of the window as highlighted in the following screenshot:

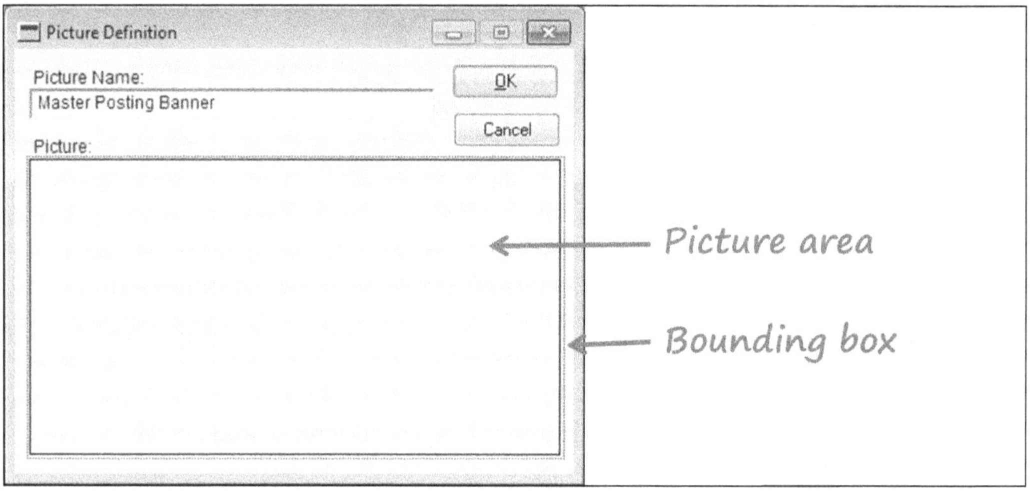

6. When you select the picture area, a bounding box will outline it. The bounding box appears as a dotted line going all the way around the rectangle.
7. From the **Edit** menu, choose **Paste** to transfer the WordArt object you previously copied to the clipboard. Once you paste the image, you cannot undo it. If you make a mistake, just click on **Cancel**, and try again.
8. After you paste your WordArt object, click on the **OK** button. Your **Picture Definition** window will close, and you will return to the **Pictures** window.
9. In the **Pictures** window, scroll down until you find the **Master Posting Banner** item, and select it.
10. When you click on the **OK** button, you will close the **Pictures** window and your image will appear in the heading of the **Master Posting** window similar to the following image:

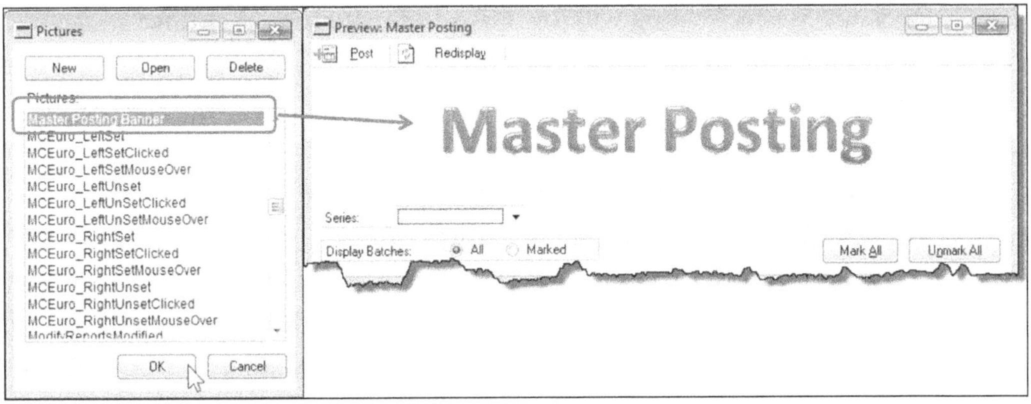

11. Close and save your window, and then return to Dynamics GP.
12. Change the **ID** field in the **Alternate/Modified Forms and Reports** settings to match the value in the following screenshot:

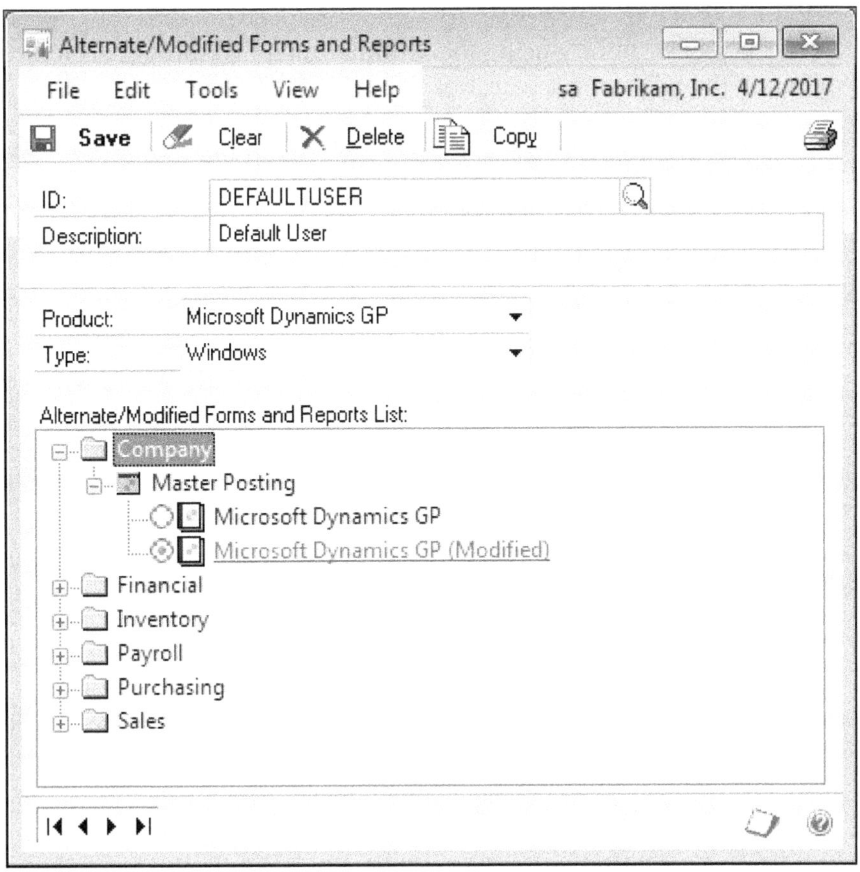

13. Open the **Master Posting** window via **Administration** | **Routines** | **Master Posting**.
14. Admire your new window!

Changing the Picture

Now you are going to change the banner by changing the picture. You won't open the window again; you will simply edit the picture stored in the picture library. When you change the picture library, you will change the old picture to the new one on every window that was previously showing the old picture.

Let's do it; follow these steps to change the existing **Master Posting Banner** picture:

1. Go back to your WordArt object and change it to, say, `Super Posting`.
2. Copy your modified WordArt object to your clipboard.
3. Launch Modifier using the application menu via **Microsoft Dynamics GP** | **Tools** | **Customize** | **Modifier**.
4. Open the picture library using the Modifier menu **Resources** | **Pictures...**, as shown in the following screenshot:

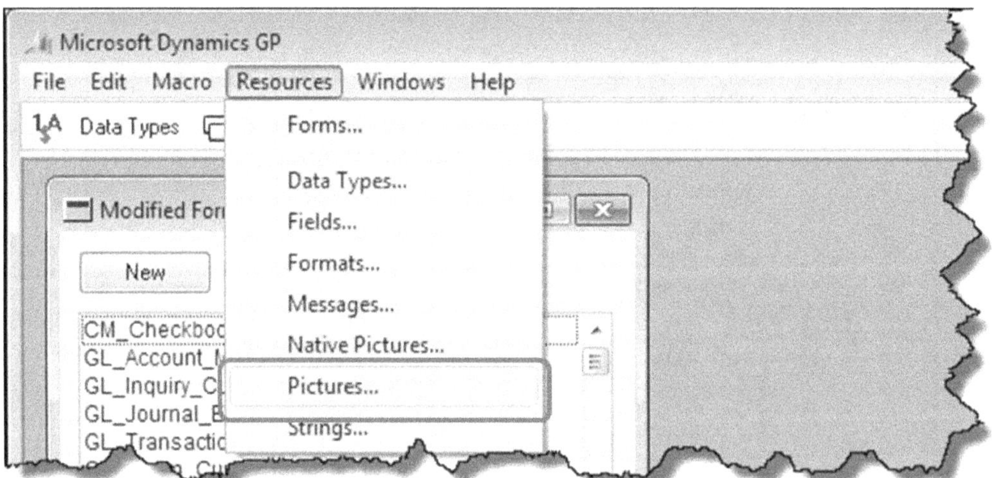

[Be careful not to open **Native Pictures...**.]

5. Find the **Master Posting Banner** item in the picture library, and click on the **Open** button.
6. Click in the **Picture:** area and notice the bounding box surrounding the picture. Your window will look similar to the following window:

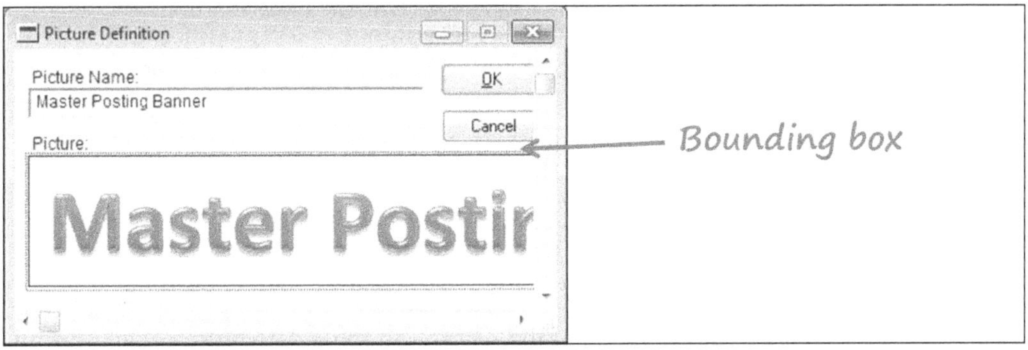

7. From the **Edit** menu, choose **Paste** to replace the picture with the new WordArt object you previously copied to the clipboard.
8. Click on the **OK** button to close the **Picture Definition** window and switch back to Dynamics GP.
9. Open the **Master Posting** window through **Administration** | **Routines** | **Master Posting**.
10. Marvel at how easy it is to change the picture using the Modifier.

Changing global resources

Global resources are available to all objects in the dictionary. If you change a global resource, you will change each object in the dictionary that uses that resource. You just learned how changing the picture you added to the picture library will change the existing picture on the **Master Posting** window.

This section describes how to change existing resources using Modifier.

Pictures and native pictures

You learned in the previous section how to change pictures. The method to change native pictures is the same. So what's the difference between a picture and a native picture? A little Dynamics GP history is the root of this difference.

Use *Alt* + *F10* to launch Modifier. Open the **Native Pictures** window from the menu bar through **Resources** | **Native Pictures**.

Open the **AboutPictureDynamics** item from the **Native Pictures** window. In the **Native Picture Definition** window, you will see the **Native Picture Name**, and a **Synchronized With** option.

Back when Dynamics GP supported the Mac, you needed to create a picture that was compatible with each operating system. You needed to synchronize the Mac pictures with the Windows pictures to be sure they were assigned the same internal ID. By having the same internal ID, you could be sure the correct picture was displayed on each platform.

Not a problem today, but that's why you find most of the old window pictures in the native pictures library.

Today, you will change the picture on the lookup button. It currently sports the image of a looking glass; let's change it to the red X from the delete button.

First, we need to figure out the name of the picture on the delete button. This should be easy because the name of the image on a lookup button is stored in the **DataType** property for the button.

The following are the steps to change the image on the lookup button:

1. Use the *Alt* + *F10* shortcut key to launch the Modifier.
2. Open the **GL_Transaction_Entry** form from the **Modified Forms** window.
3. Open the **GL_Transaction_Entry** window from the **Form Definition** window.
4. Select **Delete Button** on the window layout.
5. On the **Properties** window of **Delete Button**, double-click on the **DataType** object property.
6. Select the ellipses button next to the **Static Values** field on the **Data Type Definition** window.
7. Widen the **Button Items** window so that you can see the **Image Name**. The entire image name is **WindowToolbar_Delete_PB_Up**.
8. Close all the windows opened in Modifier (do not close Modifier).
9. Open the **Pictures** window from the menu bar via **Resources** | **Pictures**.
10. Scroll down and open the **WindowToolbar_Delete_PB_Up** picture.
11. Click on the picture to place the bounding box around the picture. From the **Edit** menu, select **Copy**.
12. Use the same method described in steps 2 through 7 to discover the name of the picture on the lookup button. The name of the picture is **Field_Lookup_PB_Up**.

13. Open the **Pictures** window from the menu bar by going to **Resources** | **Pictures**.
14. Open the **Field_Lookup_PB_Up** picture.
15. Click on the lookup button picture.
16. From the **Edit** menu, select **Paste**. The picture should change to the red X from the **Delete Button** field.
17. Close the **Picture Definition** window and return to Dynamics GP.

Changing a picture resource does not create a modified window. The lookup button image on all windows will now sport the red X instead of the magnifying glass.

Open the **Payables Transaction Entry** window; your window should look similar to the following screenshot:

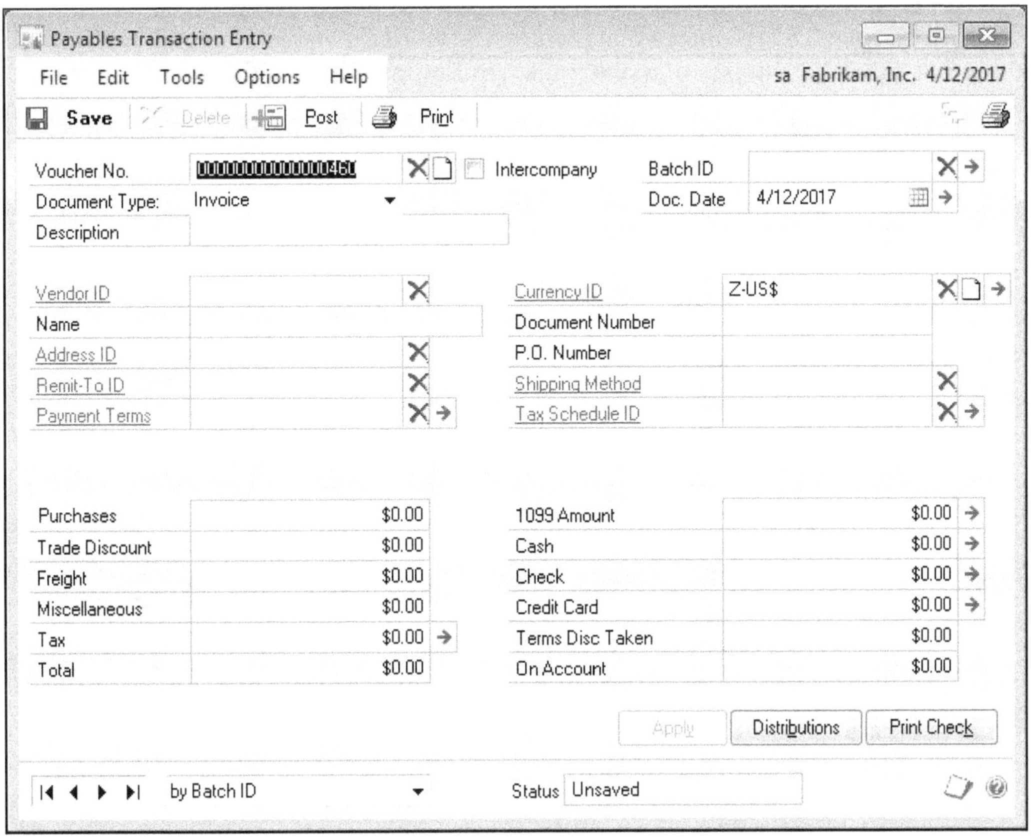

Next, let's look at string resources.

Strings

A string resource describes any static text, wherever it appears in the user interface. If you want `Customer` to read `Client`, change each instance of the word `Customer` to `Client` in the **Strings** window. This sounds simple, but it is not. When you open the **Strings** resource window, notice that there is no search and replace option anywhere in sight. While you are looking for search and replace, see if you can find a revert function—you will find neither.

You can create a *resources report* that will provide you with a list of each of the strings. But this report will just give you a road map; it will not provide a quick access method.

In order to print the resource report, select **File | Generate Resource Reports** from the menu bar. You will create a text report that lists each resource in the `Dynamics.dic` file. Nearly 70,000 resources are listed. When you search the string resources, you will discover 281 resources that include the word *Customer*. Now, all you have to do is find those 281 phrases in the **Strings** window, and change the word `Customer` to `Client`—not an easy task, but doable.

We'll pick something easier. Let's change the word on the delete button from **Delete** to **Go Away**. If you open the **DataType** property for the delete button, you'll see the caption is actually `&Delete`. Therefore, you need to change the string `&Delete`, not **Delete**. Open the **Strings** window and change `&Delete` to `&Go Away`.

Now, when you open any window that contains a delete button, it will look similar to the following screenshot:

Formats

Formats control how data is displayed. Formats put the dash in the phone number; determine how many decimals show up on the invoice, control field alignment, and so on. Changing a format changes every field whose data type includes that format; so be very careful before you change formats.

Your Dexterity skills can come in handy if you are contemplating changing formats or data types. You can use Dexterity to generate the references table, and then check to see which fields you would be changing if you modified a format or data type.

In addition to changing existing formats, you can create your own unique formats and assign them to existing data types. You should exercise the same caution when creating new formats as you do when modifying existing ones; many fields could be using the data type you changed.

The following screenshot shows the **Format Definition** window; you can create and/or modify formats in this window. By selecting a format type, you activate the corresponding options in the window:

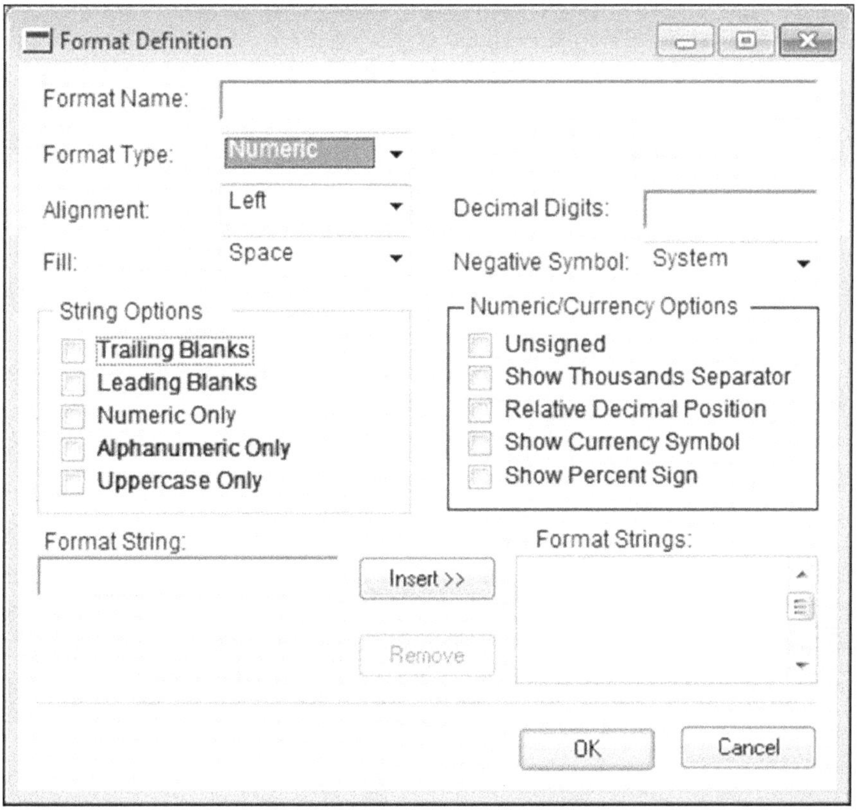

In order to see how this works, let's modify the format associated with the **Customer Number** field such that a dash is included after the eighth character.

To easily find the correct format definition, open the **Customer Maintenance** window in Dynamics GP by going to **Sales** | **Cards** | **Customers**.

Launch Modifier using the *Alt* + *F10* shortcut key. The following steps show how to modify the format:

1. Select the **Customer Number** field on the **RM_Customer_Maintenance** window.
2. Double-click on the **DataType** property under the **Object** tab of the **Properties** window, to open the **Data Type Definition** window.
3. Click on the ellipses button on the **Data Type Definition** window as seen in the following screenshot:

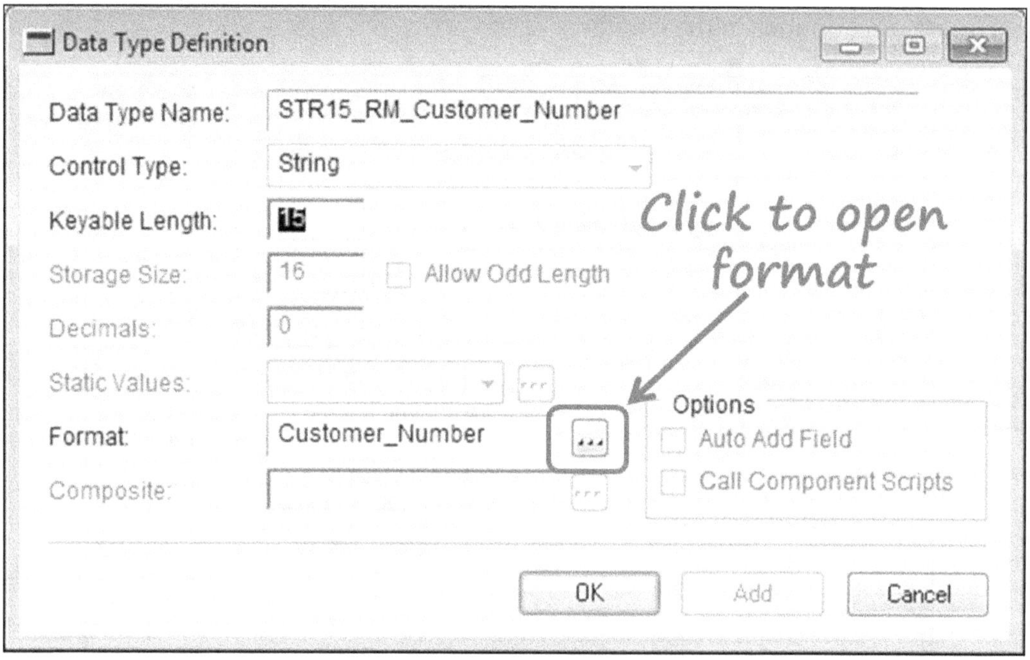

4. When you click on the button, you will open the **Format Lookup** window. The **Customer_Number** format will be selected when the **Format Lookup** window opens.

5. Click on the **Open** button on the **Format Lookup** window to open the **Customer_Number Format Definition** window displayed in the following screenshot. It's here that you'll add the format to include the dash.

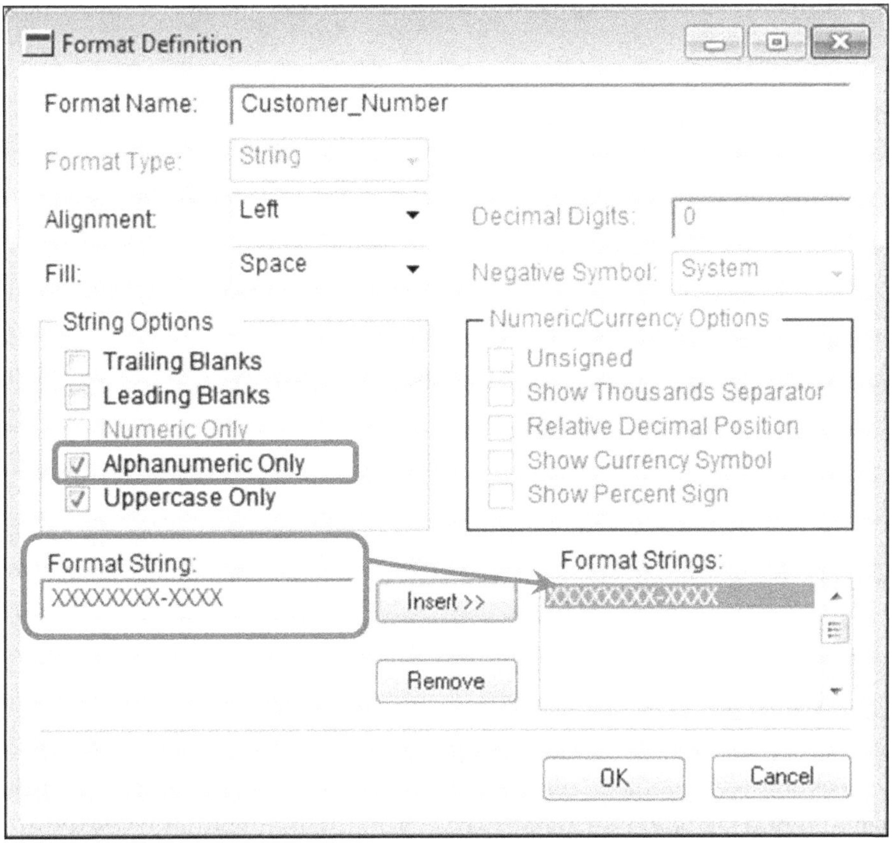

You need to make the following two changes to the **Customer_Number** format:

1. Check the box next to **Alphanumeric Only** to prevent users from entering special characters in the **Customer Number** field.
2. Create a format string that is eight capital xs and a hyphen, followed by four capital xs.

Switch back to Dynamics GP and open the **Customer Maintenance** window by going to **Sales** | **Cards** | **Customers**.

Notice that you have hard-coded a dash in the **Customer ID** field, and changed the maximum number of characters to 12. Your **Customer Maintenance** window should look similar to the window in the following screenshot:

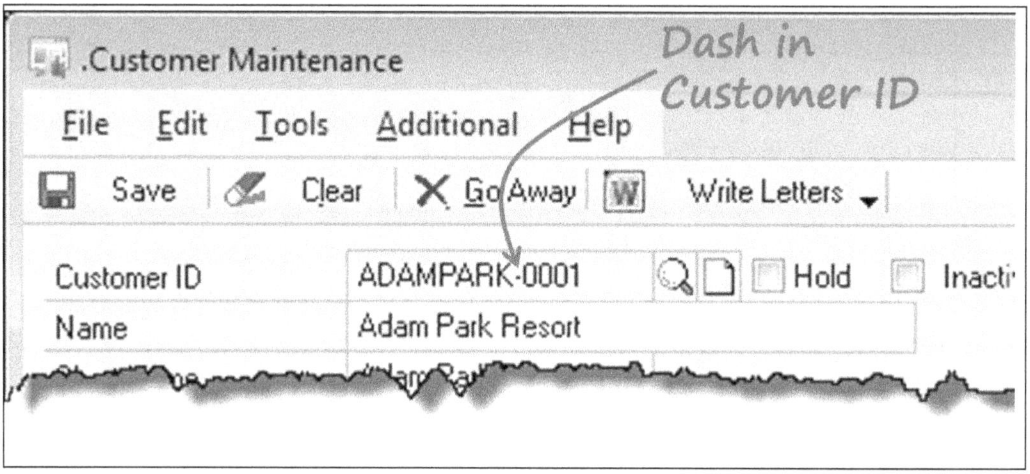

[Be sure you to fill your format with enough capital Xs to accommodate the whole field length, or you will shorten the number of characters in the field.]

Data types

We've already covered how you can change the format attached to a data type in the *Formats* section, so we won't go over it again here. There's one last point we need to look at and that is changing the data type itself. We'll use the **Item Description** field as an example.

The **Item Description** field in the **Item Maintenance** window is 100 characters long. However, the SOP invoice form can only display about 50 characters. Moreover, the longest field the Report Writer can handle is 80 characters. Let's say you want to limit the keyable length of the item description to 60 characters.

You can do it using modifier.

First, you need to find out what data type **the Item Description** field is using. That is simple enough to do by opening **the Item Maintenance** window in Modifier, and checking the **DataType** property of the **Item Description** field.

If you double-click on the property, the **Data Type Definition** window will open. As you can see in the following screenshot, the **STR100** data type is attached to the **Item Description** field:

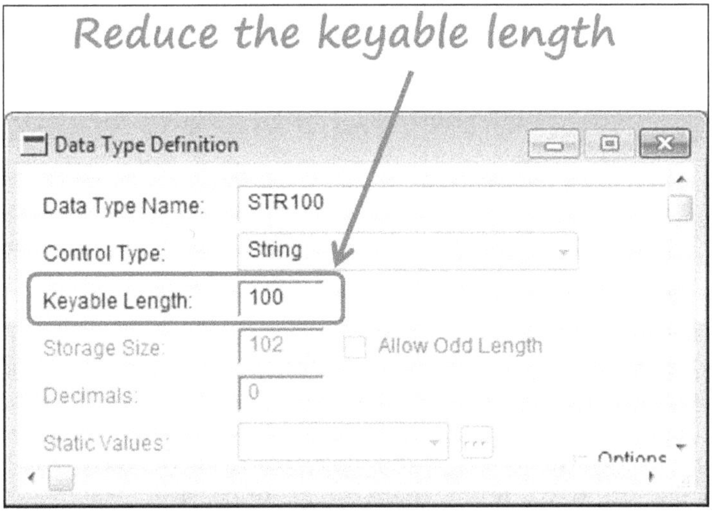

You can't change which data type is attached to the field, but you can change the properties of the data type. Using our example, if you want to limit the number of characters in the **Item Description** field to `60`, simply change the **Keyable Length** from `100` to `60`. You can use this method to reduce the length of a field, but you cannot use this method to increase the length of a field.

The final resource we'll discuss is messages.

Messages

Message resources are words or phrases that are associated with an ID. For instance, the phrase appearing in the following dialog box is message 1,003:

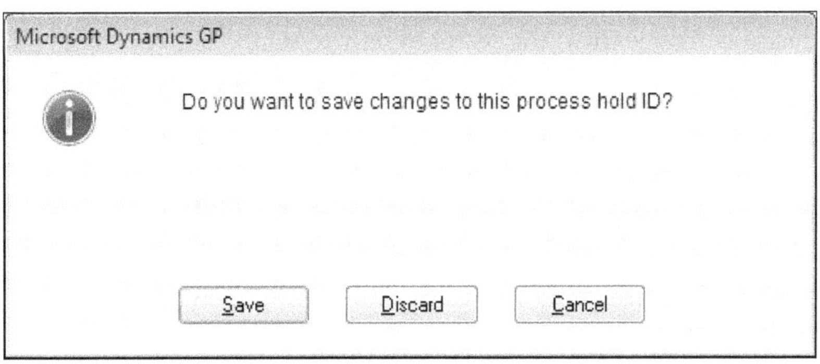

By changing the text associated with the aforesaid message, you can change the dialog. Let's look at the **Messages** window for message 1,003. Notice there is no lookup window, no print icon, or any other way to see the list of messages with their associated numbers.

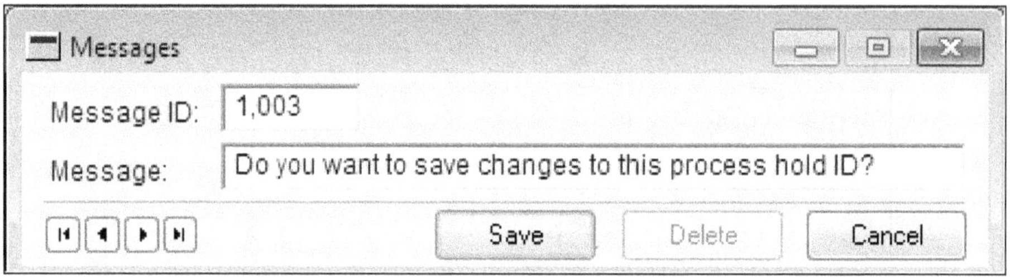

In order to determine what number goes with which message you need to print a resource report from Modifier. Print this report from the application menu bar by going to **File | Generate Resource Report**.

The resource report is quite long. For the `Dynamics.dic` file, the report is nearly 67,000 lines long. Within that report there is a list of nearly 14,000 message resources. A search tool is definitely needed. From your point of view, as a developer you can modify resources to achieve more appropriate dialogs or menu names.

For instance, two different developers may add a menu to the **Customer Maintenance** window named **Contacts**. If the developers have used messages to name their windows instead of literal strings, you can modify the message to clear up any ambiguity. You can change existing messages using Modifier, but you cannot create new ones.

Summary

You now have a pretty good understanding of the Modifier tool and all that you can do with it. You had opportunities to work hands-on with the product, and see for yourself how it works. You made several changes to the windows themselves, even adding fields and pictures. You learned the secret of how to set the big line item boundary properly, add fields from the **AutoLink** table, and even learned how to properly line up the column labels. You made tons of changes to the window fields themselves, including changing the static text attached to checkboxes and adding tooltips.

Beyond windows, you learned how to make changes to global resources that affect the entire application. You can even add new pictures to the picture library!

In the next chapter, we are going to add the VBA component to the Modifier, and turn our changes into action-capable objects.

8
Creating Customizations with VBA

This chapter will introduce you to the basics of building customizations for Dynamics GP with **Visual Basic for Applications (VBA)**. You will get a feel for how you can enhance the functionality of Dynamics GP windows. You will learn how to work with multiple dictionaries within your VBA project. This chapter will also introduce how you can change field values, populate data fields, and store and retrieve additional data from a special table in the company database.

Key points in this chapter include the following:

- How to work with windows and window fields
- The special characteristics of scrolling windows
- How to read and manipulate modal dialogs
- How to store and retrieve data using the Dynamic User Object Store
- Important considerations when deploying Modifier/VBA customizations

VBA overview

Visual Basic for Applications (VBA) is an event-driven programming language that has full core-language parity with the popular Microsoft Visual Basic product. To become better acquainted with VBA, you can access many free, online tutorials and sample applications. The following sites will get you started:

- `http://tinyurl.com/VBA-MSDN`
- `http://dynamicsgpblogster.blogspot.com/p/code.html`
- `http://support.microsoft.com/kb/163435`
- `http://blogs.msdn.com/b/developingfordynamicsgp/archive/tags/vba/`
- `http://www.gpwindow.com/DEVELOPMENT/VBA/`

Components

The heart of the VBA program is the **Visual Basic Editor** (**VBA Editor**). The VBA Editor in Dynamics GP is the same one that you will see in any of the Microsoft Office products. To launch the editor from within Dynamics GP, use the following navigation:

Microsoft Dynamics GP | **Tools** | **Customize** | **Visual Basic Editor**

Alternatively you can choose to use the hot key combination *Alt* + *F11*.
The following screenshot represents the VBA Editor window:

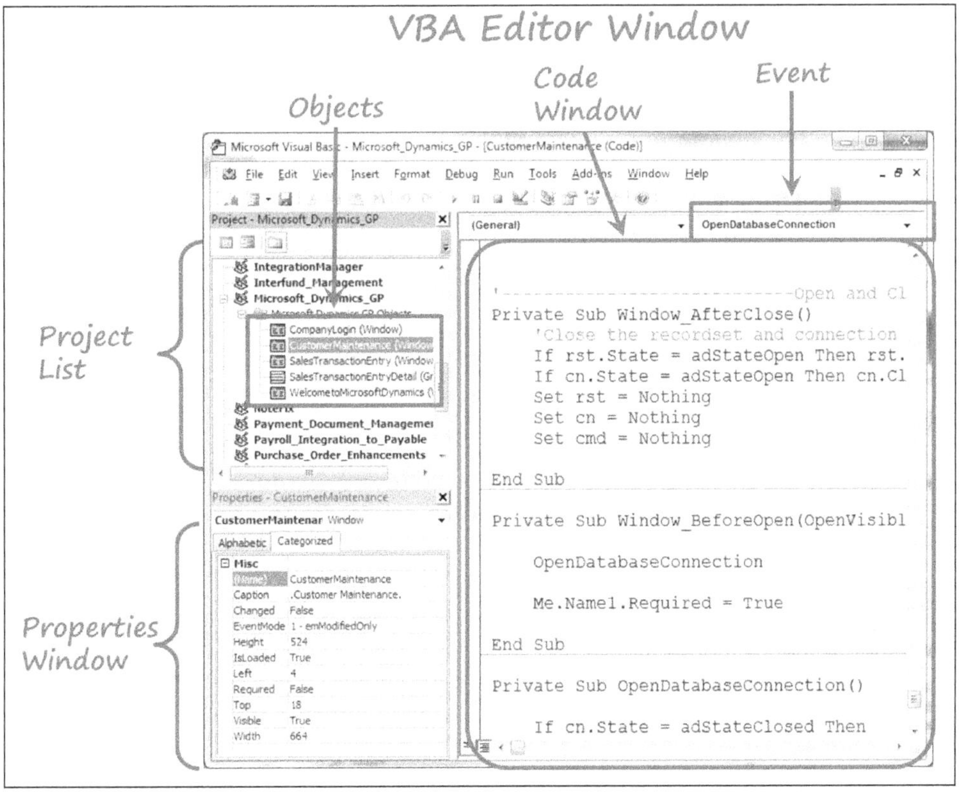

You will create a VBA project the first time you open the VBA Editor. One file is created for each product in your launch file (`Dynamics.set`), regardless of whether you have included any objects from that product in your VBA customization.

The files created are local to the workstation and are created in the same folder as the Dynamics executable (`Dynamics.exe`). You cannot share a single group of `.vba` files like you can for reports or forms dictionaries. Each workstation creates its own set of files.

The files are named after the dictionary they represent with the extension of `.vba`. For example, the file containing code for `Dynamics.dic` is `Dynamics.vba`; the file containing code for `HR.dic` is named `HR.vba`.

A project includes the following elements:

- Dynamics GP objects
- Object properties
- Object methods
- Object events
- User forms
- Variables

You define which Dynamics GP objects are included in the project and then use VBA code to manipulate the object's properties, methods, and events.

Objects

Dynamics GP exposes objects to VBA. VBA code can only be applied to exposed objects. Programmable objects for Dynamics GP include the following:

- Windows
- Reports
- Grids (scrolling windows)
- **Dynamics User Object Store** (**DUOS**) objects

The following diagram represents the Dynamics GP object hierarchy and how they relate to one another:

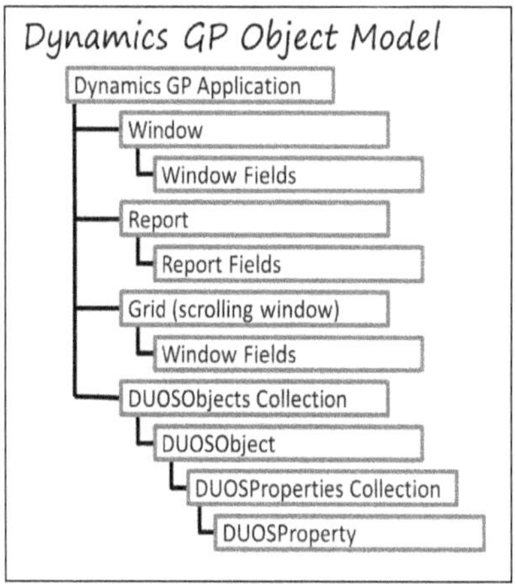

Knowing the hierarchy is important because you have to navigate down the tree in order to work with the objects further down.

For example, let's say you wanted to change the value of the **PO Number** field on the **Payables Transaction Entry** window; your fully qualified code would look something like the following code snippet:

```
'set the value of the PO Number field using the fully qualified name

Microsoft_Dynamics_GP.PayablesTransactionEntry.PONumber.Value = "123"
```

Notice how each object in the hierarchy is referenced as you move down the tree. The previous code translates to:

```
Project_Name.Window_Name.Field_Name.Property_Name.Property_Value
```

Rarely will you have to name each object in the hierarchy as we previously have, but you get the idea.

As you can see, the Dynamics GP object model is quite small. If you have ever worked with VBA in any of the other Office products, you will appreciate the simplicity of Dynamics GP's model. In contrast, Microsoft Excel has over 250 different objects in its model where Dynamics GP has only ten.

Properties

A **property** defines a characteristic of an object. The list of properties varies according to the type of object selected.

The following table contains the *Field* properties:

Property	Definition
`Caption`	The **caption** is the text you see on the screen as the label to the field. For instance, `Vendor ID` is the caption of the `VendorID` field. This is sometimes called the field prompt.
`Empty`	The **Empty** property tells you whether the field holds a value; `True` means a value is present.
`Enabled`	If **Enabled** is set to `False`, you cannot change the value of the field but you can still see it. In addition, the field's Linked Prompt is grayed out on the window.
`Height`	The **Height** property tells us the height of the object, in pixels.
`Left`	The **Left** property denotes the horizontal position of the field, in pixels.

Property	Definition
`Locked`	The **Locked** property specifies whether you can change the value of a field. It works just as a disabled field except the visual characteristic of the field's Linked Prompt is not changed.
`Name`	The **name** is what you use in VBA code to refer to the object. The name comes from the `Linked Prompt` of the field. If the field does not have a Linked Prompt, then the field's name is used (excluding spaces).
`Required`	The **Required** property indicates whether the field requires data before the record can be saved.
`Tab stop`	The **Tab stop** property controls whether you can gain focus to the field using the *Tab* key. This property does not allow you to set the tab sequence. You can set the tab sequence using the Modifier product.
`Top`	The **Top** property specifies the vertical position of the field, in pixels.
`Value`	**Value** refers to the contents of the field. The value property is the default property so you do not have to include the word in your code. `PONumber.value = "123"` is the same as `PONumber = "123"`.
`ValueSeg`	The **ValueSeg** holds the value of a specific segment of a composite field.
`Visible`	The **Visible** property determines whether you can see the field on the window. If a field is not visible, neither the field nor the Linked Prompt will be visible on the window.
`Width`	The **Width** property tells you how wide the field is, in pixels.

Many of the window properties are the same as their equivalent field properties; therefore, the following table will only show properties with unique characteristics to windows.

The following table contains *Window* properties:

Property	Definition
`Caption`	The `Caption` property sets the value of the window title.

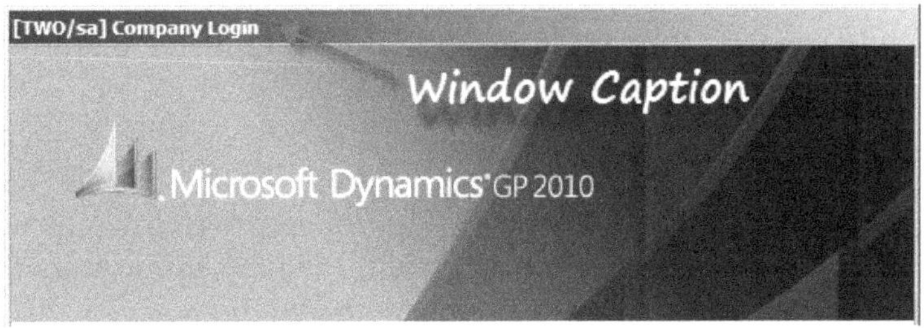

Property	Definition
`Changed`	The **Changed** property returns whether any fields on the window have changed. If the value of `Changed = True`, then one of the fields have changed.
`EventMode`	The **EventMode** property controls whether the VBA window events will occur for the original or modified version of the window. This can also be set so that window events never occur for the window.
`IsLoaded`	The **IsLoaded** property indicates whether the window is open. It is possible for a window to be loaded and not be visible.
`Required`	The `Required` property returns whether all of the required fields on the window contain data.

While you have a lot of flexibility using the object properties, you cannot override certain settings made at the Dexterity level. For instance, if a field is *required* in Dynamics GP, you cannot make it *unrequired* using VBA; nor can you make it invisible.

For fields that are not required at the Dexterity level you can use the `Visible` property to make the field invisible. To make the `Shipping Method` field invisible, use the following code:

```
ShippingMethod.Visible = False
```

Using the `Caption` property, you can change the field prompt of a field. To change the prompt for the `CustomerID` field, use the following code:

```
CustomerID.Caption = "Patient ID"
```

In short, a property tells you something about an object or causes a characteristic of the object to change.

Methods

Methods are actions you can perform for a given object. Methods include actions such as opening or closing a window, moving a field, hiding a field, or bringing focus to a field. You *set* a property and you *call* a method.

You would use the `close` method to close a window; the window is performing the action of closing:

```
PayablesTransactionEntry.Close
```

By contrast, you would use the `caption` property to change the window's title. You're changing the appearance of the window; the window isn't doing anything. Hence, you use a property instead of a method:

```
PayablesTransactionEntry.Caption = "Enter Vendor Invoices"
```

When you use the VBA Editor, you can tell which item is a method and which is a property thanks to a visual queue. The pointing finger indicates a property and the green box indicates a method. Take a look at the following screenshot to see what the different queues look like.

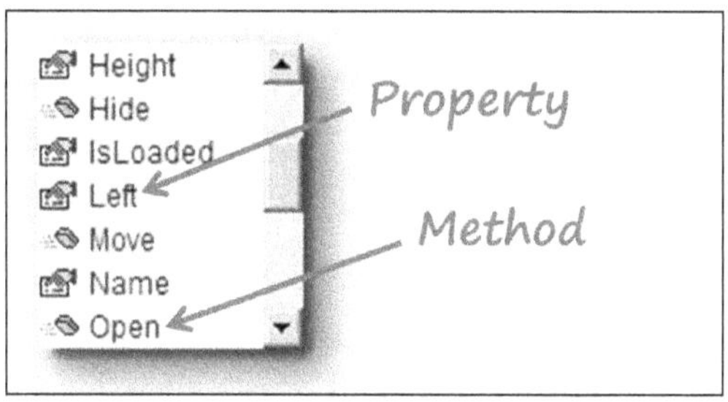

The methods available include the following:

- **Window** methods
    - `Activate`: This method makes a window the frontmost window
    - `Hide`: This method makes a visible window invisible
    - `PullFocus`: This method removes focus from a window and specifies if the data of the currently focused field is valid
    - `Move`: This method changes the location of a window
    - `Open`: This method opens the window and loads it into memory
    - `Show`: This method makes a hidden window visible again
    - `Close`: This method closes a window

- **Grid** methods
    - `Hide`: This method makes a visible grid invisible
    - `Move`: This method changes the location of a grid
    - `Show`: This method makes a hidden grid visible again
- **Field** methods
    - `Move`: This method changes the location of a field
    - `Focus`: This method moves the cursor to the subject field
    - `FocusSeg`: This method moves the cursor to a specified segment of a composite field
- **Beep**: This method causes a sound to be played through the computer's speakers

As you can see, not all objects support the same methods.

To view the full complement of objects, properties, and methods, open the **Object Browser** from the toolbar as shown in the following screenshot:

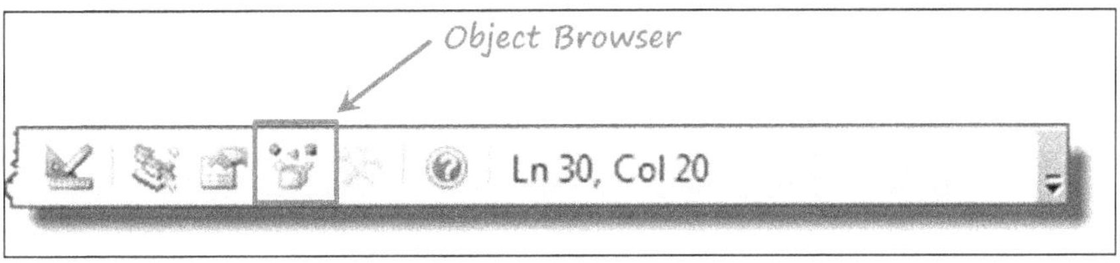

Events

Events are actions, typically initiated by the user, to which your code can respond. In other words, events specify when the VBA code should execute. Events include actions such as opening or closing a window, pushing a button, or selecting a radio button. Remember we said that VBA is an event-driven programming language; these are the kinds of events we were talking about.

VBA code can execute either before or after any Dexterity sanScript code which may be attached to the same event. Therefore, a window's `BeforeClose` event isn't something that happens before the window closes; the user has already closed the window. It determines whether the VBA code runs before or after the Dexterity code. The window's POST event would contain the Dexterity code. For example, VBA code attached to the `BeforeClose` event runs before any Dexterity script attached to that same event (the window POST event in this case). In many of the *Before* events, you are able to set a `CancelLogic` parameter to prevent the Dexterity script from running.

It's critical that you understand what *before* and *after* means in the context of VBA. The code you attach to the `BeforeOpen` event isn't running before the window opens; it's running before the Dexterity code attached to the window's PRE event executes. The user has already opened the window.

Similarly, the `BeforeClose` event runs after the window closes, but it runs before the Dexterity code runs for the window's POST event.

The *after* and *before* clauses refer to when the VBA code runs with respect to the Dexterity code. The software does not anticipate the user's wanting the window to open or close; rest assured, the user has already requested that the window close or open. The only issue is whether your VBA script runs *before* the Dexterity sanScript or your VBA script runs *after* the Dexterity sanScript.

This is the single most misunderstood concept related to Dynamics GP VBA.

VBA code can respond to the following events in Dynamics GP:

- **Window** events
    - `Before` or `After Activate`: This event occurs when the window becomes the front-most window
    - `Before` or `After Close`: This event occurs when the window closes
    - `Before` or `After Modal Dialog`: This event occurs when a modal dialog is called to be presented to the user
    - `Before` or `After Open`: This event occurs when the window opens
- **Field** events
    - `Changed`: This event occurs any time a field's value changes and the cursor moves off the field. It does not matter what initiated the change; the fact that the field's value changed causes the event.
    - `Before` or `After Got Focus`: This event occurs when the user moves the cursor into the field.
    - `Before` or `After Lost Focus`: This event occurs when the user moves the cursor out of the field.
    - `Before` or `After User Changed`: This event occurs whenever the user changes the value of a field and moves the cursor off that field.

- **Scrolling window** events (called **Grid** events in VBA)
    - `Before` or `After Line Changed`: This event occurs when the value of a field on the line changes, and the user advances to the next line.
    - `Before` or `After Line Got Focus`: This event occurs when a line is entered.
    - `Before` or `After Line Lost Focus`: This event occurs when a line is exited.
    - `Before Line Populate`: This event occurs each time a new line of data on a grid is displayed. There is no "After Line Populate" event.

In the pages that follow, you will create VBA code in Dynamics GP that will respond to **window**, **field**, and **scrolling window** events. We will not be working with **Report** events.

UserForms

UserForm is a VBA form created by the user that is not part of the Dexterity application dictionary. You can put buttons and controls, among others, on a UserForm just like you can in any other Visual Basic application. The key difference, of course, is that this window is not part of Dynamics GP's business logic.

You create a UserForm from the navigator's **Insert** menu, as shown in the following screenshot:

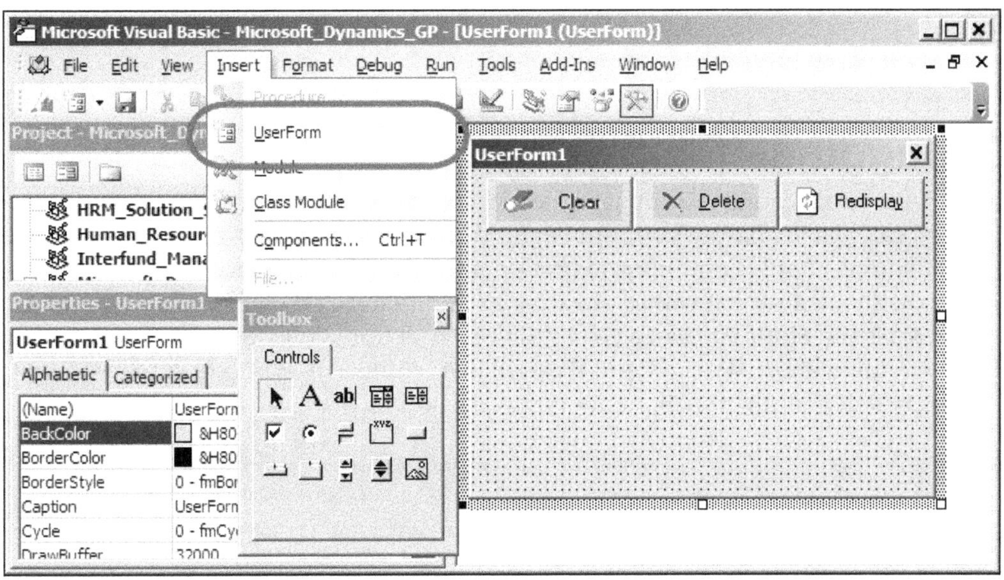

While you can use VBA code to interact with the Dynamics GP objects and create some very powerful customizations, there is no way for you to make your VBA UserForm look and behave like a native Dynamics GP window. If you copy the images from the buttons, for instance, you can make your button *resemble* Dynamics GP, but that's about it. The window itself behaves differently in that it is always the front-most window. You can move it out of the way, but you can't open another window on top of it.

An additional property available to a UserForm is the **ShowModal** property. If you want to require the user to dismiss the UserForm before using any other part of the application, you can set UserForm's **ShowModal** property to **TRUE**. For example, an error message presented that forces you to select **OK** or **Cancel** before you can do anything else, is a modal window. The default setting for a UserForm is **Modal**.

Modules

You will normally use **Modules,** also known as **Standard Code Modules**, or just **Code Modules**, to hold custom functions that you can call from other parts of your application. Using modules is a good way to organize your code. You could put any calls to Windows APIs, DLLs, or other external functions in separate modules.

You can call functions or procedures in one module from any other module or procedure in the application. You can think of modules as places to store global bits of code. You can have as many modules as you want, but it's a good idea to organize them so they can be more portable and thus, reusable.

For example, you could put your table operations in one module and your data translation procedures in another.

Debugging

The VBA module contains a robust collection of source-level script debugging tools that you can use to analyze your application. Using the debugging tools, you can set breakpoints, apply conditions to breakpoints, step through scripts one line at a time, examine and set the values of variables and fields, and control where the code should start and stop. You can also check your code for syntax errors before you run it in your application. Bear in mind though, just because its syntax is correct doesn't mean it will deliver the functionality you were hoping for.

You can access the debugging tools from the **Debug** menu as shown in the following screenshot:

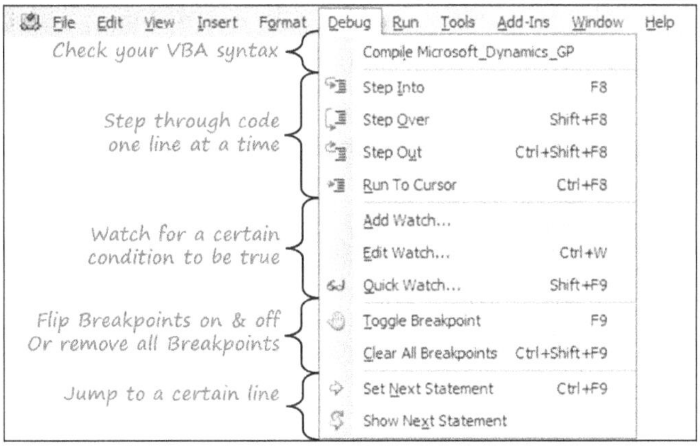

The following list will gives on each item in the **Debug** menu:

- **Compile**: This checks to verify that your syntax is correct and that there are no obvious errors such as an unknown variable, a `CASE` statement without an `END`, or an `If` block with no `End If`.
- **Step Into\Over\Out**: These control the execution of the code when it is stopped at a breakpoint.
- **Run To Cursor**: This tells VBA to execute the code until it reaches the line on which the cursor is positioned. VBA enters break mode when it reaches this line. **Run To Cursor** is similar to putting a temporary breakpoint on a line of code.
- **The Watches**: These provide a means for you to monitor a specified variable or expression. You can cause code execution to enter break mode based on the value of that variable or expression. You can have many watches active at the same time.
- **Toggle** and/or **Clear All Breakpoints**: This will place or remove a breakpoint on a line of code. Place your cursor on the line and then press *F9* to place or remove a breakpoint on that line. Press *Ctrl* + *Shift* + *F9* to remove all breakpoints.
- **Set Next Statement**: This changes the path of execution of the code. At breakpoint, you can identify the particular line where you want the execution to start. You can skip over entire sections of code or pick a previous line to have the code run again.
- **Show Next Statement**: This will display a yellow arrow to mark the line that will be executed next when you continue execution of the code.

Setting options

Before we start writing code, we need to change at least one option for the navigator itself. To access the **Options** window, go to **Tools** | **Options**.

The **Options** window will open. On the **Editor** tab, check the box next to **Require Variable Declaration**. This setting will automatically insert `Option Explicit` in new code modules. This setting will help reduce errors from misspelled variables in your code. If you don't mark this, and the VBA compiler encounters a name that it doesn't recognize, it will create a new variable for you. Therefore, if you declare a variable named `str_Name`, and then later misspell it as `sr_Name`, a new variable named `sr_Name` would be created for you. If the `Option Explicit` statement is at the start of the code module, an error will be thrown when the unknown `sr_Name` variable is encountered by the compiler. This can save you hours of debugging time later.

Windows and window fields

Now we are ready to set VBA loose! For our project we will add a button to the **Vendor Maintenance** window,which, when pushed, will open the **Vendor Credit Summary** window. We are also going to add a **Go To** button to the **Vendor Maintenance** window to provide quick navigation to the following windows:

- **Payables Transactions Entry**
- **Payables Transaction Inquiry**
- **PO Document Inquiry**
- **Vendor Yearly Summary**
- **Vendor Period Summary**
- **Vendor Summary**

First, we need to use the Modifier to add our new buttons to the window. The easiest way to do this is to open the window you want to modify (in our case the **Vendor Maintenance** window), and then select *Ctrl* + *F10* on the keyboard. The Modifier will open with the **Vendor Maintenance** layout window displayed.

Now, you are going to put the two new buttons on this window and then you will put the VBA code behind them so that they will do something when you push them. At the end of the project, the **Vendor Maintenance** window should look similar to the following screenshot. The trailing period in the window title tells you that the window has been included in a VBA project:

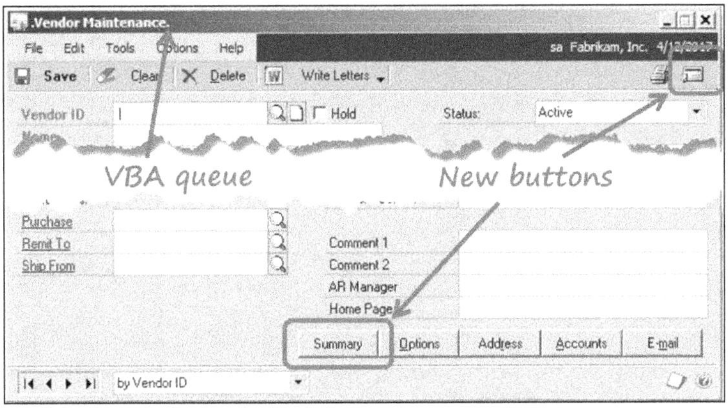

Creating the Summary button

In this section you will create a button with the the caption **Summary** on it and position it on the **Vendor Maintenance** window. When the button is pushed, the **Vendor Credit Summary** window will open. Refer to the following screenshot for more guidance regarding each step of the creation of the **Summary** button:

1. Select the push button icon from the **Toolbox** window and your mouse cursor will change to a plus sign (+) with the image of an **OK** button next to it. Click wherever you want the button to be created; you can move it and resize it later.
2. Position your button between the **Withholding** and **Options** buttons at the bottom of the window. You will need to move the **Withholding** button to the left to make room.
3. Double-click on the **Field** line of the **Properties** window and the **Local Field Definition** window will open.
4. On the **Local Field Definition** window, change the **Field Name** value to **SummaryPB**.

5. Click on the ellipses button next to the **Static Values** field to open the **Button Items** window.
6. Change the **Caption** to **Summary**. This text will appear on the face of the new button.

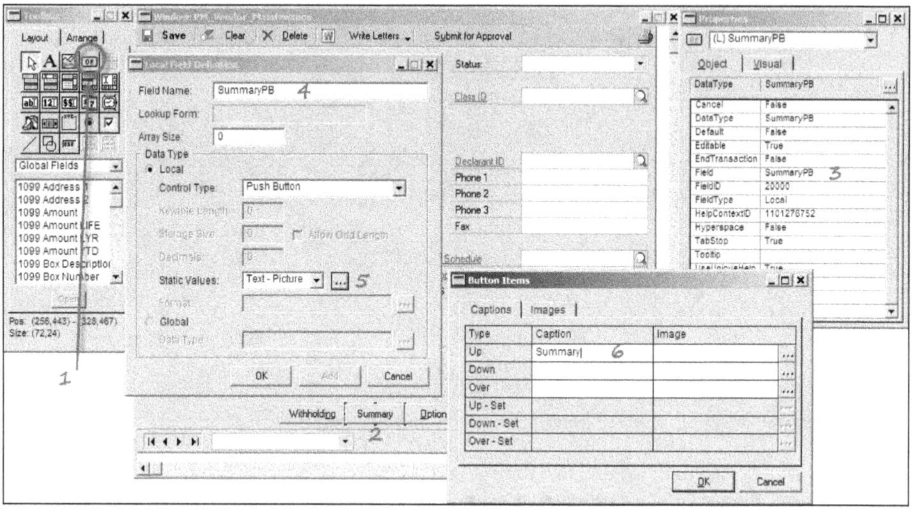

Creating the Go To button

In this section, you're going to add a **Go To** button to the **Vendor Maintenance** window.

1. Select **Local Fields** on the **Toolbox** window.
2. Push the **New** button at the bottom of the **Toolbox** window.
3. In the **Local Field Definition** window, name the field `GoToPB` and set the **Control Type** to `Button Drop List`.
4. Select the ellipses button next to the **Static Values** field to open the **Button Items** window.

5. On the **Captions** tab, enter the following values:

Type	Caption	Image
Up	&Go To	GoTo_1_Up
Down	&Go To	GoTo_1_Down
Over	&Go To	GoTo_1_Up

6. On the **Drop Items** tab, unmark the **Sort List** checkbox and add the following values:

Index	Drop Item
1	Payables Trx
2	Inquiry - PM Trx
3	Inquiry - PO Docs
4	Inquiry - Yearly Summary
5	Inquiry - Period Summary
6	Inquiry - Vendor Summary

7. Change the object properties of `(L)GoToPB` to the following values:

SetChangeFlag	False
Tooltip	Go To

8. Change the visual properties of `(L)GoToPB` to the following values:

Appearance	3D Highlight
BackColor	System-Toolbar
Position-Left	596
Size-Width	30
Style	Graphic Only

9. Select the **File** menu and then select **Microsoft Dynamics GP** to exit the Modifier and go back to Dynamics GP.
10. Change the **Alternate/Modified Forms and Reports ID** to display the modified **Vendor Maintenance** window. Use the following navigation to access the **Alternate/Modified Forms and Reports** window:

    Microsoft Dynamics GP menu | **Tools** | **Setup** | **System** | **Alternate/ Modified Forms and Reports**.

Your window should look substantially similar to the following screenshot:

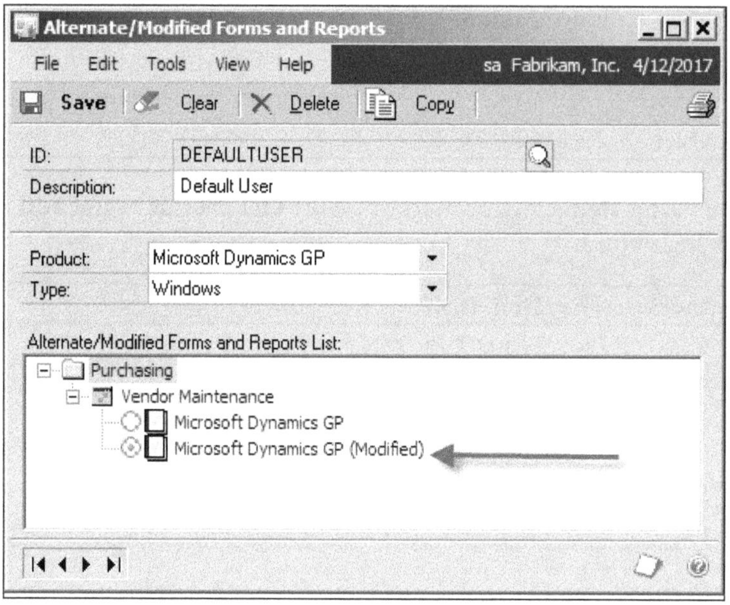

Next, you are going to create your VBA project by adding Dynamics GP objects.

Adding objects to the project

Initially, there are no objects in your VBA project. You have to individually add each window and window field that you want to use in your VBA customization.

First, you add the window to VBA, and then you add the individual fields by clicking on them one at a time. Once you've added a field to the project, the only way to remove it is to remove the entire window. Removing the window will also remove all of the fields you have added for that window, including any code you've added to those fields. Therefore, you need to be careful when adding fields or removing windows.

If you put some extra fields into the project, it probably will not be a problem, but the cleaner you can keep your project, the better. You will never know how future changes might affect your VBA project. If Dynamics GP takes a field that you have in your VBA project off the window, you will get an error at startup. Therefore, only adding the objects you need is the best practice.

Adding the Vendor Maintenance window

You are going to add VBA code to the **Vendor Maintenance** window, so you need to add it to your project. Open the **Vendor Maintenance** window. Notice the leading period in front of the window title; this is a subtle visual queue that indicates that you have modified the window. However, it might be a little too subtle. Before you start coding your VBA, you may want to change the `Title` property of the window so that it is more obvious that you are dealing with a modified window.

A better practice may be for you to put a more obvious prefix to the window title. Something like one of the following prefixes would be suitable:

- `m`: Use this is if the window is modified.
- `v`: Use this if the window is part of a VBA project.
- `mv`: Use this if the window is both modified and part of a VBA project.

In this case, you would name it **mv Vendor Maintenance** to indicate that the window is both modified and included in a VBA project. The leading period is a very slight change and can easily be overlooked.

You should see your new button to the left of the **Options** button at the bottom of the window, and the new **Go To** button in the upper right-hand corner. Right now, the new buttons are just eye candy. They do not *do* anything when you push them.

The **Withholding** button that you saw in the Modifier will not be visible unless you have set up **Withholding** in the **Company Setup Options** window. Since **Withholding** is not normally set up, you moved the **Withholding** button over to the left so that you wouldn't have a gap in your buttons.

Select the *Ctrl* + *F11* keyboard combination to add the current window to the VBA project. Alternatively, you can use the following navigation:

Tools | Customize | Add Current Window to Visual Basic

With the window added, you can now go about adding the fields.

Adding additional windows and window fields

You need to add the **Vendor ID** field from the **Vendor Maintenance** window, the **Vendor Credit Summary** window, and the **Vendor ID** field from the **Vendor Credit Summary** window. To add the window fields to your project, select *Shift* + *F11* and then click on each field you want to add.

To get started, include the following objects in your VBA project:

Window	Window fields
Vendor Maintenance	`Vendor ID`
Vendor Maintenance	`SummaryPB (your new button)`

Window	Window Fields
Vendor Credit Summary	`Vendor ID`

With your objects added, it is time to launch the VBA Editor window and go about the business of writing your code. Select *Alt* + *F11* on the keyboard or use the following navigation to launch the VBA Editor:

Tools | Customize | Visual Basic Editor

Using methods and properties

First, you are going to attach code to your button so that it opens the **Vendor Credit Summary Inquiry** window for the displayed vendor when it is pushed. Pushing the button triggers the `Changed` event; you will use the `Open` method to open the window. With the window open, you will set the `Value` property to be equal to the **Vendor ID** displayed on the **Vendor Maintenance** window. Follow these steps to accomplish this course of action:

1. In the VBA Navigator, select the **Project - Microsoft_Dynamics_GP**.
2. Move down the object list and double-click on the entry **VendorMaintenance (Window)**.
3. Select the **Summary** object and the **Changed** procedure in the code window.

4. Type the code as it appears in the following screenshot. The `Value` property is the default property and therefore you do not need to state it when setting the value of **Vendor ID**. If you need more than one line to complete a line of code, use the underscore and space for your continuation indicator.

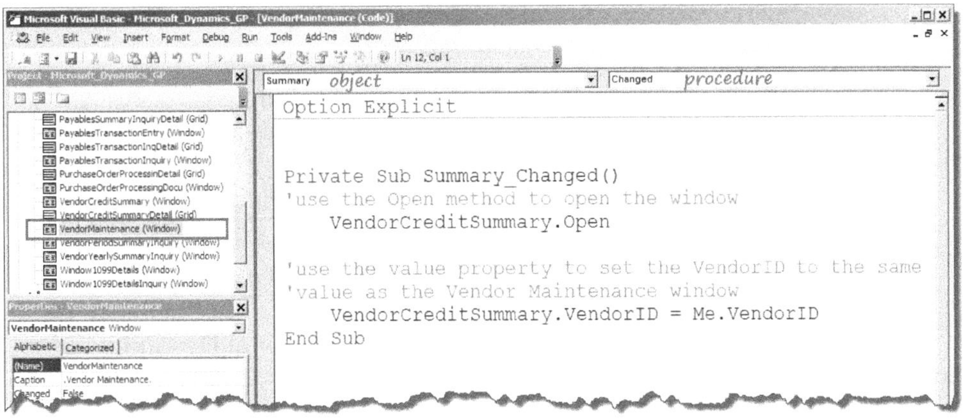

Now that you have the feel of typing code in the VBA Navigator, add the remaining objects needed for the **Go To** button.

Add the following additional objects to your VBA project:

Window	Window fields
Payables Transaction Entry	`Vendor ID`

Window	Window fields
Payables Transaction Inquiry - Vendor	`Vendor ID` Redisplay button Work checkbox History checkbox

Window	Window fields
Purchase Order Processing Document Inquiry	`Vendor ID` Redisplay button Open Purchase Orders checkbox Receipts Received checkbox

Window	Window fields
Vendor Yearly Summary Inquiry	`Vendor ID`
	Calculate button
	Summary View

Window	Window fields
Vendor Period Summary Inquiry	`Vendor ID`

Window	Window fields
Vendor Credit Summary Inquiry	`Vendor ID`
	Calculate button

You will need to populate a number of different fields in several Dynamics GP windows to make your **Go To** button functional.

Setting field values

Since you want the windows you open from your **Go To** menu to display data for the current vendor, you will need to set the value of `Vendor ID` when the window opens. For all but two of the windows, you will need to set the value for more than one field.

When you need to push a button or mark a checkbox, set the value of the field to `1`. Enter the following code in the `Changed` event for the **Go To** button:

```
Private Sub GoTo_Changed()

    With Me.GoTo

        Select Case .Value

        Case 1 'Payables Transaction Entry
            PayablesTransactionEntry.Open
            PayablesTransactionEntry.VendorID = Me.VendorID

        Case 2 'Payables Transaction Inquiry - Vendor
            PayablesTransactionInquiry.Open
            PayablesTransactionInquiry.VendorID = Me.VendorID
            PayablesTransactionInquiry.Include = 0
            PayablesTransactionInquiry.Include2 = 0
            PayablesTransactionInquiry.Redisplay = 1
```

```
        Case 3 'Purchase Order Processing Document Inquiry
            PurchaseOrderProcessingDocu.Open
            PurchaseOrderProcessingDocu.StartVendorID = Me.VendorID
            PurchaseOrderProcessingDocu.EndVendorID = Me.VendorID
            PurchaseOrderProcessingDocu.OpenPurchaseOrders = 1
            PurchaseOrderProcessingDocu.ReceiptsReceived = 1
            PurchaseOrderProcessingDocu.Redisplay = 1

        Case 4 'Vendor Yearly Summary Inquiry
            VendorYearlySummaryInquiry.Open
            VendorYearlySummaryInquiry.VendorID = Me.VendorID
            VendorYearlySummaryInquiry.SummaryView = 3
            VendorYearlySummaryInquiry.Calculate = 1

        Case 5 'Vendor Period Summary Inquiry
            VendorPeriodSummaryInquiry.Open
            VendorPeriodSummaryInquiry.VendorID = Me.VendorID

        Case 6 'Vendor Credit Summary Inquiry
            PayablesSummaryInquiry.Open
            PayablesSummaryInquiry.Ranges = Me.VendorID
            PayablesSummaryInquiry.Ranges1 = Me.VendorID
            PayablesSummaryInquiry.Calculate = 1

        End Select

    End With

End Sub
```

You should now have a **Vendor Maintenance** form with a **Go To** navigation button that looks and acts like the native **Go To** buttons. Take a look at the **Go To** button on the **Item Maintenance** screen and compare it to yours. You will not see any difference between the button you created in VBA and the button created with Dexterity.

Cross-dictionary access

VBA has no limitations as to what dictionaries you can access. When you look at your VBA customization, you will see a project for each dictionary in your `Dynamics.set` file, shown as follows:

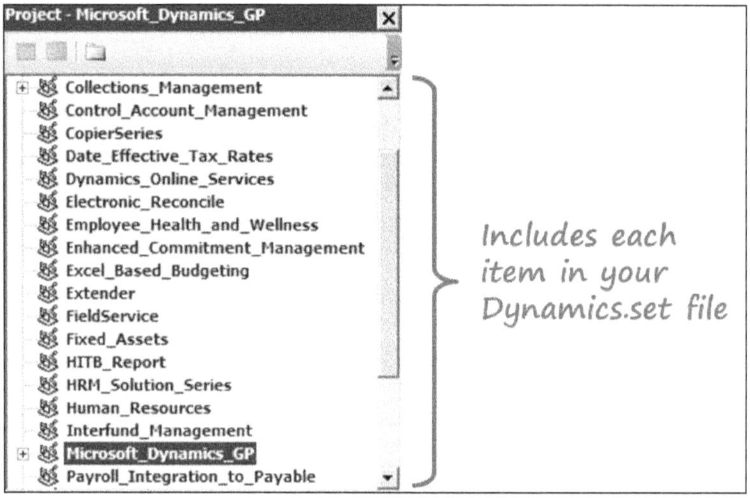

You can tell which projects have objects added to the application by the presence of a plus (+) sign next to the name of the project. In the previous **Project** screenshot, only the **Microsoft_Dynamics_GP** and the **Collections_Management** projects include any objects.

If you want to build cross-dictionary applications with VBA, you need only to include a reference to that project.

Referencing the Collections module

Let us build a customization that will provide a push button on the **Customer Inquiry** window that will open the **Collection Main** window for the selected customer.

First, add a new button to the **Customer Inquiry** window:

1. Select the push button icon from the **Toolbox** window and your mouse cursor will change to a plus sign (+) with the image of an **OK** button next to it. Click on the control area to add the button to the top of the window.
2. Double-click on the **Field** line of the **Properties** window and the **Local Field Definition** window will open.
3. On the **Local Field Definition** window, change **Field Name** to `Collection`.

4. Click on the ellipses button next to the **Static Values** field to open the **Button Items** window.
5. Change the **Caption** field to `Collections`. This text will appear on the face of the new button.

   Your **Customer Inquiry** window should look similar to the following screenshot:

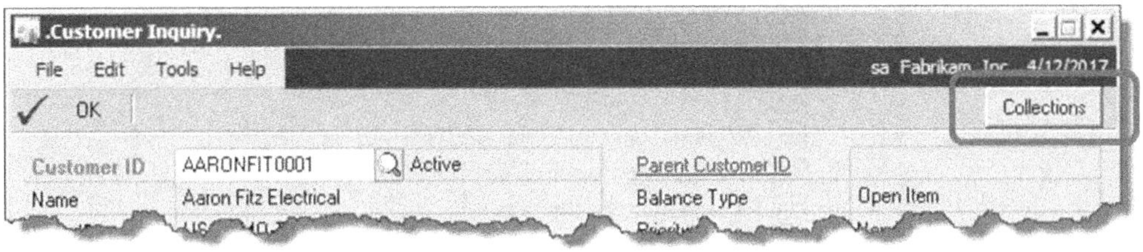

To set up the reference to the `Collections Management` module, perform the following steps:

1. With the **Microsoft_Dynamics_GP** project highlighted, select **Tools | References** from the VBA Navigator menu.
2. In the **References** window, mark the box next to the **Collections Management** item to set up the reference.

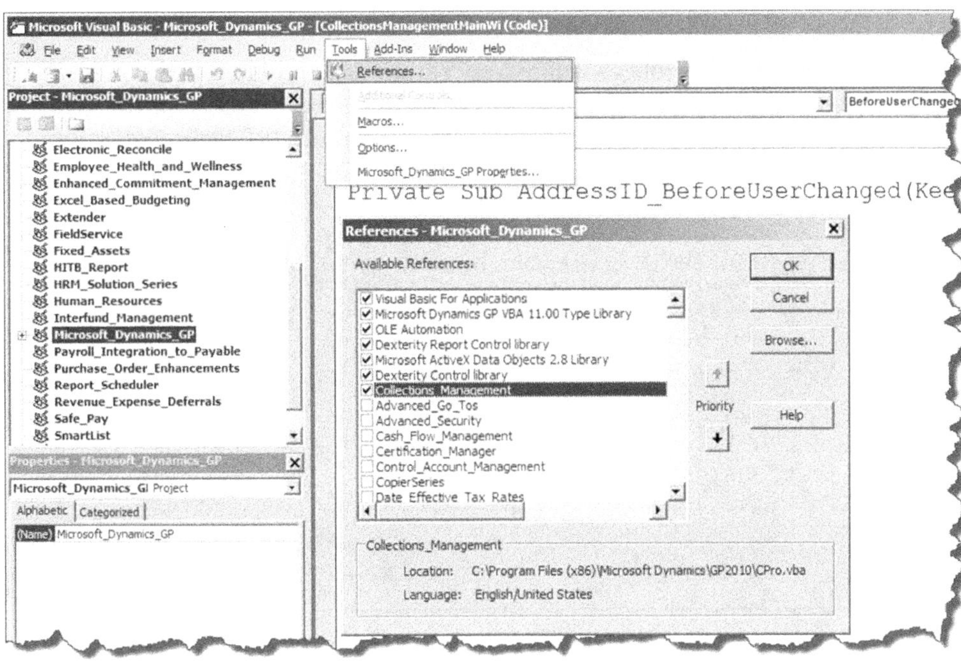

This will add a reference object to the **Microsoft_Dynamics_GP** project, as shown in the following screenshot:

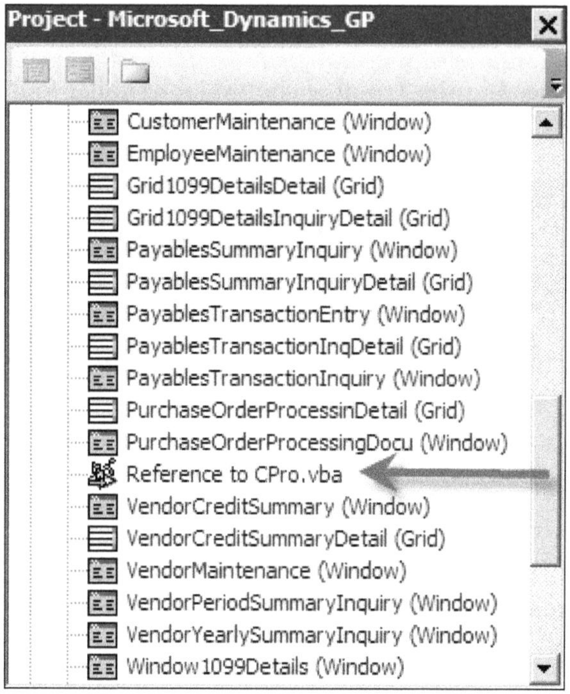

One limitation on VBA references is that they cannot be circular. Once you have set up a reference to **Collections Management** from **Microsoft_Dynamics_GP**, you cannot then set up a reference to **Microsoft_Dynamics_GP** from **Collections Management**. With the reference set, you use standard VBA coding to act upon objects in the **Collections Management** dictionary.

In the `Changed` event for the `Collections` button, type in the following code:

```
Collections                                  Changed
Option Explicit

'---open the Collection Main window from Customer Inquiry---
Private Sub Collections_Changed()
    CollectionsManagementMainWi.Open
    CollectionsManagementMainWi.CustomerID = Me.CustomerID
End Sub
```

Since we are calling across projects, we would need to fully qualify the object names if they were the same in both projects. For example, if the Dynamics dictionary also contained a window named `CollectionsManagementMainWi`, we would need to open it using the following code:

```
Collections                                   Changed
Option Explicit

'--- fully qualified object name ---
Private Sub Collections_Changed()
    Collections_Management.CollectionsManagementMainWi.Open
    Collections_Management.CollectionsManagementMainWi _
        .CustomerID = Me.CustomerID
End Sub
```

continuation character

If you need more than one line for a VBA statement, the continuation character is a space plus the underscore, as shown in the previous screenshot.

Scrolling windows

Scrolling windows are called **Grids** in VBA. Scrolling windows are used extensively throughout the application. The body of the **Sales Transaction Entry** and **Purchase Order Entry** windows are scrolling windows. A single screen can contain a number of separate scrolling windows. The following screenshot shows two screens that contain scrolling windows:

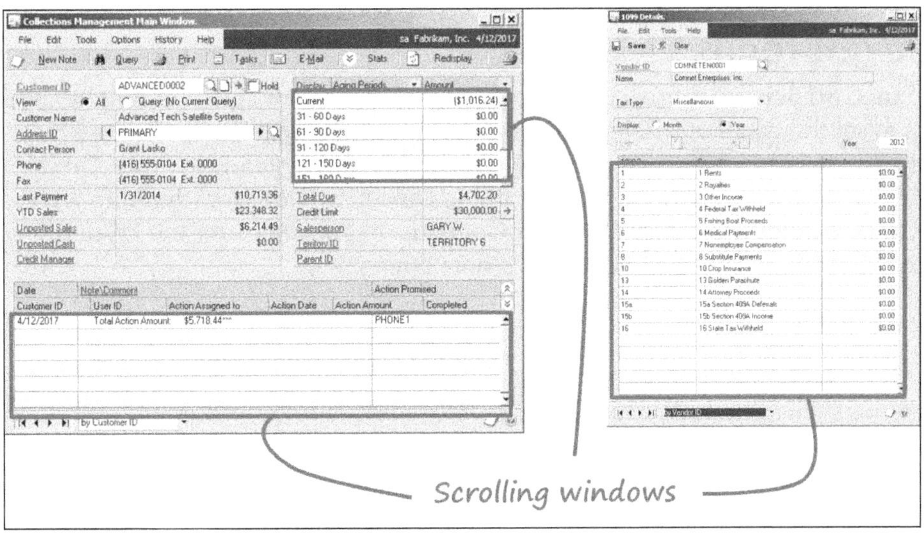

Adding a scrolling window to the project

When you add a window containing a scrolling window to a project, you automatically add a separate object for each scrolling window. You do not have to do anything extra; the VBA system does it for you. In your project, you should have several scrolling windows, called grids, listed as objects. Your project will look something like the following screenshot:

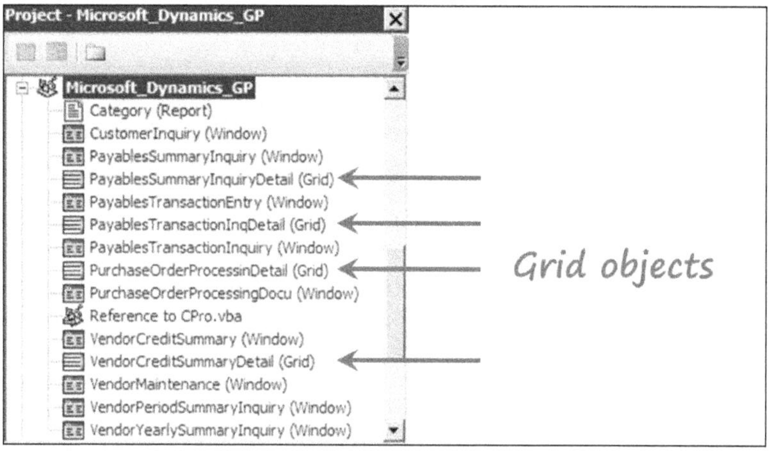

Grid events

The fields on a grid have the same events as the fields on a window. The grid itself, however, has some unique events that apply only to grids.

Line got focus

The **LineGotFocus** events occur as soon as you enter a new line. You might use this event to set the value of a field on the new line. The **BeforeLineGotFocus** event executes before any Dexterity code that runs when you enter a new line. Unlike the *Field Got Focus* events, you cannot cancel the Dexterity code from running when switching to a new line. The **AfterLineGotFocus** event runs after the Dexterity code that normally runs when switching to a new line.

Line lost focus

The **LineLostFocus** events occur as soon as you leave the line. You might also use this event to check the values of fields on the line, or to save additional data from your customization. The **BeforeLineLostFocus** event executes before any Dexterity code that normally runs when you leave a line. The **AfterLineLostFocus** event executes after any Dexterity code that runs when you exit a line.

When you switch from line to line, two events happen each time. The LineGotFocus events run for the new line and the LineLostFocus events run for the old line.

Line change

The **LineChange** events occur when you have changed the value of a field in the line and then leave the line. The **BeforeLineChange** event is useful if you want to check the value of a field and then not allow the user to advance to a new line if you don't like the value. You accomplish this by setting the value of the `KeepFocus` parameter to `True`. The `BeforeLineChange` event runs before any Dexterity code that normally runs when you change lines. Rows of data are normally saved by the Dexterity code when you change lines, so if you prevent this code from running, you can control the information that can be written to the tables.

The following code illustrates how you could check the quantity field on a purchase order line and not allow the user to advance if that quantity ordered is more than ten:

```
Grid                                    BeforeLineChange
Option Explicit

Private Sub Grid_BeforeLineChange(KeepFocus As Boolean)
    '---Check the Order Quantity field ---
    If QTYOrdered > 10 Then
        '--do not allow the line change--
        KeepFocus = True
        MsgBox "You are not authorized to order a qty greater than ten"
    End If
End Sub
```

Filtering records

Using VBA, you can filter the records allowed to fill the scrolling window. You can use the **BeforeLinePopulate** event to create the filter.

BeforeLinePopulate

This event happens each time you display a new record on a line. When you first open the grid, the `BeforeLinePopulate` event runs repeatedly until the grid is full. When you scroll down the grid and display new lines of data, the `BeforeLinePopulate` event executes for each new line of data displayed. You can use this event to filter the items filling the scrolling window by using the `RejectLine` parameter.

For example, if you want to restrict the records populating the **Vendor Lookup** window to vendors belonging to the class "USA-US-M," you could use the following code:

```
Grid                                    BeforeLinePopulate

Option Explicit

Private Sub Grid_BeforeLinePopulate(RejectLine As Boolean)

    ' Display only vendors belonging to the USA-US-M class
    If VendorsDetail.VendorClassID <> "USA-US-M" Then
        RejectLine = True
    End If

End Sub
```

This particular task is a little tricky because the **Vendor Class ID** field is not actually displayed on the **Vendor Lookup** window. To access the field so that you can add it to your VBA project, you must first modify the window and drag the field into a visible area.

Follow these steps to accomplish this task:

1. Open the **Vendor Maintenance** window and then open the **Vendor Lookup** window.
2. With the lookup window open, select *Ctrl + F10* from the keyboard. The Modifier will open with the lookup window's layout displayed.
3. Double-click in the middle of the scrolling window. The layout of the scrolling window will open.
4. If the field names are not showing, select **Layout** from the menu bar and be sure a checkmark appears next to both the **Show Field Names** menu item and the **Show Invisible Fields** menu item.
5. Select the **Vendor Class ID** field, change its `Visible` property to `True`, and then drag it up into the visible area of the scrolling window.

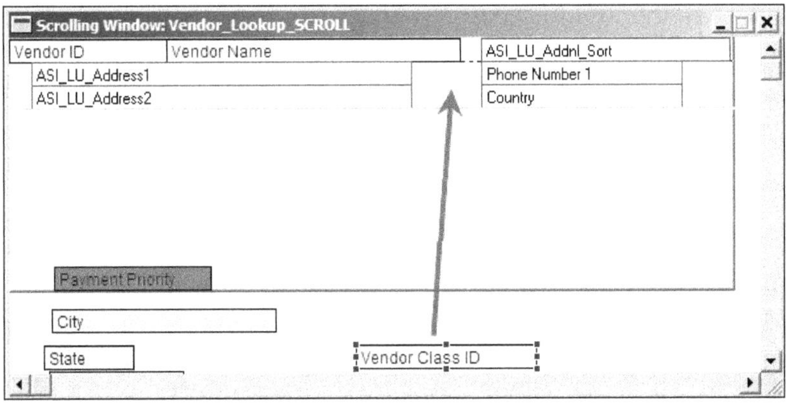

6. Switch back to Dynamics GP by selecting *Alt* + *Q* on the keyboard; be sure to save your modifications.
7. Change your **Alternate/Modified Forms and Reports** values so that it uses the modified version of the **Vendors** window in the **SmartList** dictionary. The following window shows the security window with the correct item selected:

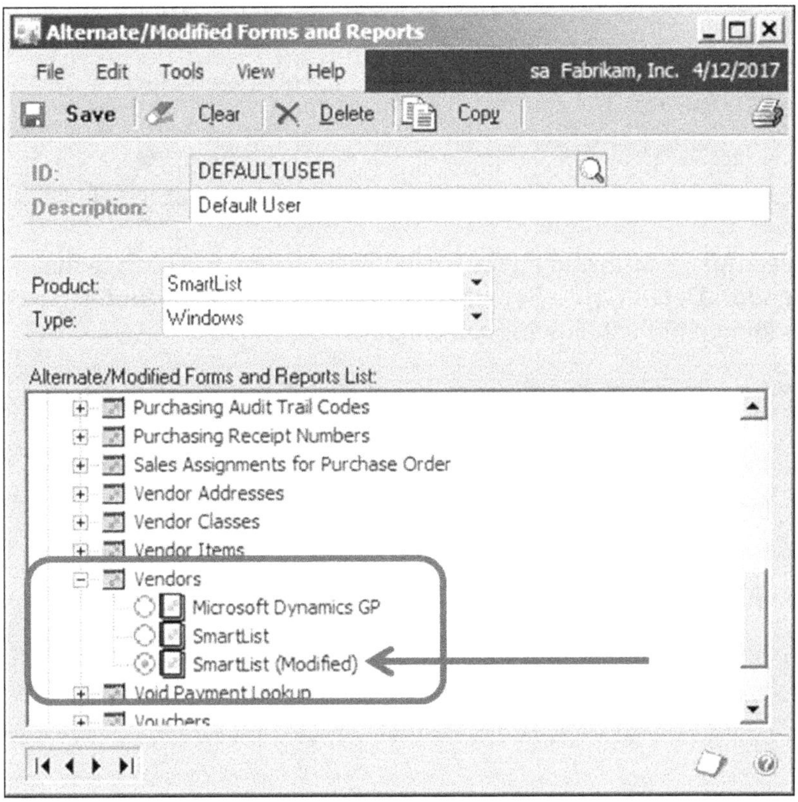

8. Now, open the **Vendor Lookup** window and press *Ctrl* + *F11* on the keyboard to add the lookup window to your VBA project.
9. Select *Shift* + *F11* and add `Vendor Class ID` to your VBA project. The **Vendors** lookup window will look something like the following screenshot:

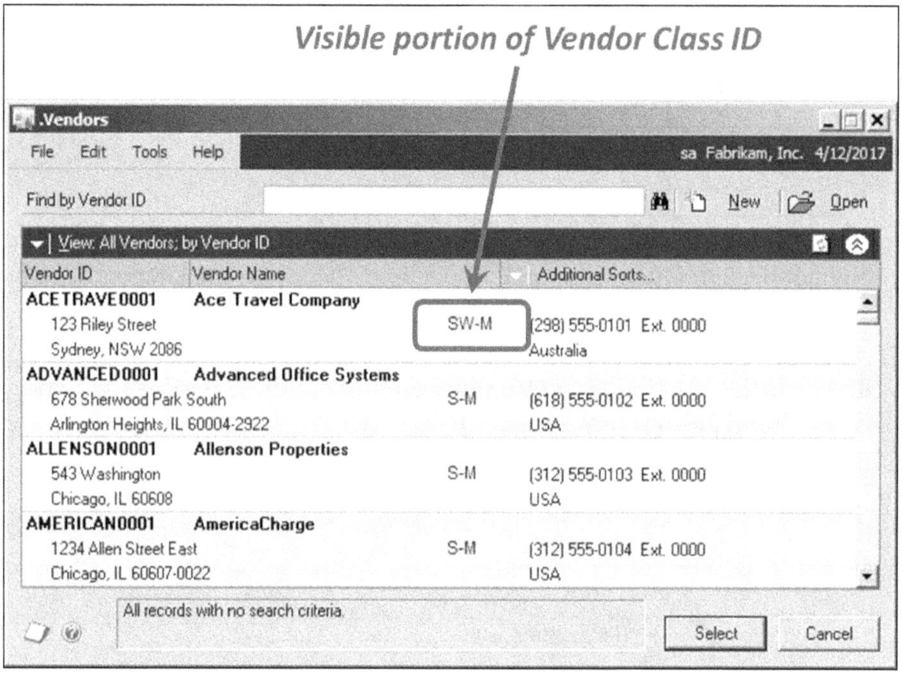

10. Select *Alt* + *F11* to launch the Visual Basic Editor and change the **VendorsDetail Grid**'s **EventMode** property back to **0 - emOriginalOnly** (as shown in the following screenshot):

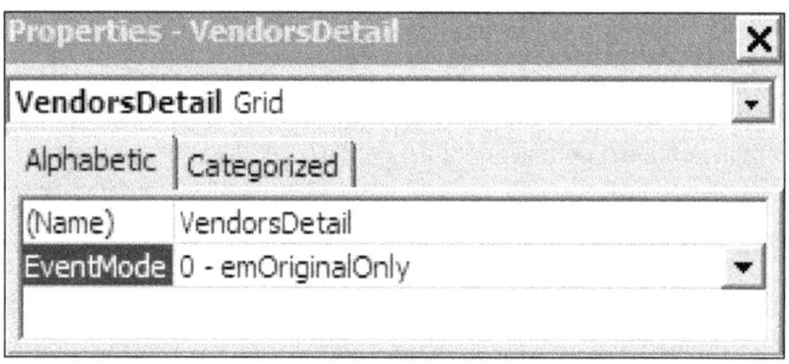

11. Change the **Alternate/Modified Forms and Reports** security back to the unmodified **SmartList** under the **Vendors** window (as shown in the following screenshot):

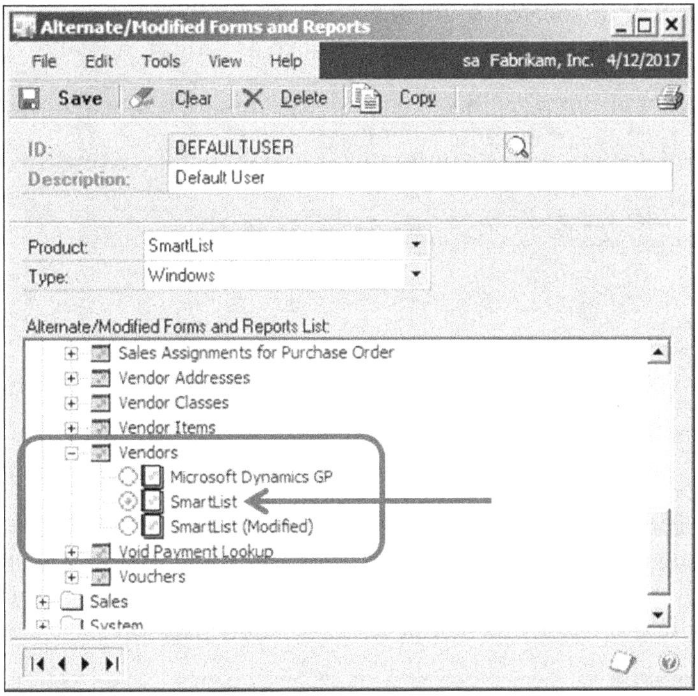

Using the `RejectLine` statement is the best way to apply your own record filtering. There is no "After Line Populate" event.

Fun with dialogs

Several dialog boxes open at various times throughout Dynamics GP. Some of these dialogs are hard to understand while others are simply a nuisance. Using VBA, you can change the text of a dialog, automatically answer the dialog before it is presented, or monitor your user's response to a dialog.

BeforeModalDialog

The **BeforeModalDialog** event is a window event that executes when a modal dialog opens but before it is visible to the user. This event is useful if you want to change the wording of a dialog or automatically answer it. You could also complete other actions based on which dialog was being displayed, such as setting the value of variables.

You can have a lot of fun with this event by customizing dialogs for individual users.

The following dialog is a prime candidate for you to dismiss using this event:

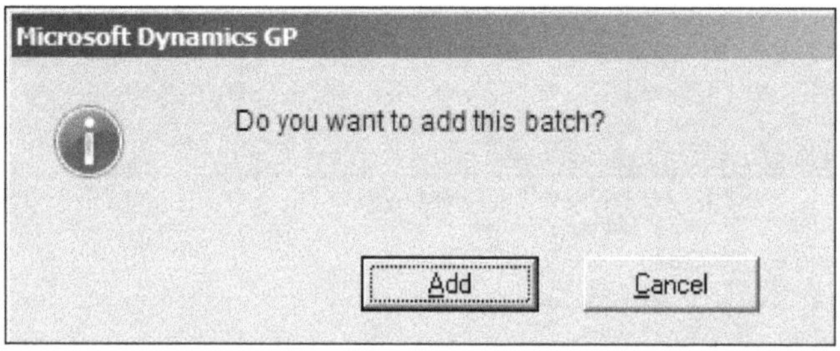

The following code will check the text of the dialog, and if it matches, will automatically push the **Add** button. The **Batch Entry** window will open and the user will never know the dialog existed. The prompt string test is case sensitive, so you need to enter the text exactly as it appears in the dialog window.

```
Option Explicit

Private Sub Window_BeforeModalDialog(ByVal DlgType As DialogType, _
    PromptString As String, Control1String As String, _
    Control2String As String, Control3String As String, _
    Answer As DialogCtrl)
    '--Check the value of the dialog prompt ---
    If PromptString = "Do you want to add this batch?" Then
        '-- If the prompt matches, push the Add button --
        Answer = dcButton1
    End If
End Sub
```

Something else you can do is change the prompt itself so that the dialog is more understandable or informative. The following dialog can be confusing because it does not define what it means by **record**:

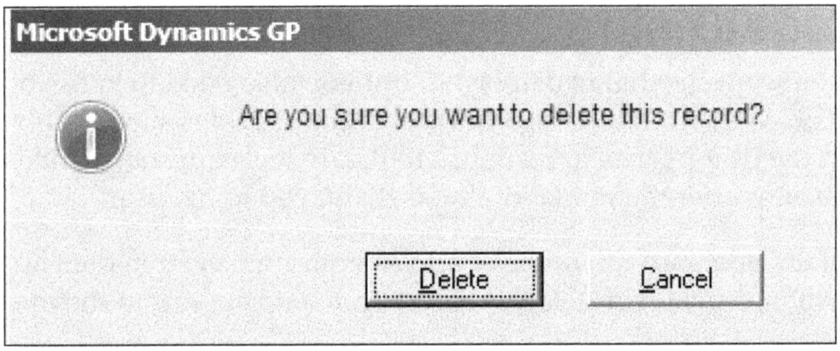

You will get this dialog if you push the **Delete** button on the **Edit Payables Checks** window. This dialog can be especially confusing to a new user. The user doesn't know if they are deleting the vendor record, the voucher record, the payment record, or what.

Alternatively, suppose the following dialog displayed instead:

You can change the dialog box prompt with the following code attached to the `BeforeModalDialog` window event:

```
Option Explicit

Private Sub Window_BeforeModalDialog(ByVal DlgType As DialogType, _
    PromptString As String, Control1String As String, _
    Control2String As String, Control3String As String, _
    Answer As DialogCtrl)
'--Check for the value of the prompt string --
If PromptString = "Are you sure you want to delete this record?" _
   Then
   '--Set the prompt string to a new value --
   PromptString = "Do you really want to delete this payment document?"
 End If
End Sub
```

AfterModalDialog

Sometimes you want the dialog displayed, but you also need to know how the user responded. The **AfterModalDialog** window event provides a means for you to check which button the user pushes in response to the dialog's question. Code attached to this event executes after the modal dialog is displayed to the user.

This is especially necessary if you have added your own custom data and need to know if you should save it or delete it based on what happens to the master record.

The following code will monitor modal dialogs on the **Customer Maintenance** window and return a message stating how the user responded to the dialog:

```
Option Explicit

Private Sub Window_AfterModalDialog(ByVal DlgType As DialogType, _
    PromptString As String, Control1String As String, _
    Control2String As String, Control3String As String, Answer As DialogCtrl)
    '-- Check the dialog displayed--
    If PromptString = "Do you want to save changes?" Then
        '-- Determine which button was selected
        If Answer = dcButton1 Then MsgBox "The Save button was selected"
        If Answer = dcButton2 Then MsgBox "The Discard button was selected"
        If Answer = dcButton3 Then MsgBox "The Cancel button was selected"
    End If

        '-- Check the dialog displayed--
    If PromptString = "Do you want to save changes to this customer?" Then
        '-- Determine which button was selected
        If Answer = dcButton1 Then MsgBox "The Save button was selected"
        If Answer = dcButton2 Then MsgBox "The Discard button was selected"
        If Answer = dcButton3 Then MsgBox "The Cancel button was selected"
    End If
End Sub
```

The Dynamic User Object Store

It is likely that your customization will involve collecting additional data. The common scenario is that you add additional fields using the Modifier and then you use VBA to store that data. However, where will you store it?

While you could create your own tables, there may be an easier method. The company database includes a table called the **Dynamics User Object Store** or **DUOS** table. The DUOS table is a powerful object that is woefully underused. Over the next few pages, you will learn how to store additional data without adding any tables to your database.

Architecture

The DUOS object is comprised of the **DUOSObjects** collection, the **DUOSObject** object, the **DUOSProperties** collection, and the **DUOSProperty** object.

The `DUOSObjects` collection is essentially a user-created set of data. You define the collection and can have as many collections as you wish. Common collections include customer, vendor, item, or employee. Inside the `DUOSObjects` collection is the `DUOSObject`. The collection is a set of data; the object is a single object record within the collection. If you had a collection of customer records, a specific customer would be an object in the collection.

The `DUOSProperties` collection is basically a list of the new fields you have created for each object. The `DUOSProperty` object represents the field value.

The following illustration represents the hierarchy of the DUOS:

The following table contains two object collections, four objects, four properties collections, and six property values. The `Objects` collection represents a category of records. In our example they are your vendors and items. The objects are individual Vendor IDs and Item IDs. The `Properties` collection contains the field name of the new fields you have created for the objects. `Property` is the individual value for the new field:

Collections	Objects	Properties	Property
Vendors	`ADVANCED0001`	`Home_Page`	`www.fitz.com`
Vendors	`BEAUMONT0001`	`AR_Manager`	`Sally Cooper`
Vendors	`DOLECKIC0001`	`Home_Page`	`www.baker.com`
Items	`100XGL`	`Manufacturer`	`AAMCO`
Items	`100XGL`	`Target_User`	`Home`
Items	`24XIDE`	`Manufacturer`	`Carllite`
Items	`256DRAM`	`EDI`	`No`

As you can see from the previous table, you can have an unlimited number of new properties for any object. Your object collections do not have to be your master files. You can create properties for anything you can uniquely identify, such as an *Item Class* or a *Document Type*.

Declaring the objects

Saving data to the DUOS table first involves declaring the collection and object so that you can access it. You would usually do this in the module's *General Declarations* section so that the DUOS objects will be available to all procedures within the module. The following code will declare the collection and object. This is the first step whenever working with DUOS objects:

```
Option Explicit
'Declaring the collection
Dim VendorCollection As DUOSObjects

'Declaring the object
Dim VendorObject As DUOSObject
```

Retrieving data

After declaring the object and collection, you need to name the collection that will contain the objects. You will use the `DUOSObjectsGet` method to create the collection. The following code creates a collection called **Vendor Information** in the `Changed` event for the `Vendor ID` field. This also demonstrates how to retrieve the fields and populate the window:

```
Private Sub VendorID_Changed()
    Set VendorCollection = DUOSObjectsGet("Vendor Information")

    If VendorID.Empty = False Then
        'The vendor ID field holds a value, retrieve DUOS
        Set VendorObject = VendorCollection(VendorID)

        'Populate the window fields
        HomePage = VendorObject.Properties("Home Page")
        ARManager = VendorObject.Properties("AR Manager")
    End If
```

Saving data

To save your DUOS data you need to monitor the **Save** button and watch for any dialogs Dynamics GP may present that will ask you to save the data. For example, if a vendor record is changed and you try to close the window or browse to another field without saving the data, the following modal dialog will appear:

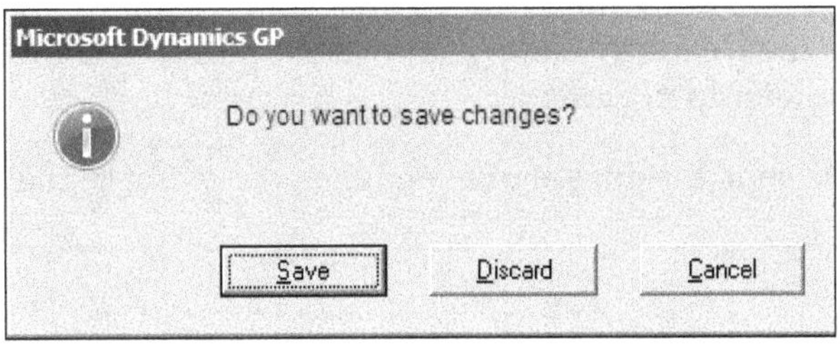

Using VBA's `AfterModalDialog` event you can find out how the user responded to this question. If they pushed the **Save** button, then you need to continue monitoring the progression of the save. If the save is successful, you will need to save your DUOS data as well. The circled part of the following code will return the user's response to the previous dialog:

```
Private Sub Window_AfterModalDialog(ByVal DlgType _
As DialogType, PromptString As String, _
Control1String As String, Control2String As String, _
Control3String As String, Answer As DialogCtrl)

    If PromptString = "Do you want to save changes?" Then
        If Answer = dcButton1 Then
            'User want's to save changes
            'Set the name and value for properties
            VendorObject.Properties("Home Page") = HomePage
            VendorObject.Properties("AR Manager") = ARManager
        End If
    End If

    If PromptString = "Are you sure you want to delete this record?"
        Then
        If Answer = dcButton1 Then
        'The user want's to delete the record
            WanttoDeleteVendor = True
        End If
     End If

End Sub
```

When you click on the **Save** button, Dexterity code examines the record for the presence of required fields. If the required fields have values, Dynamics GP will save the record. By using the `BeforeUserChanged` event of the **Save** button, you can check for required fields first and then save your data if the requirements are met. You do not want to use the `AfterUserChanged` event because the Dexterity code normally clears the window upon a successful save. Remember, the user has already pushed the **Save** button, but your VBA code gets to run first. By running first, your code can even stop the Dexterity code from ever running by setting the value of the `CancelLogic` parameter to `True`.

The following is the code for the `BeforeUserChanged` event of the **Save** button:

```
Private Sub Save_BeforeUserChanged(KeepFocus As Boolean, _
CancelLogic As Boolean)
    If VendorMaintenance.Required Then
        'Set the objects property names and values
        VendorObject.Properties("Home Page") = HomePage
        VendorObject.Properties("AR Manager") = ARManager
    End If
End Sub
```

Deleting data

You need to delete your DUOS data when the user deletes the corresponding record in Dynamics GP. You don't want to leave orphaned records in your database. Just as you did when saving a record, you need to respond to events that could result in a deleted record. The deleting process, like the saving process, has some obstacles in the way. Several dialogs could pop up and deny deletion of the record. It is important that you explore the user interface to familiarize yourself with those dialogs and what causes them to present.

Since you are using vendors, it makes sense to try to delete a few vendors to see how the system reacts. As soon as you click on the **Delete** button, you are presented with the following modal dialog:

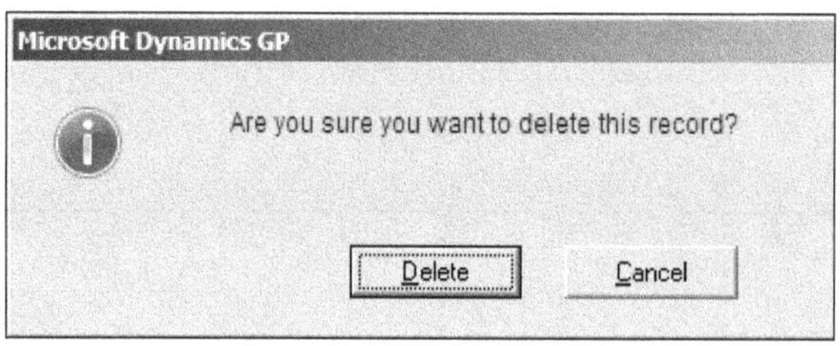

You will need an `AfterModalDialog` event to capture the user's answer. However, you are not out of the woods yet. Even after you confirmed that the user wants to delete the record, Dynamics GP may not allow it and may throw up this message:

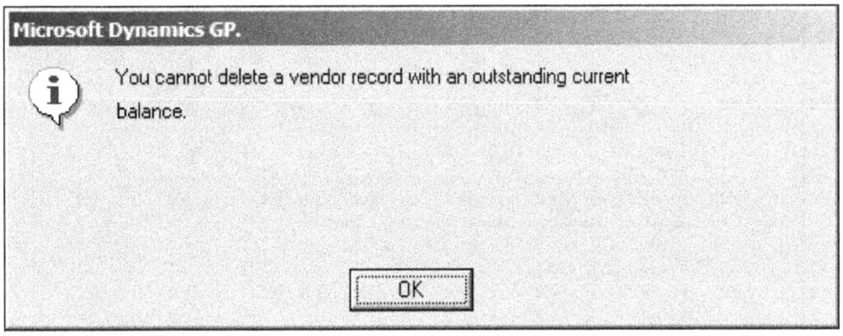

Initially, you may think that the above is a modal dialog, but it is not. It is a regular Dynamics GP window and you need to include it in your VBA project. Select *Ctrl + F11* when the window opens to add this window to your project.

This window is a modal window, but it is not a modal dialog. If you look at the modal dialogs and compare them to this window, you will notice the information icon on this window is different.

To summarize, in order to confirm a delete, you need to monitor three things:

1. The **Delete** button – when the user tries to delete the record.
2. The user's response to the *Are you sure* modal dialog – if yes, then the user wants to go ahead with the delete.
3. Dynamics GP's decision as to whether the record can be deleted – Dynamics GP may forbid the deletion of the record.

You will need three global variables to monitor the process. The circled code in the following code window shows the declaration of the variables:

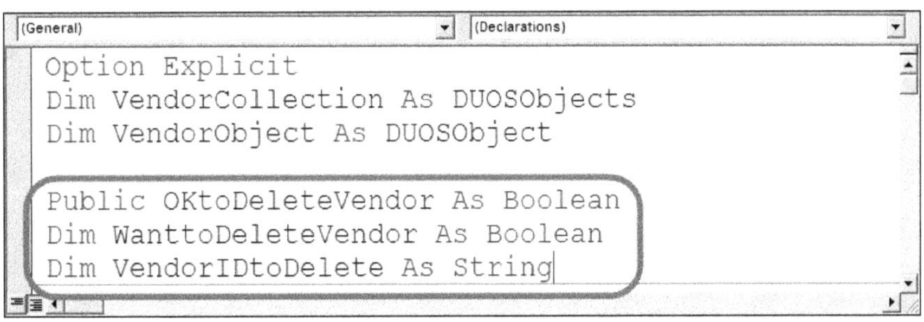

The variables are initialized in the `VendorID_Changed()` event. The circled code in the following window shows the variables being initialized:

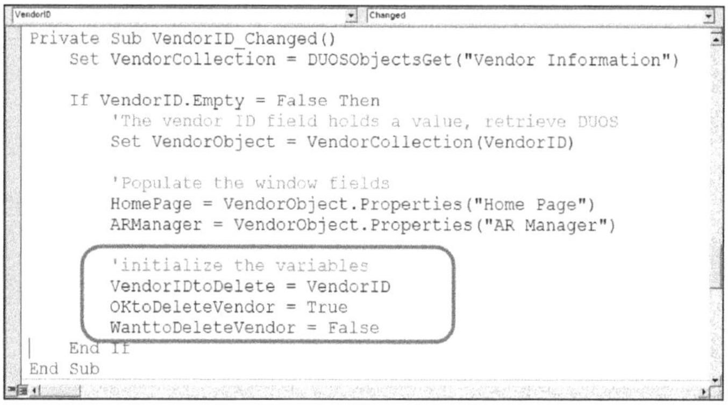

The `OKtoDeleteVendor` variable starts out with a value of `True`. This value changes to `False` if the Dynamics GP window opens and informs you that you cannot delete the vendor.

The `WanttoDeleteVendor` variable starts out with a value of `False`. The value changes to `True`, if you answer **Yes** to the modal dialog that asks **Are you sure you want to delete this record?**.

Both the `OKtoDeleteVendor` and the `WanttoDeleteVendor` variables must have a value of `True` before you can delete the vendor.

The circled code in the following window returns the user's response to the **Are you sure . . .** modal dialog. If the answer is **Yes**, it changes the value of the `WanttoDeleteVendor` variable to `True`.

```
Private Sub Window_AfterModalDialog(ByVal DlgType _
As DialogType, PromptString As String, _
Control1String As String, Control2String As String, _
Control3String As String, Answer As DialogCtrl)

    If PromptString = "Do you want to save changes?" Then
        If Answer = dcButton1 Then
            'User want's to save changes
            'Set the name and value for properties
            VendorObject.Properties("Home Page") = HomePage
            VendorObject.Properties("AR Manager") = ARManager
        End If
    End If

    If PromptString = "Are you sure you want to delete this record?"
        Then
        If Answer = dcButton1 Then
        'The user want's to delete the record
            WanttoDeleteVendor = True
        End If
     End If

End Sub
```

The following code checks to see if the window opened telling you that you couldn't delete the vendor because it had open transactions. If it does open, the `OKtoDeleteVendor` variable is set to `False`.

```
Option Explicit

'This window is the window that presents the dialog
'telling you that transactions exist for this vendor.
'This window's Window_AfterOpen() event runs BEFORE
'the Delete_AfterUserChanged() event.

Private Sub Window_AfterOpen()
    'Check to see if the Vendor Maintenance window is open
    If VendorMaintenance.IsLoaded Then
        'If this window opens, that means there are
        'transactions and it is not OK to delete the record,
        'that is why this statement is in the AfterOpen event
        VendorMaintenance.OKtoDeleteVendor = False
     End If
End Sub
```

The following window is the final step in the delete process. You put this code on the **Delete** button's `AfterUserChanged` event. The user has pushed the button to initiate the delete. This code checks both variables for a value of true. If it passes the test, the `Remove` method deletes the record. This is shown in the following screenshot:

```
Delete                                  AfterUserChanged

Private Sub Delete_AfterUserChanged()
    'User chose yes the the 'Are you sure' dialog because
    'WanttoDeleteVendor is true, and there were
    'no outstanding trx because the OKtoDeleteVendor is
    'also true
    If WanttoDeleteVendor And OKtoDeleteVendor Then
        'Delete the DUOS records
        VendorCollection.Remove (VendorIDtoDelete)
    End If
End Sub
```

Deploying a Modifier/VBA customization

Modifier/VBA customizations are very easy to deploy. You export your customizations into a `.package` file and then import the package file into each of the receiving workstations. You cannot put `.vba` files on a shared drive like you can for reports; each workstation must have its own set of `.vba` files.

Creating package files

To create a package file, go to the **Customization Maintenance** window using the following navigation:

Microsoft Dynamics GP | **Tools** | **Customize** | **Customization Maintenance**

On the **Customization Maintenance** window, highlight those customizations you want to distribute and click on **Export**. You can highlight several at a time using the *Shift* and *Ctrl* buttons on your keyboard:

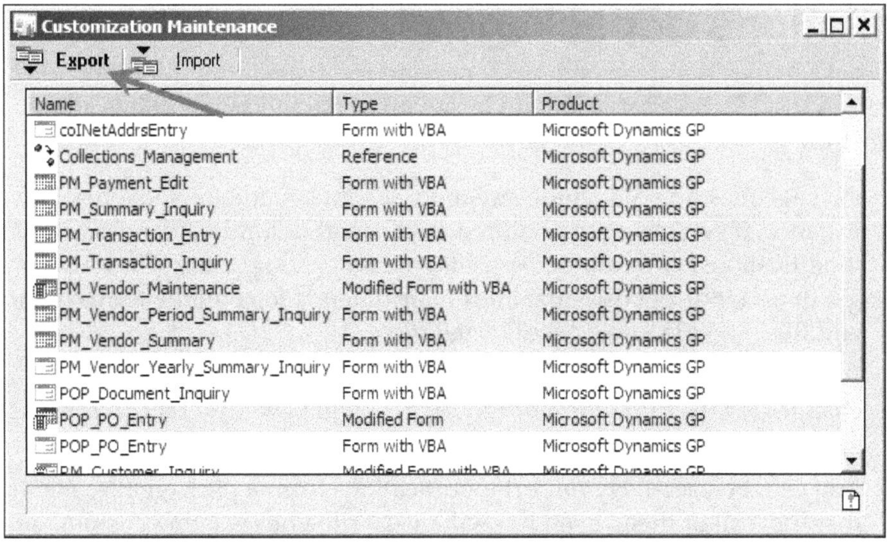

You can see what kind of customization is listed by looking at the **Type** column. The **Product** column tells you from which dictionary the objects originated.

When you create your package file keep in mind that you cannot pick and choose which customizations you import; it's all or nothing. Also, if the receiving workstation does not have the dictionary installed for each of the customizations in the package, none of the customizations will import.

Rather than exporting all of the customizations into a single package file, consider exporting them by type and dictionary for additional protection.

In addition to a vehicle for distribution, package files can serve as a backup for your VBA project. You will have invested many hours in your customizations; be sure there is a regular backup plan to protect your work.

Don't forget to import your package files occasionally to test the integrity of the file. A backup that cannot be restored is worthless.

Limitations of packages

Using package files is a clean and quick method for distributing Modifier/VBA customizations. However, you should be aware of the following points about package files:

- Package files do not include any modifications made to global resources. For instance, if you changed or added any format definitions, or made any string modifications, you cannot export them to a `.package` file. The only way to get those kinds of customizations distributed is to copy the forms dictionary and the `.vba` file to the client installation.
- Importing customizations from a package file will overwrite any existing changes for the same form or report. You will be warned if this is the case and you can cancel the import.
- You cannot selectively import modifications from a package file. You have to import all of them, even if it will overwrite your existing customizations.
- All windows belong to a form. Several related windows can be part of the same form. Even if you modify only one window, when you create the package file you export the form, which brings *all* of the windows with it. Even if your existing customization modified a different window on the form, the entire form is overwritten.
- You have to import VBA customizations on each workstation, even if the dictionaries are shared. The `.vba` file must be in the same folder as the runtime engine.

Finally, in order to use a Modifier/VBA customization, you need to either own the Modifier module or have the Customization Site Enabler registered. Any version of Dynamics GP purchased under the new Business Ready Licensing structure automatically includes the Customization Site Enabler.

Editing packages

Package files are text files and present some opportunities for editing to *tweak* the customization or make mass changes via search and replace. The following is the section of a package file that includes the list of items from the **Go To** button you added to the **Vendor Maintenance** window:

```
}
~StaticItems
{
        TextItem        "00001"
        {
                Item    "Payables Trx"
        }
        TextItem        "00002"
        {
                Item    "Inquiry - PM Trx"
        }
        TextItem        "00003"
        {
                Item    "Inquiry - PO Docs"
        }
        TextItem        "00004"
        {
                Item    "Inquiry - Yearly Summary"
        }
        TextItem        "00005"
        {
                Item    "Inquiry - Period Summary"
        }
        TextItem        "00006"
        {
                Item    "Inquiry - Vendor Summary"
        }
```

The ability to edit the `.package` file for a modified report gives you a way to copy calculated fields and copy report formats. For instance, you can copy the SOP Open Invoice Form to the SOP History Invoice Form rather than redoing all of your modifications.

Known issues with Windows 7

You may run into a couple of issues if you have MS Office 2010 installed with Dynamics GP 10 or Dynamics GP 2010. Two different scenarios have been identified related to the VBA component:

- **Scenario 1**: Dynamics GP crashes and creates a Dr. Watson fault bucket 1474386816
- **Scenario 2**: Dynamics GP freezes up and displays a warning dialog stating **File not found: VBA6.DLL**

Should you experience either of these situations, you can find some helpful tips at the following sites:

- `http://tinyurl.com/VBA6-DLL-Error`
- `http://tinyurl.com/importing-VBA`
- `http://community.dynamics.com/product/gp/f/32/p/54186/97749.aspx`

Your VBA customizations will work just as they always have using the Dynamics GP 2013 rich client. By rich client, we mean the regular Dexterity written workstation installation that you use today. What is ground-breaking about Dynamics GP 2013, is that the first Web client has been introduced. The Web client is a breakthrough for Dynamics GP, and one that has long been anticipated, but it comes at a price. Your VBA customization will not work on the Web client. The screen modifications you have made using the Modifier work just fine, but the VBA code will not execute.

Do not despair, you can still make your window changes using the Modifier, but you'll be giving life those changes with Visual Studio Tools instead of VBA. The window events are very similar to the VBA events you use now. You can find a table in *Appendix B* containing a side-by-side matrix comparing VBA, VS Tools, and Dexterity events. As you will see in the table, the events across all three tools are very similar. *Appendix B* is available as a free download from the following link :`http://www.packtpub.com/sites/default/files/downloads/0264_EN_AppB_Event_Matrix.pdf`

Go through the exercises in *Chapter 10, Creating Customizations using VS Tools*, if you want to get a feel for programming using that tool. As you go through the exercises, bear in mind that you will be using your modified window and not the WinForms described in the chapter. Dynamics GP 2013 does not recognize WinForms.

Perhaps this would be a good time to dive into Dexterity; you'll find good information and exercises to start you along the path of Dexterity in *Chapter 3, Getting Started with Dexterity*, *Chapter 4, Building the User Interface*, and *Chapter 5, sanScript - Making It Work*. Dexterity works perfectly with all aspects of the new Web client.

Summary

In this chapter we explored the VBA module and some of the things you can do with VBA to customize Dynamics GP. We now know the components of the VBA application and how to use the VBA Editor. We have worked through the objects in the Dynamics GP object module and have used several of their properties, methods, and events. We know that you can create new user forms with VBA and attach code that will interact with Dynamics windows and reports.

Our projects included adding a new button for navigation to the **Vendor Credit Summary** window and using an existing Dynamics GP object to add a **Go To** button to the **Vendor Maintenance** window. We walked through how to create, save, retrieve, and remove data from the DUOS table.

We also know how to deploy our VBA customizations along with some issues to keep in mind when creating and importing package files.

In the next chapter we will learn how to create robust customizations without writing a single line of code.

9

Code-free Customization

Dynamics GP has an impressive assortment of end-user tools available that you can use to produce an exciting array of enhancements. You can develop these enhancements without writing a single line of code.

This chapter will introduce you to some of the more inspiring code-free tools. We will be covering the following tools:

- SmartList Builder
- Excel Report Builder
- Drill-Down Builder
- Extender

At the end of this chapter, these tools will be added to your repertoire to make Dynamics GP behave the way you want it to.

Overview of tools

In the beginning, you had only Dexterity to create customizations. A few years later, VBA was embedded into the product, providing access to a stress-free cross-dictionary functionality. After that, Continuum entered the scene for Visual Basic and Delphi programmers. Continuum provided even more access to third-party dictionaries and was an efficient tool for smaller-scale customizations.

In Version 7 of Dexterity, COM objects were supported. Now you could integrate web services with Dynamics GP and call them. Using COM, Continuum was repurposed as an **Application Programming Interface** (**API**) for Dynamics GP objects.

Then, Visual Studio Tools (VS Tools) came along. VS Tools opened the black box that was the Dynamics dictionary, and the ability to customize Dynamics GP spread like a virus. Suddenly, there was a long list of options for you to choose from for custom development.

On the home front, tools were being acquired that were targeted at power users instead of programmers. At last, you could customize Dynamics GP without knowing a programming language.

SmartList Builder

SmartList Builder gives you the ability to create new SmartList objects. A SmartList object is different from a favorite. A **favorite** is a subset of an object. Favorites are what end-users create from an object.

Average Days to Pay and **Customer Balance** are the names of two prebuilt SmartList favorites that ship with Dynamics GP. To reveal these prebuilt favorites, and the objects from which they were built, launch SmartList from the menu bar using **Microsoft Dynamics GP | SmartList**. Expand the **Sales** folder. Inside the **Sales** folder you'll find the **Customers** folder. The **Customers** folder is a SmartList object. Now, expand the **Customers** folder. Inside the **Customers** folder are several prebuilt favorites including **Average Days to Pay** and **Customer Balance**.

If you wanted to build a favorite that included information regarding unposted batches, you could not accomplish it using the prebuilt objects. To satisfy this requirement, you would need to build your own SmartList object based on the Posting Definitions Master table. SmartList Builder is the tool you would use to build that SmartList object.

Excel Report Builder

Excel Report Builder is a component of SmartList Builder. It takes SmartList Builder a step further by producing refreshable Excel spreadsheets. While you can export data from SmartList to Excel, you cannot refresh the exported data with the current database information. The exported data is static. It is no longer connected to the database.

Excel Report Builder allows you to create a spreadsheet using live information that you can update with current information whenever you choose. You can even use this tool to take a SmartList object you created in SmartList Builder, and turn it into a refreshable Excel spreadsheet.

Imagine being able to format and organize a spreadsheet once, and then use it over and over again with the push of a button. No additional exporting or formatting required.

Drill-Down Builder

Using Drill-Down Builder, you can provide drill-down capability to a report. Drill-down capability means that you can double-click on a linked object and open Dynamics GP to look at the details behind the data. You are not limited to Excel reports using this tool. You can place a drill-down object anywhere in a report that will accept a hyperlink connection. So now you can take Word documents, board reports, charts, graphs, and of course spreadsheets, and transform them into interactive documents.

Extender

Extender gives you the power to add a near-unlimited amount of additional information to existing Dynamics GP windows. In addition, you can create brand-new windows that store information not related to an existing window. But Extender goes far beyond just windows.

SmartList Builder

We'll start with SmartList Builder. You can use SmartList Builder to create custom views for the database using a point-and-click interface. No SQL knowledge is required, unless you need advanced reports with custom business logic. These custom SmartList objects meld seamlessly with the out-of-the-box SmartList objects. The SmartList Builder's module satisfies the need for a user-friendly, on-demand reporting tool. SmartList Builder is likely the most popular end-user tool out there, even more popular than Modifier/VBA.

Your accountants will love SmartList Builder.

A **SmartList** is a simple report presented in a column-and-row format resembling Microsoft Excel. You can use SmartList objects wherever you need a simple display of the data; you can even download SmartList data into Microsoft Excel (Excel). Accounting managers often use this kind of report as a simple tool for managing receivables and payables.

In this section, you are going to create a SmartList object that you can use to manage inventory quantities in different warehouses. From this SmartList object, you will be able to jump to the **Item Maintenance** window, and the Item Transaction Inquiry window. These jumps are known as Go Tos.

Your job as the developer is to build the query that pulls the base dataset. It is from this dataset that your users develop SmartList favorites. A SmartList **favorite** is a saved set of parameters your user can run repeatedly.

Getting Started with SmartList Builder

The very first thing to do is to purchase SmartList Builder. SmartList Builder is not included in the Business Essentials package, or the Advanced Management package. You also need to make sure you install SmartList Builder. To check if it's already installed, or to add the feature, go to the control panel's **Programs and Features** section, select **Microsoft Dynamics GP 2010** and then click on the **Change** button in the upper half of the window. The **Select Features** window will open. Scroll down the feature list until you find the **SmartList Builder** item. If the feature is installed you will not see a red **X** mark across it. The following screenshot shows what the window will look like if **SmartList Builder** has been installed:

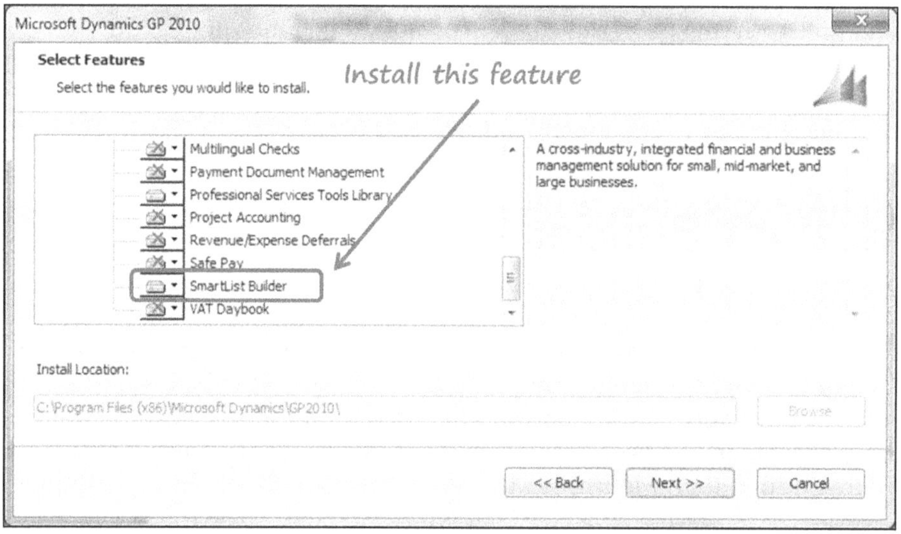

Once you've installed it, you may want to download the SmartList templates provided by Microsoft from the CustomerSource/PartnerSource website. You can access those templates in the **Reports Library** section of the website as shown in the following screenshot:

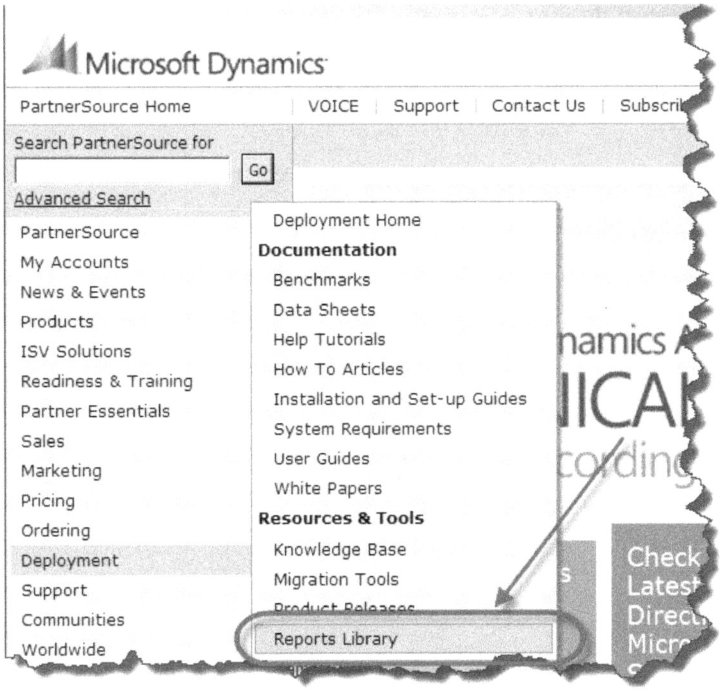

From the **Reports Library** page, select **Microsoft Dynamics GP | Microsoft Business Solutions-Great Plains**.

From the **Microsoft Dynamics GP/Great Plains Reports Library** page, scroll down until you see the **SmartList Templates** link. Click on the **SmartList Templates** link:

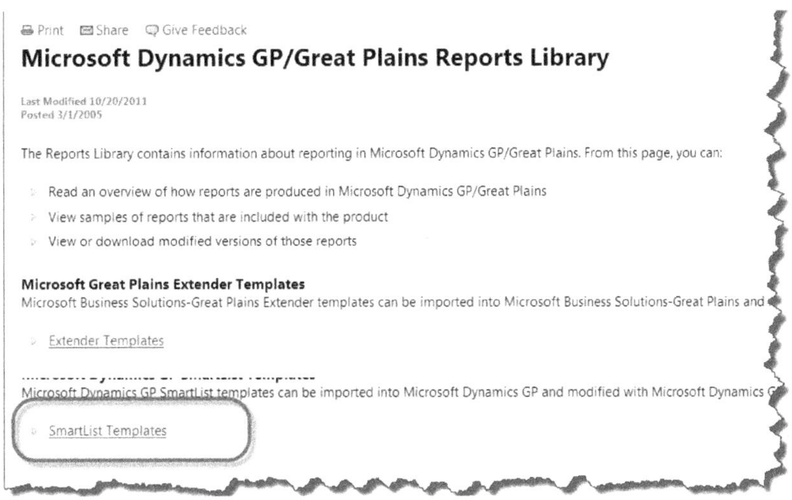

The **SmartList Templates for Microsoft Dynamics GP 10.0** window will open. Several templates are available for download that you can use to replace the out-of-the-box SmartList objects. Download the templates you use in your business:

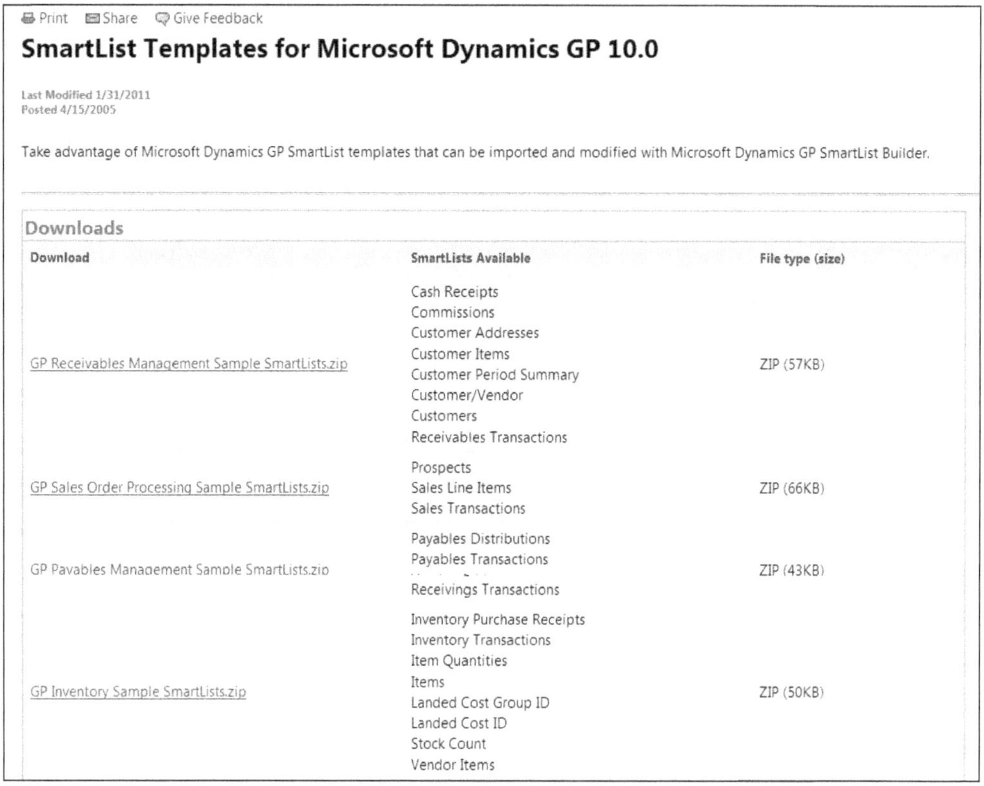
Print Share Give Feedback

SmartList Templates for Microsoft Dynamics GP 10.0

Last Modified 1/31/2011
Posted 4/15/2005

Take advantage of Microsoft Dynamics GP SmartList templates that can be imported and modified with Microsoft Dynamics GP SmartList Builder.

Downloads

Download	SmartLists Available	File type (size)
GP Receivables Management Sample SmartLists.zip	Cash Receipts Commissions Customer Addresses Customer Items Customer Period Summary Customer/Vendor Customers Receivables Transactions	ZIP (57KB)
GP Sales Order Processing Sample SmartLists.zip	Prospects Sales Line Items Sales Transactions	ZIP (66KB)
GP Payables Management Sample SmartLists.zip	Payables Distributions Payables Transactions Receivings Transactions	ZIP (43KB)
GP Inventory Sample SmartLists.zip	Inventory Purchase Receipts Inventory Transactions Item Quantities Items Landed Cost Group ID Landed Cost ID Stock Count Vendor Items	ZIP (50KB)

These templates benefit you because you cannot edit the default SmartList objects. The downside is that you will have to recreate any favorites that come with Dynamics GP. It is easy to do though; all you have to do is look at the existing favorites and duplicate the settings.

Importing the templates

Once downloaded, you need to import the templates into your current installation. You only need to do this once because SmartList objects are system-level objects rather than company-level objects. Extract the ZIP files you downloaded and put the extracted files in your `Documents` folder. You should have several `.xml` files after you have extracted the ZIP file's contents.

Navigate to the SmartList import window: **Microsoft Dynamics GP** | **Tools** | **Customize** | **SmartList Builder** | **Import**.

The window as shown in the following screenshot will open:

Click on the **Import** button and choose one of the downloaded `.xml` files, and then click on the **Open** button. Click on the **Import** button and the import will begin. It may take a little while because there are quite a few objects included in the `.xml` file. Repeat this process for each of the `.xml` files you want to import. Sadly, you cannot import more than one file at a time.

When you are ready to deploy the imported SmartList objects, change the **Product** field from **SmartList Builder** to **Microsoft Dynamics GP** and select the appropriate series. This process will take the imported SmartList objects out of the **Additional SmartLists** folder and move them into their appropriate series. Your window should look substantially similar to the one shown in the following screenshot:

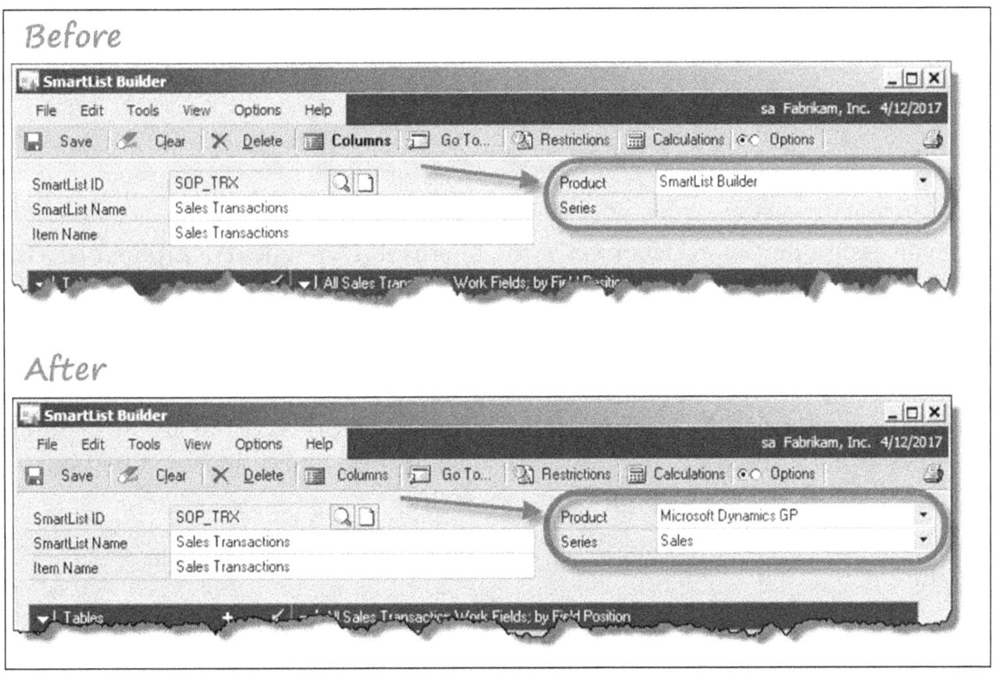

When you are ready to put your new SmartList objects into service, take the security away from the existing SmartList objects and assign it to the imported SmartList objects. By doing this security swap, you will be allowed to replace the out-of-the-box SmartList objects with SmartList Builder's SmartList objects.

If, upon opening the SmartList navigator, you have any SmartList object recorded under the **Additional SmartLists** folder (shown in the following screenshot), you have failed to change the **Product** field's value from SmartList Builder to Microsoft Dynamics GP.

Once you assign the SmartList object to the Microsoft Dynamics GP product and the correct Series, it will move down to its appropriate position as explained previously.

Creating a SmartList object

You will be creating a SmartList object that provides information about the inventory quantities in three specific warehouses. Your completed SmartList object will be similar to the following screenshot:

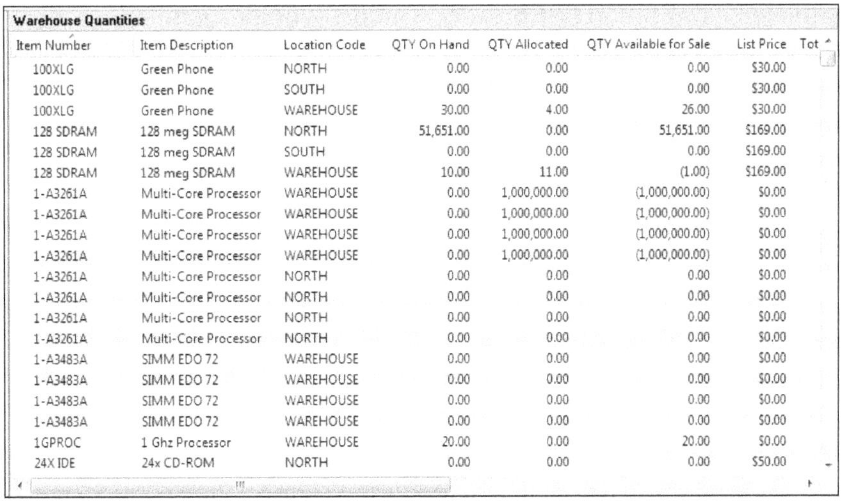

Warehouse Quantities

Item Number	Item Description	Location Code	QTY On Hand	QTY Allocated	QTY Available for Sale	List Price	Tot
100XLG	Green Phone	NORTH	0.00	0.00	0.00	$30.00	
100XLG	Green Phone	SOUTH	0.00	0.00	0.00	$30.00	
100XLG	Green Phone	WAREHOUSE	30.00	4.00	26.00	$30.00	
128 SDRAM	128 meg SDRAM	NORTH	51,651.00	0.00	51,651.00	$169.00	
128 SDRAM	128 meg SDRAM	SOUTH	0.00	0.00	0.00	$169.00	
128 SDRAM	128 meg SDRAM	WAREHOUSE	10.00	11.00	(1.00)	$169.00	
1-A3261A	Multi-Core Processor	WAREHOUSE	0.00	1,000,000.00	(1,000,000.00)	$0.00	
1-A3261A	Multi-Core Processor	WAREHOUSE	0.00	1,000,000.00	(1,000,000.00)	$0.00	
1-A3261A	Multi-Core Processor	WAREHOUSE	0.00	1,000,000.00	(1,000,000.00)	$0.00	
1-A3261A	Multi-Core Processor	WAREHOUSE	0.00	1,000,000.00	(1,000,000.00)	$0.00	
1-A3261A	Multi-Core Processor	NORTH	0.00	0.00	0.00	$0.00	
1-A3261A	Multi-Core Processor	NORTH	0.00	0.00	0.00	$0.00	
1-A3261A	Multi-Core Processor	NORTH	0.00	0.00	0.00	$0.00	
1-A3261A	Multi-Core Processor	NORTH	0.00	0.00	0.00	$0.00	
1-A3483A	SIMM EDO 72	WAREHOUSE	0.00	0.00	0.00	$0.00	
1-A3483A	SIMM EDO 72	WAREHOUSE	0.00	0.00	0.00	$0.00	
1-A3483A	SIMM EDO 72	WAREHOUSE	0.00	0.00	0.00	$0.00	
1-A3483A	SIMM EDO 72	WAREHOUSE	0.00	0.00	0.00	$0.00	
1GPROC	1 Ghz Processor	WAREHOUSE	20.00	0.00	20.00	$0.00	
24X IDE	24x CD-ROM	NORTH	0.00	0.00	0.00	$50.00	

You will be using the following features of SmartList Builder while developing your SmartList object:

- Adding and relating tables
- Designating fields to be included in the default view
- Setting field options
- Specifying the column ordering of the default view
- Designating additional fields that the user can add
- Creating calculated fields
- Creating a restriction
- Building a Go To

To begin building your SmartList object, open the **SmartList Builder** window using the following navigation: **Microsoft Dynamics GP | Tools | SmartList Builder | SmartList Builder**.

Complete the fields at the upper half of the window with the following values:

Field Name	Value
SmartList ID	`WHSE_QTY`
SmartList Name	`Warehouse Quantities`
Item Name	`Warehouse Quantities`
Product	`Microsoft Dynamics GP`
Series	`Inventory`

Once completed, your **SmartList Builder** window will look similar to the following screenshot:

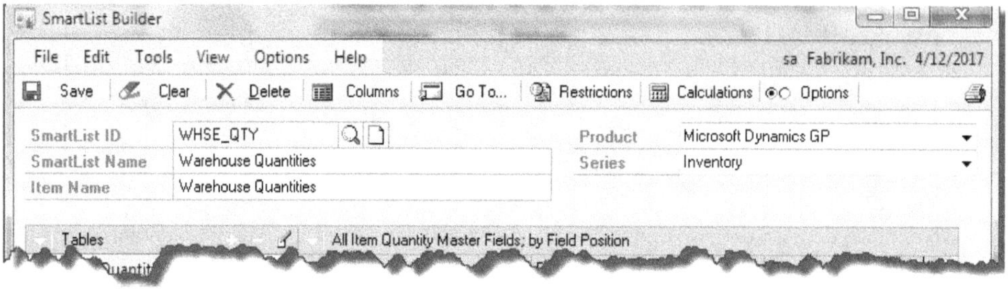

Adding tables

Now you need to add the tables from which the fields you just listed will originate. Selecting the correct table is probably the trickiest part of building a SmartList object. The following are a few hints to help you through the maze of tables:

- **Cards**: The data appearing in the Cards section is typically stored in a table containing the word Maintenance or Master; sometimes it is abbreviated to MSTR.
- **Unposted transactions**: Unposted transactions are typically found in tables containing the word Work.
- **Posted transactions**: Posted transactions are regularly found in tables containing the word Open or History.

The first table you add to a new SmartList object is the base table. All future tables you add will be related directly or indirectly to the base table. Therefore, choosing the correct base table is paramount. It's usually pretty safe to pick the table that contains the bulk of the data you want to display.

To add a table, click on the plus sign (+) and select **Microsoft Dynamics GP Table** from the button-drop list as shown in the following screenshot:

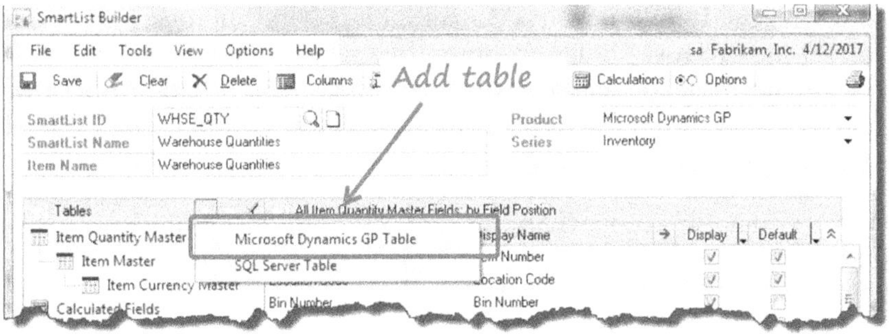

The **Add Table** window will open. Select **Microsoft Dynamics GP** as the value of the **Product** field, **Inventory** as the value of the **Series** field, and **Item Quantity Master** as the value for the **Table** field. Your window should look like the window shown in the following screenshot:

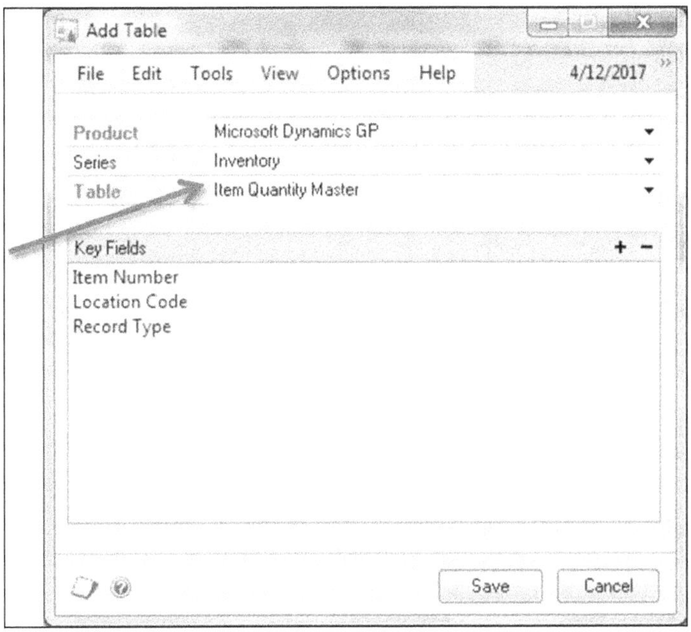

Click on the **Save** button to save the table and close the window.

As you also want to include the item's description, you need to add the **Item Master** table to the SmartList object. To add the additional table, click on the **Item Quantity Master** table and then click on the plus sign (+) again.

Like you did with the **Item Quantity Master** table, select **Microsoft Dynamics GP** as the value for the **Product** field, and **Inventory** as the value for the **Series** field. This time, select **Item Master** as the value for the **Table** field. This **Add Table** window contains a field that the first **Add Table** window didn't. The **Link Method** field is the extra field.

Whenever you add a table other than the base table, you need to identify how that table is related to the other table. You accomplish this with the **Link To Table** field and the **Link Method** field. You have two choices of link methods: **Equals** and **Left Outer**. You will be using the **Equals** link method because you want to include only records that are in both the **Item Quantity Master** table and the **Item Master** table. Your **Add Table** window will look like the following screenshot once completed:

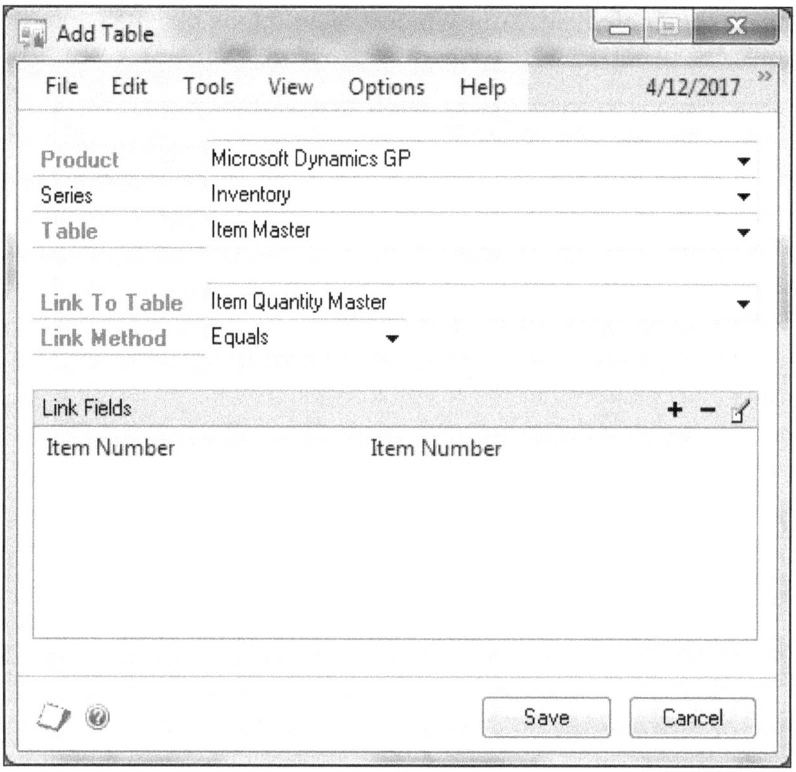

Click on the **Save** button to save the table just added and close the window.

The final table houses the **List Price** field. To retrieve the list price, you need to add the **Item Currency Master** table.

With the **Item Master** table selected, click on the plus (+) sign once more. Complete the **Add Table** window using the information shown in the following screenshot:

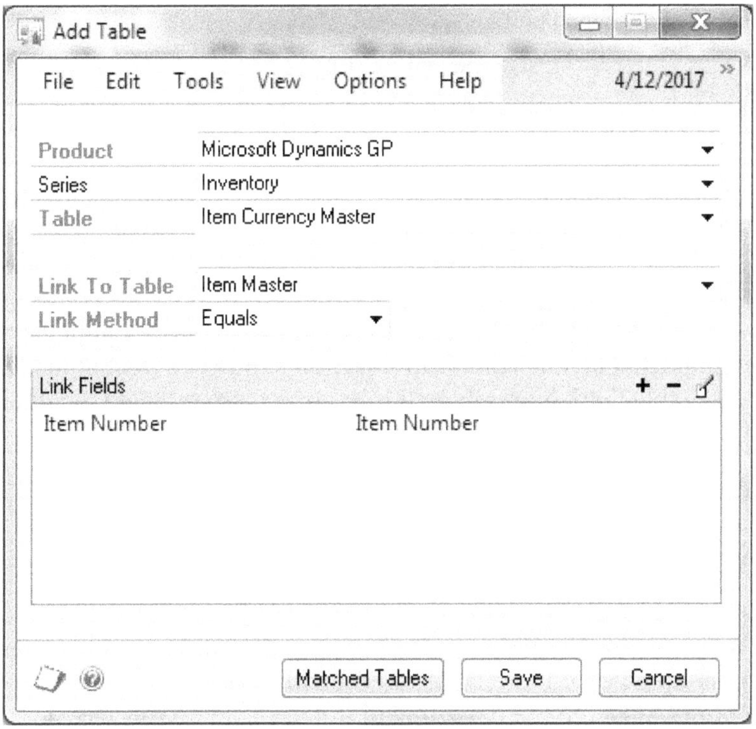

Click on the **Save** button to save the table just added and close the window.

Now that you have all of the tables added, you need to decide which fields to display by default. You also need to decide which additional fields should be available for the user to add at his or her discretion.

Fields

You want your SmartList object to display the following fields by default:

Field Name
Item Number
Item Description
Location Code
QTY on Hand
QTY Allocated
QTY Available for Sale
List Price

Field Name
List of On Hand QTY
Standard Cost
Current Cost
Item Type
Item Shipping Weight
Item Class Code

After you select the default fields, you need to choose the optional fields that your user can pick to be displayed. Your field selection should be deliberate. Do not just select every available field. Not only will it be confusing, but also the data may be wrong for the context in which you want to use it.

For example, it is not necessary for you to display the Item Number field from both the Item Master table and the Item Quantity Master table. This information is redundant and can be confusing.

Using the column of checkboxes on the far right-hand side, indicate which fields appear on the SmartList object's default view. The default view is the base list of fields that come up when you first open a SmartList object, before you select any favorite. A field is considered a default field if it appears on this base list.

Using the column of checkboxes to the left-hand side of the default list, you can indicate which fields are available for the user to add at their discretion. User-selected fields are displayed in the SmartList object's views created by the user, which are called favorites. The following screenshot is of the **SmartList Builder** window showing the fields available for display or those that are default from the Item Master table:

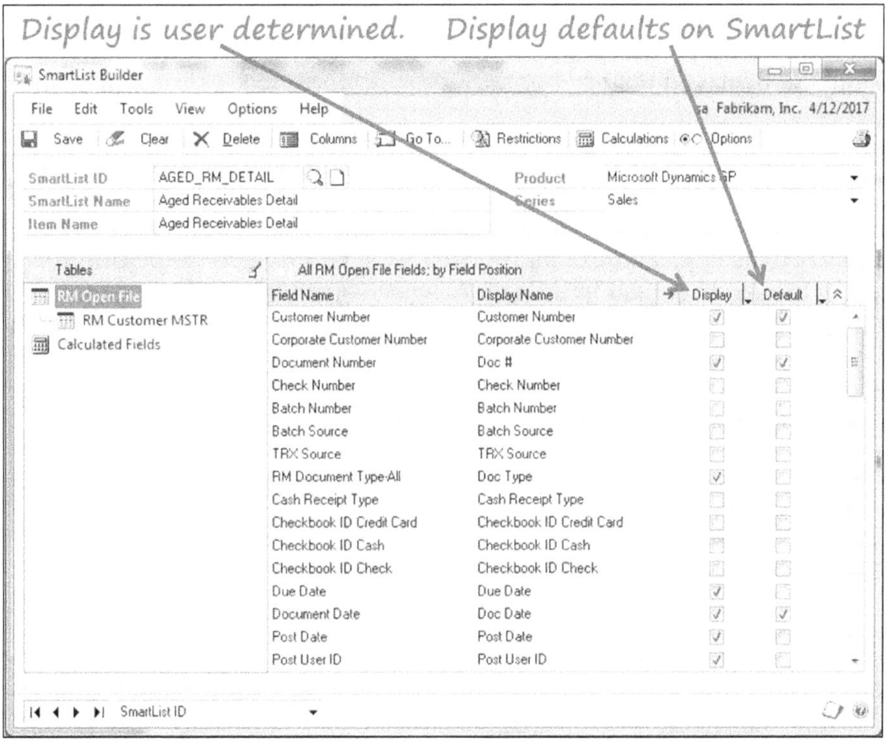

You must select at least one field in the **Default** column in order to save the SmartList object. Now, go back to your list of fields and consider from which table each field should originate. Mark the following fields from the tables indicated, as a default for your SmartList object:

Field Name	Originating Table
Item Number	**Item Quantity Master**
Item Description	**Item Master**
Location Code	**Item Quantity Master**
QTY on Hand	**Item Quantity Master**
QTY Allocated	**Item Quantity Master**
List Price	**Item Currency Master**
Standard Cost	**Item Master**
Current Cost	**Item Master**
Item Type	**Item Master**
Item Shipping Weight	**Item Master**
Item Class Code	**Item Master**

When you check the fields, there are multiple options available for you to change the display characteristics of the field. These changes apply to both the default fields and the optionally displayed fields.

Field options

You can set field options either individually or globally by field type. When selections are available, a blue expansion arrow gets activated in the column header. See the arrow circled in the following screenshot. By selecting this arrow, you will open a supplementary window that displays various options depending on the type of field you have selected. The name of the supplemental window is **Set Field Options** and is shown in the following screenshot to the left-hand side of the **SmartList Builder** window:

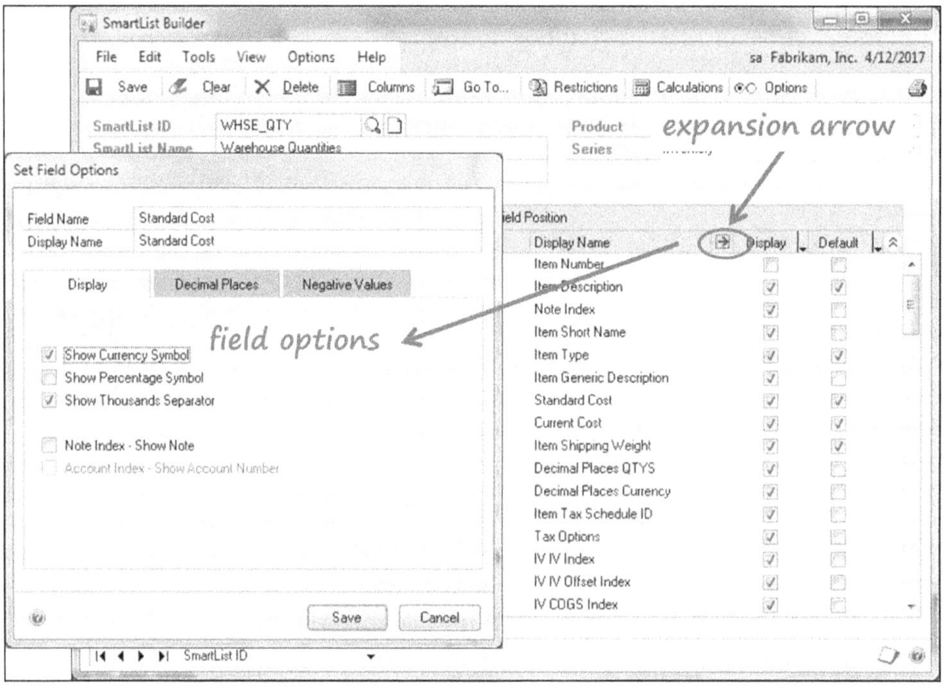

If you want to apply the same settings for all fields of a particular type, follow the menu navigation: **Options** | **Edit Field Settings**.

If you select, for example, the **Integer Fields** option, you'll open the **Set Field Options** window specifically for integers. Listed on the left-hand side of the window are all the integer fields available for your SmartList object. All the fields are displayed, not just the fields you checked. You can individually select fields using the checkboxes, or select all the fields at once. On the right-hand side of the window are the available field options for integer fields. The field options you select on the right-hand side will apply to the checked fields on the left-hand side.

The **Set Field Options** window for integer fields is displayed in the following screenshot:

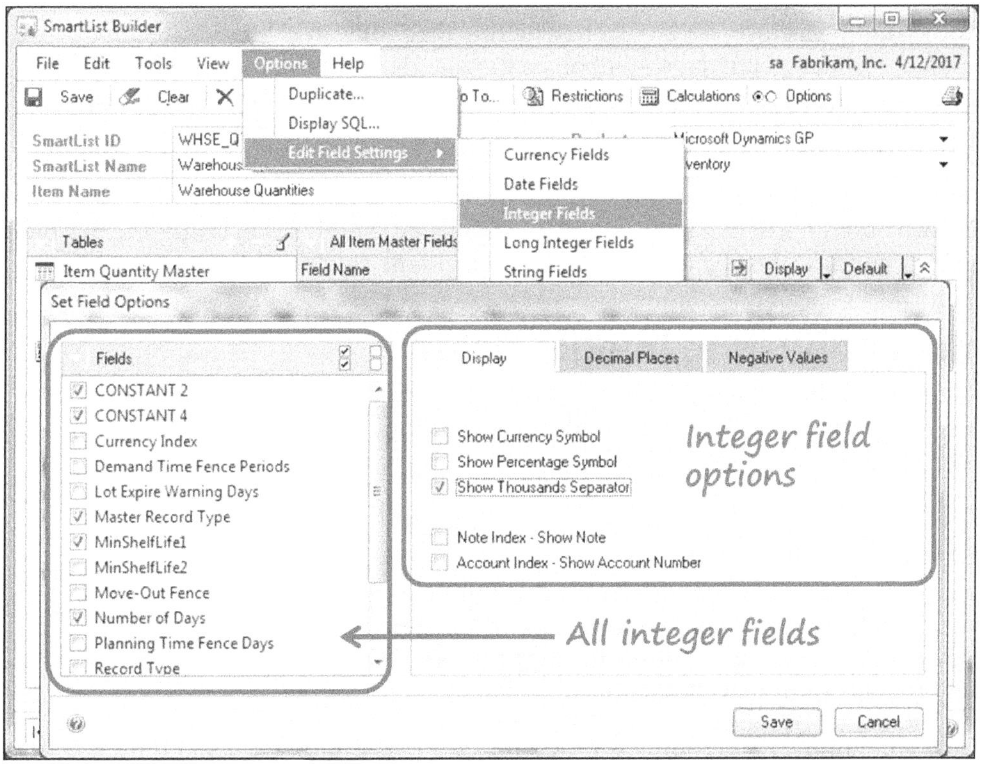

Several field types have some unique options. We'll take a closer look at the following field types:

* Currency fields
* Date fields
* Integer and Long Integer fields
* String fields

Currency fields

You have the ability to choose the value to appear as negative based on a number of factors. For instance, you could select the document type field as the determining item. You can then choose which values in that field should result in a negative number. For example, if you choose the document type as the determining field, you may select the values representing credit memos and payments to display as negative numbers.

The Record Note index is a currency field; you can choose to display the note's text instead of the index number itself.

Another unique setting for a currency field is that you can choose to display the currency symbol, percentage symbol, or thousands separator.

Date fields

Dynamics GP uses 1/1/1900 as the default bound for all date fields. If no date value has been entered into the field, the value displayed is 1/1/1900. Using the **Set Field Options** window, you can choose to display a blank field instead of 1/1/1900.

Integer and long integer fields

Integer fields share the same characteristics as currency fields, with one exception. The Account Number index is an integer field. Similar to the note index, you can choose to display the account number instead of the account index.

String fields

You can use a mask when displaying a string field. You can apply the same mask to a few fields, or no field, by using the checkboxes. The phone number and social security number masks are built-in, but you can design any type of mask you want using the **Set Field Options** window.

When you build a mask, you use the capital X as a placeholder. Field data will fill the capital Xs and any other value will appear as static text in your output. For example, let's say you want your output to always contain nine characters with a space, hyphen, and a space between each set of three characters, for example: 123 - 456 - 789; your mask would look like this: **XXX - XXX - XXX**.

The following screenshot shows the **Set Field Options** window using the mask just described:

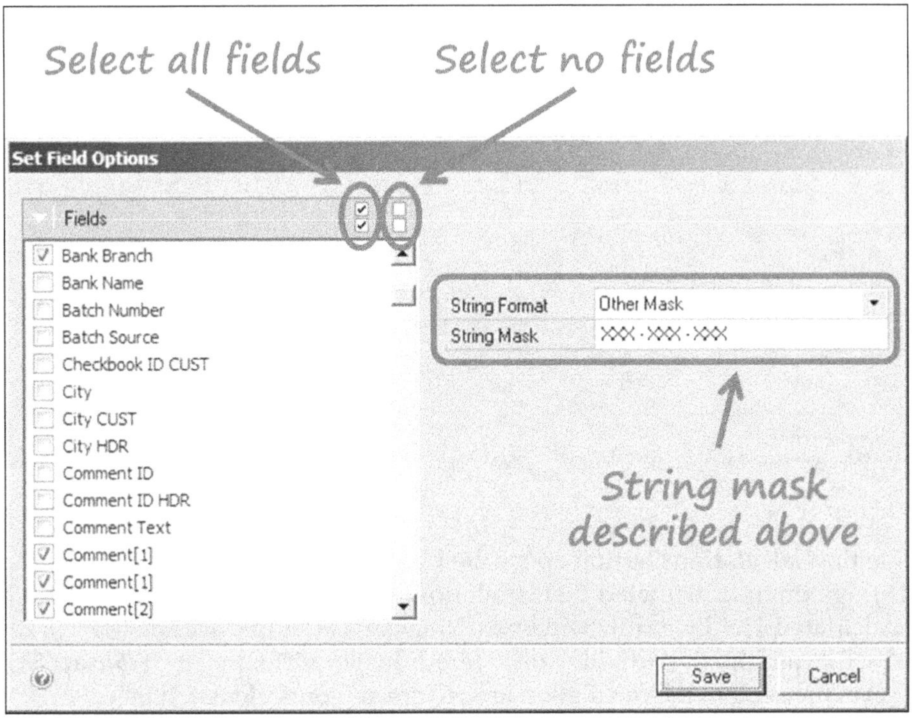

Should you choose a phone number field to appear on your SmartList object, you will probably need to assign the phone number mask to it. If you do not, the phone number will appear as a simple string of 10 numbers without any hyphen or parenthesis.

If you want the field values to appear just as they do in the database, select **None** as the value of the **String Format** field.

Calculated fields

You may need a field that doesn't exist in any table, as you do in this project. You will create this field using a formula. This type of field is called a **Calculated Field**. You will need several calculated fields for the SmartList objects you are going to build in this section.

The first step in creating a calculated field is to click the **Calculations** button on the **SmartList Builder** window, as shown in the following screenshot:

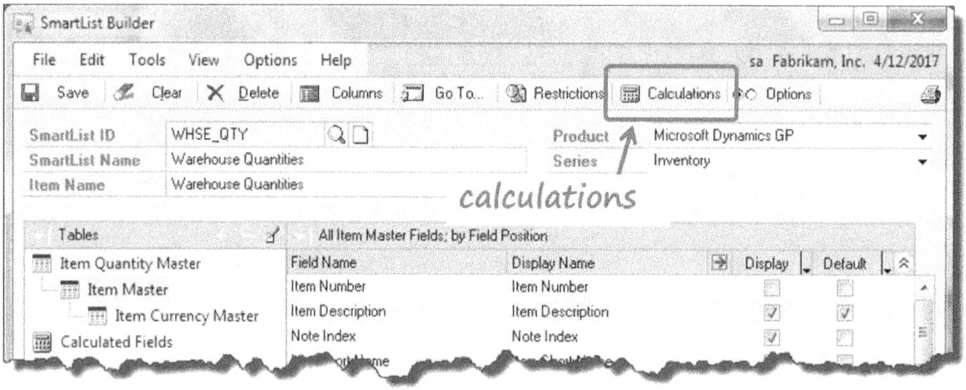

Clicking the **Calculations** button opens the **Calculated Fields** window. Press the plus (**+**) sign on the **Calculated Fields** window to create a new calculated field. The **Add Calculated Field** window will open. You can use simple arithmetic, + (add), - (subtract),* (multiply), / (divide), or % (modulus) or virtually any **Transact SQL** (**T-SQL**) expression in the calculation area to create your calculated field.

On the right-hand side of the **Add Calculated Field** window, you have a button-drop menu that you can use to toggle between functions and table field names. You will only see a few of the more popular T-SQL functions listed in the **Functions** list as a matter of convenience. This in no way is intended as a limitation on the T-SQL functions or statements you can use. You can use any valid T-SQL statement to build your calculated field, not just the ones listed.

The screenshot that you see next shows the windows you will use when creating one of your calculated fields:

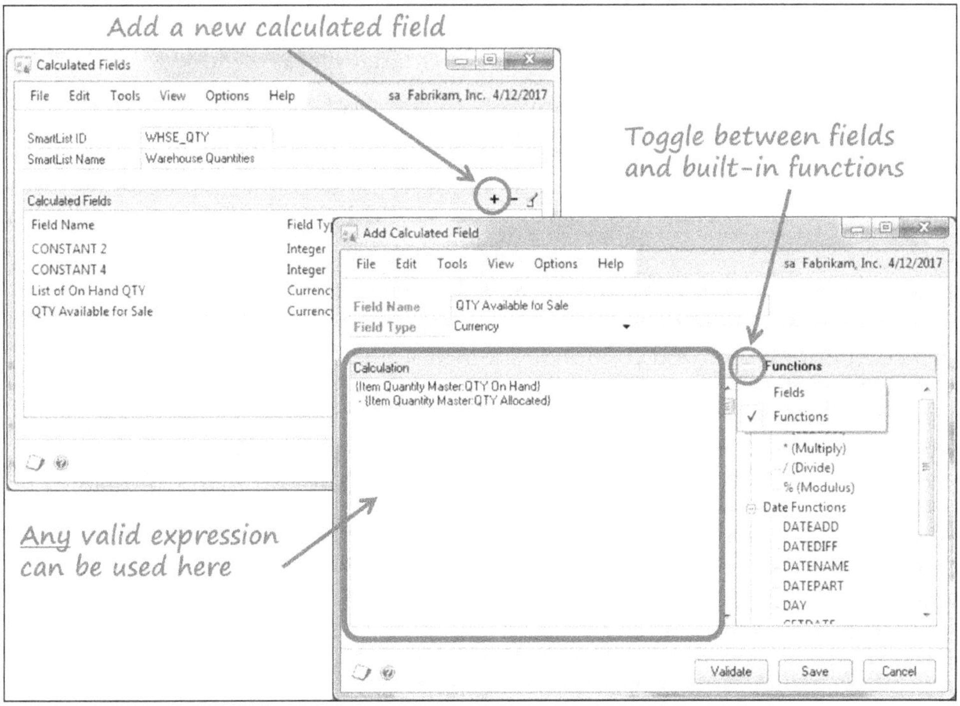

The formula shown in the screenshot subtracts the allocated quantity from the quantity on hand, resulting in the quantity available for sale.

You can place the calculated fields in the SmartList object just like table fields. You cannot, however, include a calculated field within another calculated field.

To create the calculated field, you can either type the values into the calculation area directly, or double-click on the field or function in the list on the right-hand side to move the field name or function into the calculation area.

Create the four calculated fields you need for this project using the following instructions:

Calculated field 1: QTY Available for Sale

The `QTY Available for Sale` field subtracts the allocated quantity from the quantity on hand. Follow these steps to create the `QTY Available for Sale` calculated field:

1. Open SmartList Builder.
2. Open the **WHSE QTY** SmartList object in the **SmartList Builder** window.

3. Click on the **Calculations** button in the upper half of the window.
4. Click on the plus (+) button in the **Calculated Fields** window.
5. Use the following table to complete the fields of the **Add Calculated Fields** window:

Field Name	Field Type
`QTY Available for Sale`	Currency

6. Toggle to the **Fields** option using the drop-down arrow at the upper-left corner of the second column in the **Add Calculated Fields** window.
7. Expand the **Item Quantity Master** table in the field tree.
8. Scroll until you find the **QTY On Hand** field. Double-click on the **QTY On Hand** field. The fully qualified field's name will appear in the calculation area.
9. Scroll until you find the **QTY Allocated** field. Double-click on the **QTY Allocated** field. The fully qualified field's name will appear in the calculation area.
10. Place your cursor in between the two fields and insert the minus sign (hyphen). The complete formula in the calculation area will be shown on a single line as follows:

```
{Item Quantity Master:QTY On Hand}
- {Item Quantity Master:QTY Allocated}
```

Calculated field 2: List of On Hand QTY

The `List of On Hand QTY` field calculates the total of the `On Hand QTY` field multiplied by the `List Price` field. Follow these steps to create the `List of On Hand QTY` calculated field:

1. Open SmartList Builder.
2. Open the **WHSE QTY** SmartList object in the **SmartList Builder** window.
3. Click on the **Calculations** button in the upper half of the window.
4. Click on the plus (+) button in the **Calculated Fields** window.
5. Use the following table to complete the fields of the **Add Calculated Fields** window:

Field Name	Field Type
`List of On Hand QTY`	Currency

6. Toggle to the **Fields** option using the drop-down arrow at the upper-left corner of the second column in the **Add Calculated Fields** window.
7. Expand the **Item Currency Master** table in the field tree.
8. Scroll until you find the **List Price** field. Double-click on the **List Price** field. The fully qualified field's name will appear in the calculation area.
9. Expand the **Item Quantity Master** table in the field tree.
10. Scroll until you find the **QTY On Hand** field. Double-click on the **QTY On Hand** field. The fully qualified field's name will appear in the calculation area.
11. Place your cursor in between the two fields and insert the multiplication sign (*). The complete formula in the calculation area will be as shown on a single line as follows:

```
{Item Currency Master:List Price}
* {Item Quantity Master:QTY On Hand}
```

Calculated field 3: CONSTANT 2

The `CONSTANT 2` field creates a calculated field that always returns the value of 2. You will use this field in creating the **Item Transaction Inquiry** Go To in the next section. Perform the following steps to create the `CONSTANT 2` calculated field:

1. Open SmartList Builder.
2. Open the **WHSE QTY** SmartList object in the **SmartList Builder** window.
3. Click on the **Calculations** button in the upper half of the window.
4. Click on the plus (+) button on the **Calculated Fields** window.
5. Use the following table to complete the fields of the **Add Calculated Fields** window:

Field Name	Field Type
`CONSTANT 2`	**Currency**

6. Type the digit `2` in the calculation area. The complete formula in the calculation area will be shown as follows:

```
2
```

Calculated field 4: CONSTANT 4

The `CONSTANT 4` field creates a calculated field that always returns the value of 4. You will use this field in creating the **Item Transaction Inquiry** Go To in the next section. Perform the following steps to create the `CONSTANT 4` calculated field:

1. Open SmartList Builder.
2. Open the **WHSE QTY** SmartList object in the **SmartList Builder** window.
3. Click on the **Calculations** button in the upper half of the window.
4. Click on the plus (+) button in the **Calculated Fields** window.
5. Use the following table to complete the fields of the **Add Calculated Fields** window:

Field Name	Field Type
`CONSTANT 4`	**Currency**

6. Type the digit `4` in the calculation area. The complete formula in the calculation area will be as follows:

```
4
```

Now that you have created the calculated fields, go back to the **SmartList Builder** window and mark the following calculated fields as default for the **WHSE QTY** SmartList object:

- **QTY Available for Sale**
- **List of On Hand QTY**

Restrictions

A restriction is a filter. You are going to add a restriction to your SmartList object to limit your data to three specific locations, instead of all locations.

To create a restriction, click on the **Restrictions** button in the **SmartList Builder** window as shown in the following screenshot:

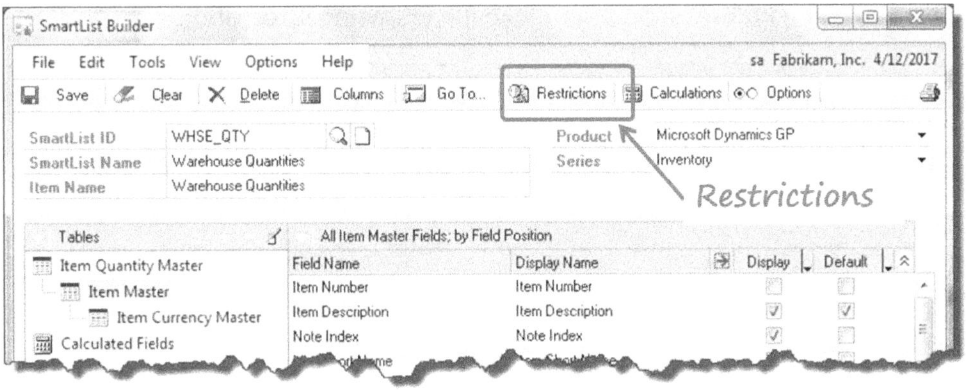

Upon clicking the **Restrictions** button, the **Restrictions** window will open. In the **Restrictions** window, click on the plus (**+**) sign to add a new restriction.

The **Add Restriction** window will now open. Use the data in the following table to complete the fields in this window. For the individual list values, type them into the **Value** field one at a time and then click on the **Add** button after each value.

Field Name	Value
Table	`Item Quantity Master`
Field	`Location Code`
Restriction	`Is Not Equal To`
Value	`Is Equal To One of List`
List Values	• `NORTH` • `WAREHOUSE` • `SOUTH`

Your completed windows should look like the following screenshot:

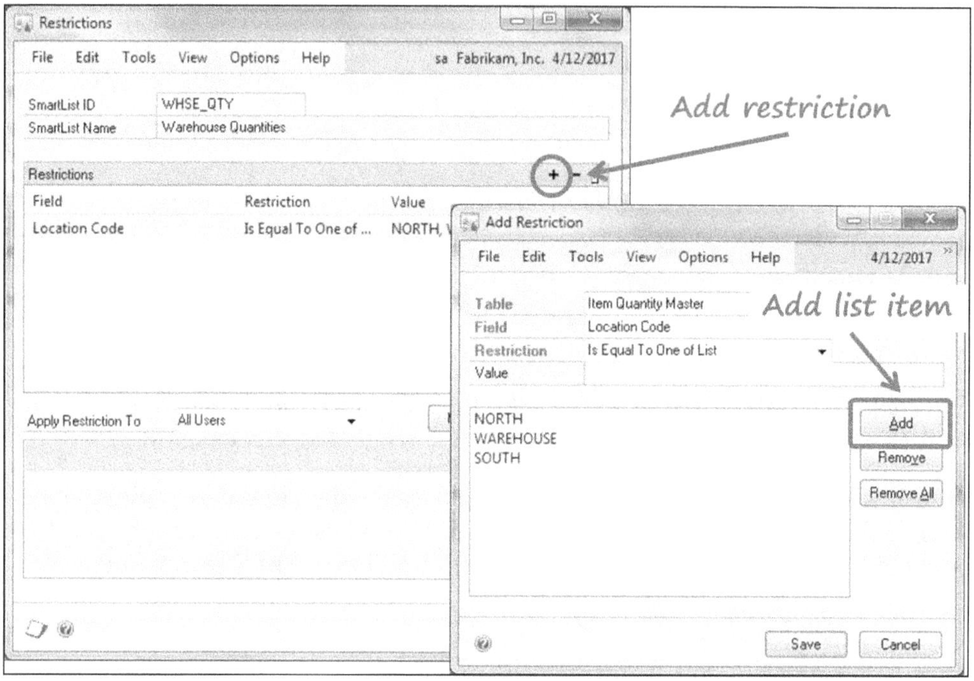

Click on the **Save** button in the lower half to close the **Add Restriction** window. Click on the **OK** button in the lower half of the **Restrictions** window to accept the restriction and close the window.

Go Tos

A **Go To** provides a link between a SmartList object and any of the following:

- A Dynamics GP form/window
- Another SmartList object
- A website
- A file
- A procedure
- An Extender resource
- A drill-down

You can designate a default Go To that will launch when you double-click on a line in your SmartList object. The other Go Tos you create are available from the Go To menu on the **SmartList** navigator screen as shown in the following screenshot:

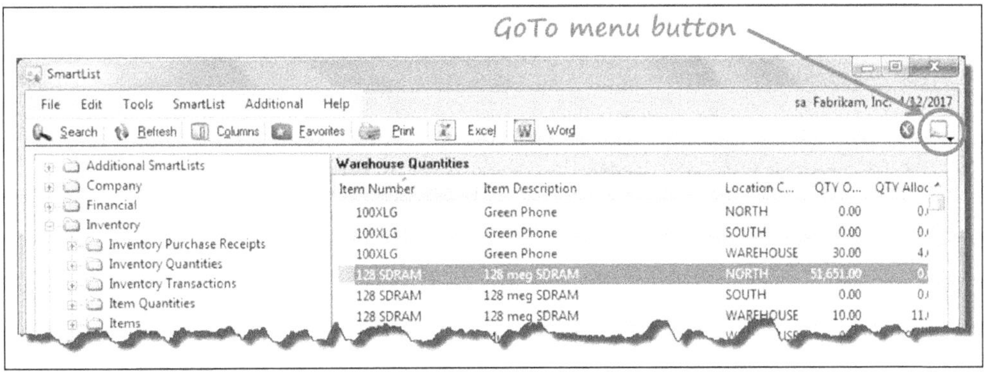

Your Go Tos will provide links to the following:

- The **Item Maintenance** window
- The **Item Transaction Inquiry** window (default **Go To**)

To get started, click on the **Go To...** button on the **SmartList Builder** window, as shown in the following screenshot:

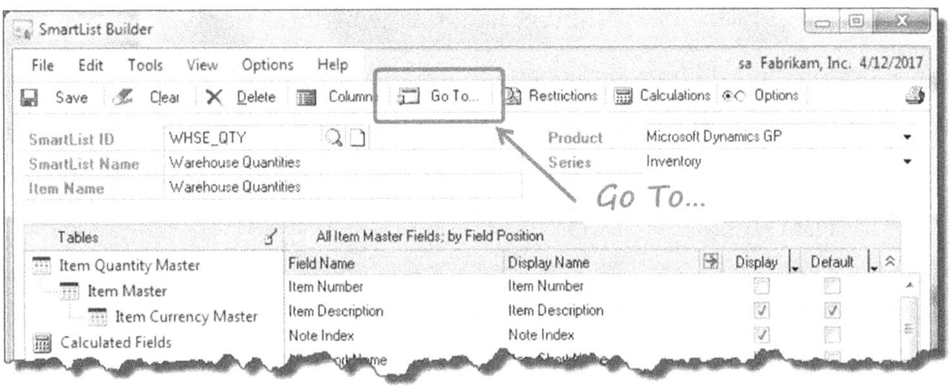

Go To: Item maintenance

On the **Go To** window, click on the **Add** button on the right-hand side and then select the **Open Form** option from the button-drop menu that opens.

Start with the simpler Go To of opening the **Item Maintenance** window. Complete the following steps to add the first Go To:

1. Open the **SmartList Builder** window.
2. Click on the **Go To...** button to open the **Go To** window.
3. Click on the **Add** button and then select the **Open Form** menu item to open the **Add Go To - Open Form** window.
4. Complete the **Add Go To - Open Form** window with the values from the following table. The **Add Go To - Open Form** window values are as follows:

Field Name	Value
Description	`Item Maintenance`
Product	`Microsoft Dynamics GP`
Series	`Inventory`
Form	`Item Maintenance`

Task: Item Number field

Complete the task for the **Item Number** field following these steps:

5. Click on the **Add** button to open the **Add Task** window.
6. Complete the **Add Task** window with the values from the following table. The **Add Task** window values are as follows:

Field Name	Value
Task Type	`Set a field value and run the field script`
Field	`(L) Temp Control Number`
Table	`Item Master`
Value	`Item Number`

7. Click on the **Save** button.

Upon completion, your windows should look like the windows shown in the following screenshot:

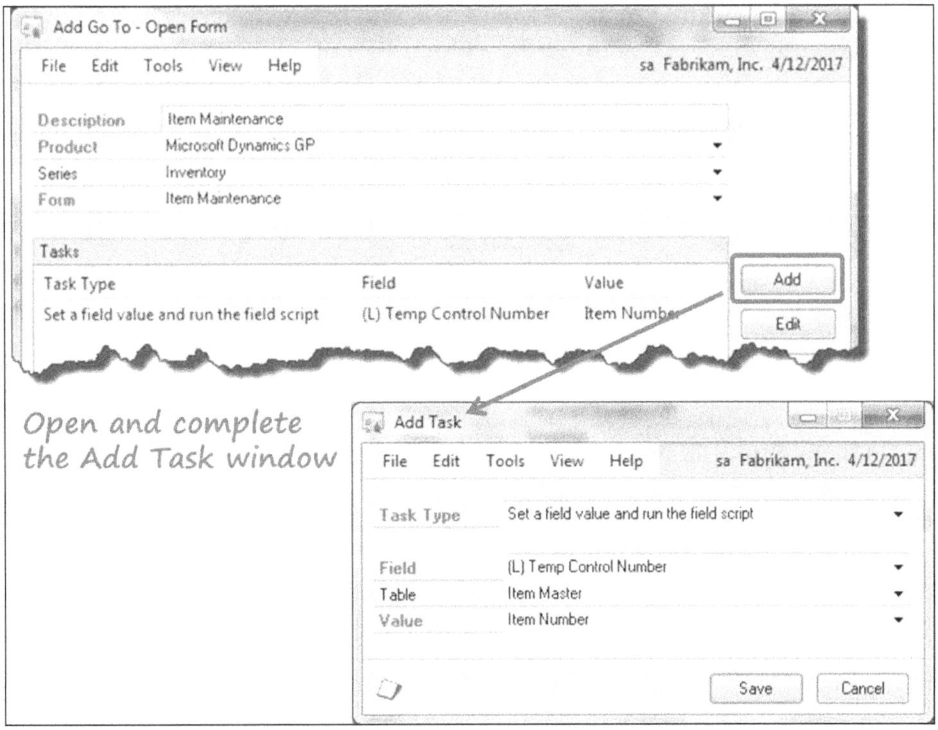

The first Go To is complete. Let's move on to the second, more advanced Go To.

Go To: Item transaction inquiry

The second Go To will open the **Item Transaction Inquiry** window. Complete the following steps to add the second Go To:

1. Open the **SmartList Builder** window.
2. Open the **WHSE QTY** SmartList object in the **SmartList Builder** window.
3. Click on the **Go To...** button to open the **Go To** window.
4. Click on the **Add** button on the **Go To** window to open the **Add Go To - Open Form** window.

5. Complete the **Add Go To - Open Form** window with the values from the following table. The **Add Go To - Open Form** window values are as follows:

Field Name	Value
Description	`Transaction Inquiry`
Product	`Microsoft Dynamics GP`
Series	`Inventory`
Form	`Item Transaction Inquiry`

When this Go To is executed, the Item Transaction Inquiry window will be populated with values pertaining to the selected item number. If item number 128 SDRAM at the WAREHOUSE site ID had been selected, the window would look similar to the window shown in the following screenshot:

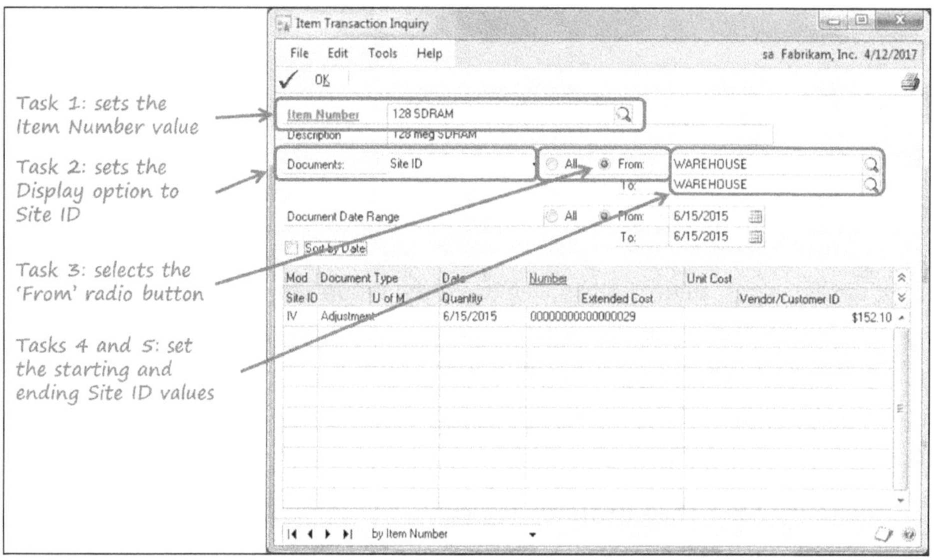

There are five tasks required to complete this Go To. Each task is defined separately.

Task 1: Item Number field

This task sets the value of the **Item Number** field in the **Item Transaction Inquiry** window. Complete the task for the **Item Number** field by following these steps:

1. Click on the **Add** button to open the **Add Task** window.
2. Complete the **Add Task** window with the values from the following table. The **Add Task** window values are as follows:.

Field Name	Value
Task Type	`Set a field value and run the field script`
Field	`Item Number`
Table	`Item Quantity Master`
Value	`Item Number`

3. Click on the **Save** button.

Task 2: (L) Display Options field

This task sets the value of the **(L) Display Options** field in the **Item Transaction Inquiry** window. The value `4` means the documents are queried according to **Site ID**. Complete the task for the **(L) Display Options** field by following these steps:

1. Click on the **Add** button to open the **Add Task** window.
2. Complete the **Add Task** window with the values from the following table. The **Add Task** window values are as follows:

Field Name	Value
Task Type	`Set a field value and run the field script`
Field	`(L) Display Options`
Table	`Calculated Fields`
Value	`CONSTANT 4`

3. Click on the **Save** button.

Task 3: (L) Display By field

This task sets the value of the display options' radio button group in the **Item Transaction Inquiry** window. This task changes it from the default value of **all** to the value of **from**. Complete the task for the **(L) Display By** field by following these steps:

1. Click on the **Add** button to open the **Add Task** window.
2. Complete the **Add Task** window with the values from the following table. The **Add Task** window values are as follows:

Field Name	Value
Task Type	`Set a field value and run the field script`
Field	`(L) Display By`
Table	`Calculated Field`
Value	`CONSTANT 2`

3. Click on the **Save** button.

Task 4: (L) Start Location field

This task sets the value of the first site in the range of sites to be queried in the **Item Transaction Inquiry** window. Complete the task for the **(L) Start Location** field by following these steps:

1. Click on the **Add** button to open the **Add Task** window.
2. Complete the **Add Task** window with the values from the following table. The **Add Task** window values are as follows:

Field Name	Value
Task Type	`Set a field value and run the field script`
Field	`(L) Start Location`
Table	`Item Quantity Master`
Value	`Location Code`

3. Click on the **Save** button.

Task 5: (L) End Location field

This task sets the value of the last site in the range of sites to be queried in the **Item Transaction Inquiry** window. Complete the task for the **(L) End Location** field by following these steps:

1. Select the **Add** button to open the **Add Task** window.
2. Complete the **Add Task** window with the values from the following table. The **Add Task** window values are as follows:

Field Name	Value
Task Type	`Set a field value and run the field script`
Field	`(L) End Location`
Table	`Item Quantity Master`
Value	`Location Code 2`

3. Click on the **Save** button.

The second Go To is complete.

In order to give additional users access to your new SmartList object, you need to add it to the security model of Dynamics GP.

Granting security to a SmartList Builder object

Having built your new SmartList object, you'll want others to be able to enjoy it too. As only you have access to the object just after you create it, you need to grant security to others.

Security for SmartList Builder objects, in general, can be a little tricky. In this section, we'll keep it simple and grant security to this object by adding it to the existing **DEFAULTUSER** task.

Open the administration area page from the shortcut bar. Go to the **Setup** content pane and expand the system menu. Select the **Security Tasks** menu item from the system menu to open the **Security Task Setup** window. Open the **DEFAULTUSER** task.

Complete the **Security Task Setup** window using values from the following table. The **Security Task Setup** window values are as follows:

Field	Value
Task ID	`DEFAULTUSER`
Task Name	`Default User`
Task Description	`Default User`
Series	`Company`
Product	`SmartList`
Type	`SmartList Object`
Series	`SmartList Objects`

The **Access List:** section in the lower half of the **Security Task Setup** window identifies the SmartList objects available. To grant access to your new SmartList object, check the **Warehouse Quantities** checkbox. With that done, any user who has a security role assigned to him or her that includes the **DEFAULTUSER** task will have access to the new SmartList object.

Your completed **Security Task Setup** window should look like the following screenshot:

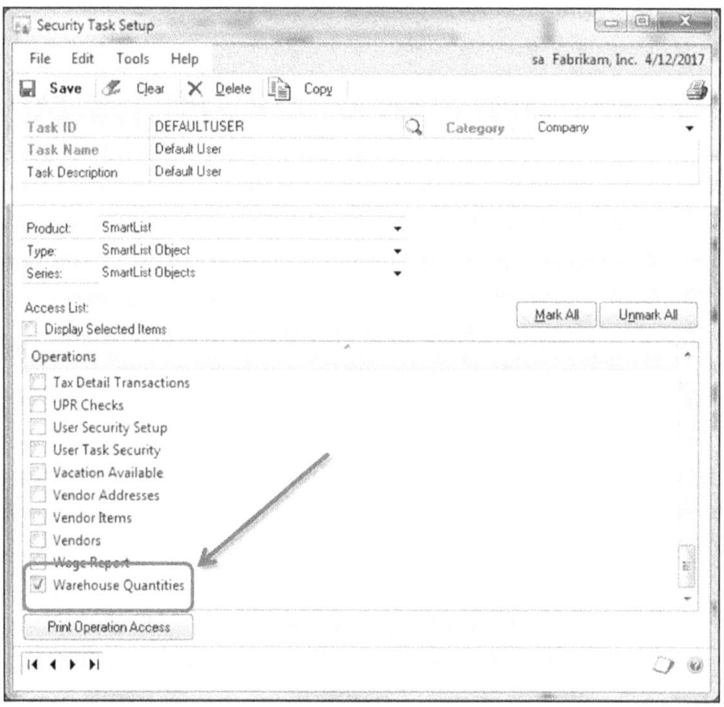

Now that security is set up, your users can try out your new SmartList object and be delighted at how clever you are!

Excel Report Builder

Excel Report Builder is a lot like SmartList Builder except the product is a refreshable Excel spreadsheet. It's refreshable because it is linked to the company's database instead of the static data you get from downloading information from a SmartList object. It's a kind of Excel SmartList object.

Let's build one!

You'll use the **Warehouse Quantities** SmartList object that you just built and turn it into an Excel report using Excel Report Builder.

Open the SmartList Builder module using the following navigation: **Microsoft Dynamics GP** | **Tools** | **SmartList Builder** | **SmartList Builder**.

The **SmartList Builder** window will open. Open the **WHSE_QTY** object, click on the **Options** menu item from the menu bar, and then click on the **Duplicate** button.

The **Duplicate** window will open. Select the **Excel Report** option for the **New List Type** field and then click on the **Duplicate** button as shown in the following screenshot:

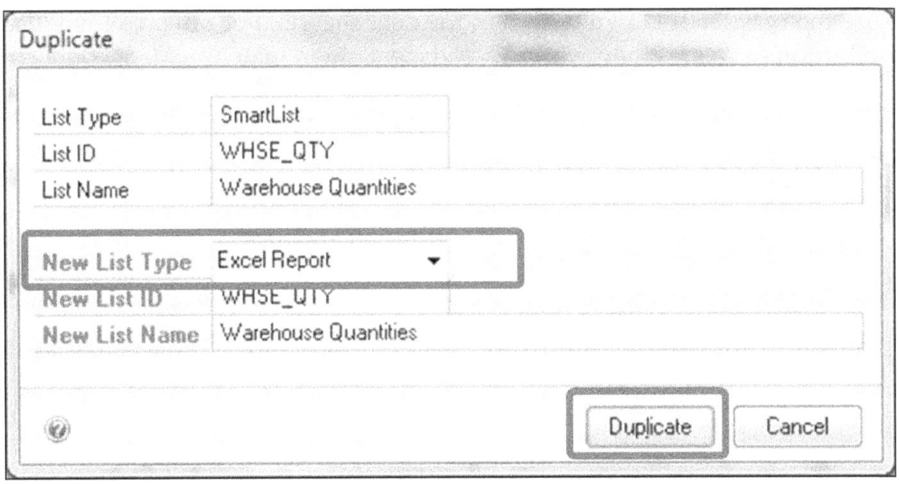

Now, launch Excel Report Builder by following this navigation: **Microsoft Dynamics GP** | **Tools** | **SmartList Builder** | **Excel Report Builder** | **Excel Report Builder**.

As you can see, the **Excel Report Builder** window is reminiscent of the **SmartList Builder** window. The steps to build an Excel report are very similar to those for building a SmartList object. You can even copy one type of list to another, as you just did.

Look at the upper half of the **Excel Report Builder** window and you will see two new buttons: the **Drill Down** button and the **Publish** button. In addition, two new fields were added: the **Report Type** and **View Name** fields. Something is also missing: the **Default** field. With Excel Report Builder, each of the fields you had in the SmartList's **Display** column will get copied to the Excel report. The following screenshot shows the **Excel Report Builder** window:

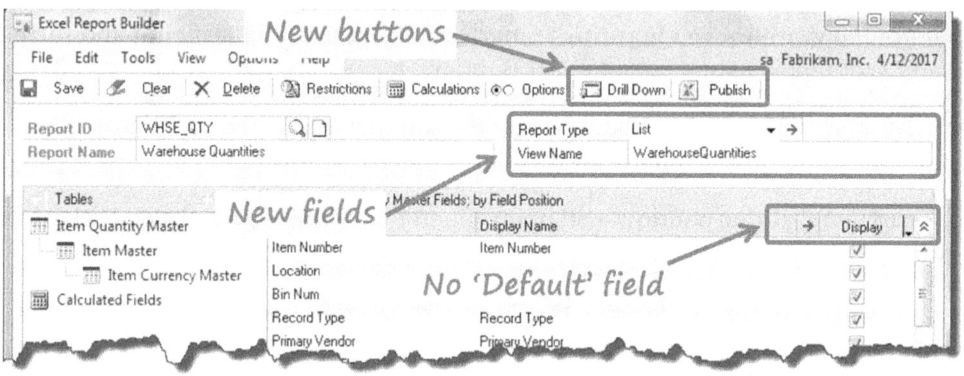

You'll use the new **Drill Down** button in the next section. For now, you will use the **Publish** button to create a new Excel spreadsheet and a matching data connection.

Click on the **Publish** button and the **Publish Report** window will open. For best results, you may want to publish the report and data connection to a network so that everyone can use them.

Don't forget to assign permissions to the appropriate database roles so that your users can access the objects. You created the database roles automatically when you installed SmartList Builder. Use the **Permissions** button to open the **Permissions** window. The following screenshot shows the **Publish Report** and the **Permissions** windows:

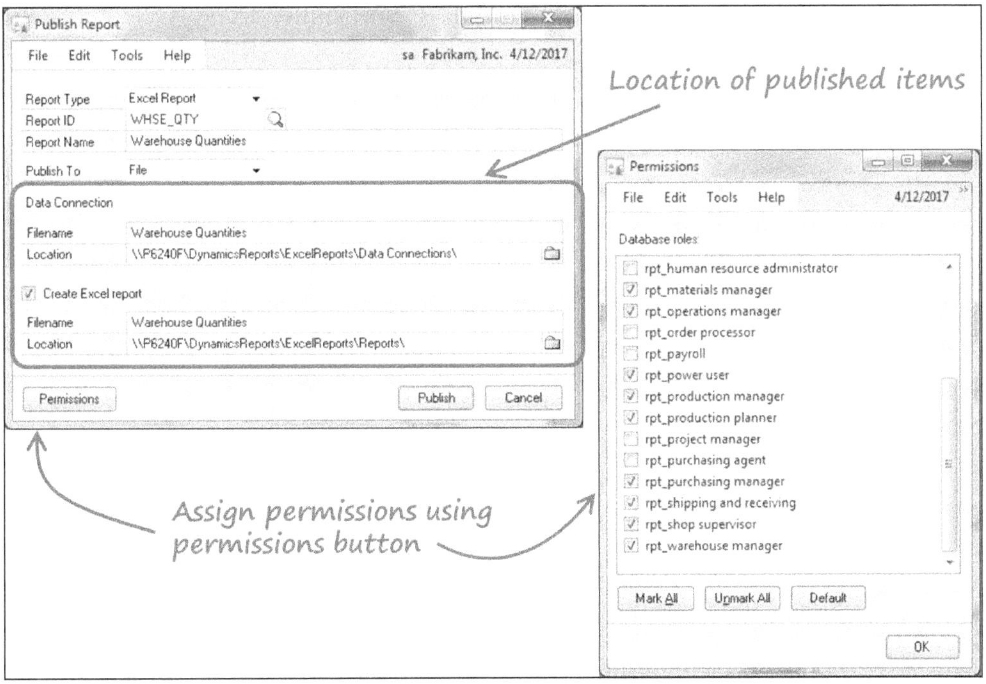

After providing locations for the data connection as well as the Excel report and assigning permissions, click on the **Publish** button to create the report and data connection.

To open the report in Excel, look in the location where you published the report. You should see a new Excel file. The name of the Excel file will be the name of your report, prefixed with the database name of the company for which you published it. For example, the published report for Fabrikam was named `TWO Warehouse Quantities.xlsx`.

Open the Excel file; it should look similar to the report shown in the following screenshot:

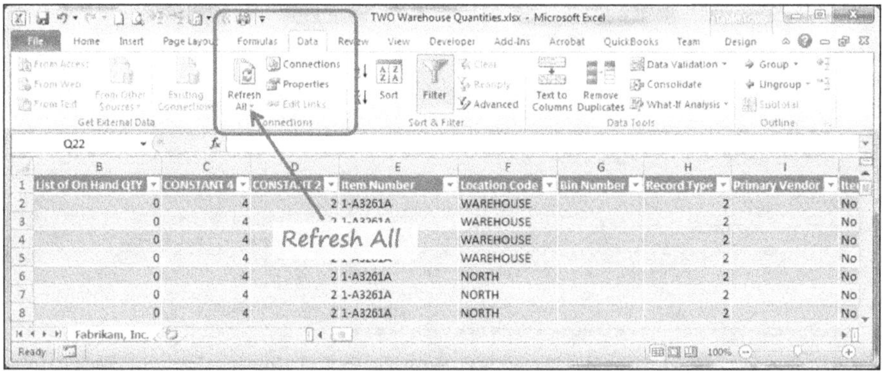

You can refresh the data, by selecting the **Data** tab and then clicking on the **Refresh** or **Refresh All** button.

Close Excel, if you've opened it.

In the next section, you will build a drill-down using Drill Down Builder. You will then add that drill-down to your Excel report.

Drill Down Builder

Drill Down Builder can take Excel Report Builder to the next level. Excel Report Builder provides a way for you to create Excel spreadsheets that are linked to the live databases. Drill Down Builder provides you with a way to create URLs that can be used by your Excel report, or any other external program, to drill down into the Dynamics GP application.

Open Drill Down Builder using the following navigation: **Microsoft Dynamics GP | Tools | SmartList Builder | Drill Down Builder**.

On the **Drill Down Builder** window, you can create three different types of drill-downs.

- **Form**: A Form drill-down opens a Dynamics GP window and sets the value(s) in the window.
- **SmartList**: A SmartList drill-down opens a SmartList object and sets search parameters.
- **Extender**: An Extender drill-down opens an Extender form or Detail form and sets the value(s) of the ID fields.

Drill-downs are comparable to the Go Tos you created for a SmartList object.

For this project, you will create a Form drill-down that will open the **Item Inquiry** window.

Complete the **Drill Down Builder** window as follows:

Field	Value
Drill Down ID	`ITEM_INQ`
Description	`OpenItemInquiry`
Drill Down Type	`Form`
Product	`Microsoft Dynamics GP`
Series	`Inventory`
Form	`Item Inquiry`

In the **Parameters** section, click on the plus (**+**) button and add the following parameters to the **Add Parameter** window:

Field	Value
Parameter Name	`ItemNumber`
Field Type	`String`

Click on the **Save** button.

In the **Actions** section, select the **Add** button and add the following actions to the **Add Action** window:

Field	Value
Action Type	`Set field value and run the field script`
Field	`Item Number`
Parameter	`ItemNumber`

Click on the **Save** button.

Upon completion, your **Drill Down Builder** window should look like the following screenshot:

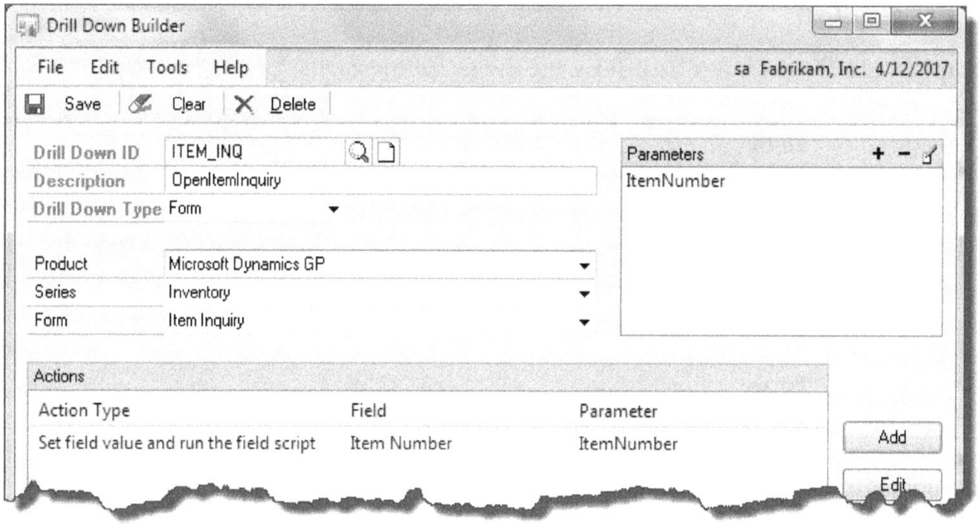

Click on the **Save** button in the **Drill Down Builder** window to save the drill-down and then close the window.

The next step is to add this drill-down to your Excel report. Launch Excel Report Builder following this navigation: **Microsoft Dynamics GP** | **Tools** | **SmartList Builder** | **Excel Report Builder** | **Excel Report Builder**.

Using the following screenshots as a guide, perform the following steps:

1. Open the **WHSE_QTY** Excel report that you created in the last project:

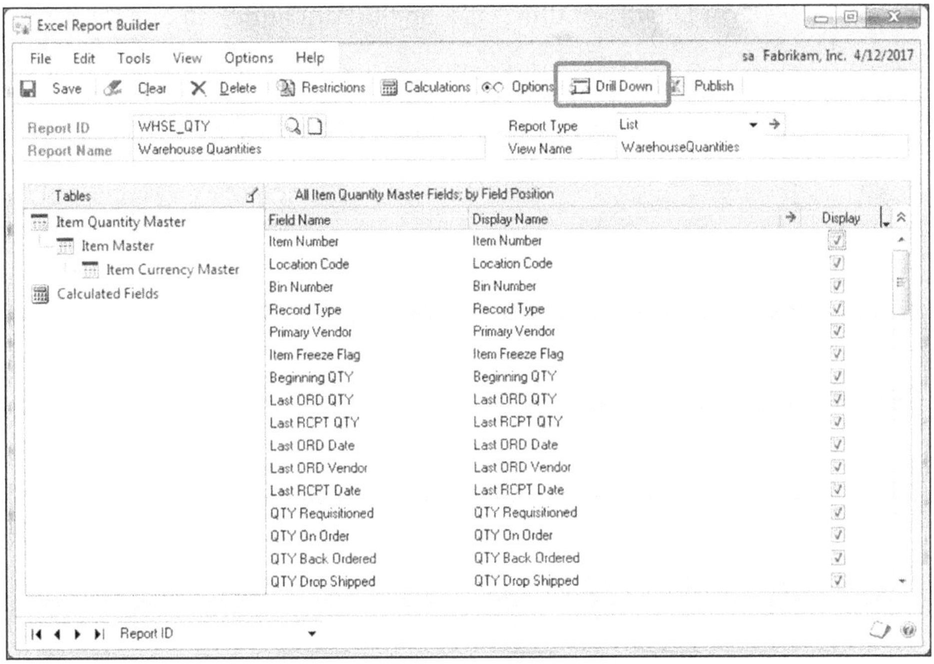

2. On the **Excel Report Builder** window, click on the **Drill Down** button in the upper half of the window to open the **Drill Down** window:

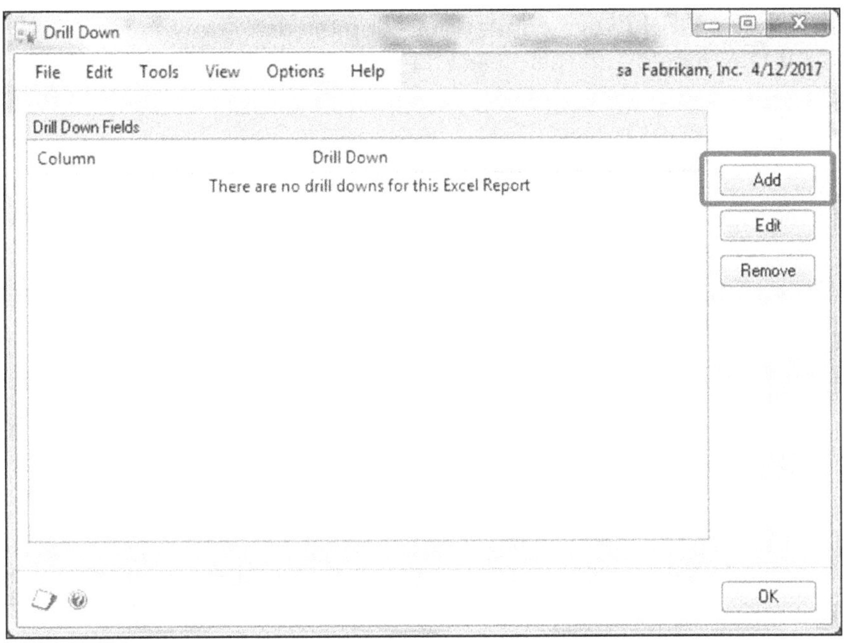

3. In the **Drill Down** window, click on the **Add** button to open the **Add Drill Down** window:

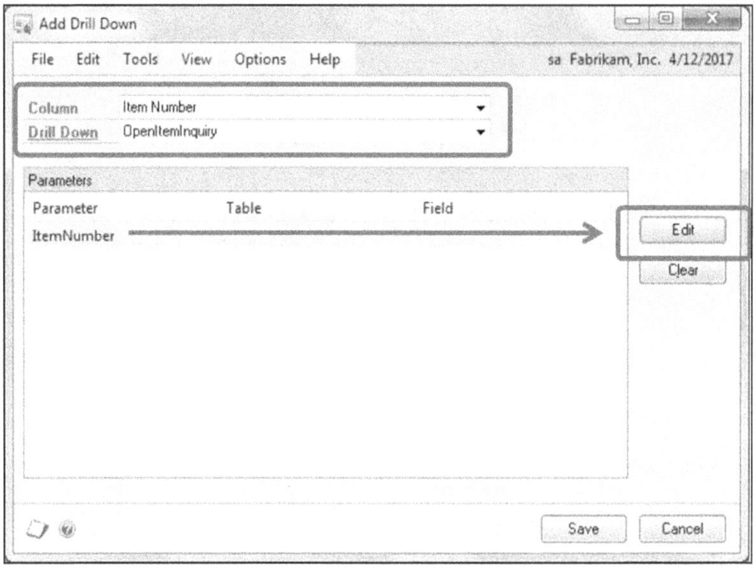

4. In the **Add Drill Down** window, click on the drop-down menu next to the **Column** field and then select the **Item Number** option.
5. In the **Add Drill Down** window, click on the drop-down menu next to the **Drill Down** field and then select the Drill Down you just created; it would be the **OpenItemInquiry** option.
6. In the **Add Drill Down** window, select the **ItemNumber** parameter in the **Parameters** scrolling window.
7. In the **Add Drill Down** window, click on the **Edit** button to open the **Map Drill Down Parameter** window:

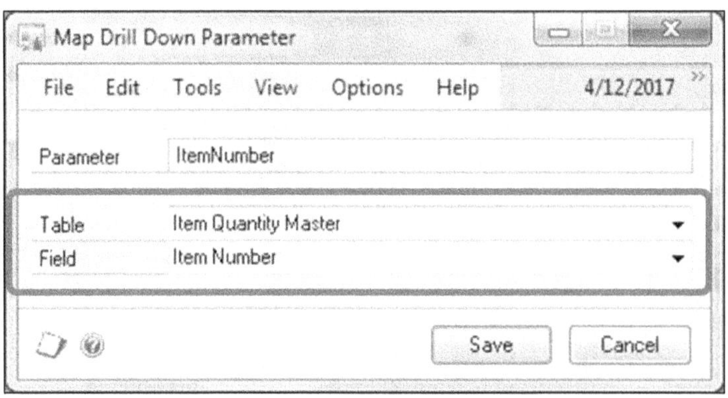

8. In the **Map Drill Down Parameter** window, click on the drop-down menu next to the **Table** field and then select the **Item Quantity Master** table as its value.
9. In the **Map Drill Down Parameter** window, click on the drop-down menu next to the **Field** field and then select the **Item Number** field as its value.
10. Click on the **Save** button in the **Map Drill Down Parameter** window.
11. The **Add Drill Down** window should now look like the following screenshot:

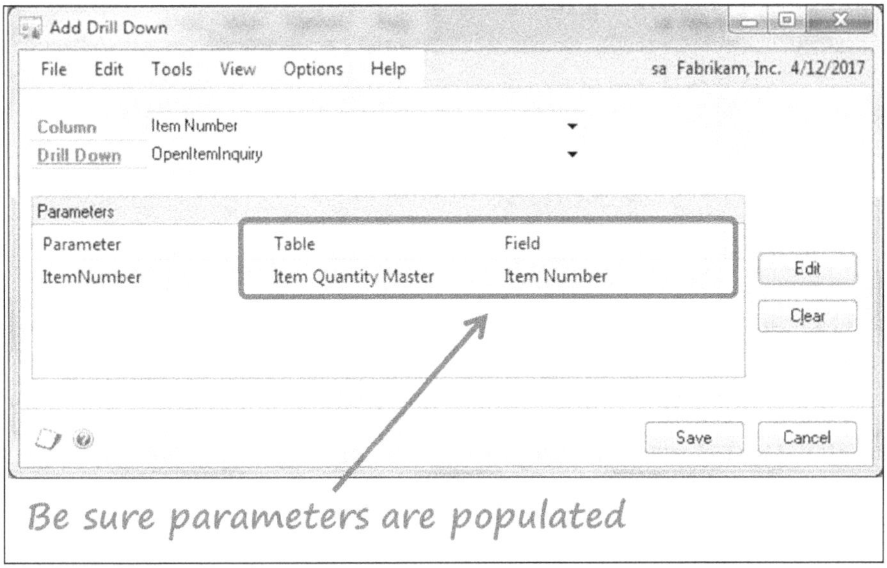

12. Click on the **Save** button in the **Add Drill Down** window.
13. Click on the **OK** button in the **Drill Down** window.
14. Publish the report again and overwrite any data connection if asked.
15. Click on the **Save** button in the **Excel Report Builder** window.

Open the Excel report in Microsoft Excel. Click on a cell in the **Item Number** column and then move your mouse cursor over the contents. Your report should look similar to the following screenshot:

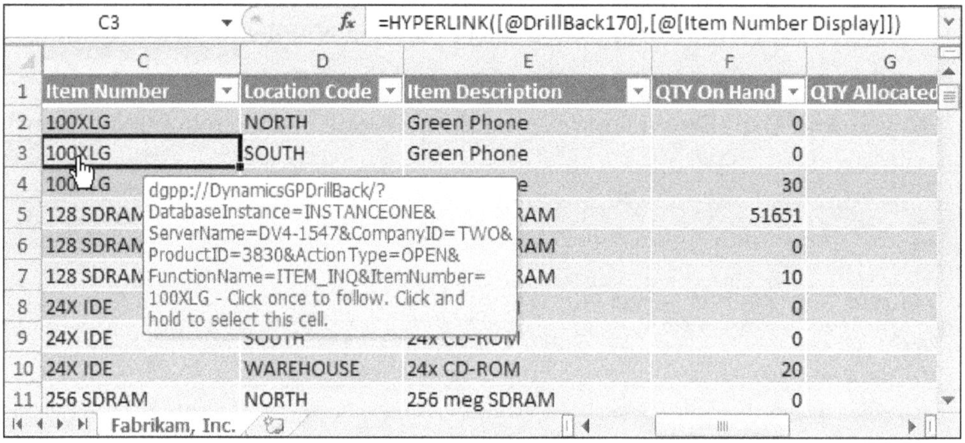

What's shown in the tooltip is the full hyperlink, not the shortcut that's in the formula bar. The full link is in a `DrillBack` column that is hidden so that it isn't visible on the report. In this worksheet, it's in column B.

If you click the mouse when the select finger cursor is displayed, you will open the **Item Inquiry** window for the selected item. Excel will not minimize; you will need to switch over to Dynamics GP to see the window.

Imagine what you can do with this feature! You can use this URL in any application and provide a drill-back functionality into any Dynamics GP window. To test this theory, copy one of the complete URLs, not the shortcut. Create a hyperlink in a Microsoft Word document using the copied URL as the address. You should be able to press *Ctrl* + click on the hyperlink to drill back into the Dynamics GP application just like you did with the Excel report.

The hyperlink URL for the **100XGL** item in the sample company looks like this:

```
dgpp://DynamicsGPDrillBack/?DatabaseInstance=INSTANCEONE&ServerName=
DV4-1547&CompanyID=TWO&ProductID=3830&ActionType=OPEN&FunctionName=IT
EM_INQ&ItemNumber=100XLG
```

If you edit this hyperlink by changing the `ItemNumber` parameter appearing at the end, it will point the hyperlink to a different item.

This feature surely deserves a Wow! And don't forget: you did this with zero code.

In the next section, you are going to continue with codeless customization by creating new Dexterity windows, without Dexterity!

Extender

Extender is one of the most remarkable modules available for Dynamics GP. Using Extender, you can build an unlimited number of additional data-entry windows that you can tie to existing Dynamics GP windows. You can even tie additional windows to the rows in a scrolling window. Furthermore, you can create your own standalone forms that do not need to be associated with an existing Dynamics GP window. You get all of this functionality through a point-and-click interface.

Overview

Your Dynamics GP system tracks a plethora of information, but still we crave for more. The native user forms often do not provide enough fields to house all of the data you need to store. For example, would you like to have line item notes on the Sales Transaction Entry window? How about an Inventory options window that displays different sets of information based on the kind of item you selected?

All of the features just mentioned, and more, are available if you use the Extender module. As a bonus, Dynamics GP maintains the upgrade path for future releases of Extender! You will not have to worry about changes to the data tables or business logic with each new release. Your customization is upgraded automatically.

Extender editions

Extender comes in two flavors:

- Extender Standard
- eXtender Enterprise

Extender Standard

Extender Standard is the module you buy from Microsoft. This module allows you to create Extender resources quickly and easily.

An Extender window can hold a maximum of 15 data fields, and you can attach an unlimited number of windows to an existing Dynamics GP window.

Extender provides a means to add objects using a point-and-click interface. You can create the following object types using Extender:

- **Forms**: A form is a data-entry screen that is independent of other Dynamics GP windows.

- **Detail forms**: A detail form is a data-entry screen containing a scrolling window that is independent of other Dynamics GP windows.
- **Windows**: A window is a single data-entry screen that is related to an existing Dynamics GP window.
- **Detail windows**: A detail window is a scrolling window that is related to an existing Dynamics GP window.
- **Window groups**: A window group is a set of windows that open according to the conditions you assign.
- **Notes windows**: A note window is an enhanced version of the Dynamics GP note window. The Extender note window is better because you can create multiple notes for a single window that are date stamped and classified.
- **Lookups**: A Lookup is a window you can use on an Extender window to select and validate existing Dynamics GP data.
- **Imports**: An import is a tool you can use to populate your new Extender fields with data.
- **Menus**: An Extender menu is used to add navigation to your Extender forms using the Dynamics GP menu structure. You can also use an Extender menu to build your own navigation path to existing Dynamics GP windows.
- **Dialogs**: An Extender dialog is something you would create if you wanted to send a message to the user. Any response from the user can be stored in table `EXT30200` in the company's database. The basic Extender product cannot invoke actions based on dialog responses. eXtender Enterprise can.

You will do a project a little later where you'll work with several of the objects just listed.

eXtender Enterprise

This flavor of eXtender comes from eOne Solutions at `http://www.eonesolutions.net`. It does everything that a basic Extender will do, and you can add code to it! eOne Solutions calls these code parts logic scripts. Essentially, what they have done is create a point-and-click Dexterity. You can even download several logic script templates and samples from the eOne Solutions website to get you started.

Working with Extender

To help you get a feel of Extender, you are going to create some Extender objects and see how they interface with Dynamics GP. Creating each of the Extender objects is beyond the scope of this book, but you will be working with quite a few.

You will be working with the following objects:

- Forms
- Windows
- Detail windows
- Window groups
- Lookups
- Menus
- Notes

These objects will be included in a variety of projects.

Your first project will be the Event form project. You have a client that plans different party events. They need some new forms and fields to track information about the events they plan.

In this project, you will create a new form that will serve as the **Event Maintenance** window. You will also create some Lookups that you will use with fields on the **Event Maintenance** window. You will then create extra windows that will include additional information about the event that you didn't want on the main window. Finally, you need a way to open this new window, so you will add a line to the existing **Cards** menu to handle navigation.

Your second project will be the Customer Window Group project. In this project, you will create three different windows. Each window will capture different information about a customer according to the customer's class.

Your third project will be the Customer Contacts project. In this project, you will create a Detail window to store multiple contacts for each customer.

Your final project will be the Line Item Note project. In this project, you will add a note to each line of the Sales Transaction Entry window.

Event form project

This project will work with the following Extender objects:

- Form
- Lookup
- Menu added to the existing **Cards** menu.
- Extra windows

At the end of the day, your Extender form will look substantially similar to the following screenshot:

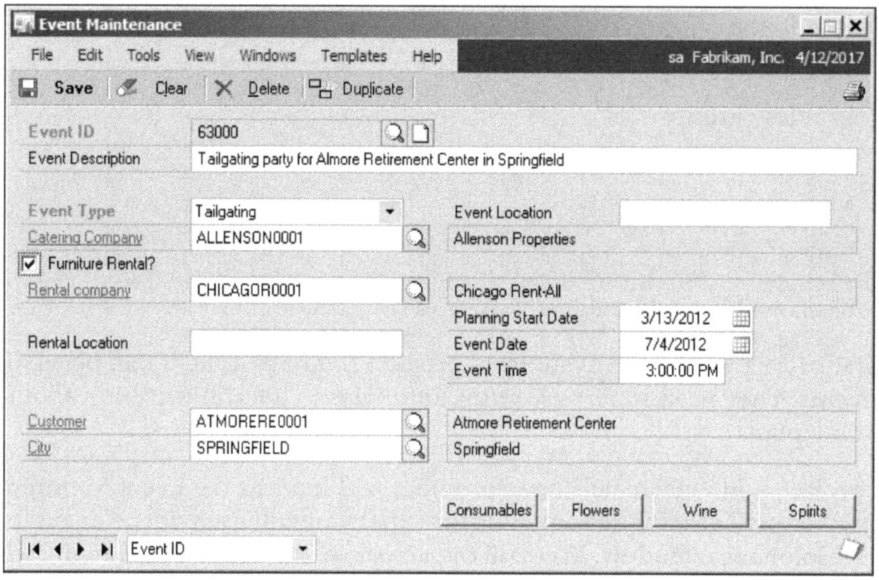

Let's create the new form!

Open the **Extender** window using the following navigation: **Microsoft Dynamics GP | Tools | Extender | Extender**.

Select the **Forms** object from the list on the left-hand side and then click on the **New** button or select **Forms** from the **New** button drop-down list:

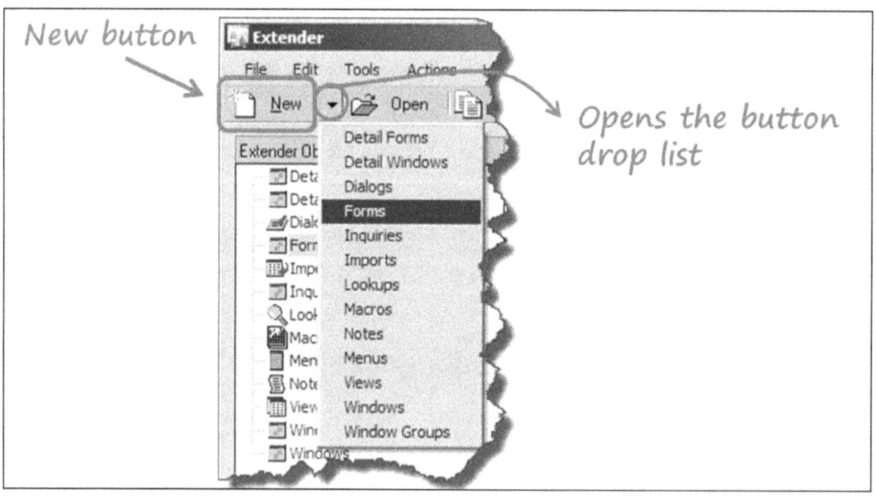

When the **Extender Forms** window opens, create the new form by using the field values from the window as shown in the following screenshot:

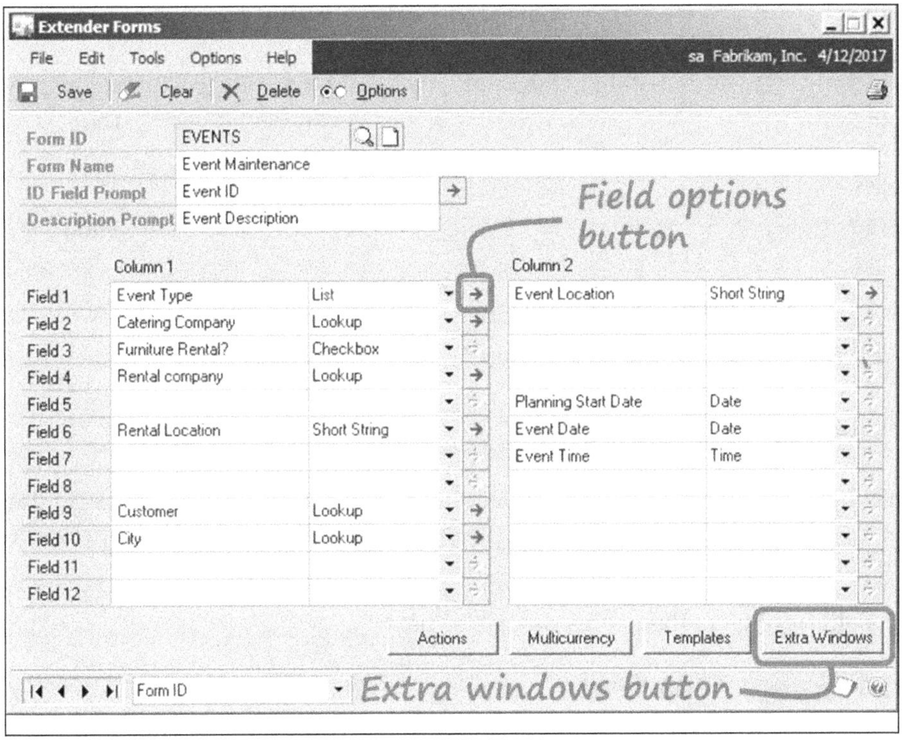

Enter the following values in the header fields:

Field Name	Value
Form ID	`EVENTS`
Form Name	`Event Maintenance`
ID Field Prompt	`Event ID`
Description Prompt	`Event Description`

Enter the following values in the body fields:

	Column 1		Column 2	
Field ID	**Value**		**Value**	
Field 1	`Event Type`	**List**	`Event Location`	**Short String**
Field 2	`Catering Company`	**Lookup**		

	Column 1		Column 2	
Field ID	**Value**		**Value**	
Field 3	`Furniture Rental?`	**Checkbox**		
Field 4	`Rental Company`	**Lookup**		
Field 5			`Planning Start Date`	**Date**
Field 6			`Event Date`	**Date**
Field 7			`Event Time`	**Time**
Field 8				
Field 9	`Customer`	**Lookup**		
Field 10	`City`	**Lookup**		
Field 11				
Field 12				

You need to set some field options. Select the desired field and then click on the blue arrow that will become active if options are available. Refer to the preceding screenshot for the location of the blue arrow button.

Set the field options to the values described next.

Field 1: Event Type field

The **Event Type** is a list field. You will identify the items on the list using the **List Field Settings** window. Access this window by clicking on the blue expansion arrow to the right-hand side of Field 1. Follow the screenshot to create the list:

Field 2: Catering Company field

The **Catering Company** field is a Lookup. A lookup field can be associated with existing lookups in Dynamics GP, such as customers or vendors. Additionally, you can create your own lookups using data from your Extender windows. For this lookup, select **Vendor** in the **Lookup Settings** window.

Field 3: Rental Company field

The **Rental Company** field is also a Lookup field. It too will be associated with the existing vendor lookup. Select **vendor** in the **Lookup Settings** window.

Field 4: Rental Location field

The **Rental Location** field is a **Short String** field. You may also have seen the Long String field in the list of field types. A short string can have a maximum length of 30 characters. A long string allows for up to 255 characters. A unique feature of the short string is that you can apply a mask to the string, limit the type of data that can be entered, and also limit the number of characters. When defining a mask, you use the capital X as a placeholder. Your data flows into the placeholder positions, and any other static values included in the mask will print as entered. For example, if you wanted your short string to represent a social security number, your mask would look like this: XXX-XX-XXXX.

Field 9: Customer field

The **Customer** field is a Lookup field; associate it with the customer lookup.

Field 10: City field

The **City** field is also a Lookup field, but there is not a pre-existing lookup associated with the vendor's city. You will build an Extender Lookup to use for this field.

Field 1, Column 2: Event Location field

The **Event Location** field is also a short string. Like the **Rental Location** field, you will not apply a mask.

Creating the City Lookup object

To create the City Lookup object, select the **Lookups** option from the **Extender Objects** list and then click on the **New** button. Complete the **Extender Lookup Maintenance** window using the following screenshot as your guide. Associate this new lookup with the **City** field after you save it:

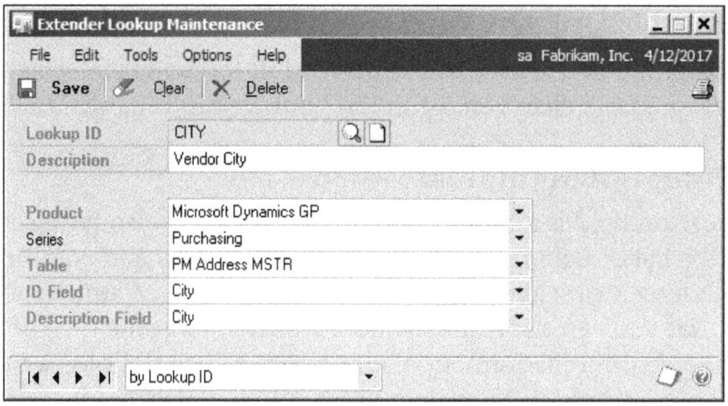

Extra windows

You create the **Consumables**, **Flowers**, **Wine** and **Spirits** buttons appearing in the lower half of your **Event Maintenance** form using the **Extra Windows** button at the lower right-hand corner of the **Extender Forms** window.

Click on the **Extra Windows** button on the **Extender Forms** window to open the **Extra Windows** window. Click on the plus (+) button to open the **Extra Window Setup** window. Create your four extra windows using the following screenshot as your guide:

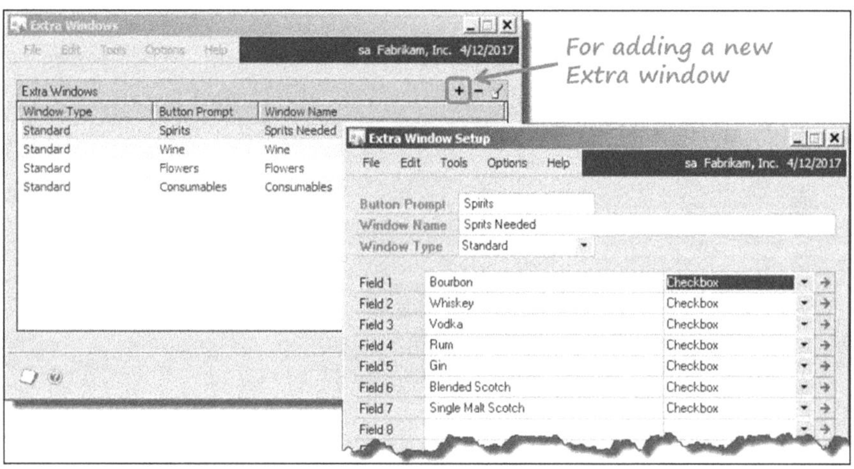

The last thing you need to do is to create a way to open your new form. You are going to use an Extender Menu object to add a selection to the existing **Cards** menu.

Extender menu

Select the **Menus** option from the **Extender Objects** list, and then click on the **New** button. In the **Extender Menus** window fill in the details about your menu, including the text you want to appear on the menu and the menu group you would like to join. Use the following screenshot as a guide to create your menu:

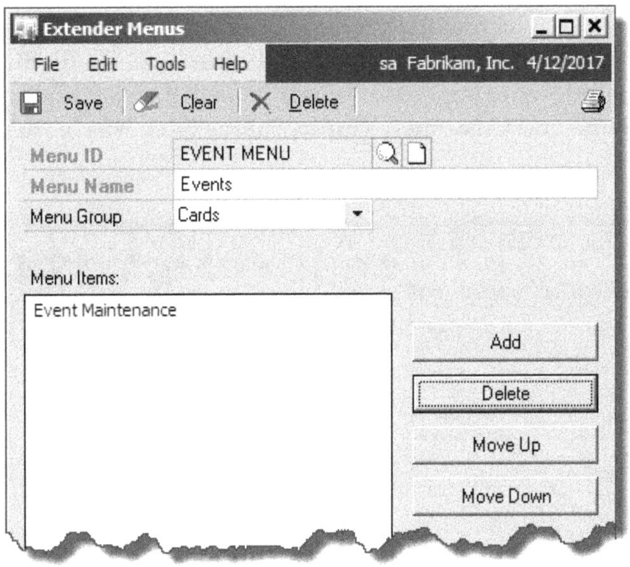

Restart or launch Dynamics GP, navigate to the **Cards** menu, and select **Event Maintenance** to open your **Event Maintenance** window. Try out your lookups and extra windows and be amazed at how simple it was to create this masterpiece!

Customer window group

This project will show you how window groups work. You use a window group when you need different Extender windows to open, based on the criteria you set. For instance, imagine your inventory includes both plastic plants and real plants. For your real plants, you might want to include information such as hardiness zone, watering needs, soil requirements, and the like. For plastic plants, all you need are its dimensions and weight. To collect the same information for the plastic plants and the real plants wouldn't make any sense. A window group will solve this dilemma. You would have one window for plastic plants and a different window for real plants. The window that would be displayed would depend on the value in a certain field, for example the item's class field.

In this project, you will use a customer class as your differentiator. You'll create three different windows and associate each of them with a different Customer class. When you view a customer in the **Customer Maintenance** window, you will see that only the window you associated with that customer's class will be available. Additionally, you will set a hot key that you can use to open the Extender window.

Let's get started.

First, you need to create the three windows. From your **Extender Objects** list click on the **Window** button and then click on the **New** button. You're going to create three windows, each of which will pertain to a unique customer class. In this project, you will name your windows **Churches**, **Distributors**, and **Radio Stations**. Create each of the three windows using the following screenshot as your guide. Note that both **Customer Name** and **Customer Number** fields have been added to the **Key Fields** section:

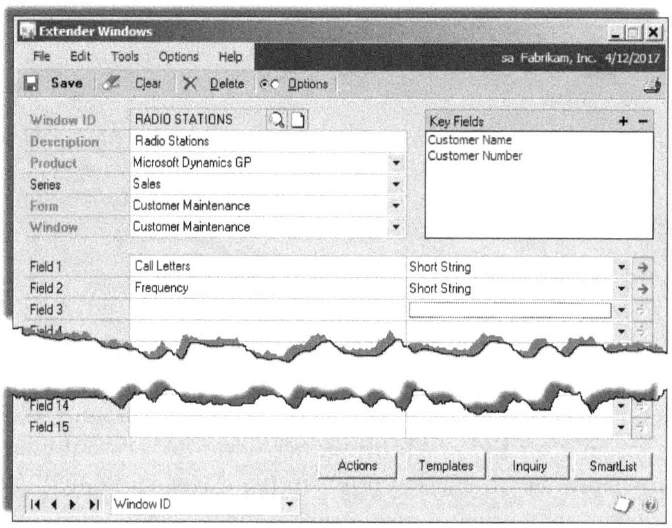

> The customer's name is optional, but if you include it, the **Customer Name** field will display on the Extender window.

If you're using the sample company, you will need to create the three customer classes that correspond with your new windows. Add a few customers to each class so that you have some data to test.

You're going to use two additional features of Extender in this project. The first feature will cause the window to open and close automatically; the second feature marks a field as required. To set up these additional features click on the **Options** button in the upper half of the **Extender Windows** window. In the **Options** window, check the boxes next to **Auto Open** and **Auto Close**; choose the **Field Exit** option as the value for the **Open Method** field and the **Customer Name** option as the value for the **Field** field. Check the box next to the **Call Letters** checkbox to make **Call Letters** a required field. Complete the window using the following screenshot as your guide:

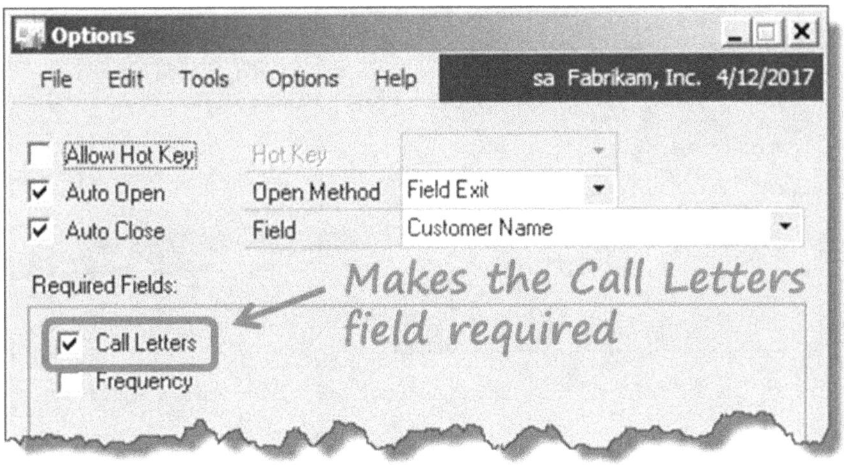

With the preceding settings, your Extender window will open automatically whenever you exit the **Customer Name** field on the **Customer Maintenance** window. When you tab off the last field on the Extender window, it will close automatically and return focus to the **Customer Maintenance** window. In addition, you will not be able to save the data on the Extender window if the **Call Letters** field is blank.

After you complete all three windows, you can move on to building the window group.

To create the window group, click on the **Window Groups** button from the **Extender Objects** list and then click on the **New** button. The **Window Group Maintenance** window will open. Click on the plus (+) button in the **Windows** section to add your windows to the group. Once your windows are added, select one of the windows and click on the plus (+) button in the **Conditions** section. The condition tells Dynamics GP which window(s) to make available according to the criteria you specify. You can make more than one window available.

You want to assign each window to a specific customer class. Add conditions to each of your windows. You are not limited to one window per condition nor one condition per window. You have a lot of flexibility on how you want the windows in your window group to behave. Our conditions will associate each of your windows to a specific customer class. The conditions you will enter will be similar to the ones in the **Add Condition** window shown in the following screenshot:

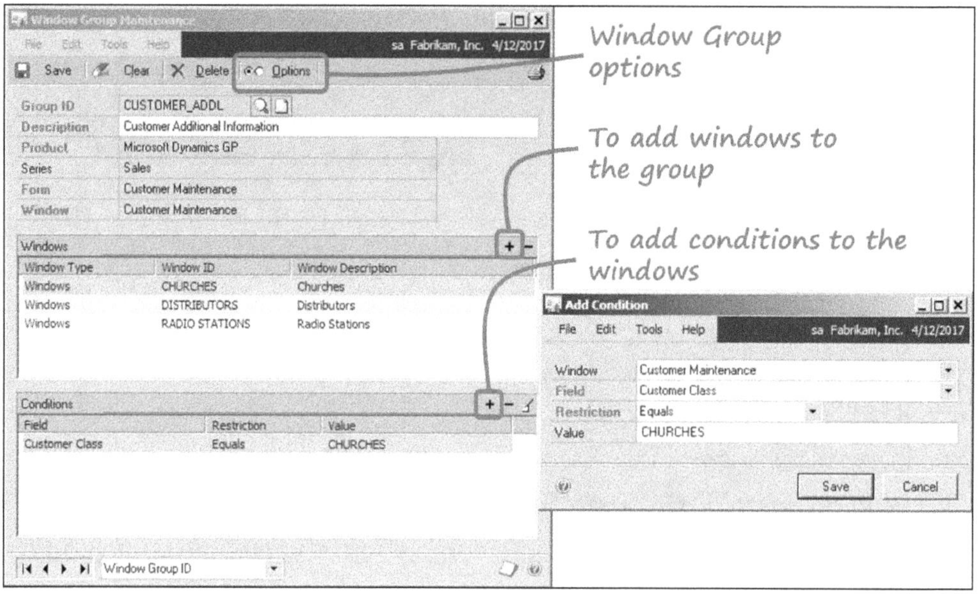

Define a hot key for opening your Extender window to reduce the quality time you spend with your mouse. Also, define an open method so that the Extender window will automatically open when you leave the **Customer Name** field on the **Customer Maintenance** window.

To define these choices, click on the **Options** radio button in the upper half of the **Window Group Maintenance** window. Use the following screenshot as your guide for completing the **Options** window:

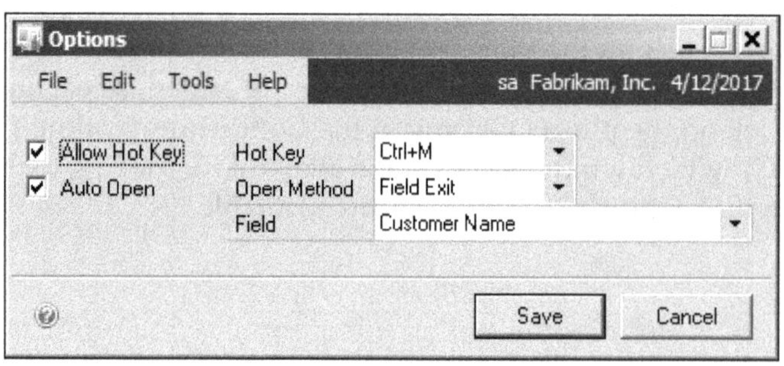

Restart Dynamics GP after you have saved your changes. Now, open a customer that you have associated with one of the classes of your window group. When you go to the additional menu and select the name of your window group, only the window for that customer's class will open. None of the other windows in the group will be available.

All of this without a single line of code.

Customer contacts

This project will work with the following Extender object:

- Detail window

Often, you need to keep track of multiple people associated with a customer. You may have an accounts payable person as well as your sales representative, who you interact with on an on-going basis. You would like to keep those individuals' contact information with customer IDs, but Dynamics GP does not offer sufficient fields to accomplish this task. With Extender, it's easy to add a window that will track as many contacts as you need.

The Extender object you will use to solve this dilemma is called a Detail window. Essentially, a scrolling window can hold ten fields on two rows per record. Let's build one so that you can see how this works.

From the **Extender Objects** list, click on the **Detail Windows** button and then click on the **New** button. Use the following screenshot as a guide for completing the **Detail Windows** window. Use the **R1** and **R2** checkboxes to indicate whether you want the field to appear in the first or second row of your scrolling window:

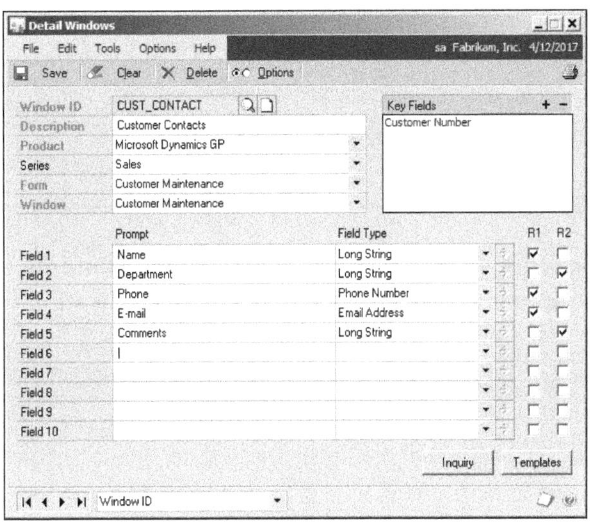

Save your new window and then open the **Customer Maintenance** window and pull up a customer. In the additional menu, you will now have a menu item that matches with the **Description** field you entered in the **Detail Windows** window, shown in the following screenshot:

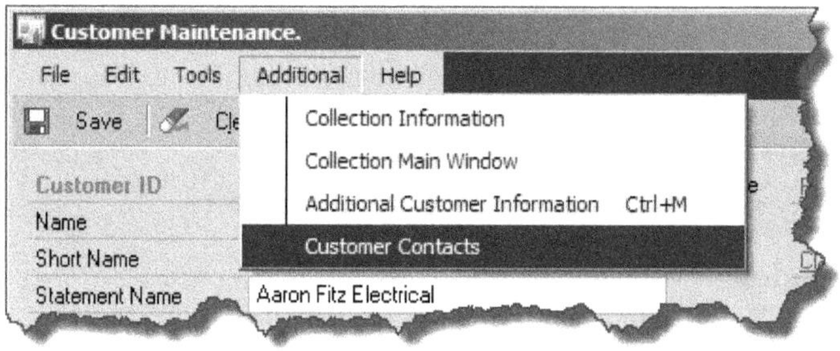

When you select the **Customer Contacts** item, a window similar to the one shown in the following screenshot will open:

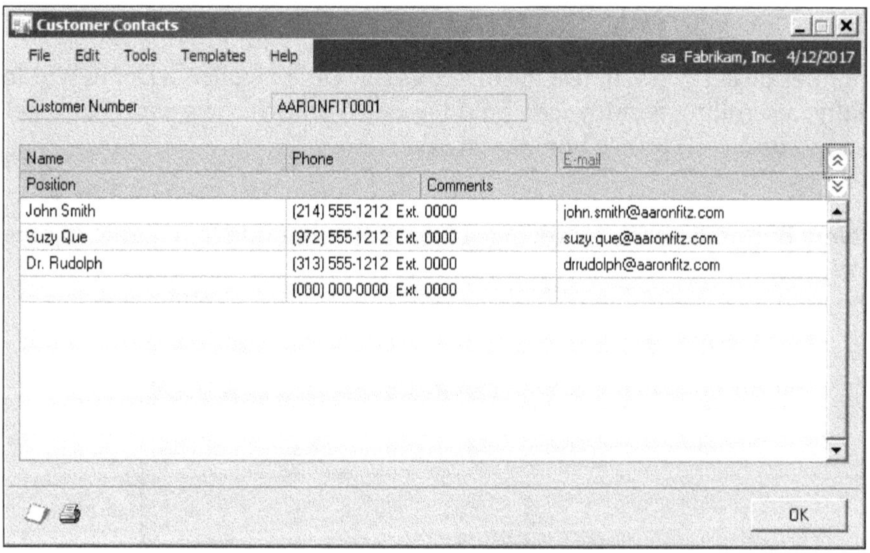

If one of your fields is not visible, go back to the **Details Window** window in Extender and make sure you checked one of the boxes indicating whether you wanted the field in row one or row two.

Line item note

Our final project uses the Notes object to add a note to the line item of a Sales Order Processing document. An Extender note is far more versatile than the native Dynamics GP note. You can attach multiple notes to a window instead of the single note that is allowed using the out-of-the-box version. In addition, Extender will automatically stamp your note with the date and time, as well as your user ID. If you compare the native Dynamics GP note (shown in the following screenshot) to the Extender note (also shown in the following screenshot), there is really no contest. Extender wins!

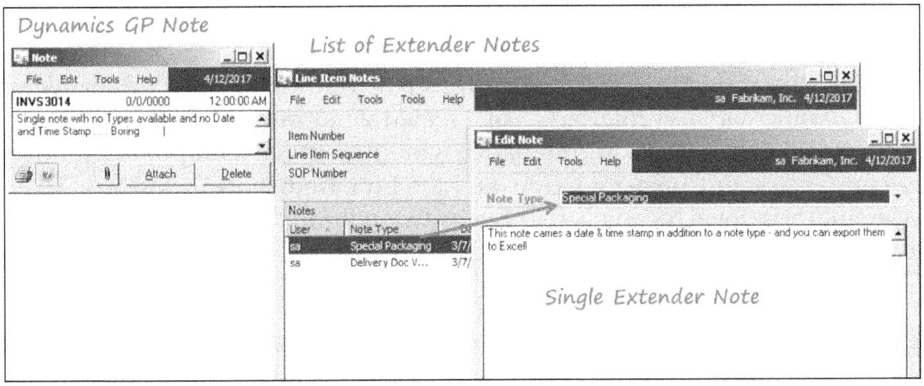

From the **Extender Objects** list click on the **Notes** button and then click on the **New** button. The **Extender Notes** window will open. Complete the **Extender Notes** window using the following screenshot as your guide:

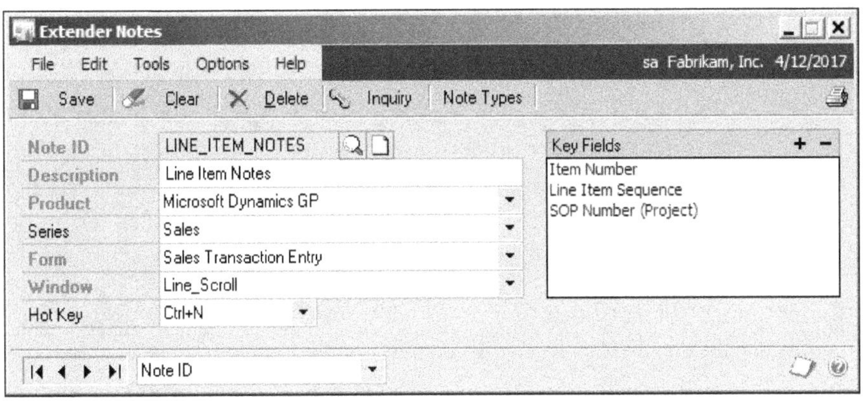

Summary

This chapter focused on the wide variety of enhancements you can create without any coding. You created animated SmartList objects that allowed your users to navigate to other windows and forms by merely clicking on a selected line item. You created Drill Down instances that could integrate Dynamics GP with any program that can accept hyperlinks.

Your SmartList objects morphed into Excel reports that you could link to live company databases. Now you can have refreshable spreadsheets that you only need to format once. You created data connections that you could use in any Microsoft Office application.

Beyond reporting, you moved into creating new windows that melded seamlessly into the Dynamics GP user interface. You created windows that could stand alone, such as the new **Event Maintenance** window, as well as support windows, such as the **Customer Contact** window and the **Line Item Notes** windows. These supplemental windows became a part of your standard user experience as if they had been part of the native code.

You can create cross-dictionary actions, such as opening any window in any dictionary using a SmartList object GoTo. You can use Extender to achieve more cross-dictionary integration by adding companion windows to any window in any dictionary.

The tools you have added in this chapter give you a broad variety of possibilities for a wide-range of enhancements.

In the next chapter you'll learn about using Visual Studio Tools to create Dynamics GP customizations.

10

Creating Customizations with VS Tools

When Microsoft introduced **Visual Studio Tools (VS Tools)** for Microsoft Dynamics GP, the development opportunities for Dynamics GP increased vastly. Now C# (C Sharp) and VB.NET (Visual Basic .NET) programmers could build .NET-based integrated applications using tools they were already familiar with. Dexterity didn't rule the roost anymore.

In this chapter, you will get to work with Visual Studio Tools. We'll explore several areas, which include the following:

- The architecture of the VS Tools add-in for Dynamics GP
- Installing VS Tools for Dynamics GP
- Using the tool to create a custom entry window
- Accessing resources in the Dynamics dictionary
- Building an assembly to access resources in a third-party dictionary
- Providing navigation to open your custom window
- Basic table operations
- Working with ranges
- Building and deploying your application

That's a lot to touch on in one chapter, but we can do it! Roll up your sleeves and we'll dig into VS Tools for Dynamics GP.

Architecture

Special runtime components give VS Tools the ability to work with Dynamics GP resources. You must install these runtime components before you can use VS Tools to create a Dynamics GP integration project. You will install the runtime components when you install VS Tools for Dynamics GP SDK. The runtime components include:

- Dexterity Shell
- Dexterity Bridge
- Application assemblies
- AddIns folder

Dexterity Shell

The **Dexterity Shell** assembly (`Microsoft.Dexterity.Shell.UI.dll`) provides the means for you to develop windows using VS Tools that look like native Dynamics GP windows. The assembly adds the following three additional properties to WinForms you created:

- **AutoSetDexColors**: This property is used to match the color of Dynamics GP windows and controls
- **ControlArea**: This property denotes the area located at the top of the window containing the button bar (refer to the following screenshot)
- **StatusArea**: This property denotes the area located at the bottom of the window containing the browse buttons and help icon (refer to the following screenshot)

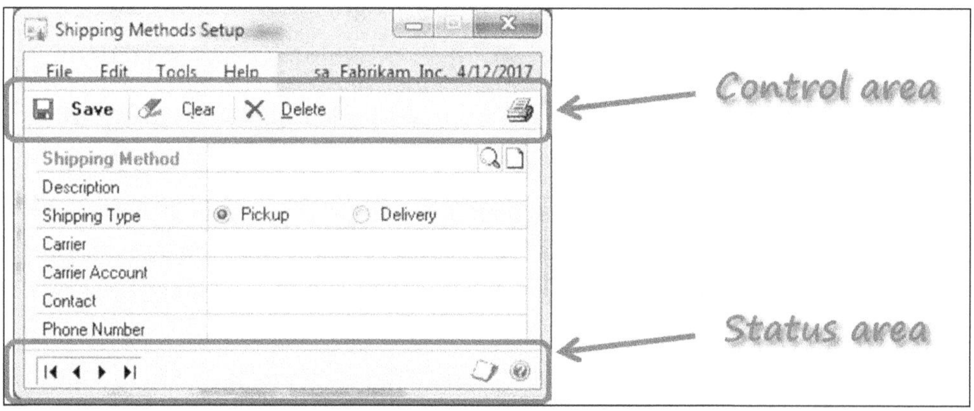

Every integration project must provide a reference to this assembly.

Dexterity Bridge

The **Dexterity Bridge** assembly (`Microsoft.Dexterity.Bridge.dll`) provides VS Tools access to the resources and events stored in the application dictionaries (the `.dic` files). Every integration project must provide a reference to this assembly.

Application assemblies

An **application** assembly provides VS Tools access to the resources and events of a single dictionary. Each dictionary that your integration project uses will have a corresponding application assembly.

The name of the application assembly will follow the name of the dictionary. For example, the name of the resource dictionary for fixed assets is `fam.dic`. Therefore, the name of the application assembly for fixed assets is `Application.FA.dll`.

Similarly, the name of the application assembly for the core modules is `Dynamics.dic`. The name of the application assembly for the core modules is `Application.Dynamics.dll`.

Every integration project must provide a reference to the appropriate application assembly or assemblies.

VS Tools for Dynamics GP ships with prebuilt application assemblies for each of the modules that come with Dynamics GP. You will select the application assemblies you want to include within the integration project when you install the runtime for VS Tools.

If you are working with an application provided by a third-party developer, either the developer will provide you with an application assembly or you will build one using the DAG tool. We will discuss the DAG tool a little later in this chapter.

Add-ins folder

We consider this a runtime component because it's where Dynamics GP looks for assemblies created with VS Tools. If it finds any, it will try to launch them as part of the startup routine.

Installing VS Tools

Before we can use VS Tools for Dynamics GP, we need to install the following:

- Visual Studio 2005 or higher
- Microsoft Dynamics GP version 9.0 or higher
- Visual Studio Tools for Dynamics GP SDK for the corresponding version of Dynamics GP

We presume that you have the first two software of the preceding list installed. In this chapter, we will be using Visual Studio 2010 and Dynamics GP release 2010 R2.

Download it

To install VS Tools, the first thing you need to do is download it. It can be a little difficult to find on your own. Using the following steps will lead you right to it:

1. To find the download link, navigate to `http://msdn.microsoft.com/en-us/dynamics/gp`
2. Click on **Downloads** from the landing page, which is shown in the following screenshot:

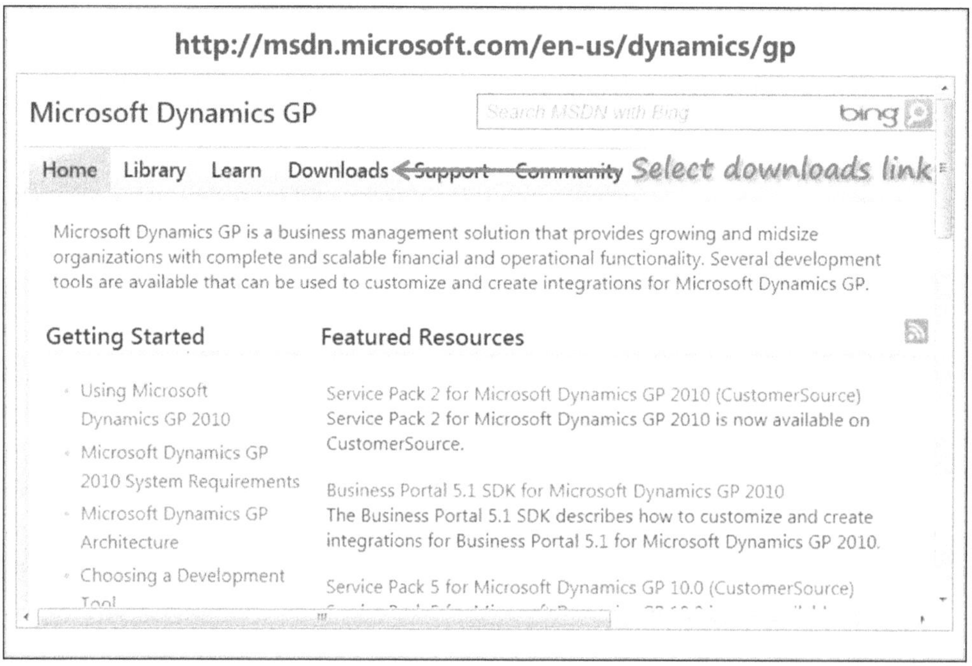

3. On the **Microsoft Dynamics GP** page that opens, click on the **Software Development Kits (SDKs)** link.
4. On the **Software Development Kits** page that opens, scroll down and click on the **Visual Studio Tools for Microsoft Dynamics GP 2010 SDK** link.
5. Clicking the **CustomerSource** link will launch the CustomerSource login page. Log in and the SDK download article will open. Click on the **Downloads** link.
6. From the **Downloads** section of the CustomerSource article, click on the most recent download:

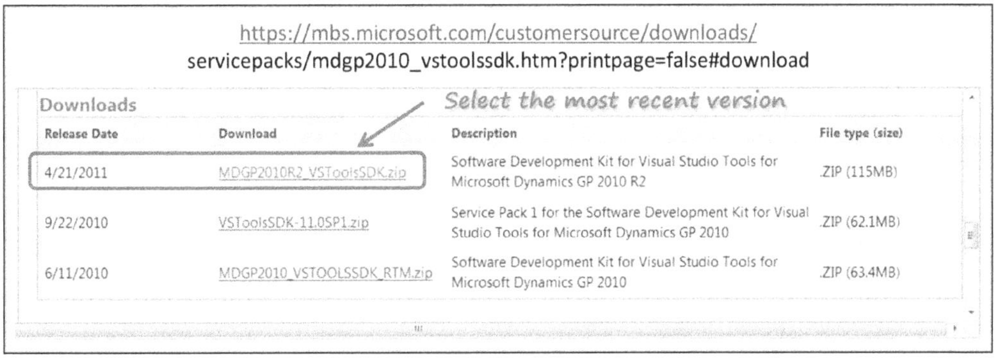

7. Extract the files from the `.zip` file and then navigate to where you extracted them.

At last! You are now ready to install VS Tools.

Run the installation

Execute the `Microsoft_DynamicsGP11_VSToolsSDK_x86_en-us.msi` installation file. In the **Select Features** window (shown in the following screenshot), select the assemblies you wish to install. Be sure to install the assembly generator. Install the Visual Studio templates that match the version of Visual Studio that you intend to use. We are using Version 2010 in this chapter, so we're not installing the templates for versions 2005 or 2008. We are also using the default installation location:

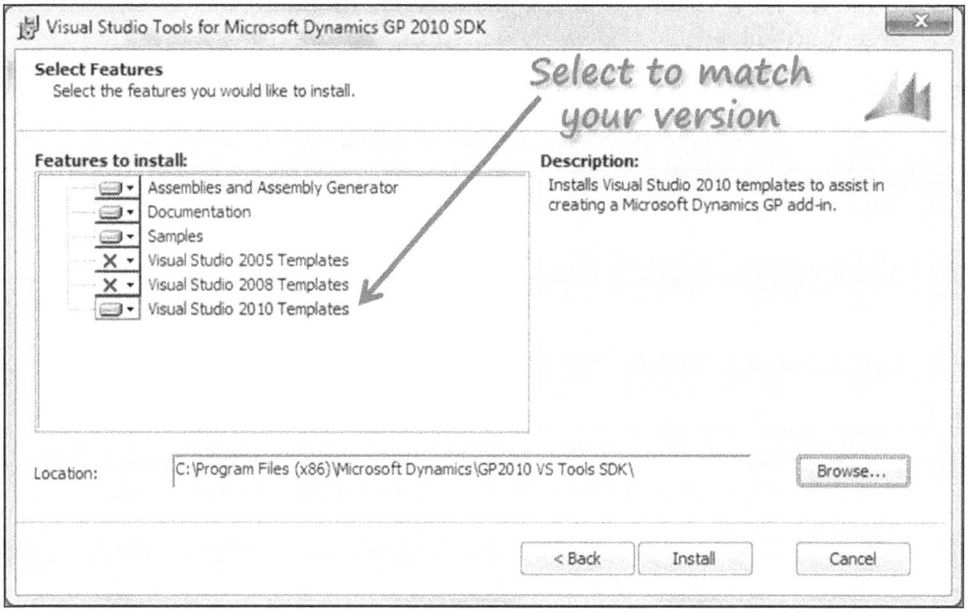

The installation reconfigures Visual Studio so that it recognizes the new templates. This process will take a few minutes to complete, so be patient.

The programmer's guide for VS Tools can be found at the location specified by you in the **Location** field within the **Select Features** window. You should probably print this programmer's guide; you will find it very helpful as you begin to use these new templates.

Vendor Quick Entry project

Your first VS Tools project will be to create a new window that you can use to quickly set up a new vendor. The window will include a much smaller number of fields than the regular **Vendor Maintenance** window. This project will show you how to create a WinForm and add controls to the new form. At the end of the project, your new window will look similar to the window shown in the following screenshot:

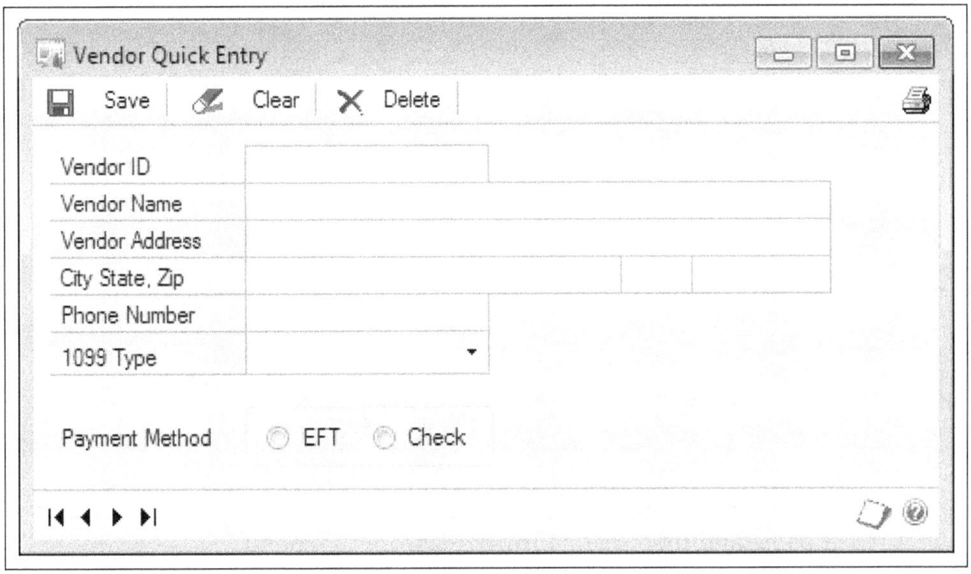

Creating the new project

Launch Visual Studio and click either the **New Project** link or select **File** | **New** | **Project** from the menu bar. Use the following steps to create your new project:

1. In the **New Project** window, set the .NET framework version drop-down menu to **.NET Framework 4**.
2. Expand the **Visual C#** option in the **Installed Templates** section.
3. Select the **Dynamics GP** template. If the Dynamics GP template does not appear in your list of installed templates, it means VS Tools did not get installed properly. Reinstall the templates and try again.
4. Name the project `Vendor_Quick_Entry`, accept the default location, and name the solution `Vendor_Quick_Entry`.
5. Mark the **Create directory for solution** checkbox.

6. Click on the **OK** button to create the project. The new project window should look similar to the window shown in the following screenshot, but you're not done yet:

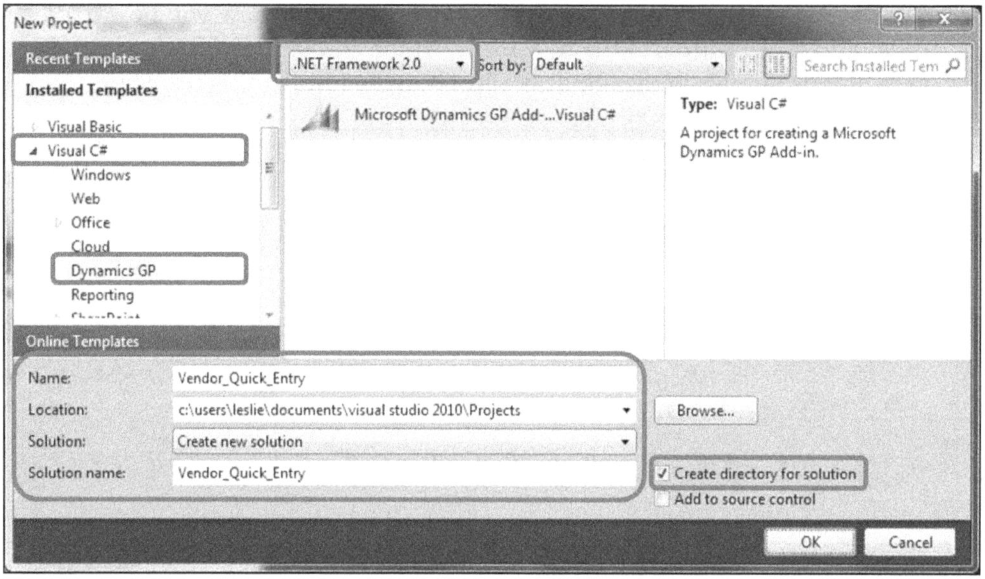

7. In the **Solution Explorer** pane, right-click on the **Vendor_Quick_Entry** project name and then select **Properties** from the menu.
8. Set the target framework to **.NET Framework 2.0**, as shown in the following screenshot. Respond with a yes to the warning dialog that will be presented to you:

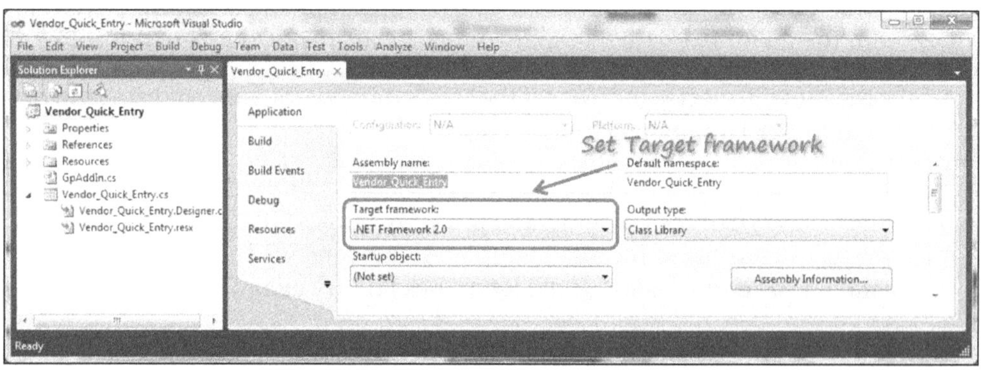

Now you're done with creating the project. Next we'll create a window.

Adding the new window

With your project created, your next step is to create the new `Vendor_Quick_Entry` window and add controls to it. To create the new window, you need to add a new form to the **Vendor_Quick_Entry** project. Perform the following steps to add a new form:

1. Right-click on the **Vendor_Quick_Entry** item in the **Solution Explorer** pane.
2. From the flyout menu, select **Add** and then **Component**.
3. The **Add New Item - Vendor_Quick_Entry** window will open. Select **Microsoft Dynamics GP Form** from the list of installed templates. Name the new form `Vendor_Quick_Entry`.

Alternatively, you can create the new window using these instructions:

1. Select **Project** from the menu bar.
2. Select **Add Windows Form** from the menu. Your **Add New Item - Vendor_Quick_Entry** window should look similar to the following screenshot:

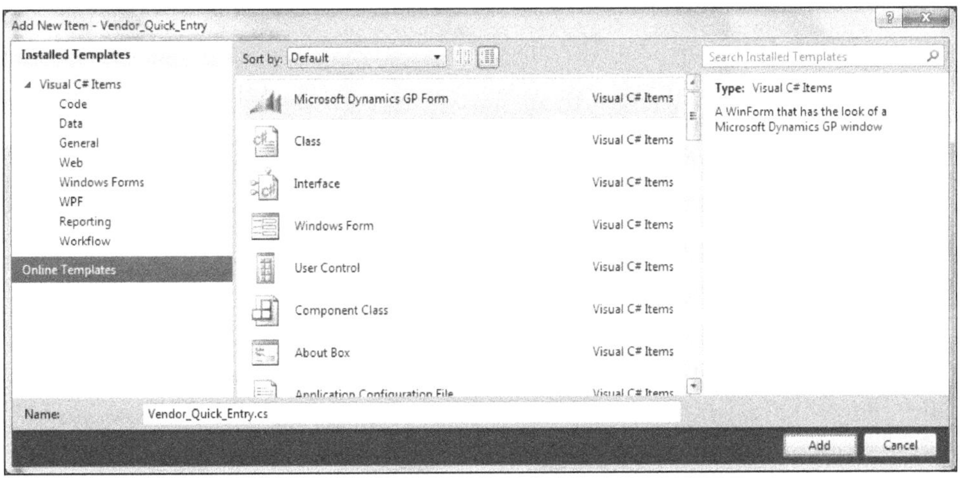

3. Select **Add**, and the newly created **Vendor_Quick_Entry** window will open.
4. Set the following properties for the **Vendor_Quick_Entry** window:

Category	Property	Value
Appearance	Text	Vendor Quick Entry
Layout	Size	500,275
Location	X	230
Location	Y	230

5. At this point, your new window should now look similar to the following screenshot:

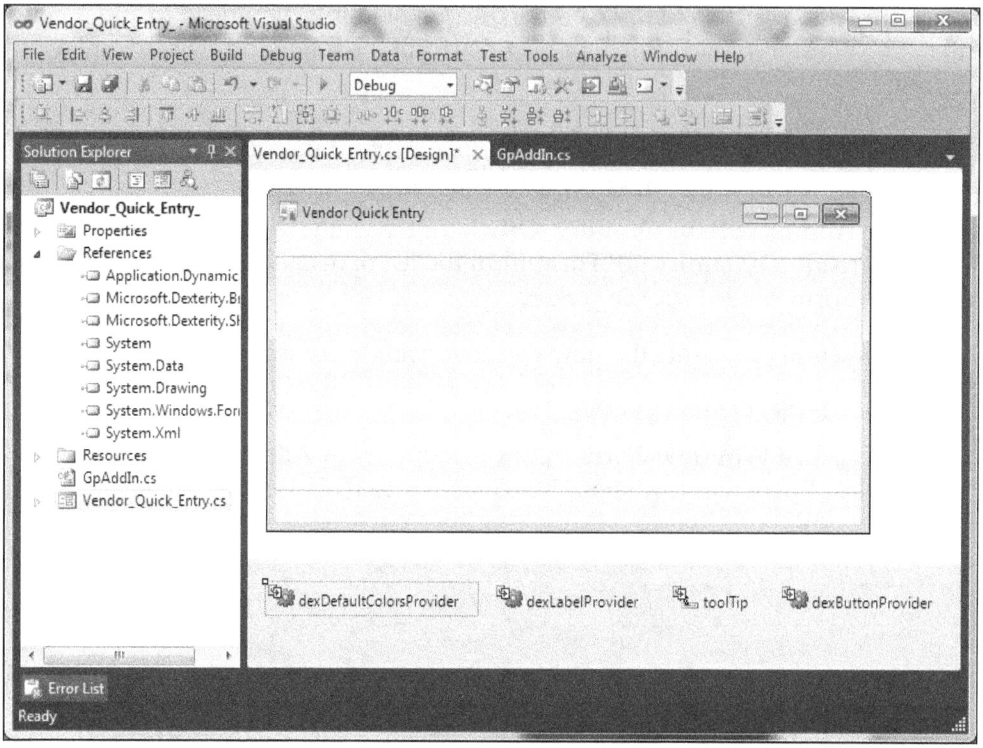

The next step in our project is to add controls to the window.

Window controls

To add a control to a window, simply select the control from the **Toolbox** pane and drag it out onto the form. If your **Toolbox** pane is not displayed, click on the toolbox icon to open it. The names of the controls are listed in the **Toolbox** pane. The **Toolbox** pane and toolbox icon are identified in the following screenshot:

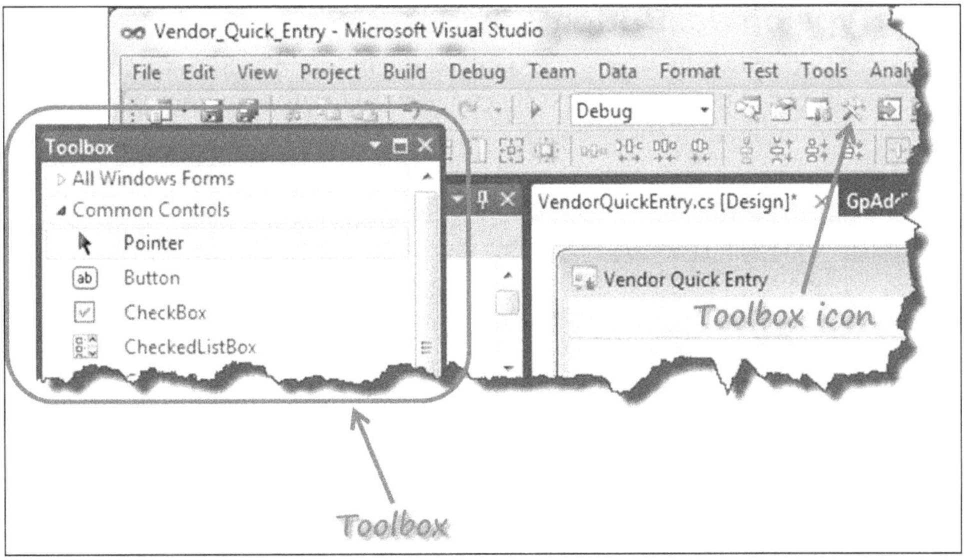

You're going to add the following types of controls to your new window:

- Button
- TextBox
- Label
- RadioButton and GroupBox
- ComboBox

Many controls in the VS Tools Dynamics GP template have additional properties specific to Dynamics GP integration projects. These additional properties are listed in the **Dexterity** section of the **Properties** window.

Button

A **button** control appears as a push button in the window. Pushing the button causes a `Click` event in Visual Studio. The button control includes two Dexterity properties:

- ButtonType on dexButtonProvider
- AutoSetDexColors

ButtonType on dexButtonProvider

Five types of buttons are available. The table in the following screenshot shows each button type and how the button looks in a window. These are merely examples; many more button images are available.

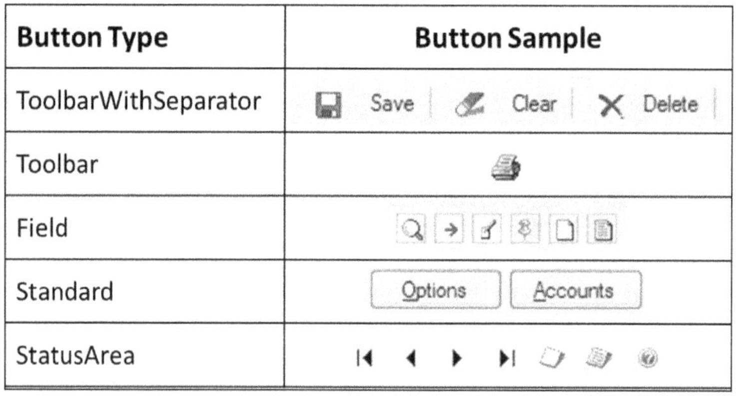

Button Type	Button Sample
ToolbarWithSeparator	Save Clear Delete
Toolbar	
Field	
Standard	Options Accounts
StatusArea	

The **ToolbarWithSeparator** button type describes the buttons that appear at the top of the window in the Control area.

The **Toolbar** button type describes the buttons in the Control area that are typically just a single icon with no words, such as the printer button.

The **Field** button type describes the buttons that you would normally attach to fields. The lookup button, expansion button, and record notes button are all examples of **Field** button types.

The **Standard** button type is the generic Visual Studio button; it is not a special button designed for the Dynamics GP template, and it looks the same even if you are not working on a Dynamics GP customization.

The **StatusArea** button type is for buttons you place at the bottom of the window in the Status area. The so-called VCR buttons and the help icon are StatusArea buttons.

VS Tools provides many, but not all, of the standard images for Dynamics GP buttons. They come as `.png` files; you installed these files when you installed VS Tools SDK. The images and their names are shown in the following screenshot:

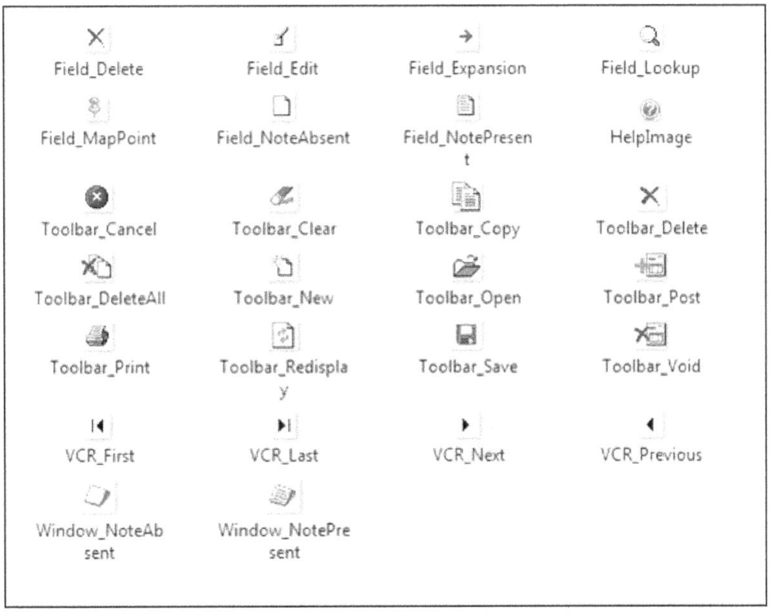

Any images that you did not get as part of VS Tools, such as the Print Preview icon for example, can easily be extracted from Dexterity's Picture resource or Native Picture resource library.

AutoSetDexColors

When set to `True`, the AutoSetDexColors property will cause a Visual Studio button control to take on the colors that match the colors in Dynamics GP. The default setting is `True`.

TextBox

The TextBox control is equivalent to the following Dynamics GP controls:

- String
- Integer
- Currency
- Date
- Time

Since there are no separate controls dedicated to each of the data types that were just listed, you need to set the format for the TextBox control. You'll use a small bit of code to accomplish this. For example, in order to change the format of a TextBox control to display currency, you could use the `DataBindings` property.

The following example code will format `textBox1` to display the currency as $1,234.56:

```
textBox1.DataBindings.Add(new Binding ("Text", data_source_name,
"table_name.field_name", true, DataSourceUpdateMode.OnValidation,
0 "C"));
```

The last parameter (`"C"`) is what formats the field to the currency format.

The following table lists the string-formatting parameters and their results:

Name	Parameter	Result
Currency	`C`	`$1,234.56`
Scientific Exponential	`E`	
Percentage	`P`	`45.6%`
Fixed Decimal	`F?`	`?` is number of decimal places For example: `F4` = `1234.5678`
ShortDate	`d`	`M/d/yyyy`
LongDate	`D`	`MMMM dd, yyyy`
LongDate ShortTime	`f`	`dddd,MMMM dd, yyyy HH:mm aa`
LongDate LongTime	`F`	`dddd,MMMM dd yyyy HH:mm:ss:aa`
Month and Day	`M`	`MMMM dd`
General	`G`	Date and Time format comes from your system locale settings.

Visual Studio will build the code for you using the `DataBindings` property and the **Formatting and Advanced Binding** window. You will first need to add a data source before you can use the `DataBindings` property. Use the **Data** menu to add the data source.

Additionally, you can read an article dedicated to the formatting types at `http://msdn.microsoft.com/en-us/library/fbxft59x(v=vs.80).aspx`

The TextBox control includes the following Dexterity properties.

AutoSetDexColors

When set to `True`, the AutoSetDexColors property will cause a Visual Studio TextBox control to take on the colors that match the colors in Dynamics GP. The default setting is `True`.

Label

A Label control is equivalent to the static text that appears next to a field on a Dynamics GP window.

The Label control includes the following Dexterity property.

LinkField on dexLabelProvider

Using this property, you can provide the name of the control to which you are linking the label. An example of this control is shown in the following screenshot.

By filling this out, you will cause an underline to appear below the linked label, which is a standard Dynamics GP user interface feature.

After you have linked the labels, they will look like the following screenshot:

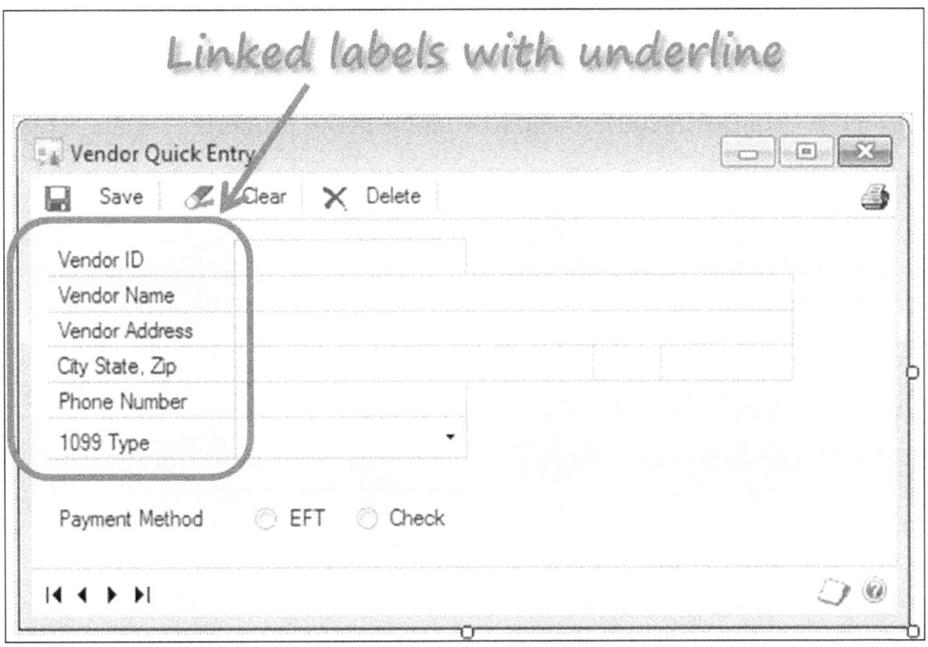

RadioButton and GroupBox

RadioButton and GroupBox controls do not have any Dexterity properties. These controls operate as you would expect any radio group to behave.

ComboBox

Use a ComboBox control when you want to present a drop-down list to the user.

In order to display an uneditable drop-down list, you need to set the **Style** property on the control.

Style

The ComboBox control has three **DropDownStyle** options:

- **Simple**: With the Simple option selected, you can always see the item list; you can also edit the text portion of the list.
- **DropDown**: Using the DropDown option, you can display the list by clicking on an arrow button on the right-hand side of the field. You can also edit the text portion of the list.
- **DropDownList**: Like the DropDown style, you can display the list by clicking on an arrow button on the right side of the field. However, the text portion is not editable when DropDownList is selected.

The ComboBox control includes the following Dexterity property:

AutoSetDexColors

When set to `True`, the AutoSetDexColors property will cause the Visual Studio ComboBox control to take on the colors that match the colors in Dynamics GP. The default setting is `True`.

Adding window controls

To build your **Quick_Vendor_Entry** window, you need to add several controls. In this project, you will add the controls one control type at a time for convenience. When you add your controls, position them on the window similarly to the preceding screenshot displayed at the beginning of this project.

TextBox controls and properties

Add seven TextBox controls to the window and then set their properties according to the information in the following table:

Control	Category	Property	Value
textBox1	Design	(Name)	txtVendorID
	Misc	AutoCompleteMode	SuggestAppend
	Misc	AutoCompleteSource	RecentlyUsedList
	Behavior	CharacterCasing	Upper
	Behavior	MaxLength	15
textBox2	Design	(Name)	txtVendorName
	Behavior	MaxLength	64
textBox3	Design	(Name)	txtVendorAddress
	Behavior	MaxLength	60
textBox4	Design	(Name)	txtCity
	Behavior	MaxLength	35
textBox5	Design	(Name)	txtState
	Behavior	CharacterCasing	Upper
	Behavior	MaxLength	35
textBox6	Design	(Name)	txtZipCode
	Behavior	MaxLength	10
textBox7	Design	(Name)	txtPhoneNumber
	Behavior	MaxLength	14

Label controls and properties

Add seven Label controls to the window and then set their properties according to the information in the following table:

Control	Category	Property	Value
label1	Design	(Name)	lblVendorID
	Appearance	Text	Vendor ID
	Dexterity	LinkField on dexLabelProvider	txtVendorID
label2	Design	(Name)	lblVendorName
	Appearance	Text	Vendor Name
	Dexterity	LinkField on dexLabelProvider	txtVendorName

Control	Category	Property	Value
label3	Design	(Name)	`lblVendorAddress`
	Appearance	Text	`Address`
	Dexterity	LinkField on dexLabelProvider	`txtVendorAddress`
label4	Design	(Name)	`lblCity`
	Appearance	Text	`City State,Zip`
	Dexterity	LinkField on dexLabelProvider	`txtCity`
label5	Design	(Name)	`lblPhoneNumber`
	Appearance	Text	`Phone Number`
	Dexterity	LinkField on dexLabelProvider	`txtPhoneNumber`
label6	Design	(Name)	`lbl1099Type`
	Appearance	Text	`1099 Type`
	Dexterity	LinkField on dexLabelProvider	`cb1099Type`
label7	Design	(Name)	`lblPaymentMethod`
	Appearance	Text	`Payment Method`

Button controls and properties

Add eleven Button controls to the window and then set their properties according to the information in the following table:

Control	Category	Property	Value
button1	Design	(Name)	`btnSave`
	Dexterity	ButtonType on dexButtonProvider	`ToolbarWithSeparator`
	Appearance	Text	`Save`
	Appearance	TextAlign	`MiddleRight`
	Appearance	Image	`Toolbar_Save`
	Appearance	ImageAlign	`MiddleLeft`

Control	Category	Property	Value
	Design	(Name)	btnClear
	Dexterity	ButtonType on dexButtonProvider	ToolbarWithSeparator
button2	Appearance	Text	Clear
	Appearance	TextAlign	MiddleRight
	Appearance	Image	Toolbar_Clear
	Appearance	ImageAlign	MiddleLeft
	Design	(Name)	btnDelete
	Dexterity	ButtonType on dexButtonProvider	ToolbarWithSeparator
button3	Appearance	Text	Delete
	Appearance	TextAlign	MiddleRight
	Appearance	Image	Toolbar_Delete
	Appearance	ImageAlign	MiddleLeft
	Design	(Name)	btnPrinter
	Dexterity	ButtonType on dexButtonProvider	Toolbar
button4	Appearance	Text	(blank out):
	Appearance	Image	Toolbar_Print
	Appearance	ImageAlign	MiddleCenter
	Design	(Name)	btnVCR_First
	Dexterity	ButtonType on dexButtonProvider	StatusArea
button5	Appearance	Text	(blank out):
	Appearance	Image	VCR_First
	Design	(Name)	btnVCR_Previous
	Dexterity	ButtonType on dexButtonProvider	StatusArea
button6	Appearance	Text	(blank out):
	Appearance	Image	VCR_Previous
	Design	(Name)	btnVCR_Next
	Dexterity	ButtonType on dexButtonProvider	StatusArea
button7	Appearance	Text	(blank out):
	Appearance	Image	VCR_Next

Control	Category	Property	Value
button8	Design	(Name)	btnVCR_Last
	Dexterity	ButtonType on dexButtonProvider	StatusArea
	Appearance	Text	(blank out):
	Appearance	Image	VCR_Last
button9	Design	(Name)	btnHelpIcon
	Dexterity	ButtonType on dexButtonProvider	StatusArea
	Appearance	Text	(blank out):
	Appearance	Image	HelpImage
button10	Design	(Name)	btnNotePresent
	Dexterity	ButtonType on dexButtonProvider	StatusArea
	Appearance	Text	(blank out):
	Appearance	Image	Window_NotePresent
button11	Design	(Name)	btnNoteAbsent
	Dexterity	ButtonType on dexButtonProvider	StatusArea
	Appearance	Text	(blank out):
	Appearance	Image	Window_NoteAbsent

Position the **btnNotePresent** and **btnNoteAbsent** controls on top of one another on the right-hand side of the status area next to the **btnHelpIcon** control.

RadioButton and GroupBox

Add one GroupBox control with two RadioButton controls to the window.
The GroupBox control is in the **Containers** section of the **Toolbox** pane.
Use the following table to set their properties:

Control	Category	Property	Value
groupBox1	Design	(Name)	grpPaymentMethod
	Appearance	Text	(blank out)
radioButton1	Design	(Name)	rbEFT
	Appearance	Text	EFT
radioButton2	Design	(Name)	rbCheck
	Appearance	Text	Check

ComboBox

Add one ComboBox control to the window and then set its properties according to the information in the following table:

Control	Category	Property	Value
comboBox1	Design	(Name)	cb1099Type
	DataBindings	Items	Not a 1099 Vendor Dividend Interest Miscellaneous
	Appearance	DropDownStyle	DropDownList
	Dexterity		

Accessing dictionary resources

When you installed VS Tools, you installed assemblies for each dictionary that ships with Dynamics GP. If you followed the instructions in this chapter, you installed all of the assemblies. In order for your application to read those assemblies, you need to tell it where they are. You can accomplish this using a reference. Two references are required:

- Referencing the application assembly
- Referencing the namespace

Referencing the application assembly

When you initially created your VS Tools project, you automatically created a reference to the Dynamics assembly. You can see that reference in the following screenshot:

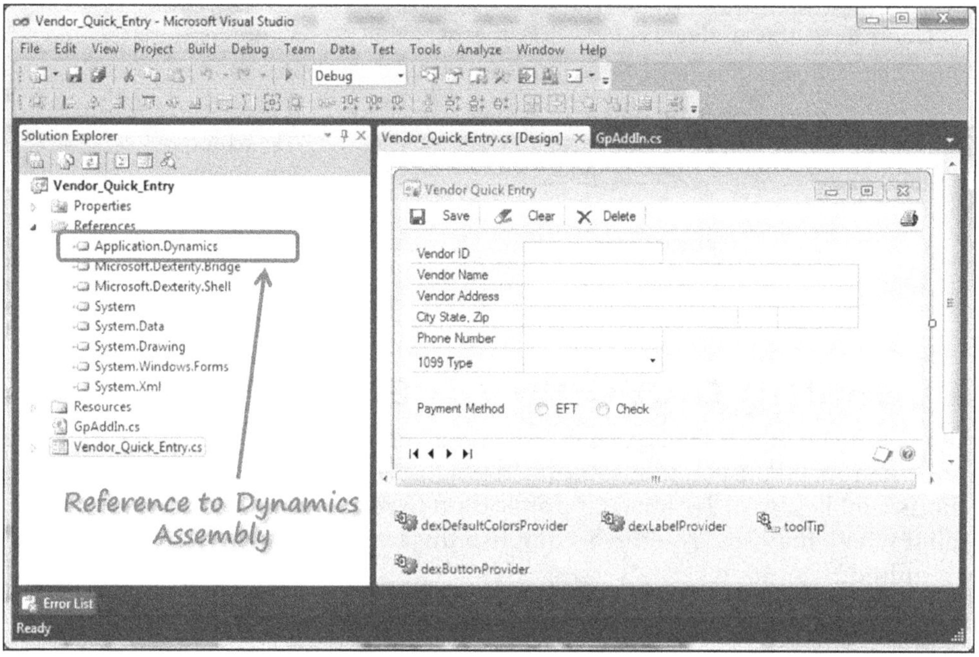

Since that reference already exists, let's imagine that your application will also be accessing resources in the Fixed Assets dictionary. Under that scenario, you would need to create a reference to the Fixed Assets assembly. Follow these steps to add the reference to your application:

1. Right-click on the **Vendor_Quick_Entry** project.
2. Select **Add Reference** from the menu.
3. The **Add Reference** window will open; click on the **Browse** tab and browse for the `Application.FixedAssets.dll` file:

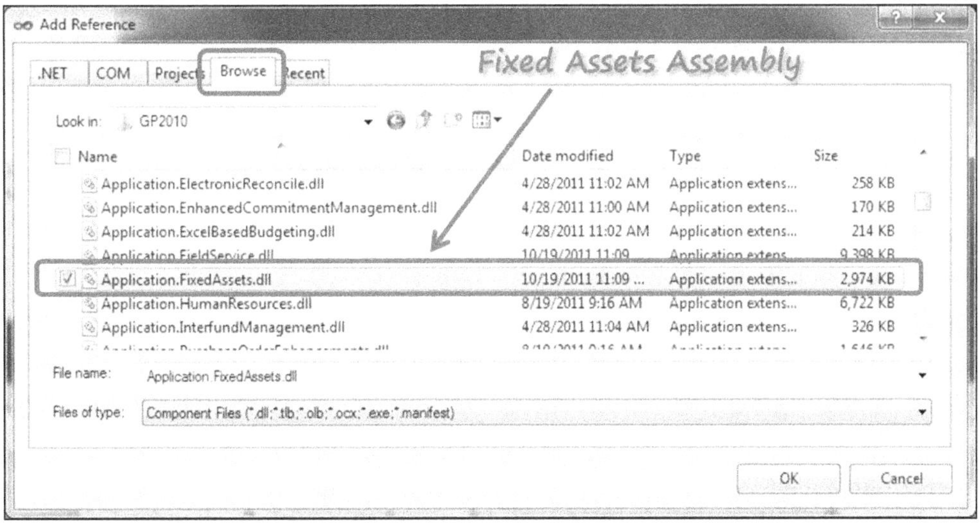

Referencing the namespace

A reference to the namespace is not required, but it makes coding much easier because you will not have to provide fully qualified references to dictionary resources. For example, if you want to open the **Vendor Maintenance** window, here's how it looks:

- Without the namespace reference:
  `Microsoft.Dexterity.Applications.Dynamics.Forms.Vendor.Open()`
- With the namespace reference:
  `Dynamics.Forms.Vendor.Open()`

Convinced?

When you created your project, the following code was automatically added to reference the listed namespaces:

```
using System;
using System.Collections.Generic;
using System.ComponentModel;
using System.Data;
using System.Drawing;
using System.Text;
using System.Windows.Forms;
using Microsoft.Dexterity.Bridge;
using Microsoft.Dexterity.Applications;
using Microsoft.Dexterity.Shell;
```

What do you do about applications that are prepared by third-party developers? Read on.

Building dictionary assemblies

If you need to access resources in a dictionary that did not ship with Dynamics GP, you need to build an application assembly yourself. You're in luck! VS Tools includes a tool called the **Dictionary Assembler Generator** (**DAG**). You should never generate an assembly for dictionaries shipped with Dynamics GP. If you didn't install an assembly you need, rerun the VS Tools installation.

The `dag.exe` file is in the `GP2010 VS Tools SDK` folder. For you to try it out, you need a third-party application dictionary. You're in luck again!

When you installed Dexterity (I'm presuming you installed Dexterity), a sample integrating application was also installed. You can find it in the following folder:

```
… \Microsoft Dexterity\Dex 11.0\Samples\Develop
```

Copy the `Develop.cnk` file and drop it into your Dynamics GP installation folder. Launch Dynamics GP to expand the `.cnk` file, thereby creating the `Develop.dic` dictionary.

Now you're ready to create the application assembly for the Develop dictionary.

Dictionary Assembly Generator (DAG)

DAG is a command-line utility that creates two files when you run it against a dictionary. It creates an application assembly file and an IntelliSense file.

The application assembly file provides access to resources in the third-party dictionary. The IntelliSense file is used by Visual Studio to list available resources that you can choose from as you use the code editor. You can still integrate with the dictionary without the IntelliSense file, but you will have to know the names of each resource that you want to access. Unless you know those resource names by heart, this is not a good plan.

Using the DAG

As this is a command-line utility, you need to run it from the command prompt. To open the command prompt, you can type `cmd` in the **Run** window or navigate to **All Programs** | **Accessories** | **Command Prompt**.

At the command prompt, navigate to the folder where the `DAG.exe` file resides. To see all of the options available for this application, enter the following command and hit the *Enter* key:

```
dag.exe /?
```

You can enter the following command at the command prompt to go directly to the folder you want (modify Program Files as appropriate):

```
cd %programfiles(x86)%.\microsoft dynamics\gp2010 vs tools sdk
```

To generate the assembly, the first thing you need is the product ID of your target dictionary. You can find the product ID in the `Dynamics.set` file. A `Dynamics.set` file is shown in the following screenshot; note that the product ID for **Sample Integrating App.** is **3333**:

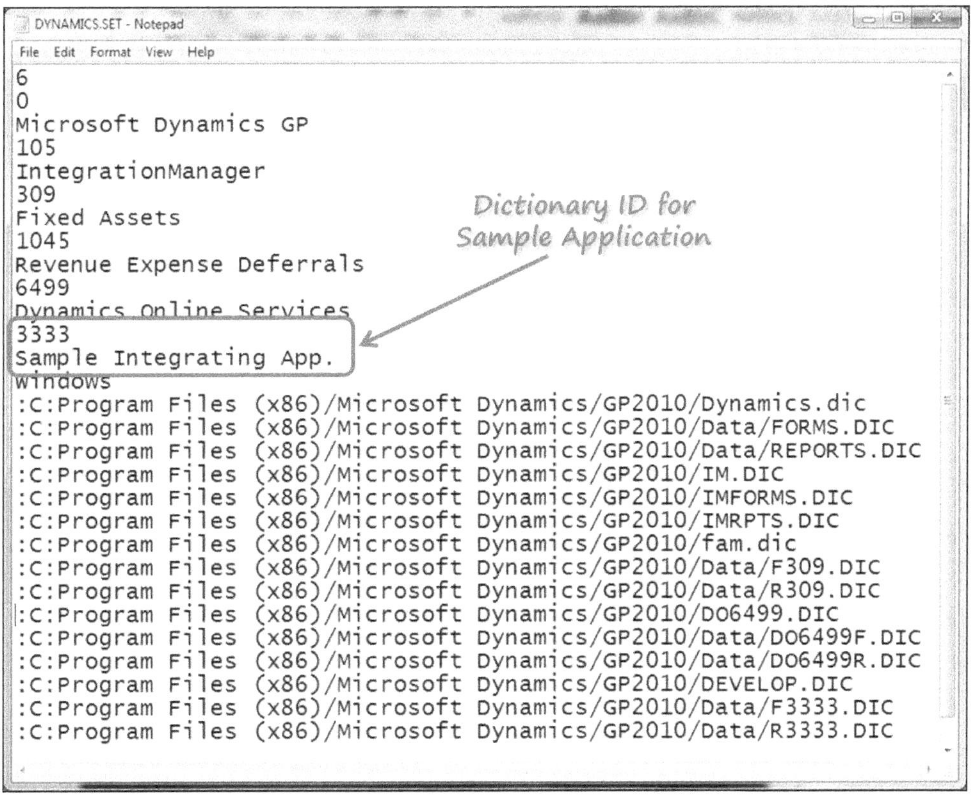

Armed with the dictionary ID, you can now create the assembly.

Using DAG, you can create an assembly for the main dictionary or an assembly for the forms dictionary. To generate the desired assembly, use the following commands:

- Main dictionary:

  ```
  dag.exe 3333 /M
  ```

- Forms dictionary:

  ```
  dag.exe 3333 /F
  ```

Creating the AddIn assembly

It's time to create the assembly for the sample integration application. Copy the `dag.exe` file to the folder containing your `Dynamics.set` file. Navigate to that folder using the command prompt and execute the following command:

```
dag.exe 3333 /M
```

When you have completed the process, open your `GP2010` folder and look for the following two files:

- `Application.SampleIntegratingApp.dll`
- `Application.SampleIntegratingApp.xml`

To prove the validity of the application assembly, include a reference to it in your project and then enter the following code:

```
SampleIntegratingApp.Forms.IgLeadMaintenance.
```

As you type in each period, the IntelliSense file should take over and present a list of choices to you from the application assembly. The following screenshot shows you what the screen looks like when you enter a dot after `IgLeadMaintenance`:

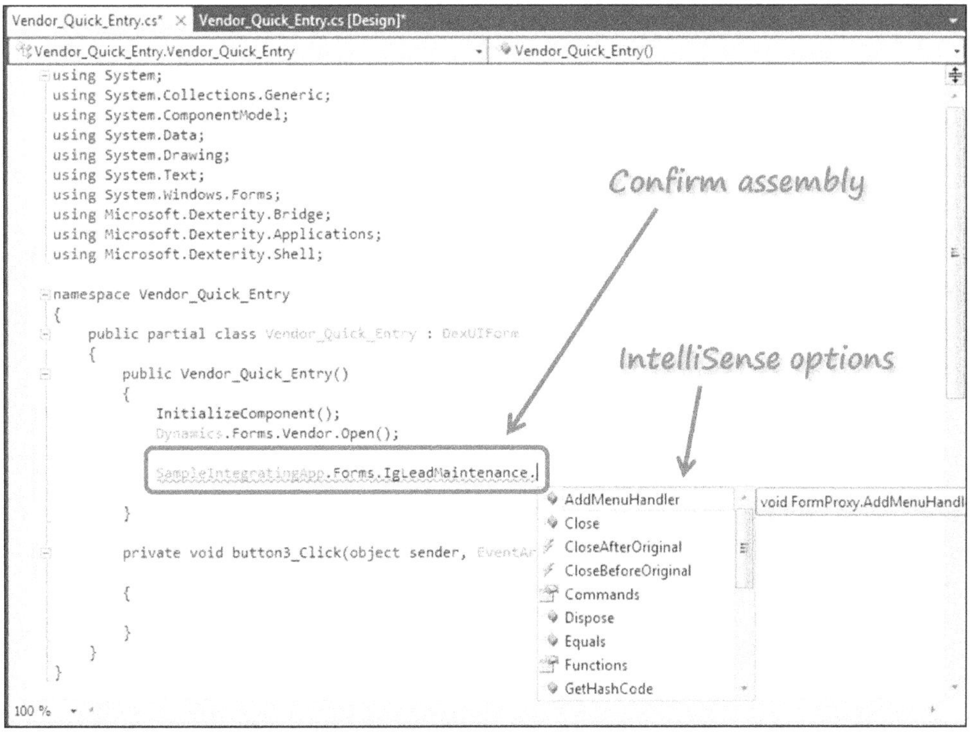

Opening your window

Now that you have your **Vendor Quick Entry** window constructed, you need to provide a method to open it in Dynamics GP. You're going to use the **Additional** menu on the **Vendor Maintenance** window to open your **Vendor Quick Entry** window.

Code the action

The following screenshot shows you what the **Additional** menu will look like:

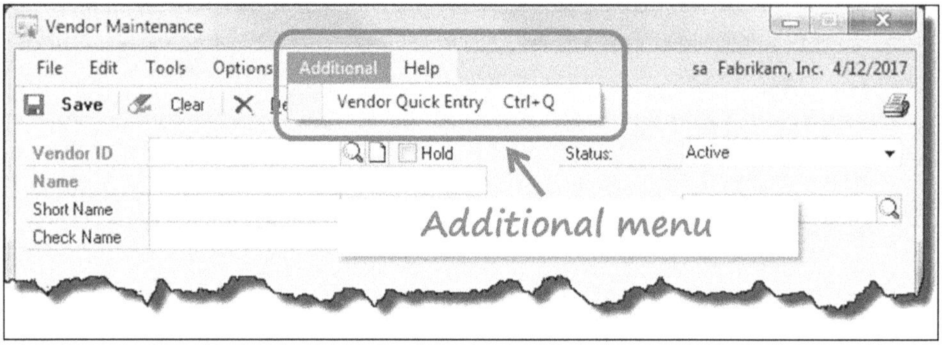

To create the **Additional** menu and add an item to it, you need to add some C# code to the `GpAddin.cs` file. To open the code window, double-click on the filename as shown in the following screenshot:

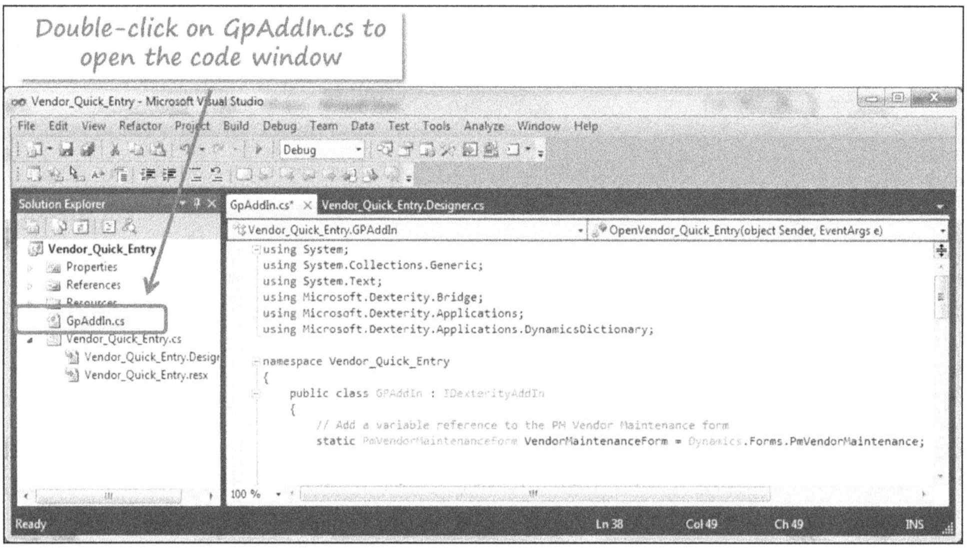

Add the following snippet to the code window:

```
using System;
using System.Collections.Generic;
using System.Text;
using Microsoft.Dexterity.Bridge;
using Microsoft.Dexterity.Applications;
```

```
using Microsoft.Dexterity.Applications.DynamicsDictionary;

namespace Vendor_Quick_Entry
{
    public class GPAddIn : IDexterityAddIn
    {
        // Add a variable reference to the PM Vendor Maintenance form
        static PmVendorMaintenanceForm VendorMaintenanceForm =
Dynamics.Forms.PmVendorMaintenance;


        // Keep a reference to the Vendor_Quick_Entry WinForm
        static Vendor_Quick_Entry Vendor_Quick_EntryForm;

        public void Initialize()
        {
            // Sets the menu text to -- Vendor Quick Entry --
            // and the Alt+ shortcut to -- Q --
            VendorMaintenanceForm.AddMenuHandler(OpenVendor_Quick_
Entry, "Vendor Quick Entry", "Q");
        }

        static void OpenVendor_Quick_Entry(object Sender, EventArgs e)
        {
            // check to make sure the form is not either empty or
already created
            if (Vendor_Quick_EntryForm == null)
                Vendor_Quick_EntryForm = new Vendor_Quick_Entry();
            else
                if (Vendor_Quick_EntryForm.Created == false)
                    Vendor_Quick_EntryForm = new Vendor_Quick_Entry();

            // make the form visible
            Vendor_Quick_EntryForm.Show();

            // make the form the window in focus
            Vendor_Quick_EntryForm.Activate();
        }
    }
}
```

With your code entered, build your assembly to create the `Vendor_Quick_Entry.dll` file.

Building and testing your assembly

Save any changes to your project, select **Build** from the menu bar, and then select **Build Vendor_Quick_Entry** from the submenu. If all goes well, you will be rewarded with the **Build succeeded** message shown in the following screenshot:

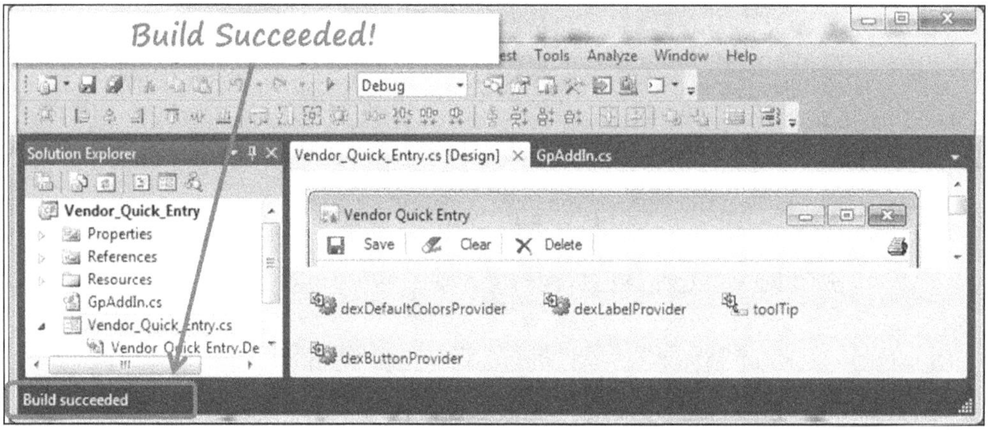

The build created a file named `Vendor_Quick_Entry.dll`. If you chose to leave your application in the default location, you will find this file in `C:\Users\User_Name\Documents\Visual Studio 2010\Projects\Vendor_Quick_Entry\Vendor_Quick_Entry\bin\Debug`.

If it isn't in the `Debug` folder, check the folder `C:\Users\User_Name\Documents\Visual Studio 2010\Projects\Vendor_Quick_Entry\Vendor_Quick_Entry\bin\Release`.

Copy the file and paste it into the `AddIns` folder at `C:\Program Files (x86)\Microsoft Dynamics\GP2010\AddIns\`.

Now launch Dynamics GP, open the **Vendor Maintenance** window, and check out the **Additional** menu selections. The **Vendor Maintenance** window should be sporting a new **Additional** menu.

Table operations

As with any database program, you need to get down to the basic table operations before you do anything fancy. Now is the time for you to put some code behind those buttons and see your window in action.

Knowing the basic table operations is the first thing you need to learn using any database programming. Learning table operations is often referred to as "knowing your CRUD." Let's do it. This is what CRUD means:

- Creating a record
- Retrieving a record
- Updating a record
- Deleting a record

Creating a record

You're going to attach your code to the `Click` event of the **Save** button. When you click on the **Save** button, you will add a record to the `PM Vendor Master` table. Open the code window by double-clicking on the **Save** button. Add the following code to the **Save** button's `Click` event:

```
//////////////////////  SAVE BUTTON //////////////

        //The SAVE button SAVES a record to the vendor master table
        private void btnSave_Click(object sender, EventArgs e)
        {
            //Variable for table operation error
            TableError err;

            //Create a reference to the PM Vendor Mstr table
            PmVendorMstrTable VendorMSTRTable;
            VendorMSTRTable = Dynamics.Tables.PmVendorMstr;


            // Release any existing lock
            VendorMSTRTable.Release();

            //set the window field values in the table
            VendorMSTRTable.VendorId.Value = txtVendorID.Text;
            VendorMSTRTable.VendorName.Value = txtVendorName.Text;
            VendorMSTRTable.VendorCheckName.Value = txtVendorName.
   Text;
            VendorMSTRTable.Address1.Value = txtVendorAddress.Text;
            VendorMSTRTable.State.Value = txtState.Text;
            VendorMSTRTable.City.Value = txtCity.Text;
            VendorMSTRTable.ZipCode.Value = txtZipCode.Text;
            VendorMSTRTable.PhoneNumber1.Value = txtPhoneNumber.Text;
```

```
            //set default values for other fields in the table
            VendorMSTRTable.VendorAddressCodePrimary.Value =
"PRIMARY";
            VendorMSTRTable.VendorStatus.Value = 1;
            VendorMSTRTable.FreeOnBoard.Value = 1;
            VendorMSTRTable.KeepCalendarHistory.Value = true;
            VendorMSTRTable.KeepGlDistHistory.Value = true;
            VendorMSTRTable.KeepPeriodHistory.Value = true;
            VendorMSTRTable.KeepTrxHistory.Value = true;
            VendorMSTRTable.Hold.Value = true;
            VendorMSTRTable.CreditLimit.Value = 1;
            VendorMSTRTable.RevalueVendor.Value = true;
            VendorMSTRTable.FreeOnBoard.Value = 1;


            //save the new vendor
            err = VendorMSTRTable.Save();

            if(err== TableError.NoError)
            {
                //If no error, clear the window after the save
                txtVendorID.Text = string.Empty;
                txtVendorName.Text = string.Empty;
                txtVendorAddress.Text = string.Empty;
                txtCity.Text = string.Empty;
                txtState.Text = string.Empty;
                txtZipCode.Text = string.Empty;
                txtPhoneNumber.Text = string.Empty;
                cb1099Type.SelectedItem = null;
            }

            else if (err == TableError.Duplicate)
            {
                //If there is a duplicate vendor error
                MessageBox.Show("Vendor already exists, choose another
vendor ID");
            }
            else
            {
               //If there is some other kind of error
                MessageBox.Show(err.ToString());
            }

            //close the vendor master table
            //VendorMSTRTable.Close();
        }
```

Retrieving a record

Retrieving a record is a two-step operation. First you need to determine which table key you will use to uniquely identify the record. Next, you need to set the values to that key, and finally you need to request the record from the table. If you use the `Change()` method, you can make changes to the record and update it. If you use the `Get()` method, it's a look but don't touch retrieval.

The following code could be placed on the `TextChanged` event of the `txtVendorID` field to retrieve a record:

```
            ///////////////  RETRIEVING A ROW  ////////////////       }
        private void txtVendorID_TextChanged(object sender, EventArgs e)
          {
             //Variable for table operation error
                TableError err;

             //Create a reference to the PM Vendor Mstr table
                 PmVendorMstrTable VendorMSTRTable;
                 VendorMSTRTable = Dynamics.Tables.PmVendorMstr;

             //set the key to use for the table
             //we are using key 1 because it contains the vendor ID
                 VendorMSTRTable.Key = 1;

             // Release any existing lock
                 VendorMSTRTable.Release();

             //set the key field in the table
                 VendorMSTRTable.VendorId.Value = txtVendorID.Text;

            //try to read the row. The Change() method will put a
   passive lock on the row.
            //If you had used the Get() method, no lock would be put on
   the row
                 err = VendorMSTRTable.Change();

             //check for table operation error
                 if (err == TableError.NoError)

                 //Display the values from the table to the window
                     txtVendorID.Text = VendorMSTRTable.VendorId.Value;
                     txtVendorName.Text = VendorMSTRTable.VendorName.
   Value;
                     txtVendorAddress.Text = VendorMSTRTable.Address1;
```

```
                    txtCity.Text = VendorMSTRTable.City.Value;
                    txtState.Text = VendorMSTRTable.State.Value;
                    txtZipCode.Text = VendorMSTRTable.ZipCode.Value;
                    txtPhoneNumber.Text = VendorMSTRTable.
PhoneNumber1.Value;
                    cb1099Type.SelectedItem = VendorMSTRTable.
Number1099Type.Value;


                //close the table
                    VendorMSTRTable.Close();
        }
```

Updating a record

Updating a record does not require any special procedure. You simply need to retrieve the record using the `Change()` method, change whichever fields you want, and then use the `Save()` method to commit the changes to the database.

Deleting a record

To delete a row from the database, attach the following code to the **Delete** button's `Click` event:

```
    ////////////  DELETE BUTTON  ////////////////////////

        private void btnDelete_Click(object sender, EventArgs e)
          {
            //Create a reference to the PM Vendor Mstr table
            PmVendorMstrTable VendorMSTRTable;
            //here we are accessing a form level table buffer to view
its contents.

            {
                //Variable for table operation error
                TableError err;

                VendorMSTRTable = Dynamics.Tables.PmVendorMstr;

                //set the key to use for the table
                //we are using key 1 because it contains the vendor ID
                VendorMSTRTable.Key = 1;

                //set the key field in the table
```

```
                VendorMSTRTable.VendorId.Value = txtVendorID.Text;

                 //try to read the row. The Change() method will put a
passive lock on the row.
                err = VendorMSTRTable.Change();

                //check for table operation error
                if (err == TableError.NoError)
                {
                    //try to remove the record
                    err = VendorMSTRTable.Remove();

                    if (err == TableError.NoError)
                    {
                        //If no error, clear the window after the save
                        txtVendorID.Text = string.Empty;
                        txtVendorName.Text = string.Empty;
                        txtVendorAddress.Text = string.Empty;
                        txtCity.Text = string.Empty;
                        txtState.Text = string.Empty;
                        txtZipCode.Text = string.Empty;
                        txtPhoneNumber.Text = string.Empty;
                        //cb1099Type.SelectedItem = null;
                    }
                    else //There was some sort of table error
                    {
                        MessageBox.Show("The following error occurred
deleting the record: " + err.ToString());
                    }
                }

                //close the table
               // VendorMSTRTable.Close();


            }
```

Clearing the window

To add functionality to the **Clear** button, you simply need to add code that will remove all of the data from the window fields. Enter the following code to accomplish this task:

```
/////////////////// THE CLEAR BUTTON /////////////////////
        // The CLEAR button simply blanks out the fields on the form
        private void btnClear_Click(object sender, EventArgs e)
        {
                        //Variable for table operation error
            TableError err;

            //Create a reference to the PM Vendor Mstr table
            PmVendorMstrTable VendorMSTRTable;
            VendorMSTRTable = Dynamics.Tables.PmVendorMstr;


            // Release any existing lock
            VendorMSTRTable.Release();

            txtVendorID.Text = string.Empty;
            txtVendorName.Text = string.Empty;
            txtVendorAddress.Text = string.Empty;
            txtCity.Text = string.Empty;
            txtState.Text = string.Empty;
            txtZipCode.Text = string.Empty;
            txtPhoneNumber.Text = string.Empty;
            cb1099Type.SelectedItem = null;

            //close the vendor master table
            VendorMSTRTable.Close();
        }
```

Working with ranges

Similar to Dexterity, working with ranges should be on the top of your list of things to master. Establishing a range allows you to work with a subset of data instead of everything in the database's table. When you define a range, the software treats the records in the range as if they were the only records in the database's table.

For example, if your range is set to include only documents dated in 2012, then the `GetFirst()` method would return the first document dated in 2012. Similarly, the `GetLast()` method would return the last document dated in 2012. In reality, the table could include documents spanning 20 years, but setting a range would make the other records invisible to your code. Pretty cool!

To set a range, you first need to decide which table key you will use to define the series of records. For example, if you only want to evaluate invoices in `PM Paid Transaction History File`, you would need to find a key that included Document Type (the `DOCTYPE` field) as one of the elements.

An investigation of **PM Paid Transaction History File** (**PM30200**) would reveal nine keys. These keys have the following components:

- Key1
    - Vendor ID
- Key2
    - Document Type
    - Voucher Number
- Key3
    - Document Number
- Key4
    - Document Date
- Key5
    - TRX Source
    - Voucher Number
- Key6
    - Voucher Number
- Key7
    - Document Type
    - Document Number
- Key8
    - Vendor ID
    - Document Date

- Key9
    - Document Type
    - Voided
    - Document Date
    - Purchases Amount

Three of the keys, Key2, Key7, and Key9, include Document Type as one of the components. You could use any of these keys to establish your range of documents, but the fewer components you have to deal with, the better. In light of that, Key2 or Key7 would be better choices than Key9. Choosing between Key2 or Key7 would depend on your preference of the secondary sorting order. Let's assume we want the secondary sorting order to be Document Number. Then Key7 would be our choice.

To create the range, you'll use the `RangeStart()` and `RangeEnd()` methods. To clear the range, use the `RangeClear()` method. Here's what the code will look like if you want to establish a range on the `PM Paid Transaction History File` file that includes only invoices:

```
////////////////// SETTING A RANGE //////////////////

 //declare a variable for table operation errors
     TableError err;

 //create a reference to the table
    PmPaidTransactionHistTable PMTrxHistory;
    PMTrxHistory = Dynamics.Tables.PmPaidTransactionHist;

 //set the table key to use for the range
 //key 7 will be used.
 //Its components include doc type and doc number
    PMTrxHistory.Key = 7;

 //specify the beginning of the range
 //specify a value for each component of the key
 //type 1 denotes an invoice
       PMTrxHistory.DocumentType.Value = 1;

 //set the minimum value for the doc number
       PMTrxHistory.Clear();
       PMTrxHistory.RangeStart();

//specify the end of the range
//specify a value for each component of the key
//type 1 denotes an invoice
```

```
            PMTrxHistory.DocumentType.Value = 1;

   //set the maximum value for the doc number
            PMTrxHistory.Fill();
            PMTrxHistory.RangeEnd();

   //retrieve the first invoice in the range
             err = PMTrxHistory.GetFirst();
```

Building and deploying the application

Once your coding is complete, it's time to build and deploy your application. This time, however, you are going to set the assembly information for your integration. Setting this information will allow you to establish a version number and a build number for your program.

Assembly information is stored in the `AssemblyInfo.cs` file. This file should be included in the **Properties** section of your project. Open the file by double-clicking on it. Fill in the various values for the assembly information similarly to the following screenshot:

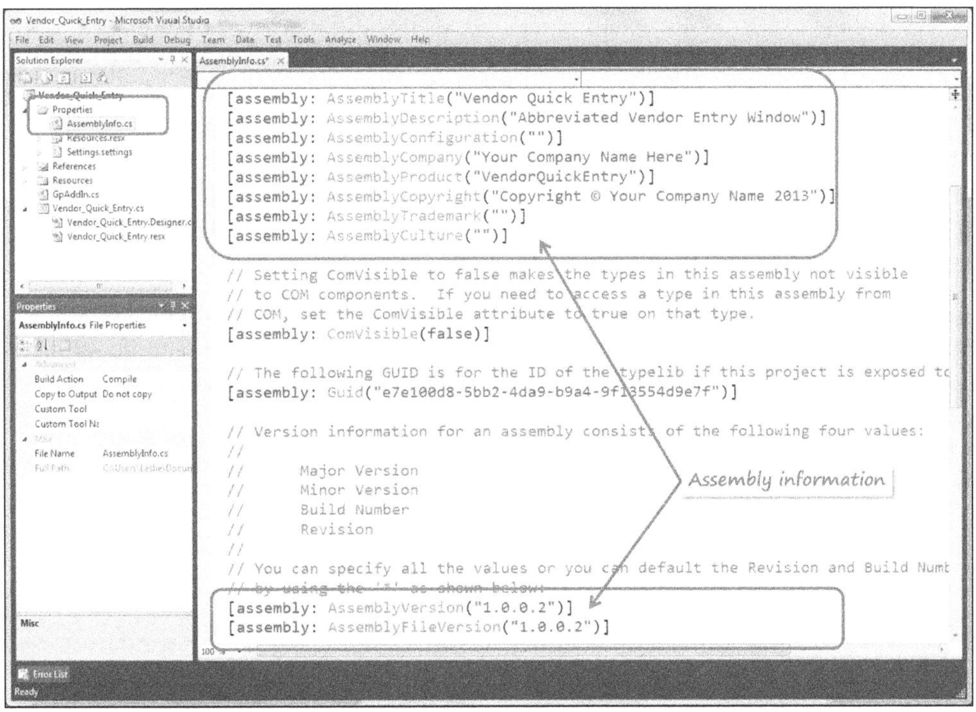

Close and save all of the objects you have opened. Select **Build** from the menu bar and then select **Build Vendor_Quick_Entry**, as shown in the following screenshot:

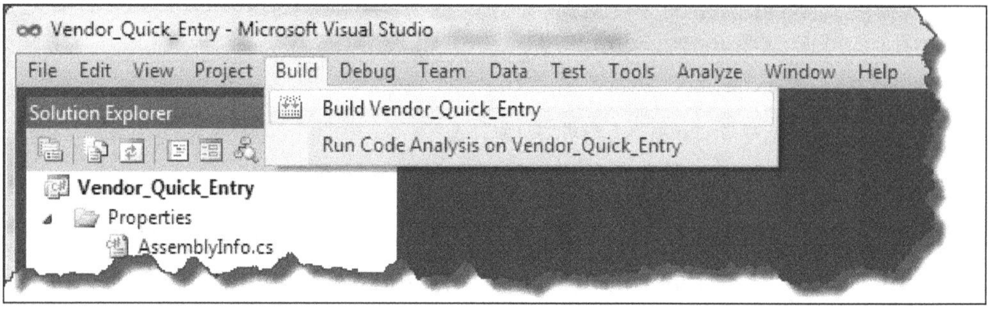

When the build has succeeded, copy the `Vendor_Quick_Entry.dll` file from your project folder to the `AddIns` folder of your Dynamics GP application.

Launch Dynamics GP and admire your new integrated customization!

Dynamics GP 2013 consideration

Your VS Tools customizations will work just as they always have using the Dynamics GP 2013 rich client. **Rich client** implies the regular Dexterity written workstation installation that you use today. Using the Web Client is a different matter. The Web Client is a ground-breaking achievement for Dynamics GP and one that has long been anticipated, but it comes at a price. Any user interface that you created using WinForms will not work with the Web Client. The screen modifications you have made using Modifier work just fine, but the WinForm events will not fire when using the Web Client.

Your code will work just fine, but you need to rework your interface and register your events against a window created with Dexterity. You have two choices here. You can use Modifier and make changes to existing windows, or you can learn Dexterity and build your own windows using its graphical forms designer. You'll find the graphical forms designer feature easy to use, and making the reference change shouldn't be that difficult.

The window, function, and procedure events in Dexterity are very similar to the VS Tools events you use today. You can find a table in *Appendix B, Event Matrix*, that contains a side-by-side matrix comparing Dexterity, VS Tools, and VBA events. As you will see in the table, the events across all three tools are very similar. *Appendix B, Event Matrix* is available as a free, downloadable chapter from the following link: `http://www.packtpub.com/sites/default/files/downloads/0264EN_Appendix_B_Event_Matrix.pdf`

To get a feel for Dexterity, go through the exercises in *Chapter 3, Getting Started with Dexterity, Chapter 4, Building the User Interface,* and *Chapter 5, sanScript – Making It Work.* Who knows, you may give up VS Tools and move your entire application to Dexterity!

Summary

We covered a lot of ground in this introduction to VS Tools for Dynamics GP. You found it on the Web, installed it, built a window that looks just like one of the native Dexterity windows, and set it in motion. But remember, this just scratches the surface. So many more things are possible. You can work with scrolling windows, call lookup lists, respond to window events, integrate with web services, call native procedures, query the SQL database, and even record macros.

Several sample applications are available online. Use the search argument `VS Tools for Dynamics GP sample applications` and you'll find many more resources to help you excel with this exciting development tool.

11
Upgrading Customizations

It is inevitable; with each enhancement we create, another release of Dynamics GP is just around the corner. In this chapter, we'll explore what you need to do to update your customizations when you upgrade to the next release of Dynamics GP. Some customizations will require few, if any, upgrade tasks, while others will require quite a number of steps.

This chapter is not intended to be a comprehensive upgrade guide, but rather an introduction to what the upgrade process looks like. We will cover the following key topics:

- How to use Dynamics GP SDK
- Upgrading a Dexterity customization
- Upgrading a customization using Modifier with VBA
- Upgrading Extender, SmartList Builder, and Excel Report Builder customizations
- Upgrading a Visual Studio Tools add-in

Determining if there were any changes to the resources that your customization uses is an important part of every customization upgrade, regardless of the type of customization. Studying the SDK is the right place to start.

Using the SDK

The SDK makes it very easy to determine those, if any, Dynamics GP objects that changed from release to release. One of the steps included in setting up your Dexterity development environment was installing Dynamics GP SDK (*Chapter 3, Getting started with Dexterity*). It's time to go over some of the goodies in Dynamics GP SDK that will help you move through the upgrade process.

Launch Dynamics GP SDK using the **Start** button. If you accepted the default locations during installation, the **Dynamics GP SDK** selection will be within the **Microsoft Dynamics** folder in the Start menu.

If it isn't in the Start menu (and you accepted the defaults), double-click on the following file to open it:

```
C:\Program Files (x86)\Microsoft Dynamics\GP 11.0 SDK\DynamicsGP_SDK.
chm
```

The opening screen of Dynamics GP SDK looks like the following screenshot:

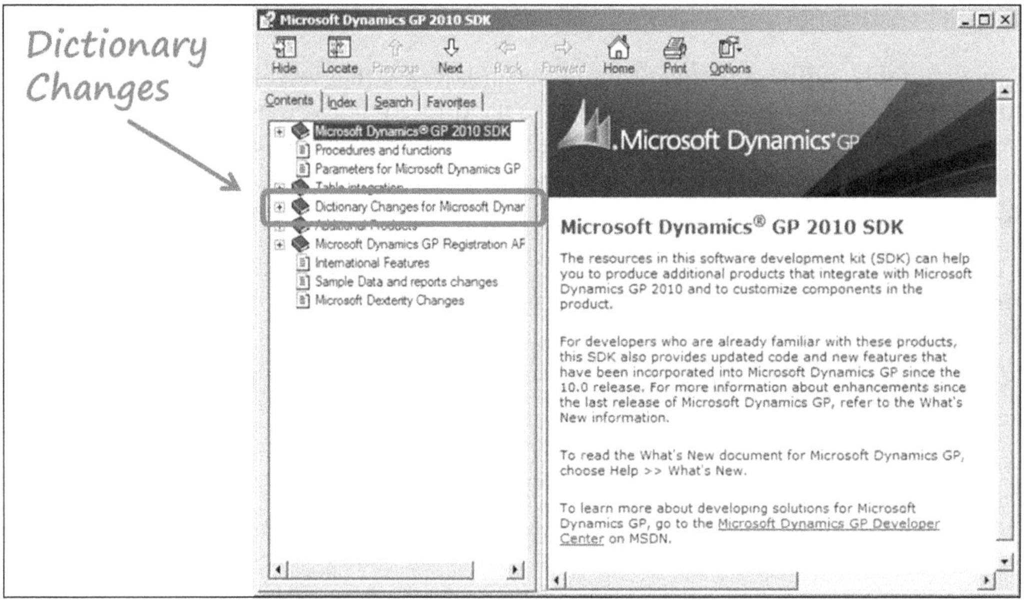

From the preceding screenshot, click on the **Dictionary Changes for Microsoft Dynamics GP** item. In the screen that opens, you'll find links to information on the following types of changes:

- Script changes
- Data model changes
- Table changes
- Form changes

Let's explore each one of these selections to understand what they reveal.

Script changes

Click on the **Script Changes** link from the list of topics. The script changes s ection provides details regarding which global procedures, global functions, form procedures, and form functions have changed since the last two releases. Each release is listed separately. You'll see one section for changes between releases 10 and 11, and another for changes between releases 9 and 10. For each release, you can find changes to scripts in the following dictionaries:

Product Name	ID	Dictionary
Microsoft Dynamics GP	0	`Dynamics.dic`
Project Accounting	258	`PA258.DIC`
Fixed Assets	309	`fam.dic`
Manufacturing	346	`ICONMFG.DIC`
Human Resources	414	`HR.DIC`
FieldService	949	`SrvcAdv.DIC`
SmartList	1493	`EXP1493.dic`

Although the dictionaries in the following table are also contained in the installation media, their script changes are not included in Dynamics GP SDK:

Product Name	ID	Dictionary
Interfund Management	1042	`IFund.dic`
Revenue Expense Deferrals	1045	`D1045.DIC`
Collections Management	1157	`CPro.dic`
Safe Pay	1235	`SFPAY.dic`
Electronic Reconcile	1428	`AREC.dic`
Cash Flow Management	1632	`CFM.DIC`
Technical Service Tools	1838	`TAUTIL.DIC`
Excel-Based Budgeting	1878	`XLBudget.dic`
HRM Solution Series	1911	`HRMSS.DIC`
Payment Document Management	2150	`PMNTDOC.DIC`
Purchase Order Enhancements	2277	`POE2277.DIC`
Control Account Management	2416	`CAM2416.dic`
Enhanced Commitment Management	2547	`ECM2547.DIC`
CopierSeries	2992	`QK2992.DIC`
VAT Daybook	3096	`vatDaybk.dic`
Advanced Security	3104	`AdvSecur.dic`

Product Name	ID	Dictionary
Extender	3107	extud.dic
Analytical Accounting	3180	AA.dic
Encumbrance Management	3258	ENC3258.dic
Report Scheduler	3278	RPTSCHED.dic
SmartList Builder	3830	SLBuild.dic
ML Checks	4067	MLChecks.dic
Grant Management	4421	GTM4421.dic
Payroll Integration to Payable	4522	PIP.dic
Advanced Go Tos	4612	advgoto.dic
Analysis Cubes for Excel	4621	WHPivot.dic
Certification Manager	4933	CLTM.DIC
Employee Health and Wellness	4955	EHW.DIC
Electronic Signatures	4965	4965MS.DIC
Audit Trails	4966	4966MS.dic
HITB Report	5597	HTB5597.DIC
Dynamics Online Services	6499	DO6499.DIC
Date Effective Tax Rates	6831	DET.dic

Your best bet for finding out about changes to the dictionaries not included in the SDK is to start a support case with Microsoft or post a question on the Dynamics community website (`https://community.dynamics.com/product/gp/f/32.aspx`). One advantage of using the Dynamics community is that it's free. On the downside, the information you receive might be wrong.

Our experience has been that the information received from the Dynamics community website is excellent and often better than what you may get from Microsoft Support. Be sure you validate anything you plan to rely on coming from the Dynamics community website, because the fine people providing suggested solutions are 100 percent volunteers.

The following screenshot is an example of what you'll see if there is a change in a global procedure:

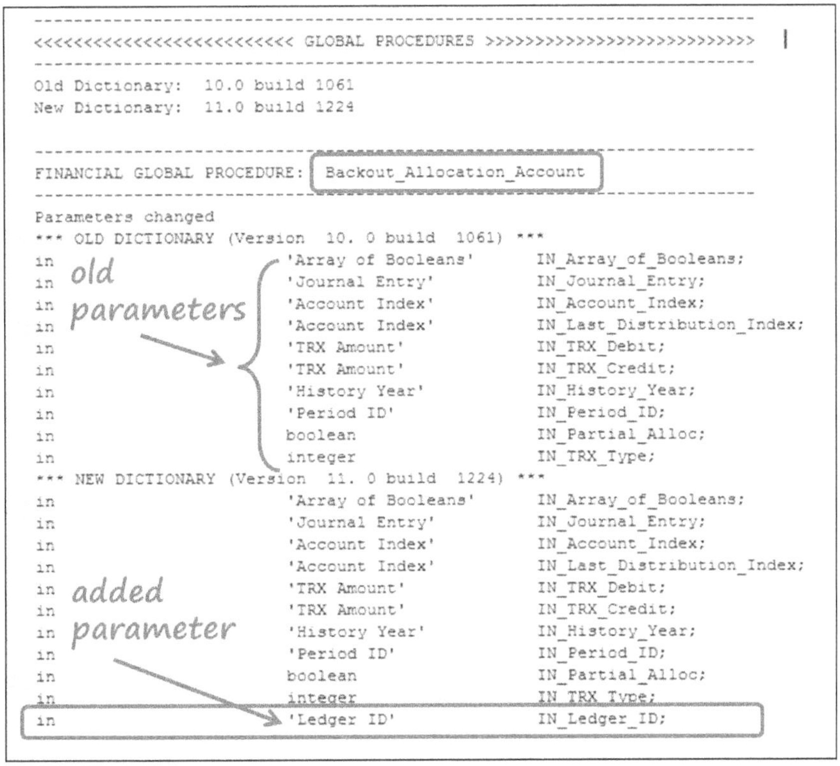

The name of the procedure in the preceding screenshot is **Backout_Allocation_Account**. There was a change to the procedure's parameters. Parameters from the old dictionary are listed in the top section. Parameters from the new release are listed in the lower section. Unfortunately, the parameter that changed has not been easily identified.

However, the changes to data model can easily be identified; we'll look at those next.

Data model changes

Select the **Data Model Changes** link from the list of topics. The data model changes section provides information on changes in the database tables.

New tables

Database tables added in the new release that did not exist in the old release are identified in the new tables section. A portion of the **Dynamics GP SDK** window documenting new tables is shown in the following screenshot:

Tables in 11.0.1200 that cannot be found in 10.00.0774

Table Technical Name	Table Physical Name
cmEFTDepositInfo	Temp
cmTransactionEFTBatch_TEMP	TEMP
cmTransactionEFT_TEMP	TEMP
CM_Checkbook_MSTR	CM00100

Deleted tables

Database tables that existed in the old release but do not exist in the new release are identified in the deleted tables section. A portion of the **Dynamics GP SDK** window documenting the deleted tables is shown in the following screenshot:

Tables in 9.00.0114 that are not in 10.00.0774

Table Technical Name	Table Physical Name
SY_Class_Normal_SETP	SY40300
SY_Security_Normal_MSTR	SY02000

New columns

Fields (columns) added to tables in the new release that did not exist in the tables of the old release are identified in the new columns section. A portion of the **Dynamics GP SDK** window documenting new table columns is shown in the following screenshot:

Columns in 11.0.1200 that cannot be found in 10.00.0774

Table Technical Name	Table Physical Name	Field Technical Name	Field Physical Name
AccountTrxTemp	TEMP	Ledger Name	Ledger_Name
BatchHeadersTmp	TEMP	Clear Recurring Amounts	ClearRecAmts
		WF Step Type	WF_Step_Type
Batch_Headers	SY00500	Clear Recurring Amounts	ClearRecAmts
Batch_Headers_DUP	SY00500	Clear Recurring Amounts	ClearRecAmts
cmCheckbookEFT	CM00101	EFT PM Post to Bank Rec in Summary	EFTPMPostBankRecSummary
		EFT RM Post to Bank Rec in Summary	EFTRMPostBankRecSummary
		RM Use EFT Numbering	RMUSEEFTNUM

Deleted columns

Fields (columns) that existed in the tables in the old release but do not exist in the tables of the new release are identified in the deleted columns section. Only one field was deleted between releases 10 and 11. The **Dynamics GP SDK** window documenting that change is shown in the following screenshot:

Columns in 10.00.0774 that are not in 11.0.1200

Table Technical Name	Table Physical Name	Field Technical Name	Field Physical Name
UPR_TEMP_Report_Destination	UPR10212	File Name	FILENAME

New indexes

Indexes added to existing tables that did not exist in the tables of the old release are identified in the new indexes section. A portion of the **Dynamics GP SDK** window documenting new table indexes is shown in the following screenshot:

Indexes in 11.0.1200 that cannot be found in 10.00.0774

Table Technical Name	Table Physical Name	Index Technical Name
cmTransactionEFT	CM20202	cmTransactionEFTIdx_BatchSourceSeries
cmTransactionEFTBatch	CM20203	cmTransactionEFTBatch_Key3
		cmTransactionEFTBatch_Status
CM_Checkbook_MSTR	CM00100	CM_Checkbook_MSTR_Key1
		CM_Checkbook_MSTR_Key2
		CM_Checkbook_MSTR_Key3

Deleted indexes

Indexes deleted from the tables that existed in the old release are identified in the deleted indexes section. A portion of the **Dynamics GP SDK** window documenting the deleted indexes is shown in the following screenshot:

Indexes in 9.00.0114 that are not in 10.00.0774

Table Technical Name	Table Physical Name	Index Technical Name
SY_Class_Normal_SETP	SY40300	SY_Class_Normal_SETP_Key1
SY_Security_Normal_MSTR	SY02000	SY_Security_Normal_MSTR_KEY1
taxReturn	TX00300	taxReturnIdx_ID
		taxReturnIdx_StartDate
taxReturnDetail	TX00304	taxReturnDetailIdx_ID
taxReturnEuInputTax	TX00302	taxReturnEuInputTaxIdx_Date
taxReturnException	TX00301	taxReturnExceptionIdx_ID
taxReturnGLException	TX00303	taxReturnGLExceptionIdx_ID

Different data types

This section documents any instances where a field was assigned a different data type in the new release. For example, if a field's data type was **STR15** in release 10 and that field's data type changed to **STR20** in release 11, this would fall into the different data types category. Only two data type changes were made between releases 10 and 11. A portion of the **Dynamics GP SDK** window documenting those changes is shown in the following screenshot:

Different Datatypes in 10.00.0774 vs. 11.0.1200

Table Name			Table Physical Name			
Field Name			Field Physical Name			
Version	Great Plains Data type		Data Type	Keyable length	Storage size	
RM_Statements_ROPT		Table Name	RM40501			
Restrict From User Defined		Field Name	RSTRFRUD			1st Change
10.00.0774	STR15	Old Datatype	String 16	15	16	
11.0.1200	STR20	New Datatype	String 22	20	22	
RM_Statements_ROPT		Table Name	RM40501			
Restrict To User Defined		Field Name	RSTRTOUD			2nd Change
10.00.0774	STR15	Old Datatype	String 16	15	16	
11.0.1200	STR20	New Datatype	String 22	20	22	

Different segments

Any changes to the segment of a composite field are identified in the different segments section. There were no segment changes between releases 10 and 11.

Different index columns

Any changes to the components of an existing index are identified in the different index columns section. For example, if you look at the following screenshot, you will see where the first and third key of the `GL_Budget_SUM_MSTR` table changed between releases 9 and 10. The first and third keys no longer include the `Period Date` column:

Different IndexCol in 9.00.0114 vs. 10.00.0774

Table Technical Name	Index Technical Name	Table Physical Name	Current set of fields	Previous set of fields
GL_Budget_SUM_MSTR	GL_Budget_SUM_MSTR_Key1	GL00201	Budget ID, Account Index, Period ID	Budget ID, Account Index, Period Date, Period ID
	GL_Budget_SUM_MSTR_Key3	GL00201	Account Index, Budget ID, Period ID	Account Index, Budget ID, Period Date, Period ID

New RW relations

This section describes any new table relationships that were created for Report Writer. These new relationships are used exclusively by Report Writer to create reports. They do not enforce referential integrity, nor are all possible relationships represented. A portion of the **Dynamics GP SDK** window documenting new RW relations is shown in the following screenshot:

Reporting_relationships in 11.0.1200 that cannot be found in 10.00.0774

Primary Table Name	Report Writer Relation Name	Secondary Table Display Name	Secondary Table Technical Name
CM_Checkbook_MSTR	Bank Master	Bank Master	CM_Bank_MSTR
CM_Deposit_TEMP	CM Checkbook Master	CM Checkbook Master	CM_Checkbook_MSTR
CM_Deposit_WORK	CM Checkbook Master	CM Checkbook Master	CM_Checkbook_MSTR

Deleted RW relations

This section describes the Report Writer table's relationships that existed in release 10 but were deleted for release 11. A portion of the **Dynamics GP SDK** window documenting the relationships that were deleted is shown in the following screenshot:

Reporting_relationships in 10.00.0774 that are not in 11.0.1200

Primary Table Name	Report Writer Relation Name	Secondary Table Display Name	Secondary Table Technical Name
cmTransactionEFTBatch	Checkbook Transaction Electronic Funds Transfer	Checkbook Transaction Electronic Funds Transfer	cmTransactionEFT

As you can see, Microsoft makes it easy to determine changes to the metadata from release to release.

Table changes

The table changes section includes the same information as the *Data model changes* section we just discussed; it's just presented to you in a different way. In some respects, you may find that data model changes are a little easier to spot using the table changes section's documents.

There are two choices per dictionary for table changes: Summary, which is a list of each of the tables that changed, and Detail, which shows the details for each changed table. A portion of the **Dynamics GP SDK** window listing the table changes section's topics is shown in the following screenshot:

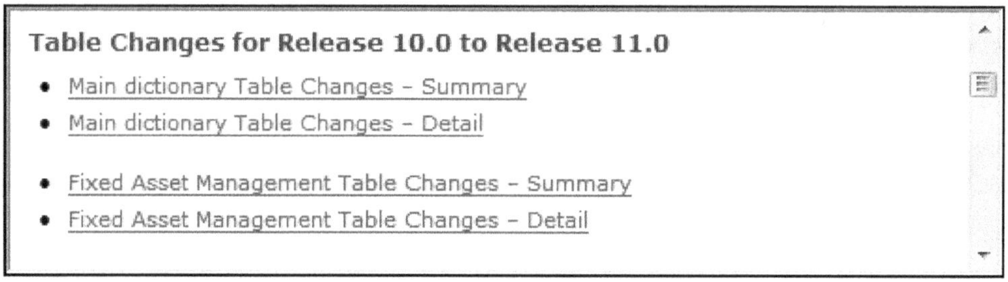

Table changes – summary

This section displays a list of all of the tables that have changed since the last release of Dynamics GP. This is actually a text file and not the HTML file that you find in the data model changes section.

As you can see in the following screenshot, a comment next to the physical name of the table identifies the new tables. Table **DD10400** included only a change to a key, which you can tell by the comment next to the table. At the bottom of the list, you can see how many tables changed out of the total number of tables in the dictionary.

The file can be searched easily when in text format. Also, you can readily copy the data to an Excel worksheet from the text file. Using information from the **Main dictionary Table Changes - Summary** file, you'll quickly be able to determine if any of the tables your customization relies on have changed.

The actual text file for summary changes to the Dynamics dictionary (assuming the default location was selected), can be found at `C:\Program Files (86)\Microsoft Dynamics\GP 11.0 SDK\Content\10.0to11.0\ CoreTableSummary_1000_1100.txt`. If you are running a 32-bit machine, look at `C:\Program Files\Microsoft Dynamics\GP 11.0 SDK\Content\10.0to11.0\ CoreTableSummary_1000_1100.txt`.

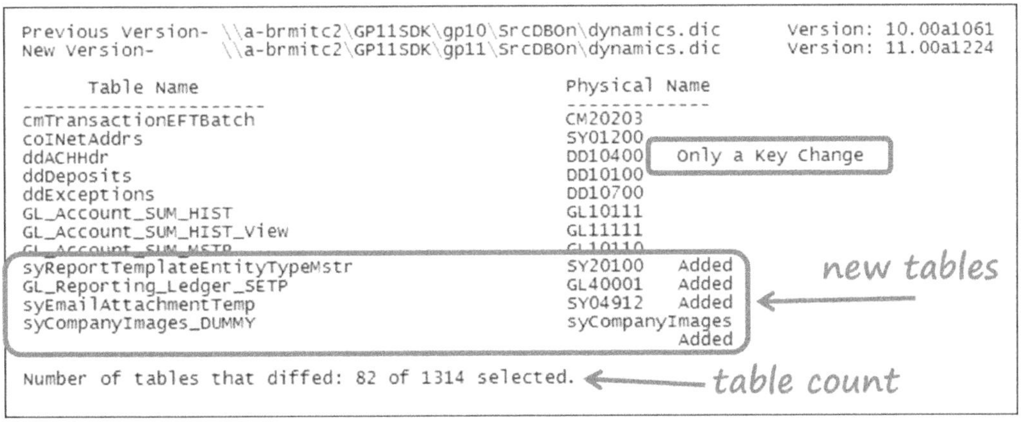

Table changes – detail

The detail report is a one-stop shop to view all the relevant changes made to a table. Instead of needing to look at each type of change separately (fields, indexes, and so on), you can see everything at once. Look at the following screenshot; it shows the detail report for the `GL_YTD_TRX_OPEN` table. You can quickly see the two fields that were added, and their specifications, in this report. You also have a good view of the new key that was added, as well as the change to an existing key's segments:

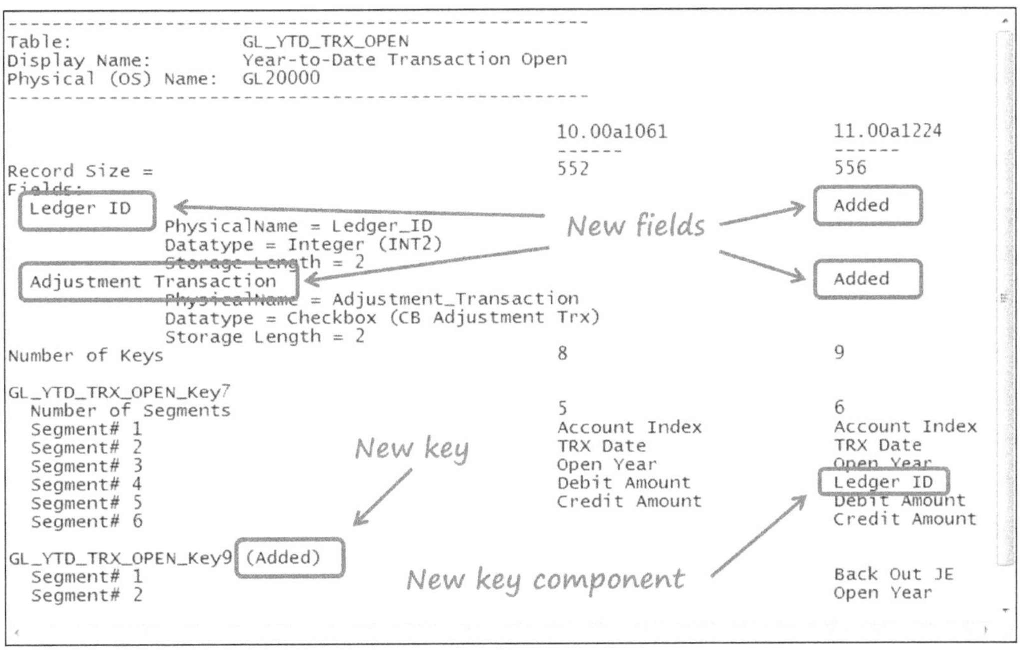

If you are upgrading a customization that uses third-party application tables, the information regarding table changes may not be readily available.

An easy way to figure out the changes is possible if you have a copy of both the old database version and the new database version. You can then use Visual Studio's Schema Compare tool or any other SQL schema compare tool to compare the changes between the tables. For details about using schema compare tools, refer to this article:

`http://www.mssqltips.com/sqlservertip/2089/sql-schema-comparison-with-visual-studio-2010/`

A tool that is often used for schema comparison is Red Gate's SQL Compare:

`http://www.red-gate.com/products/sql-development/sql-compare/`

The final element of the data model section of Dynamics GP SDK is form changes. Let's look at how you can discover the changes made to forms.

Form changes

The form changes section is determined from running the Compare utility in Dexterity Utilities. The comparison is between an unmodified release 10 dictionary as the source dictionary and a release 11 dictionary as the compared dictionary.

As part of your code update, you should also run this utility against your own code dictionaries to keep track of changes.

Look at the following screenshot. Each form resource appearing on the report has changed in some way. Not every form is included in the report; only those that have changed are included. If something was added to a form, such as a menu, a constant, a string, and so on, you would see an asterisk next to the new resource. Not all of the resources in a form are presented; just the ones that changed or new ones that were added:

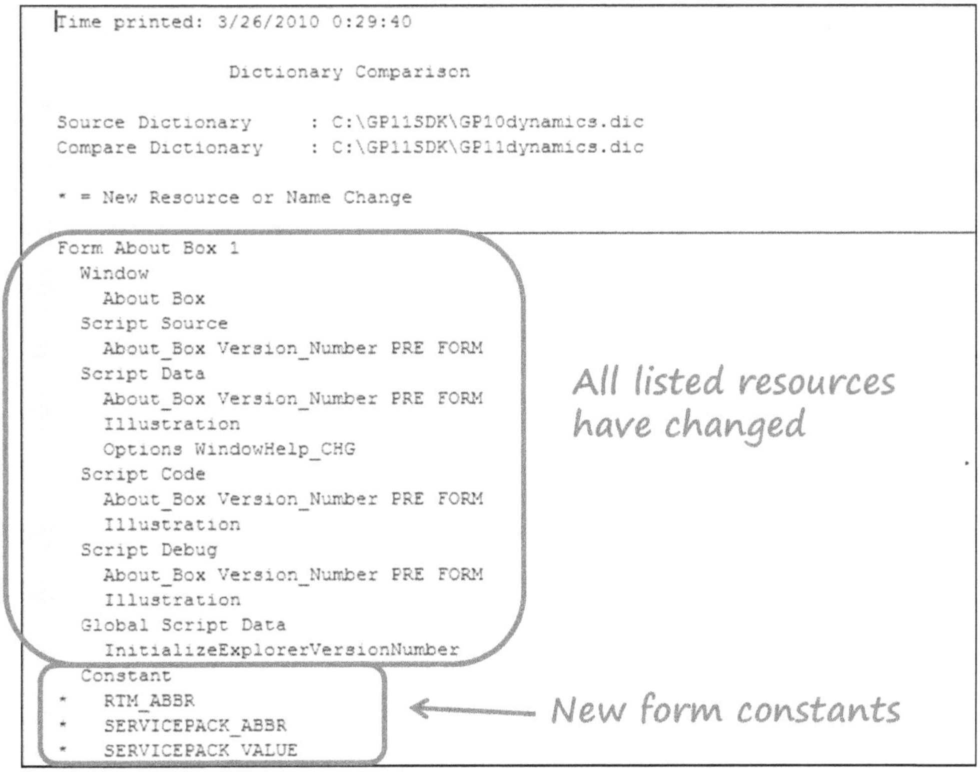

```
Time printed: 3/26/2010 0:29:40

               Dictionary Comparison

Source Dictionary       : C:\GP11SDK\GP10dynamics.dic
Compare Dictionary      : C:\GP11SDK\GP11dynamics.dic

* = New Resource or Name Change

Form About Box 1
  Window
    About Box
  Script Source
    About_Box Version_Number PRE FORM
  Script Data
    About_Box Version_Number PRE FORM
    Illustration
    Options WindowHelp_CHG
  Script Code
    About_Box Version_Number PRE FORM
    Illustration
  Script Debug
    About_Box Version_Number PRE FORM
    Illustration
  Global Script Data
    InitializeExplorerVersionNumber
  Constant
*   RTM_ABBR
*   SERVICEPACK_ABBR
*   SERVICEPACK_VALUE
```


Taken together, you can discern a great deal of information from the dictionary change section of Dynamics GP SDK. Too often, dictionary changes are not studied sufficiently and more problems are discovered after the deployment of the upgrade. I know you'll do your homework and not be one of those developers who send what is essentially a beta copy of the upgrade to their users.

Dexterity

This section describes how to upgrade your Dexterity customization. You'll go through several steps to upgrade your code; it can be tedious, but it isn't difficult. Remember, this chapter is not about troubleshooting an upgrade, nor is it a step-by-step guide; it's an introduction to the process.

The first step, of course, is to install the new release of Dynamics GP. Once you install the new release, we can begin.

Starting with release 6, you could no longer use the Developer Update utility in Dynamics Utilities to upgrade your code to the new release. Instead, you now have to use Source Code Control. You can use Visual Source Safe or Team Foundation Server. If your Source Code Control application isn't supported, or you don't have one, use the generic source code control included with Dexterity to perform the upgrade.

You will complete the following tasks in the upgrade process:

1. Set up generic Source Code Control server.
2. Create the new development dictionary.
3. Transfer third-party resources to the new release of the Dynamics dictionary.
4. Review the Microsoft Dynamics GP changes and make any necessary modifications to your application.
5. Redo any customizations made for alternate forms and reports.
6. Test your application using the Test mode in Dexterity.
7. Build a chunk dictionary for your upgraded code.
8. Test your application in multidictionary mode.
9. Repeat steps 6 through 8 until you get it right (c'mon, nobody's code is perfect the first time through).
10. Prepare the chunk file for distribution.

Setting up generic source code control

We will be using the built-in generic source code control to perform the upgrade. The first thing to do is set it up. This section explains how to do that.

Installing Dexterity Source Code Control Server (DSCCS)

The first step won't make much sense right now, but you'll see why you need this as we progress through the installation and update.

In our example, release 11 of Dexterity is installed in the `C:\DEX_GP2010` folder. Perform the following steps:

1. Create the following folder structures:
    - `C:\DEX_GP2010\SCC\PROJECT_1` (to hold the source code control objects for Project1)
    - `C:\DEX_GP2010\PROJECTS\ORIG_DICS` (to hold a copy of the unmodified original dictionary)
    - `C:\DEX_GP2010\PROJECTS\PROJECT_1` (to hold a copy of the development dictionary)
    - `C:\DEX_GP2010\SCC\TEMP` (to hold the temporary files used by the DSCCS)

   The default location for the temp folder is the user's default temp folder. The user's temp folder can normally be found in the following location:

   `C:\Users\user_name\AppData\Local\Temp\`

   When you name your folders during an actual update, you can name them anything. The names of the folders do not matter. The names used here are just for this example, not as a best practice. Develop a naming convention for yourself and use it consistently.

2. Navigate to the following folder in the Dynamics GP installation media: ...\ `Tools\Dex\DSCCS\`. Launch the file in the `DSCCS` folder. There should only be one file in that folder. The name of the DSCCS installation file is `Microsoft_Dexterity11_SourceCodeControlServer_x86_en-us.msi`.
3. Accept the default location for the installation.

4. Check the **Local System account** checkbox, as shown in the following screenshot, and then click on the **Install** button:

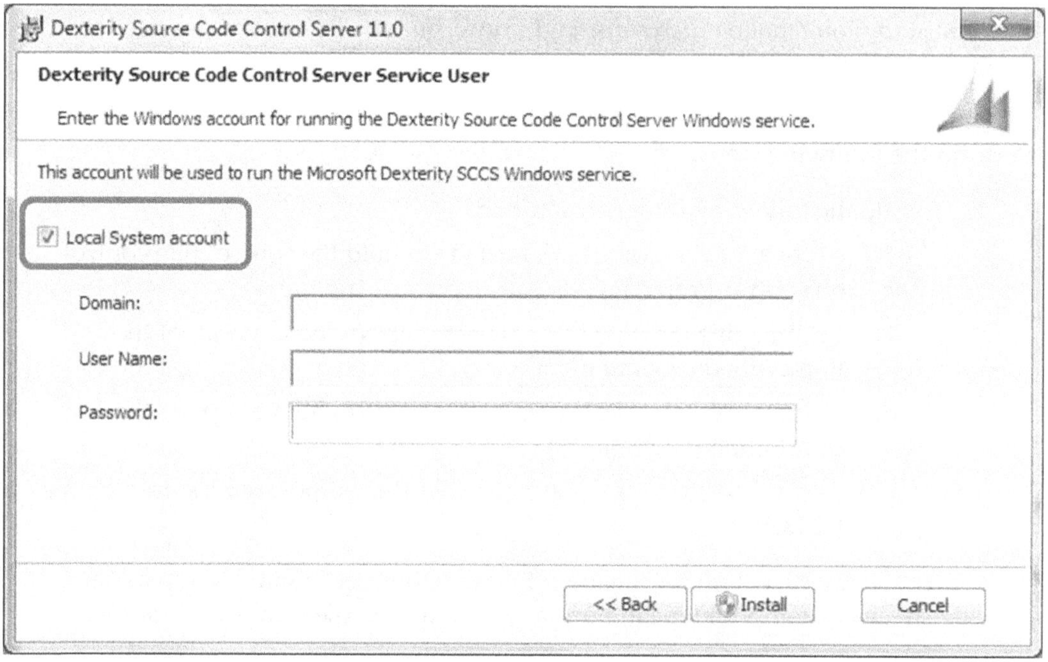

If you are on a domain, do not select the Local System account. Instead, fill out the window below with the correct domain information and the name of the user that will run the service. For the sake of simplicity, we'll just check the box.

The **Dexterity Source Code Control Server** window will open.

5. On the **Dexterity Source Code Control Server** window, select **Generic** as the value of the provider and set the **Root Directory** field to `C:\DEX_GP2010\SCC`. Your window should look like the following screenshot:

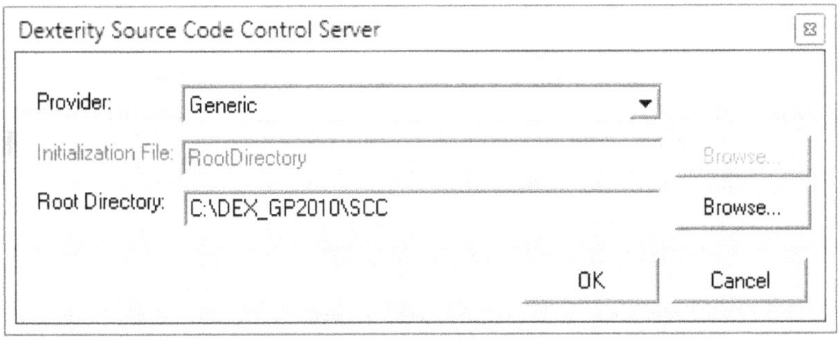

Should you ever need to change the location of the root directory, use the Dexterity SCCS (32-bit) applet in the control panel of your computer's operating system. The following screenshot shows you what the applet looks like in **Control Panel**:

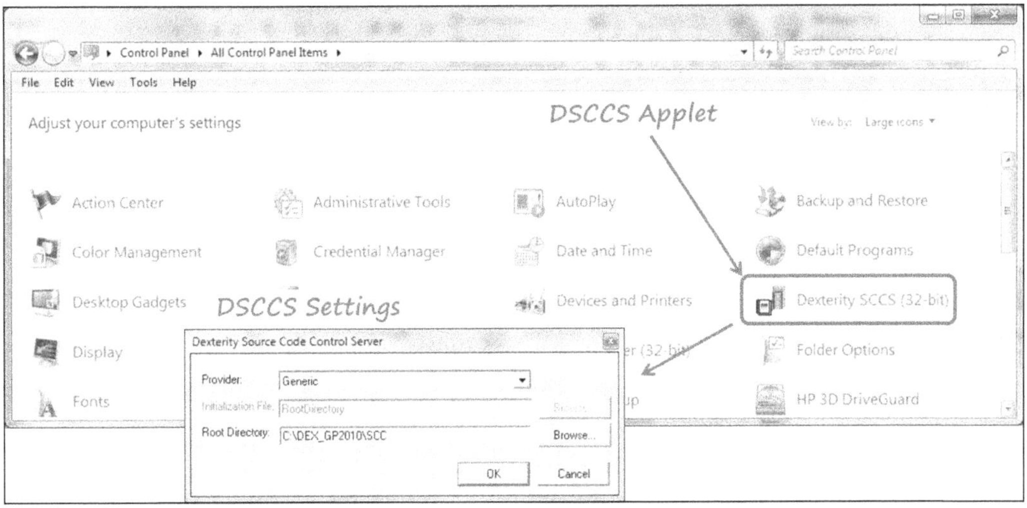

6. The finish screen will come up next. Click on the **Finish** button and that's all there is to the installation. Simple, right?

Create a separate folder in the root directory for each of your development projects. These project folders will store the resources for each project. The folder you created earlier for this project was `C:\DEX_GP2010\SCC\PROJECT_1`. Next, you'll configure the Source Code Control server to work with Dexterity.

Configuring the DSCCS

1. First, confirm that the Microsoft Dexterity SCCS service shows the status as **Started**. You do this by launching **Control Panel** | **Administrative Tools** | **Services**. The navigation is shown in the following screenshot:

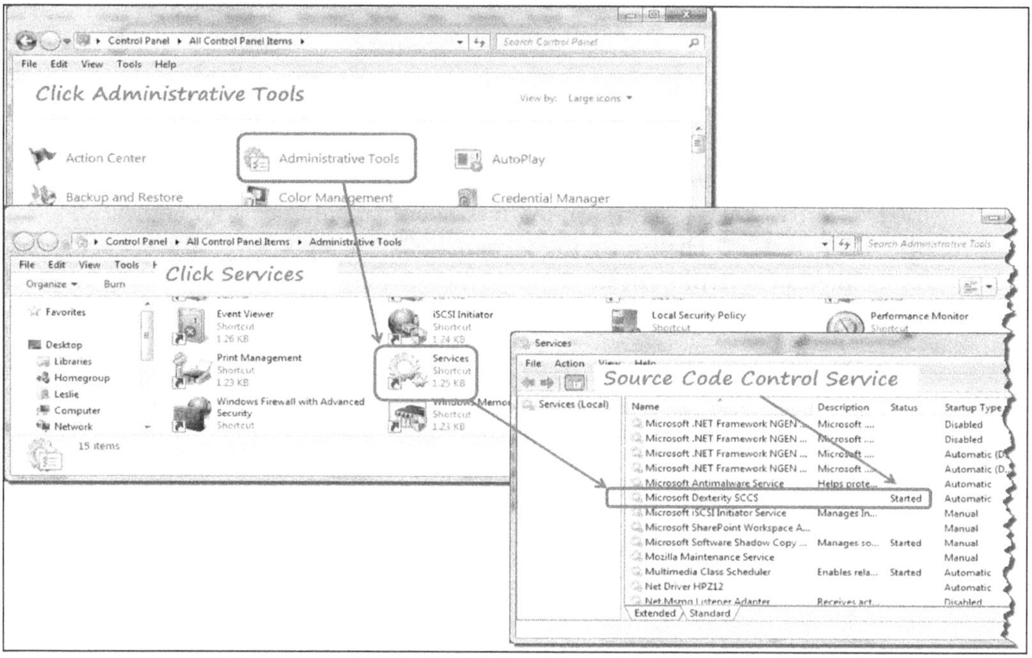

2. Copy your old development dictionary and all the associated `.dat` and `.idx` files into the `C:\DEX_GP2010\PROJECTS\PROJECT_1` folder.
3. Copy the `dex.ini` file from the new release of Dexterity and paste it into the `C:\DEX_GP2010\PROJECTS\PROJECT_1` folder. Each project must have its own `dex.ini` file because the DSCCS settings are stored in this file.
4. Create a shortcut to launch this specific project using its unique `dex.ini` file. Always use this project-level shortcut to open the development dictionary. The shortcut includes the path to `Dex.exe`, the development dictionary, and the unique `dex.ini` file.
5. Your shortcut for the project in this chapter will appear as follows: **"C:\DEX_GP2010\Dex.exe" "C:\DEX_GP2010\PROJECTS\PROJECT_1\ DEV_PROJECT_1.dic" "C:\DEX_GP2010\PROJECTS\PROJECT_1\ dex.ini"**
6. Open your old development dictionary in Dexterity using your new shortcut. Navigate to the **Options** window (**Edit** | **Options...**), click on the **Source Control** tab in the **Options** window, and perform the following steps:

i. Check the **Use Source Control** checkbox.

ii. Check all the three boxes at the bottom of the window and then complete the window with the following values:

Field Name	Value
Repository Name	`localhost.`
User Name	Leave blank.
Password	Leave blank.
Project Name	`PROJECT1` – if you click the ellipses button, the values presented are the names of any folders you have created in the root directory of the DSCCS.
Temp File Location	`C:\DEX_GP2010\SCC\TEMP` – this can be any valid folder; many people use their local temp folder. You need available disk space equal to twice the size of your development dictionary.
Original Dictionary	`C:\DEX_GP2010\PROJECTS\ORIG_DICS` – this must be an unmodified copy of the original dictionary of the release you are upgrading from.

7. Upon completion, your window should look like the following screenshot:

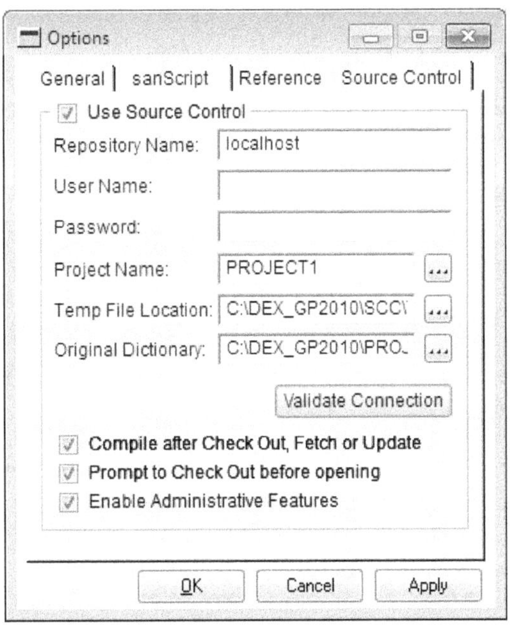

8. Click on the **Validate Connection** button. If everything is fine, you will get the **Connection Validated** message.

Resolving validation errors

If the connection cannot be validated, you may get one of the messages shown in the following screenshot (or worse, nothing at all):

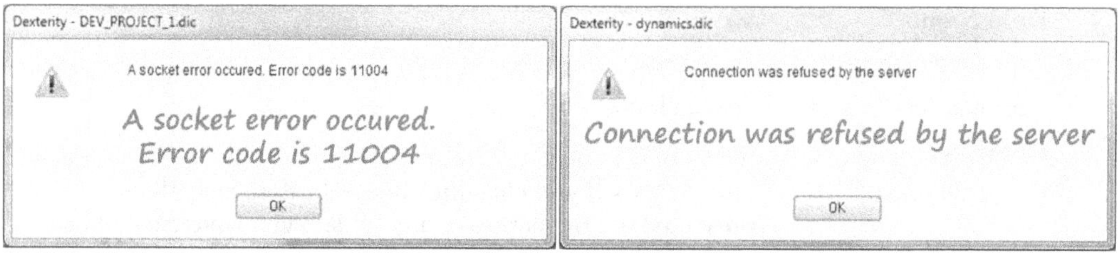

To resolve validation errors like these, check the following steps:

1. Confirm that the **Microsoft Dexterity SCCS** service is running.
2. Verify that the repository name is correct. If you are using a single machine to run Dexterity and the source code control provider, you can use `localhost` or `127.0.0.1` as the repository name.
3. Make sure that all of the folders, including the root folder, are properly set up.
4. Ensure that the firewall has port 2725 open.
5. Change the TCP/IP port number if necessary.

The service uses port 2725 by default. If any other application is trying to use that port, there will be a conflict. To change the port, you will need to edit the registry as follows:

Open the registry key with **HKEY_LOCAL_MACHINE | SOFTWARE | Great Plains Software | Dexterity Source Code Control Server | DSCCSProvider**.

Create a new DWORD named `Port`.

Set the value to the port number that you want DSCCS to use. You'll need to do this for each workstation.

The next step is to transfer your third-party resources to the new release of the Dynamics GP dictionary. This process involves checking in your old development dictionary, creating a new development dictionary including the checked-in resources, and then performing the update operation.

Checking in the old dictionary

Now that Dexterity Source Code Control is properly configured, you need to perform the initial check in of your old development dictionary. Checking in the old dictionary puts all of your resources into the repository so that they may be transferred to the new development dictionary. Only the resources you added to the old dictionary, including any alternate forms and reports, will be held in the repository.

Before you start checking in your resources, add the following line to the `Dex.ini` file:

```
ShowResIDs=TRUE
```

This `.ini` parameter will add another column to your **Dexterity Explorer** window that shows the Dexterity resource ID of each object in the opened dictionary. Your new **Dexterity Explorer** window will look like the following screenshot:

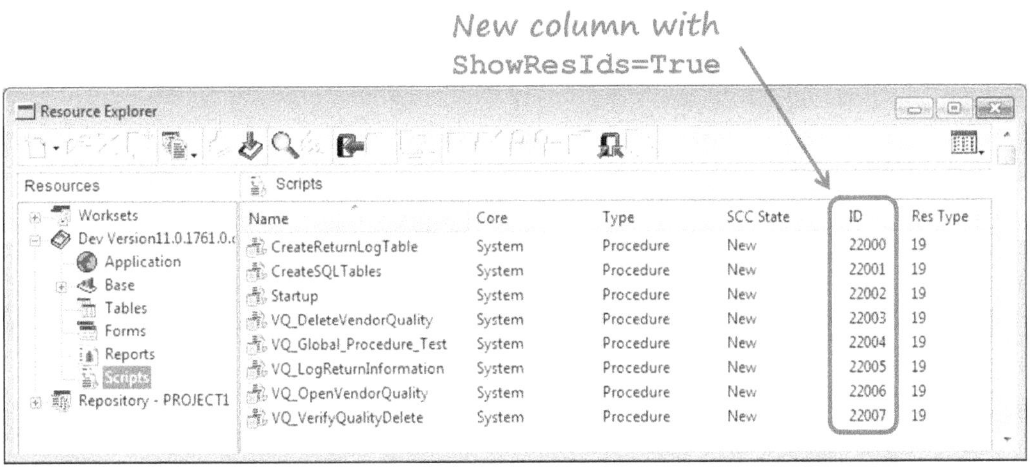

With this parameter in your `.ini` file, your resources are much easier to find because the ID in the **ID** column of your resources is **22000** or higher.

Follow these steps to check in your dictionary:

1. Open the copy of your old development dictionary using your new shortcut.
2. In the **Edit** menu, select the **Options** menu item and then click on the **Source Control** tab.

3. Add `\10-0-1841` after the name in the **Project Name** field in the **Options** window. After the change, your **Project Name** field's value will be **PROJECT_1\10-0-1841**. Do not attempt to validate it (because it won't get validated). The suffix you just added represents the build number of the old release. You can use any naming convention that pleases you.

   By changing the project name, you will create a separate project folder in your SCC folder for this initial project.

4. From the Dexterity Explorer menu, select **Source Control | Update SCC State**.
5. Open the definition for each of the alternate forms or reports that you c reated and then click on the **OK** button to close it.

   You do not need to do anything to the definitions other than simply opening and closing them. This process is known as **touching** the alternate forms and reports. When you do this, you update the **Source Code Control** (**SSC**) state of the resource. Updating the SCC state will change the state of the form or report in the main product to **New**. See the following screenshot:

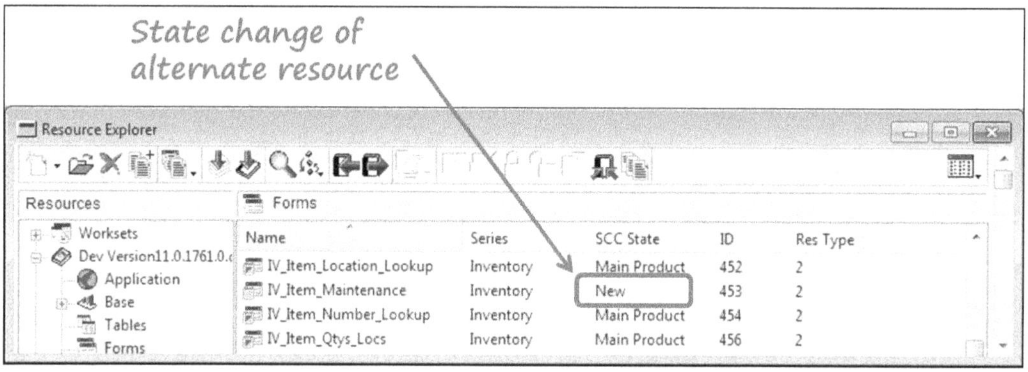

6. Next, check in the old dictionary. From the Dexterity Explorer menu, select **Source Control | Check in**. After a few moments, the **Check In to Repository: PROJECT1** window will open. Click on the **Insert All** button and add `Initial 10.0.1841 Check in` as the check-in comment. After that, click on the **Check In** button. Close the window when the check in has completed. The following screenshot shows the **Check In to Repository: PROJECT1** window:

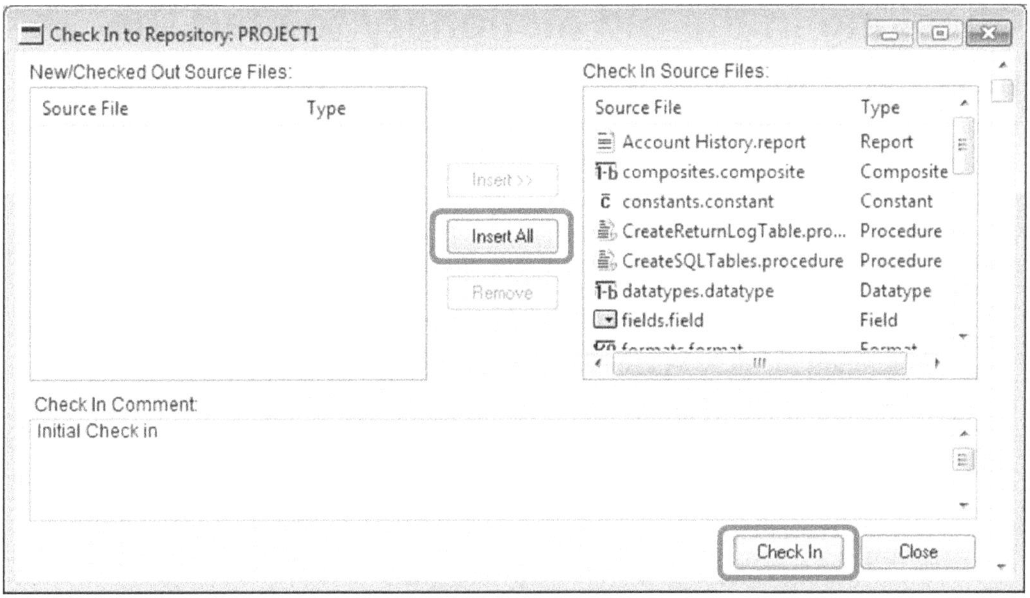

Resolve any errors displayed after the check in before moving on to the next step. Common errors are described in chapter 43 of the integration manual (`IG.pdf`) shipped with Dexterity. You can find this manual in the ...`\Microsoft Dexterity\Dex 11.0\Manuals` folder.

7. Update the index file by selecting the Explorer menu and then selecting **Source Control** | **Update Index File**. Respond with a yes to the dialog question that will be presented.

> Updating the index file ensures that all of your resources maintain the same resource ID throughout the upgrade. More information about the index file is in **Knowledge Base** (**KB**) 894699. You can get a copy of this KB at `http://support.microsoft.com/kb/894699`.

Now would be a good time to back up (copy) the repository directory. In this example, the repository directory is `C:\DEX_GP2010\SCC\PROJECT_1\10-0-1841`.

Now that you've checked in your resources, updated the index, and backed up the repository, it's time to start your new Dexterity project.

Checking in the old dictionary to start the new project

The starting point of your new Dexterity project is to first check in the old dictionary into the new project. This check in will put your resources into the repository so that you can transfer them into the new development dictionary. You will essentially do the same thing that you did in the last section, only this time you'll transfer the resources to a new dictionary. Perform the following steps:

1. Use your shortcut to open the old development dictionary.
2. In the **Edit** menu, select the **Options** menu item and then click on the **Source Control** tab.
3. Change the suffix of the project name from `\10-0-1841` to `\11-00-1935`. The complete new project name will be **PROJECT_1\11-00-1935**. Do not attempt to validate it (because it won't get validated). The suffix represents the build number of the new release. You can use any naming convention that pleases you. It's done this way merely as an example.

   When you change the project name, you will create a separate project folder in your SCC folder for your new project.

4. From the Dexterity Explorer menu, select **Source Control | Update SCC State**.
5. Open the form or report definition for each of the alternate forms or reports that you created and then click on the **OK** button to close it.

   You do not need to do anything to the alternate form or report definitions other than simply opening and closing them. This process is known as touching the alternate forms and reports. When you do this, you will change the SCC state of the resource. Updating the SCC state will change the state of the form or report in the main product to **New**.

6. From the Dexterity Explorer menu, select **Source Control | Check in**. After a few moments, the **Check In to Repository: PROJECT1** window will open. Click on the **Insert All** button. Add `Initial 11.00.1935 Check in` as the check-in comment and then click on the **Check In** button. Close the window when the check in is complete.

   Resolve any errors displayed after the check in before moving to the next step. Common errors are described in chapter 43 of the integration manual (`IG.pdf`) shipped with Dexterity. You can find this manual in the ...\ `Microsoft Dexterity\Dex 11.0\Manuals` folder.

7. Update the index file by selecting the Explorer menu and then selecting **Source Control | Update Index File**. Respond with a yes to the dialog question.

> Updating the index file ensures that all of your resources maintain the same resource ID throughout the upgrade. More information about the index file is in KB 894699. You can get a copy of this KB at `http://support.microsoft.com/kb/894699`.

With the resources now checked in to the new project's repository, it's time to create the new development dictionary.

Creating the new development dictionary

Your new development dictionary will start with a copy of the new release's `dynamics.dic` file. You'll add your resources to that dictionary and then update it. After the update, your code will reflect the new release. You can then test your code and make any changes necessary for it to work with the new release.

Follow these steps to create the new dictionary and update your code:

1. After you've made a backup copy, delete your old development dictionary.
2. Make a copy of the new `Dynamics.dic` file and paste it into your `C:\DEX_GP2010\PROJECT_1` folder. This will become your new development dictionary.
3. Rename the new dictionary to the name of the old dictionary. In this example, you'll name your new dictionary `DEV_PROJECT_1.dic`.
4. Make another copy of the new `Dynamics.dic` file and paste it into your virgin dictionary folder. Rename the new dictionary to reflect that it is an original unmodified dictionary. In this example, we'll use `C:\DEX_GP2010\ORIG_DICS\Virgin_11-00-1935.dic`.
5. Use your project shortcut to open the new dictionary in Dexterity.
6. Change Source Code Control's **Options** window to specify the location of the unmodified copy of the new dictionary just created.

> To access Source Code Control's **Options** window, use the navigation **Edit | Options**.

7. Transfer your resources to the new development dictionary using the Update operation.

> To perform the Update operation, open the **Update from Repository: PROJECT1** window using the navigation **Explorer** | **Source Control** | **Update**.

Complete the window using the following options:

Field Name	Value
Resources section:	
Update All	marked
By Tip	marked
New Resources section	
Prompt to Overwrite	marked
Reports section	
Run SCC Error Report	checked
Run Compile Error Report	checked
Compile section	
With Debug Information	marked
Options section	
Use Index File	checked

Upon completion, the window should look like the following screenshot:

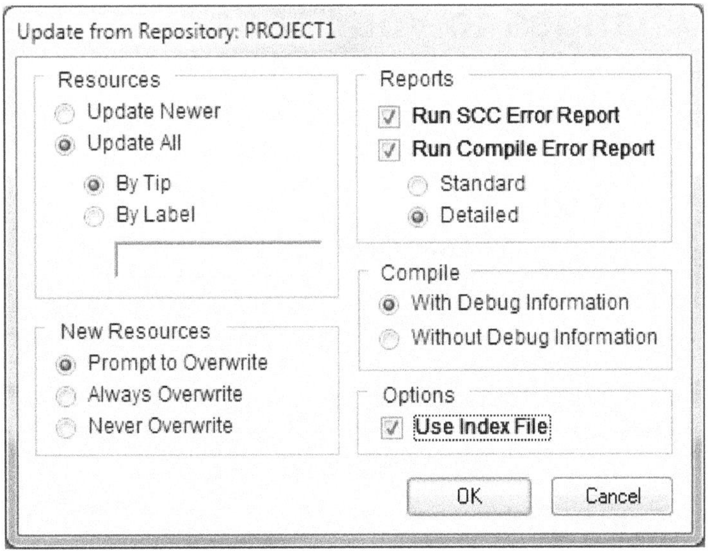

8. Click on the **OK** button to perform the update; print the reports when prompted. The **Resources to Update** window will open. Make sure all of the source file objects, other than the alternate reports, are checked and then click on the **OK** button.
9. Resolve any errors displayed after the check in before advancing to the next step. Common errors are described in chapter 43 of the integration manual (`IG.pdf`) shipped with Dexterity. You can find this manual in the `\Microsoft Dexterity\Dex 11.0\Manuals` folder.

Your development dictionary for the new release is now complete! This would be a good time to back it up. Using your newly created dictionary, move on with your upgrade.

Making changes to your code

Use Dynamics GP SDK to analyze how any changes to the Dynamics GP data model could affect your application. Specifically, you should look for the following:

- Data type changes
- New or deleted fields
- Changes to procedures or functions
- Table changes
- Functionality changes

Other changes might affect your application, such as changes to stored SQL procedures or views, but the ones just listed are the primary culprits for errors.

Data type changes

Change to a data type, such as increasing or decreasing the keyable (allowable) length of a string, will change the data type's storage size. If your table includes a field with one of the changed data types, the size of your table will change. This type of change would require you to convert the table. Likewise, if you have any reports referencing fields with a changed data type, you will need to update your report.

Also, watch out for changes to a data type's control type, such as changing a listbox to a multiselect listbox. Any data type change can cascade changes throughout an application, so be very mindful of this when you analyze a change to a data type.

Use the **Reference Information** window in Dexterity to see what objects a resource refers to or is referenced by. Before you can see this reference information, you must first build the reference table using Dexterity options.

To build the reference table, select **Edit** | **Options** from the Dexterity menu bar. In the **Options** window, click on the **Reference** tab and then click on the **Build Table** button. You might also want to consider checking the **Update Reference Information on Compile** checkbox in order to keep your reference table up to date. For the Dynamics dictionary, it will take about five minutes to build the table. The **Reference** tab of the **Options** window is shown in the following screenshot:

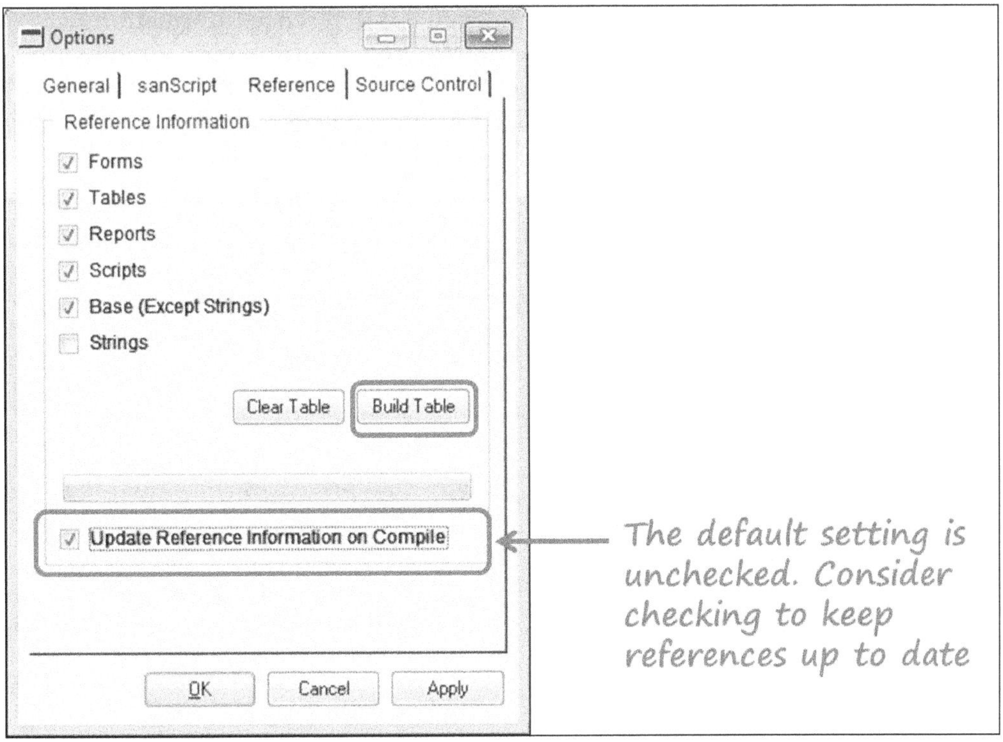

With the reference table built, you can open any Dexterity resource and click on the reference button to discover which other resources may have been affected by the change. The following screenshot identifies the location of the reference button on the **Data Type Definition** window:

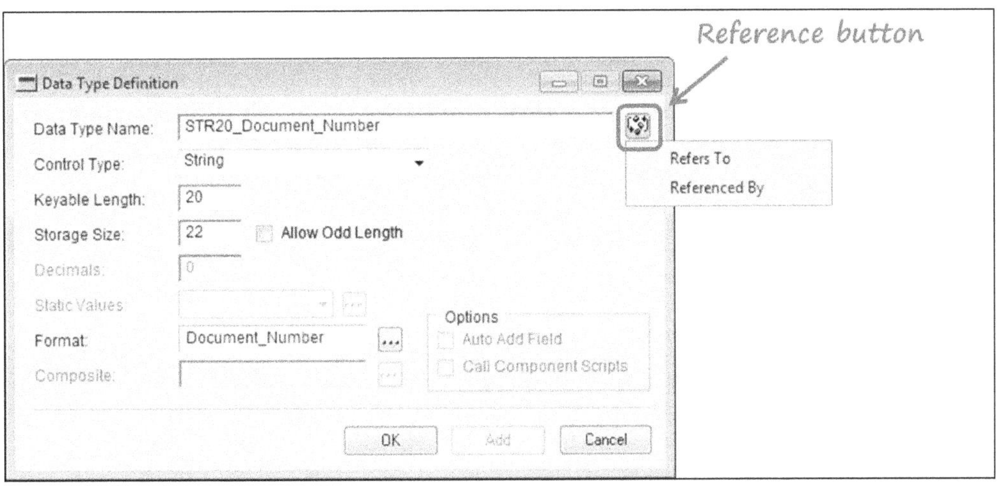

Select the **Refers To** item from the menu to open the **Reference Information** window. The following screenshot shows the **Reference Information** window for the **STR20_Document_Number** data type. The left-hand side of the window lists the format attached to the data type; the right-hand side lists each field that uses this data type:

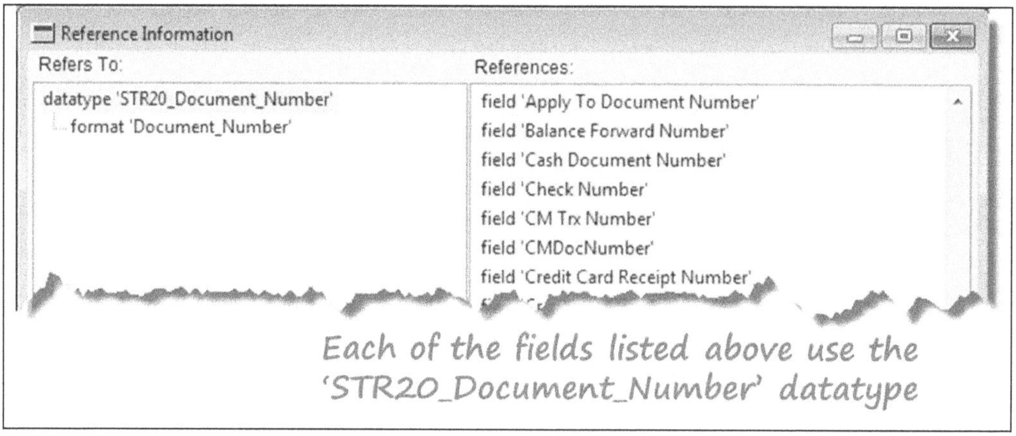

Field changes

New fields added to the new release may have the same name as one of your fields; you may need to rename your field and then modify all of the scripts that reference it. If a field has been removed, you'll get an illegal address error if any of your scripts refer to a missing field.

Procedure or function changes

Changes to procedures or functions are most often changes to the number, order, or type of its parameters. If your script calls one of these procedures or functions, it will not get compiled. You will need to modify your script to comply with the changed parameter details.

Table changes

Check tables for the following changes:

- Adding, changing, or removing fields
- Adding, changing, or removing keys
- Adding or removing a table

If any part of a table definition changes, you may need to modify your code. Any type of change to a global field can affect a table's storage size. If fields have been added to or removed from a table, the storage size will also change. Table changes will require you to convert the existing data, recreate the table, and recreate the autogenerated procedures.

If a table is removed when an update takes place and you reference the missing table, you'll get an illegal address error.

Refer to Volume 1 of the Dexterity programmer's guide for more information on how to handle table changes. This document was installed as `PRGV1.pdf` within the `Microsoft Dexterity\Dex 11.0\Manuals` folder.

Functionality changes

In addition to resource changes, a new release of Dynamics GP is likely to include a new functionality. Review this new business logic and make sure it is still consistent with the routines in your application. Each new release of Dynamics GP includes a `what's new` document that describes its new features to the user. This document might be a good place to start if you're researching functionality changes. You can find more detailed information about functionality changes on the CustomerSource/PartnerSource website.

Completing the update

So many things can go wrong during an update. Consider creating an upgrade checklist for your application to make the process more streamlined. Take into account any changes to the items just discussed, combined with the changes we are about to discuss shortly. You're close, but you're not done yet. Make sure you have a good backup of your old dictionary.

Converting the data

Changes to any global fields that you have used in one of your tables will necessitate a table update routine. For example, if you included the `Customer Name` field in one of your tables, and Dynamics GP changes the length of that field from 65 to 75 characters, your table's record length will change. At that point, data in your old table will be incompatible with the new table definition and will create errors.

To incorporate the new table definition in your application, you need to convert the existing data. Essentially, converting existing data involves transferring the data into a temporary table, dropping and recreating the table with a new table definition, and then transferring the data back into the updated table.

You'll most likely use the SQL pass-through technology to convert the existing data in your SQL tables. You can find more information on how to convert SQL tables in Volume 1 of the Dexterity programmer's guide. This document was installed as `PRGV1.pdf` within the `Microsoft Dexterity\Dex 11.0\Manuals` folder.

Recreating alternate forms and reports

If you created any alternate forms, scrutinize them closely to determine if the update operation adequately updated your forms.

If you created any alternate reports, you'll need to recreate those reports from scratch. The update operation does not attempt to update alternate reports.

Updating forms and report dictionaries

There are special considerations if your users have created any modified forms or reports. It's important that you ensure they are updated properly when your new chunk file is installed. The easiest way to deploy modified forms and reports is to export the modifications to a package file before the upgrade. After the export, rename the existing form and report dictionaries for your application.

Once the upgrade is complete, you can import the package file back into the dictionary. As the changes are imported, Dynamics GP will create new form and report dictionaries for your application. Most, but not all, of the modifications will be recovered. What won't be recovered are any modifications the user has made to global resources, such as formats or strings.

GP 2013 considerations

Looking forward, to maximize your Dexterity application's performance with GP 2013 Web Client, you'll want to clean up your code. As you update your applications, remove any unnecessary comments and scripts with no executable code. For example, you may have large chunks of code that you've commented out to keep track of code changes. The Dexterity runtime engine simply skips over these comments, but Web Client will pass the comments back and forth between the **Windows Communication Foundation** (**WCF**) GP Runtime service and the Silverlight client. More traffic means slower performance. Getting rid of the unnecessary clutter will keep messaging efficient.

Another change to prepare for is that GP 2013 allows you to name the system database to something other than `DYNAMICS` (for new installations only). Being able to name the system database gives you the opportunity to have multiple copies of Dynamics GP running on the same SQL Server instance.

You will want to search your code to make sure you don't have `DYNAMICS` hardcoded as the database name anywhere. You also need to make sure you change any code, including the constants `SQL_SYSTEM_DBNAME` or `ST_DYNAMICS`, or any constants you may have created yourself that resolve to the word `DYNAMICS`.

Replace any instances of these constants or the hard-coded name with code by using the new `GetSystemDatabaseName()` function.

Testing your application and building the update chunk

After an upgrade, it is especially important to test your application thoroughly before delivering it to your users. Be sure to test it in both multidictionary mode and Dexterity's Test mode.

To determine which modules need to be included in the updated chunk file, use the Compare Dictionaries utility in Dexterity Utilities to create the dictionary comparison report. The Compare Dictionaries utility also generates the `CHANGES.TXT` report, which will reveal changes to any of your forms and reports.

It is considered best practice to release a beta version of your upgraded application to a limited set of users for testing. If the customization is client specific, then a complete test upgrade of Dynamics GP with your customization is recommended before performing the production upgrade.

When building the update, include data conversion procedures that operate as seamlessly as possible without requiring a lot of user interaction. In addition, including a procedure that will not allow a user to go forward without updating the tables is a good idea.

Now you're done with your Dexterity upgrade!

Modifier with VBA

Modifier/VBA customizations can open a mixed bag of issues at upgrade time. Moreover, it's not just Dynamics GP upgrades you need to concern yourself with; issues such as intervening operating system updates can also ruin your day. For example, Microsoft Security Bulletin MS08-070, originally published on December 9, 2008, addressed security holes found in Visual Basic 6 Runtime. While this critical update may have fixed the runtime vulnerabilities, it broke many a VBA customization.

This section will review some hazards that could affect your Modifier/VBA customizations.

Modifier

You can only make visual changes using the Modifier tool. You can add or remove buttons and fields, move things around, hide things, change the tab order, colors, fonts, and so on. You cannot change or attach code using only the Modifier tool (without VBA), but you can completely change the look and feel of a window.

Upgrades to Dynamics GP can change the layout and content of any window. If the changes affect one of your modified windows, you may need to tweak your modified window.

For instance, imagine you have modified the **Sales Transaction Entry** window so that it displays the entire 100 characters of the item description. If the new release of Dynamics GP placed a new field in the original window at exactly the same position as your expanded description field, you would need to change your modified window to move things around.

Any changes you make to a window are stored in a forms dictionary for the appropriate application. Before starting any upgrade, it is critical that you do the following:

- Back up (copy) your modified forms dictionary
- Export form changes to a package file

Copying and pasting the dictionary's folder will suffice to back up the forms dictionary. Use the **Customization Maintenance** window to create the package file. This package file will later be imported after your update is complete.

Put these steps on your upgrade checklist; you don't want to risk losing those valuable modifications!

VBA

VBA may be the trickiest upgrade of all. You need to worry about any changes to the environment as well as changes to Dynamics GP. So really, any changes to the workstation could cause you to upgrade your VBA code. Something else to consider during an upgrade is that your VBA code itself could be causing problems with the upgrade.

So many events, such as security patches, Windows version updates, Modifier changes, adding another third-party product, and others can ruin your VBA code. These kinds of changes can happen at any time, and not just as part of a Dynamics GP release upgrade.

The good news is that with most upgrades, nothing at all will impinge on your VBA additions to Dynamics GP. On the other hand, some changes may look simple but may prove to be more challenging.

While this is in no way a complete list, here are some key changes to look for:

- Environment changes
    - Processing system (32 or 64-bit)
    - Operating system (XP, VISTA, Windows 7, Windows 8)
    - Other installed software (Microsoft Office)
    - Service packs
    - Device drivers
- Window changes
    - Moving fields
    - Deleting fields
    - Changing a field's data type
    - Changing the control type of a data type
- Report changes
    - Changes to the fields
    - Changing the sort order
    - Changing restrictions
    - Changing report sections

Environment changes

Over the last couple of years, we've seen many environment changes that proved difficult to resolve. We had a major change in operating systems with the move to Windows 7. Many problems were introduced by changing from an x86 (32-bit) processor to a 64-bit processor.

Major changes in the operating system environment, such as this, can wreak havoc on Dynamics GP VBA customizations. The change just mentioned created massive problems. VBA would simply stop working and start throwing the error message:

File not found: VBA6.DLL

At the same time, a change in **Microsoft Office 2010** (**Office 2010**) created a problem in the system registry keys that would cause Dynamics GP to crash and make a call to Dr. Watson. To further complicate matters, the order in which you installed the programs and whether or not you ever had the previous version installed made a difference. Several environment changes were at work, causing errors. It seemed as if a new slant on the same problem was creeping up everywhere!

Just to make things more interesting, new device drivers being released to support the 64-bit processor were battling with the existing device drivers and crashing the system.

If you find yourself running into odd problems, such as these, don't despair! Reach out to the web community and search for a solution; you will not be left alone in the dark. The 32-bit/64-bit, Office 2010 registry changes, and the device driver problems were all worked out in time.

You can read all about the reason behind the file not found error and how to ultimately resolve it by typing in the following search criteria in your Internet browser:

```
VBA6.DLL File Not Found GP
```

Of particular interest is a series of articles posted by Vaidy Mohan that begin with the article posted at `http://vaidymohan.com/2010/08/23/`.

The move to Windows 8, SQL Server 2013, and Office 2013 is at our doorstep, so beware!

Window changes

Because most VBA customizations rely heavily on objects and fields in the windows; a change to a window could impair your code in a big way.

For example, the upgrade from release 9.0 to release 10.0 changed the **Sales Transaction Entry** window a great deal. In release 9.0, you had buttons in the control area of the window. In release 10.0, the buttons disappeared. A single button with a drop-down list appeared in their place, called the **Action** button.

This change was fantastic for users; it cleaned up the screen and matched actions to the document type. On the developer's side, however, the change was painful. Suddenly, your code was trash if it referenced the **Save**, **Delete**, **Void**, **Post**, **Transfer**, or **Purchase** buttons. There was a lot of garbage, ah, code, that you needed to rewrite. Take a look at the following change.

The release 9.0 window looked like the following screenshot:

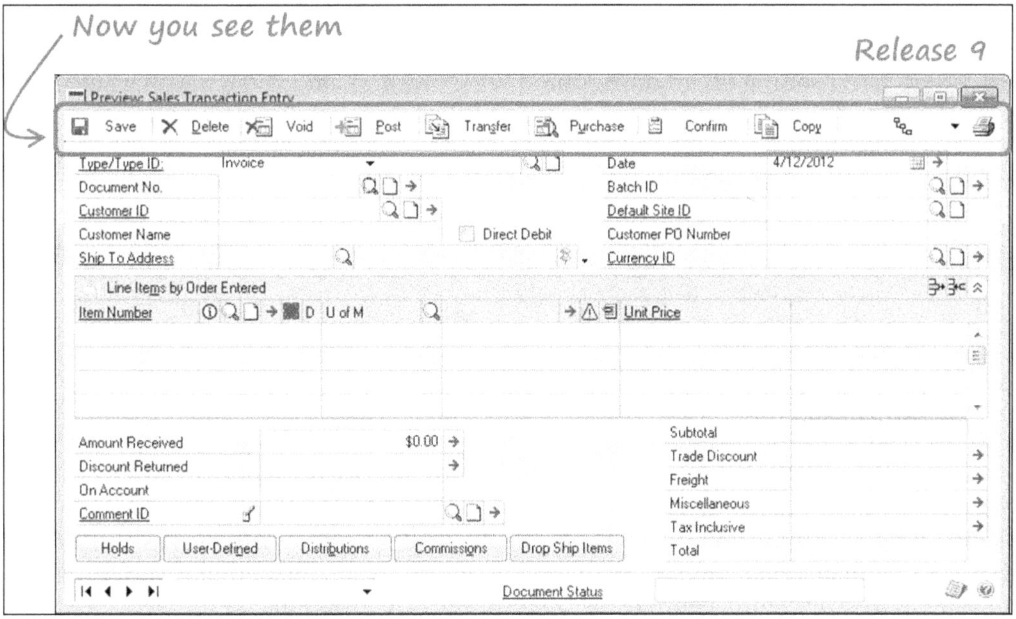

The release 10 window looked like the following screenshot:

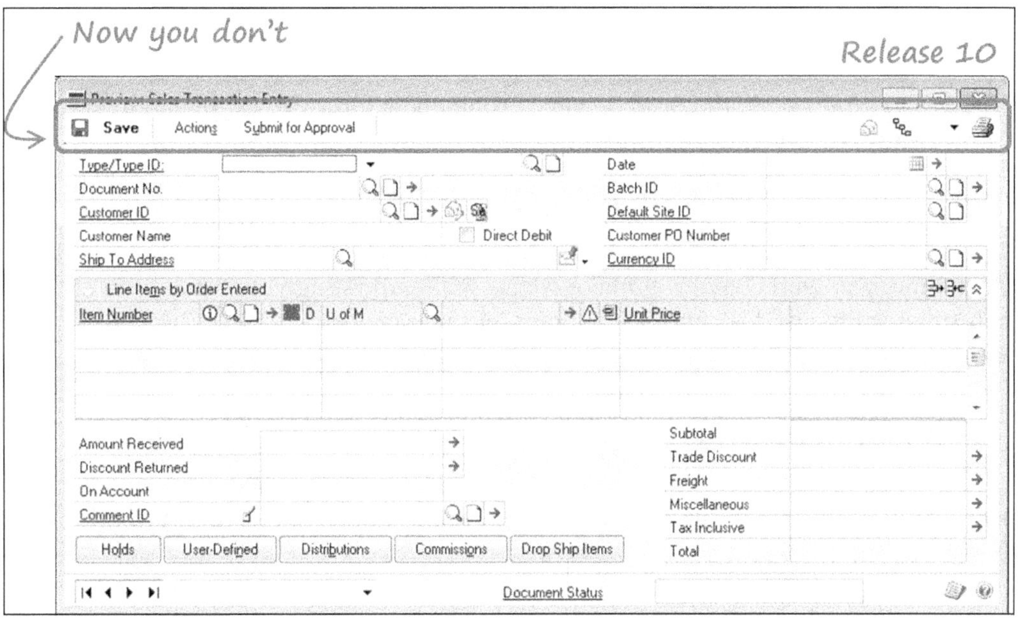

Without thoroughly testing every function and procedure of your code, this kind of change could be devastating.

Another window-change consideration involves newly modified windows. If your VBA code was running on an unmodified window, and later the window is modified (using the Modifier tool), your code will not work anymore. The reason for the breakdown is that you specify whether your VBA code runs against the original or modified version of the window. You can only pick one; it cannot run against both. If you want your code to start running against a modified version of the window, you'll need to change the **EventMode** property for the window in the VBA editor.

Report changes

As VBA can also be used with Report Writer, we'll touch on a couple of things to consider during an update.

Changes to the report fields can cause similar problems with reports as they do with windows. For example, if a currency field is changed to an integer field, some of your calculations may fail. Similar to windows, any changes to a data type or control type that your code depends on may require code modification. Studying the SDK makes it easier to identify many changes before you start the upgrade.

Also documented in the SDK are changes to the data tables themselves as well as changes to Report Writer relationships. If your code references one of the changed tables, you may need to update it. Even changes in Report Writer relationships can lead you to potential difficulties.

Changes to report sections could be very problematic for a VBA application because report sections control data grouping.

If any of your VBA code pulls in a SQL view, you'll want to double-check the SQL object to make sure it is still the same.

Most often, you won't have any major issues while updating your VBA code. You still need to test, test, and test it again, but you probably will not need to change a thing.

GP 2013 considerations

Your VBA code will still work fine with the Rich Client (read Dexterity) installation of GP 2013, but it will not be supported when using the Web Client. You might want to start learning Dexterity and VS Tools!

Extender and Builder(s)

Upgrading applications you created with Extender, SmartList Builder, or Excel Report Builder is simpler than updating a VBA project.

Somebody else, say Microsoft, will work out any code problems with the functionality of the actual Extender or Builder modules themselves!

Your job is to check the dataset delivered by your SmartList objects or Excel reports, and make sure your Go Tos go to the right place. Test your Extender windows to make sure they open at the right time, the conditions work, the linked tables work, and the fields show up in a SmartList object.

Extender

More specifically, for Extender, you need to examine the GP windows that you have used with your Extender windows to make sure that there weren't any changes that might affect your customization. You can use the SDK, as explained in an earlier section, to discover any changes.

For example, if you are using a field on the **Customer Maintenance** window to determine which Window Group window(s) should open, and that field is moved to another window in the new release, you'll need to rework your Window Group's condition statement(s).

Similarly, if you are using any of the auto-open options, you need to make sure the Dynamics GP window field you're using is still in the window and correct tab sequence to satisfy your customization's requirements.

For example, if you want your Extender window to open when the user tabs off the last field in the **Customer Maintenance** window, and the last field in that window changes, you'll need to adjust your Extender window accordingly.

In short, you need to test each of the touch points between your Extender resources and Dynamics GP. Some of the things to check include:

- Form and window names
- Fields used as key fields
- Fields used in condition statements
- Fields used for auto-open options
- SmartList integrations
- Table links

SmartList Builder and Excel Report Builder

For the Builder objects, your job is much the same as it is for Extender, except that you will need to concentrate on data table changes. Changes to forms are important too, because they may affect your Go Tos, but table information is the king of these tools.

In addition to any data model changes, if you have built any SmartList objects or Excel reports using custom views or stored procedures, you'll need to make sure those SQL objects are still returning the correct data.

For example, the Analytical Accounting tables were changed in Dynamics GP 10 SP2, which caused some custom SmartList Builder reports to stop working. This also served as a wake-up call to many, reminding us that service packs need to be tested as well, and not just major release changes. More information about this change is available on Jivtesh Singh's blog at `http://www.jivtesh.com/2009/02/aa-table-changes-with-gp-10-sp2.html`.

Visual Studio Tools (VS Tools)

Upgrading a customization created with VS Tools is a very straightforward operation. There are fewer steps than those needed to upgrade a Dexterity application.

Before getting started, be sure you have the most current release of VS Tools installed.

In this section, you'll walk through the steps to upgrade a VS Tools integration from Dynamics GP release 10 to Dynamics GP release 2010. The procedure is the same for both C# and Visual Basic.

Downloading the application

This example uses a sample application, which you can download. Download this release 10 integration from the address `https://www.box.com/s/eqvsyrsv89i6o5p3ahyq`. From the web page, select the **PostingSOPBatches.zip** item as shown in the following screenshot:

After you have downloaded the file, create a new folder named `Samples` in the `\GP2010 VS Tools SDK` folder. Extract the contents of the `PostingSOPBatches.zip` file into that folder. Delete the hidden file `PostingSOPBatches2.suo` from the extracted files.

Opening the solution

Launch VS Tools and open the solution file of the target integration. Be sure you have a backup of the solution files before you embark on this journey. The extension of a solution file is typically `.sln`.

From the **File** menu select **Open** | **Project/Solution**. Navigate to the ...`\GP2010 VS Tools SDK\Samples\PostSOPBatchDemo_10\PostSOPBatchDemo100` folder and open the `PostingSOPBatches2.sln` file.

If you created your solution with an earlier release of VS Tools, an update wizard will ask you if you want to update your solution. You do. Follow the instructions from the wizard to update the solution files to the current release. At the end of the conversion, you'll get a report that will show you any conversion errors and warnings. The following screenshot is a copy of the conversion error report you may receive. This report shows no errors and only one warning. Click on the plus (+) sign to the left-hand side of the file generating the warning to get more information.

The warning in the following screenshot is telling you that you probably want to update your .NET settings. It's not directly saying that, but if you read between the lines a bit, that's the gist of the .NET warning (I can hear your groans). Actually, this project references .NET 4.0 resources (look at the system references):

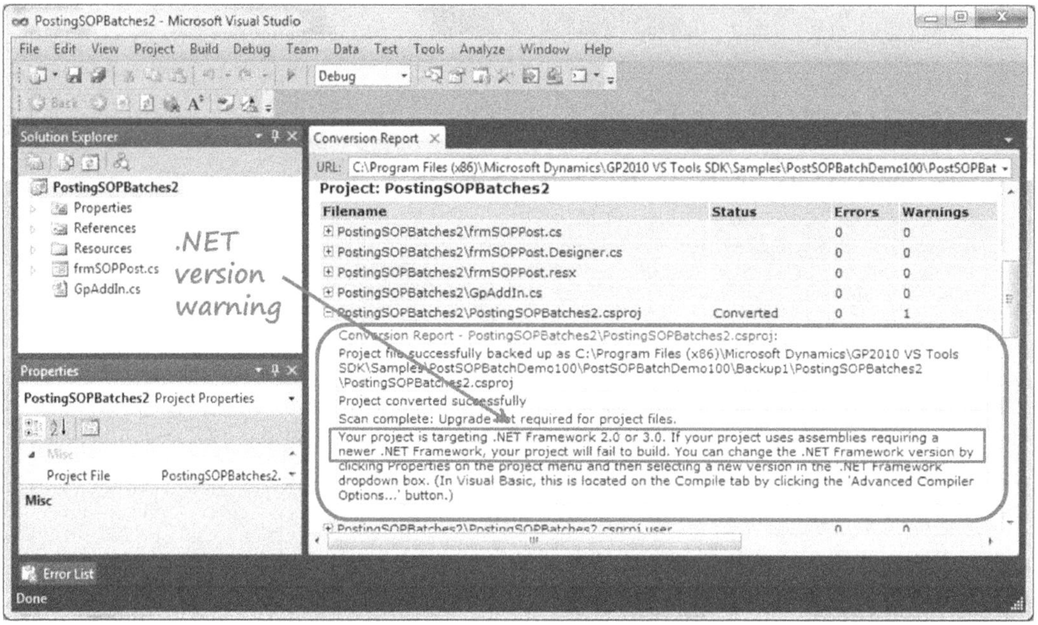

Fortunately, changing the .NET framework is easy. Close the conversion report and double-click on the **Properties** folder under the **PostingSOPBatches2** solution. In the window that opens, change the target framework to 4.0 and then answer with a yes to the dialog that pops up. The following screenshot shows the location of the framework selection field:

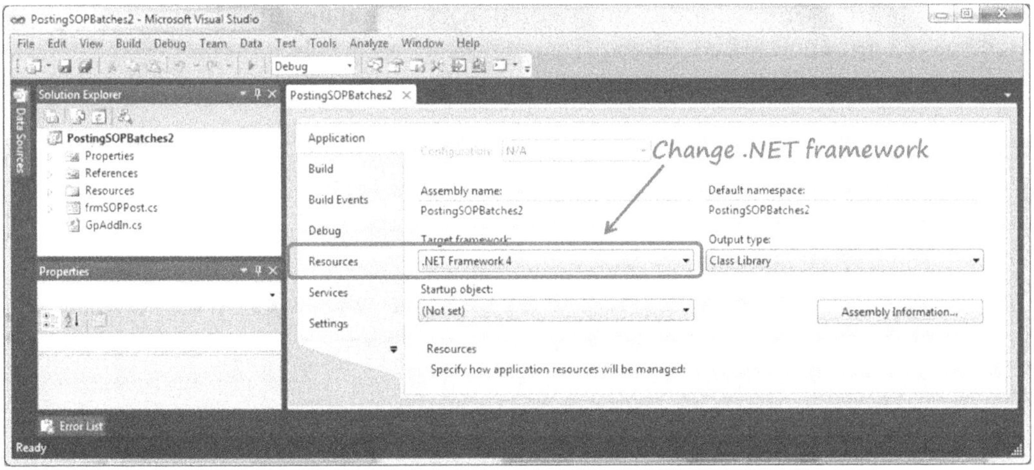

Rebuilding application assemblies

If you used a dictionary other than those shipped with Dynamics GP, then you created an application assembly. You may also have created an assembly for a forms dictionary. You need to rebuild those assemblies.

Use the DAG program to create the new assembly. You installed DAG at the same time you installed VS Tools. You'll find it in the `GP2010 VS Tools SDK` folder. It is a command-line utility; you need to run it from the command prompt.

Our sample project does not use any outside dictionaries. In order to illustrate the process, we'll use the sample integrated application you installed when you installed Dexterity.

Drop the `C:\Program Files (x86)\Microsoft Dexterity\Dex 11.0\Samples\Develop\DEVELOP.CNK` file into the Dynamics GP's root folder. Launch Dynamics GP to unchunk the file, thereby creating the `Develop` dictionary and updating the launch file. The dictionary ID for `Develop.dic` is 3333. You can discover a dictionary's ID by looking in the `Dynamics.set` file.

To generate an assembly for this (main) dictionary, you'll need to use the `/M` parameter with the DAG command. To generate an assembly for a modified forms dictionary, use the `/F` parameter.

From the command prompt, navigate to the location of the `DAG.exe` file and generate the assembly. If you installed everything in the default locations, the command to generate the assembly files is as follows:

- For a 32-bit computer:

```
dag.exe 3333 "C:\Program Files\Microsoft Dynamics\GP2010\Dynamics.
set" /M
```

- For an x64 computer:

```
dag.exe 3333 "C:\Program Files (x86)\Microsoft Dynamics\GP2010\
Dynamics.set" /M
```

You may need to use an elevated command prompt to execute the second command shown. To obtain an elevated command prompt, run the `cmd.exe` file as the administrator. The command-line shortcut is located in the **Accessories** group in the **All Programs** menu. The following screenshot shows the **Command Prompt** menu item. Right-clicking on the item will open the second menu where you select the **Run as Administrator** menu item:

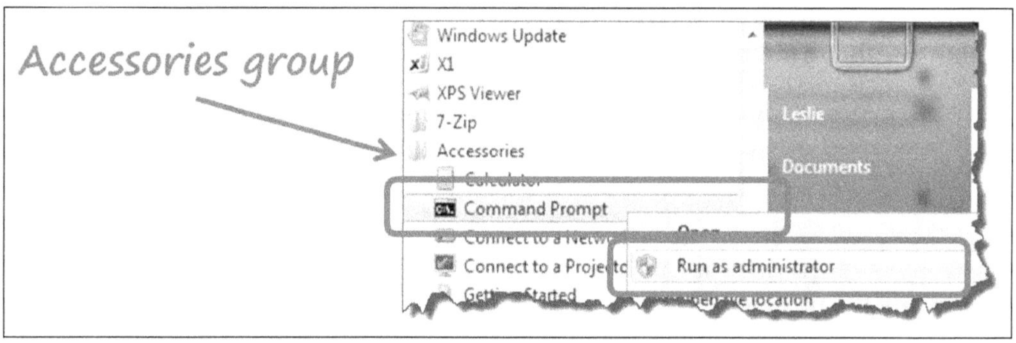

If your command was successful, you'll have two new files:

- `Application.SampleIntegratingApp.dll`
- `Application.SampleIntegratingApp.xml`

If you used the `/F` parameter and built an assembly for a forms dictionary, you'll create the following two files:

- `Application.SampleIntegratingApp.ModifiedForms.dll`
- `Application.SampleIntegratingApp.ModifiedForms.xml`

You'll find the files in the same folder as the `DAG.exe` file.

With the assemblies rebuilt, we move on to fixing the references.

Updating references to the assemblies

You need to update the assembly references because they are most likely pointing to the wrong location. Your task is to change the path so that they are pointing to the correct file. The **PostingSOPBatches2** project uses the following assemblies, which we need to change:

- Application.Dynamics
- Application.SmartList
- Microsoft.Dexterity.Bridge
- Microsoft.Dexterity.Shell

First, delete the current references. You can do this by selecting the reference, right-clicking on it, and then selecting the **Remove** menu item. Do this for each of the four references just listed.

Next, add a new reference for each of the references that you deleted. Do this by right-clicking on the **References** folder and then select the **Add Reference...** menu item. See the following screenshot as an example:

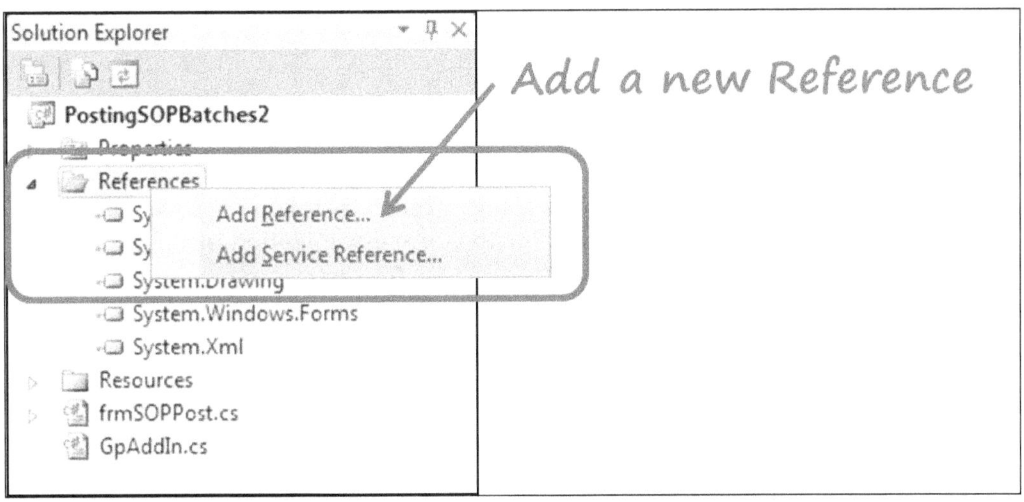

Add the following references, as indicated, from the `\Microsoft Dynamics\GP2010\` folder:

- `Application.Dynamics.dll`
- `Application.SmartList.dll`
- `Microsoft.Dexterity.Bridge.dll`
- `Microsoft.Dexterity.Shell.dll`

Fixing the code

These instructions would certainly be incomplete without this section. You should always start by reviewing Dynamics GP SDK to see if you need to make any obvious changes. Beyond that, it is test, test, test, and then test again!

Not only do you have to worry about the new features you want to include in your application, but you also need to make sure Dynamics GP hasn't done anything to mess you up.

We put "Fixing the code" as if it were a one-time event. As you know, the *Fixing the code* section happens over and over (and over) again.

Building the new solution

The final step is to build the updated solution and test it in multidictionary mode, and then fix the code again!

As we suggested in the *Testing your application and building the update chunk* section, releasing a beta version to a select number of users is considered the best practice for testing updates. That suggestion certainly applies to your VS Tools customizations as well. Likewise, if the customization is client specific, then you should consider a complete test upgrade of Dynamics GP with your customization installed before performing the production upgrade.

Once testing is complete, building the new solution is relatively painless. From the **Build** menu, simply select the **Build PostingSOPBatches2** menu item. You'll get an **Error List** section (report) informing you of any errors, warnings, or messages generated from the build. Of course, you will need to clean up any errors and decide whether you care about the warnings or messages. The following screenshot shows what the postbuild error list looks like:

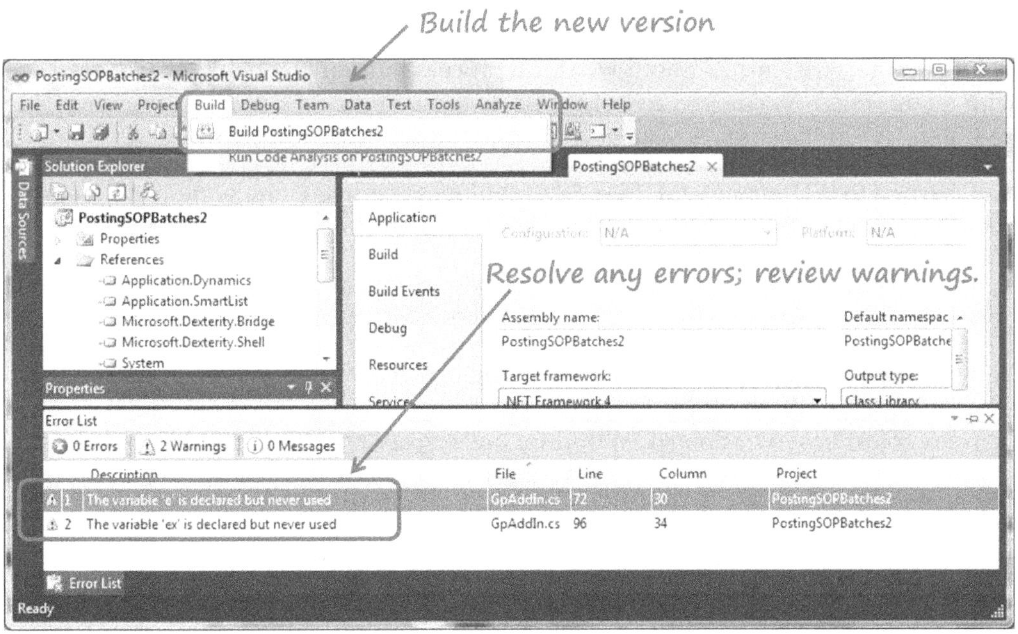

You're done! Now you can bundle up your customization and send it off for deployment.

GP 2013 considerations

Your .NET code will function just fine with the Web Client, but WinForms will not be recognized. Without a user interface, your event-driven code won't get executed. To get it up and running on the GP 2013 Web Client, you need to create your user interface using Dexterity and change your .NET code to trigger off the Dexterity window events.

You can find a table in *Appendix B, Event Matrix*, that contains a side-by-side matrix comparing Dexterity, VS Tools, and VBA events. As you will see in the table, the events across all three tools are very similar. *Appendix B, Event Matrix* is available as a free, downloadable chapter from the following link:
`http://www.packtpub.com/sites/default/files/downloads/0264EN_Appendix_B_Event_Matrix.pdf`

To get a feel for how to build your user interface using Dexterity instead of WinForms, go through the exercises in *Chapter 4, Building the User Interface*.

Summary

In this chapter, we learned about what the process looks like when you are faced with an upgrade. As you see, each type of customization goes down a different path to arrive at the same destination.

You learned that Dynamics GP SDK is a fabulous resource when you need to learn about the changes to the data model from version to version.

You did a sample upgrade of both a Dexterity customization and a VS Tools add-in. The two most feature-rich tools also require the most effort to update.

You looked at several scenarios that can bring down a VBA customization, and where you can find some help for this. Finally, the best of the bunch in terms of upgrade effort has to be SmartList Builder and Excel Report Builder. If there aren't any SQL complications or widespread metadata changes, upgrading Dynamics GP to a new release will automatically upgrade your Builder customization.

In the next chapter, you'll learn where to go from here and how to get there. The chapter is rich with information about support forums, blogs, books, web links, and training materials that will help you learn more about creating customizations for Dynamics GP.

Index

Symbols

.dic files 447
.dll files 132
(L) Display By field 416
(L) Display Options field 415
(L) End Location field 417
(L) Start Location field 416
.NET assembly 64
.NET Framework 62
.tlb files 132

A

account framework 23
Active Data Objects. *See* **ADO**
AddIn assembly
creating 470
AddIns folder 447
AddsAllowed scrolling window 230
ADO
about 71
capabilities 72
developer skills requisites 72
features 71
Advanced Security module 35
AfterActivate event 344
AfterClose event 344
AfterGotFocus event 344
AfterLineChanged event 345
AfterLineGotFocus event 345, 362
AfterLineLostFocus event 345, 363
AfterModalDialog event 344, 370-375
AfterOpen event 344
AfterUserChanged event 344, 374, 378
Allow Active Locking 165
Alternate form 119
Alternate/Modified Forms and Reports settings 126
Alternate Report 62
Alternate Window 62
AltLineColor property
about 135, 145
setting 224
Appearance property 136
application
building 483, 484
deploying 483, 484
testing 519
application assembly
about 447
referencing 466
Application.Dynamics.dll 447
Application.FA.dll 447
Application Programming Interface (API) 385
articles, Vaidy Mohan
URL 522
assembly
building 474
testing 474
AssemblyInfo.cs file 483
Auto-Chunk window
about 277
Chunk Dictionary 277
Current Dictionary 277
Product Information window 280

AutoCopy property 134
AutoLinkTable property 140, 291
automation clients 72
automation controllers 72
automation servers 72
AutoOpen property 140, 291
AutoSetDexColors property 84, 446, 457, 459, 460

B

Band events, VBA
about 112
BeforeAF 113
BeforeAH 113
BeforeBody 113
BeforePF 113
BeforePH 113
BeforeRF 113
BeforeRH 113
base resources, Dexterity
about 155
data types 155, 156
fields 158
format 157
base table 394
Beep method 343
Before Activate event 344
BeforeClose event 344
BeforeGotFocus event 344
BeforeLineChange event 345, 363
BeforeLineGotFocus event 345, 362
BeforeLineLostFocus event 345, 363
BeforeLinePopulate event 345, 364, 365
BeforeModalDialog event 344, 368, 369
BeforeOpen event 344
BeforeUserChanged event 344, 374
Big Line Item 146, 310
big line item, scrolling windows
about 222
marking 223
boolean, Dexterity control types 551
Border property 136
bottom-up approach 132
browse buttons, Vendor Maintenance window 46
BrowseOnly scrolling window
about 224
lookup windows 225
visual properties, modifying 224
Btrieve 10
built-in report writer 15
button control 455
button control, properties
AutoSetDexColors 457
ButtonType on dexButtonProvider 456
Button controls
about 462
properties 462-464
button drop list, Dexterity control types 551
Buttons, VS Tools controls
about 84
field 86
Standard 84
StatusArea 85
Toolbar 85
ToolbarWithSeparator 85
ButtonType on dexButtonProvider property 456

C

C# 445
calculated fields, SmartList objects
about 403-405
CONSTANT 2 407
CONSTANT 4 408
List of On Hand QTY 406, 407
QTY Available for Sale 405, 406
CancelLogic parameter 343, 374
Cancel property 134
capabilities, Continuum 68
capabilities, DDE \ ODBC \ ADO \ OLE Automation 72
capabilities, Dexterity 61, 62
capabilities, eConnect 78
capabilities, Extender / eXtender Enterprise 69
capabilities, Integration Manager 73
capabilities, Modifier with VBA 67
capabilities, Table Import 75
capabilities, VS Tools 64

capabilities, Web services 80
caption property 339, 341
Cards section 394
Catering Company field 435
Certified for Microsoft Dynamics.
See **CfMD**
CfMD 12
Changed property 341
Change() method 477, 478
Change script 195
change, table operations 198
char() function 219
check box, Dexterity control types 551
Chunk Dictionary
Build Number 278
Compression 279
Installation Scripts 278
Major Version 278
Minor Version 278
Module 278
Version Information 278
Chunk file
creating 119
chunk file, Dexterity application
creating 272
dictionary module, transfer 275-277
resources, extracting 273, 274
City field 435
City Lookup object
creating 436
CLARK 23
classroom training, Dynamics GP 536, 537
Click event 455
CloseBox property 141, 291
Code Modules 346
Collections module
referencing 358-361
column headings, static text 189
ComboBox control
about 460, 465
properties 465
ComboBox control, DropDownStyle options
DropDown 460
DropDownList 460
Simple 460
ComboBox control, properties
AutoSetDexColors 460
combo box, Dexterity control types 552
ComboBoxes, VS Tools controls 86
COM objects 385
Compare Dictionaries utility 519
Component Object Model (COM) 62
components, Dexterity 131
components, VBA
about 336
events 343
methods 342
objects 338, 339
properties 339, 340
Composite
about 138, 552
creating 138, 139
COM support 16
considerations, Dynamics GP 2013 484
CONSTANT 2 field 407
CONSTANT 4 field 408
Contact Department field 173
Contact Lookup window 181
Contact Person field 173
Contact Phone Master Key1 163
Contact Phone Master Key2 164
Contact Phone Master Key3 164
Contact Phone Master table
creating 163
table keys 163, 164
Contain.exe 131
Continuum
about 68, 113, 385
capabilities 68
developer skills requisites 69
features 68
ControlArea property 446 84, 142
controls
adding, to window 454, 455
control type 551
copy from table statement 203
copy from table, table operations 200
copy to table, table operations 199

core modules, Dynamics.dic dictionary
Bank Reconciliation 82
General Ledger 82
Inventory 82
Payables Management 82
Purchase Order Processing 82
Receivables Management 82
Sales Order Processing 82
System Manager 82
cross dictionary access
about 358
Collections module, referencing 358-361
CRUD 475
currency, Dexterity control types 552
currency fields 402
current table naming convention 41
Customer Contact Lookup window
creating 152
Customer Contact Maintenance form and window
creating 167
Customer Contact Maintenance window
about 181
creating 152
Customer Contact Master Key1 161
Customer Contact Master Key2 162
Customer Contact Master table
creating 160
table keys 161, 162
Customer contacts, Extender 441, 442
Customer Contacts lookup window
creating 152
customer exmaple, record retrieving
about 202, 203
browse buttons 206-208
customer zoom 203-206
Customer field 435
Customer ID field 202
Customer Number field 202
Customers and Prospects lookup window 226
Customer window group, Extender 437-441

D

DAG
about 468
using 468-470
dag.exe file 468, 469
database-level integrations
about 59
tasks 59
Database Management System (DBMS) 71
database trigger
about 97, 235
considerations 237
cross-dictionary considerations 237
registration procedure 235
DataBindings property 458
data model changes, Dynamics GP SDK
about 491
deleted columns 493
deleted indexes 493
deleted RW relations 495
deleted tables 492
different data type 494
different index columns 494
different segments 494
new columns 492
new indexes 493
new RW relations 495
new tables 492
data table naming convention, Dynamics GP
about 33
current table naming convention 41-43
module prefixes 34, 35
numbering convention 35, 36
stored procedure 39, 40
DataType
about 133
composite definition 133
Control Type 133
elements 133
format 133
Keyable length 133
static values 133
storage size 133

data type changes 514-516
Data Type Definition window 156
DataType property 134
data types
changing, Modifier used 332, 333
creating 155, 156
date, Dexterity control types 553
date fields 402
DDE
about 71
capabilities 72
developer skills requisites 72
features 71
DDE Conversation 71
debugging 346, 347
debugging tools 16
Debug menu 347
Debug mode 124
Default Button 228
DefaultDblClick property 144, 228
detail form 430
detail window 430
developer skills requisites, Continuum 69
developer skills requisites, DDE \ ODBC \ ADO \ OLE Automation 72
developer skills requisites, Dexterity 63
developer skills requisites, eConnect 79
developer skills requisites, Extender / eXtender Enterprise 70
developer skills requisites, Integration Manager 74
developer skills requisites, Modifier with VBA 67
developer skills requisites, Table Import 76
developer skills requisites, VS Tools 64
developer skills requisites, Web services 80
Developer Toolkit articles and links 548
Developing for Dynamics blog
about 547
Developer Toolkit articles and links 548
Dexterity articles and links 548
Dynamics GP blogster 549
DynDeveloper.com 550
Modifier & VBA articles and links 548
Support Debugging Tool 549
development dictionary
creating 123
development environment, integrating application
development dictionary, creating 123
Dex.ini file, modifying 122
DexSense, installing 129
Dexterity, installing 121
SDK, installing 121
SDT, installing 130
setting up 121
test mode 124, 125
user security, modifying 126-129
development process, integrating application
application, developing 118
Chunk file, creating 119
development environment, preparing 118
final product, delivering 120
software, installing 118
Development Tool
reference link 538
Dex.chm 131
Dex.dic 131
Dex.exe 131
Dex.ini file
modifying 122
DexSense
about 118, 193
installing 129
Dexterity
about 12, 60, 118, 385
built-in report writer 15, 16
capabilities 61, 62
components 131
COM support 16
debugging tools 16
design 17
developer skills requisites 63
development environment, setting up 121
development process 117
dictionary resources 132
end user prerequisites 63
exception handling 15
features 60

Form Definition window 82
Forms 82
function library 15
Graphical forms designer 16
limitations 63
multidictionary environment 61
overview 12, 13, 130
Resource Explorer 13, 148
resources 132
sanScript 14
source code control 15
Table Definition window 160
triggers 97
used, for modifying UI 81, 82
Dexterity application
chunk file, creating 272
completing 263
deploying 245
distributing 284
forms and windows 263
reports 267
resources available 283
tables 267
testing, in multi-dictionary
environment 281
Dexterity application deployment
about 245
system requisites 246
table creation routines 253
versions and builds 251
Dexterity application, distributing
chunk, sending 284
Windows Installer services 285
Dexterity application, testing
chunk file, doesn't unchunk 282
tools and techniques, testing 282, 283
Dexterity articles and links 548
Dexterity Basics training manual
downloading 43
Dexterity Bridge assembly 447
Dexterity control types
about 551
boolean 551
button drop list 551
check box 551
combo box 552
composite 552
currency 552
currency (variable) 553
date 553
drop-down list 553
horizontal list box 553
integer 553
list box 554
list view 554
long integer 554
multi-select list box 555
non-native list box 555
picture 556
progress indicator 556
push button 556
radio button 556
Radio group 556
reference 556
string 556
text 556
time 556
tiny integer 557
tree view 557
visual switch 557
Dexterity customization
code, changing 514
generic source code control, setting up 500
new development dictionary,
creating 511-513
new project, starting 510, 511
old dictionary, checking in 507-509
upgrade checklist, creating 517
upgrading 500
Dexterity design 17
Dexterity documentation
URL 283
Dexterity events
about 94
Field events 95
Form events 94
scrolling window events 96
Window events 95

Dexterity integration guide 537
Dexterity object events
Activate (Window) 559
Change (Field) 560
ContextMenu (Field) 561
ContextMenu (Line) 561
ContextMenu (Window) 561
LineChange 560
LineDelete 560
LineFill 559
LineInsert 560
LinePost 560
LinePre 559
MouseEnter 560
MouseExit 561
Post (Field) 560
Post Script (Form) 560
Post (Window) 559
Pre (Field) 560
Pre Script (Form) 560
Pre (Window) 559
Print (Window) 560
Dexterity Runtime Engine (Dynamics.exe) 31, 32
Dexterity sample applications
reference link 539
Dexterity Shell assembly
about 446
AutoSetDexColors property 446
ControlArea property 446
StatusArea property 446
Dexterity Source Code Control Server. *See* **DSCCS**
Dexterity, triggers
Database 97
Focus 98
Form 98
Function 98
Procedure 98
Dexterity WYSIWYG Form Designer window 290
DexUtils.chm 131
DexUtils.dic 131
DexUtils.exe 131
dialogs 367
dictionary assemblies
building 468
Dictionary Assembly Generator. *See* **DAG**
dictionary resources
accessing 465
application assembly, referencing 466
namespace, referencing 467, 468
dictionary resources, Dexterity
Composite 138
DataType 133
Field 134
form 140
Format 133
Scrolling window 142
Table 139
dictionary resources, VS Tools 99
documentation for Developer Toolkit
reference link 538
Dpm.exe 131
Dps.exe 131
Drill Down Builder 387, 422-428
drill downs, types
Extender 422
Form 422
SmartList 422
drop-down list, Dexterity control types 553
DSCCS
configuring 504, 505
installing 501-503
DUOS
about 114, 338, 370
architecture 371
data, deleting 374-378
data, retrieving 372
data, saving 373, 374
field information 115
key information 115
object, declaring 372
table information 114
DUOS data
deleting 374-378
retrieving 372
saving 373, 374
DUOS object
about 371
declaring 372

DUOSObjects collection 371
DUOSObjectsGet method 372
DUOSProperties collection 371
DUOSProperty object 371
DUOS table
about 114, 115, 370
information, adding 114
Dynamic Data Exchange. *See* **DDE**
Dynamics communities
about 539, 540
URL 539
Dynamics.dic dictionary 82, 83
Dynamics.exe 131
Dynamics GP
about 9, 118, 385
additional data, storing 114
additional window elements 48
Btrieve 10
classroom training 536, 537
data table naming convention 33
FairCom's c-tree Plus 10
functionality, changing 94
Microsoft SQL Server 10
native user interface 10
online training 536
project, defining 56
resources 538
Sales Transaction Entry window 48
software manuals 537
SQL table 32
SQL table names 32
tools 60
user interface, modifying 81
User Interface (UI) 43
Dynamics GP 2013
considerations 484
Dynamics GP application
components 17
launching 18, 19
runtime environment 19
Dynamics GP data model changes
data type changes 514-516
field changes 516
functionality changes 517
procedure or function changes 516
table changes 516
Dynamics GP Most Valuable Professionals (MVPs) 539
Dynamics GP MVP blog
URL 526
Dynamics GP SDK
about 487
data model changes 491
form changes 498
launching 488
reference link 538
screen 488
script changes 489-491
table changes 495
Dynamics GP Software Development Kit 118
Dynamics.set file 63, 358, 469
dynamics user group forum
about 542
URL 542
Dynamic User Object Store. *See* **DUOS**
DynDeveloper.com 550

E

eConnect
about 76
capabilities 78
components 76
developer skills requisites 79
end user prerequisites 79
features 76
interfaces 76
eConnect for Microsoft Dynamics GP 2010 Integration Service 76
Editable property 135
editable scrolling window
about 228, 229
Line Events 229
editions, Extender
eXtender Enterprise 430
Extender Standard 429
e-mail link button, Sales Transaction Entry window 48
Empty property 339
Enabled property 339

end user prerequisites, Dexterity 63
end user prerequisites, eConnect 79
end user prerequisites, Modifier with VBA 67
end user prerequisites, VS Tools 64
end user prerequisites, Web services 81
environment changes, VBA customizations 521-523
eOne Solutions
about 78, 430
URL 430
event form project, Extender
about 431-434
Catering Company field 435
City field 435
City Lookup object, creating 436
Customer field 435
Event Location field 435
Extender menu 437
Extra Windows button 436, 437
Rental Company field 435
Rental Location field 435
event handler 99
Event Location field 435
EventMode property 341
events 343
Event Type field 434
Excel Report Builder 386, 419-422, 526
exception handler 15
exception handling 15
expansion arrows, Sales Transaction Entry window 52
ExportOneLineBody switch 28-31
Export to Text File window 168
Extender
about 69, 387, 429, 525
Customer contacts 441, 442
Customer window group 437-441
editions 429
event form project 431-434
line item note 443
object types 429, 430
overview 429
working with 430
Extender Detail form 90
Extender Detail window 91, 92
Extender dialog 430
Extender drill down 422
eXtender Enterprise
about 69, 114, 430
using 114
Extender / eXtender Enterprise
capabilities 69
Detail Forms 90
Detail Windows 91
developer skills requisites 70
end user prerequisites 70
features 69
Forms 89
Notes window 93
used, for modifying UI 88
windows 90
Extender form
about 89
fields 89
Extender menu 430, 437
Extender Notes window 93
eXtender sample applications
reference link 539
Extender Standard 429
Extender window 90
Extender Windows window 88
Extract dictionary 119
Extra Windows button 436, 437

F

FairCom's c-tree Plus 10
fam.dic 447
favorite 386
feature specific requisites, Dynamics GP 2010
about 247
business analyzer 249
Business Portal 5.0 249
Business Portal 5.1 250
Charts and KPIs (Key Performance Indicators) 249
e-mail functionality 248

Unified communications 249
Word form documents 248
Word form template generator 249
Workflow in 32-bit environments 250
Workflow in 64-bit environments 250
features, SmartList Builder 393
feld changes 516
Field
about 134
lookup buttons, linking 137
prompts, linking 136, 137
field button 86
Field button type 456
Field Definition window 158
field events, Dynamics GP
about 344
After Got Focus 344
After Lost Focus 344
After User Changed 344
Before Got Focus 344
Before Lost Focus 344
Before User Changed 344
Changed 344
field events, Dexterity
about 95
change 95
ContextMenu 95
MouseEnter 95
MouseExit 95
post 95
pre 95
field events, VBA
AfterGotFocus 110
AfterLostFocus 110
AfterUserChanged 110
BeforeGotFocus 109
BeforeLostFocus 110
BeforeUserChanged 110
field events, VS Tools
about 105
Change 105
ClickAfterOriginal 105
ClickBeforeOriginal 105
EnterAfterOriginal 105
EnterBeforeOriginal 105
LeaveAfterOriginal 105
LeaveBeforeOriginal 105
ValidateAfterOriginal 106
ValidateBeforeOriginal 106
field methods
Focus 343
FocusSeg 343
Move 343
field options, SmartList Builder 400, 401
field properties
about 134, 135, 339
caption 339
Empty 339
Enabled 339
Height 339
Left 339
Locked 340
name 340
object properties 134
Required 340
Tab stop 340
Top 340
Value 340
ValueSeg 340
Visible 340
visual properties 135
Width 340
fields
displaying, SmartList object used 397, 398, 399, 400
fields, Dexterity
creating 158
Field_SetAltLineColor() function 145
fields, form and windows
adding 173
global field, adding 174
global field properties 175
local fields 178, 179
window text properties 175
field types, SmartList Builder
currency fields 402
date fields 402
integer fields 402
long integer fields 402
string fields 402, 403

field values
setting 356, 357
fill window statement 216
final product
delivering 120
Focus trigger
about 98, 233
considerations 235
cross-dictionary considerations 235
registration procedure 234
form
about 140, 429
adding, to Vendor Quick Entry project 453
form and windows, Dexterity
Customer Contact Maintenance form 168
Customer Contact Maintenance form and window 167
Export to Text File window 168
Import From Text File window 169
Replace window 169
RM_Customer_Address form 167
RM_Customer_Contact_Maintenance form 169
scrolling window 180, 181
tables, attaching 169
window fields, removing 171, 173
window properties, setting 170
form and windows
Contact Lookup window 181
creating 167
window fields 182
form and windows, Dexterity application
about 263
fields, adding 173
formats, linking 265
lookup buttons, hyperspacing 265
lookups, linking 264
maintenance 167
prompts, linking 264
tab order, setting 266
tool tips, adding 264
user interface standards, complying with 267
windows opening positions and sizes, setting 266
Format
about 133
changing, Modifier used 328, 329
Format Definition window
creating 157
format, Dexterity
creating 157
form changes, Dynamics GP SDK 498
Form Definition window 82, 181
Form drill down 422
Form events, Dexterity
about 94
Post 94
Pre 94
Form events, VS Tools
about 100
CloseAfterOriginal 100
CloseBeforeOriginal 100
OpenAfterOriginal 100
OpenBeforeOriginal 100
Forms
about 82
window objects 83
windows 83
Form trigger
about 98, 232
considerations 233
creating 240
cross-dictionary considerations 233
processing procedure, creating 240-242
registration procedure 232
forums
about 539
Dynamics communities 539
Dynamics user group 542
GPWindow 543
MSDN 546
Tek-Tips forum 541
functionality changes 517
Function events, VS Tools
InvokeAfterOriginal 106
InvokeBeforeOriginal 106
function library 15

function trigger
about 98, 239
considerations 240
cross-dictionary considerations 240
registration procedure 239

G

General Entry window modifications
about 304-306
fields, adding to scrolling window 309-312
graphic elements, adding or changing 319, 320
static text, modifying 313, 314
window fields, adding 307-309
window fields, modifying 307-309
general requisites, Dynamics GP 2010 246, 247
generic source code control
DSCCS, configuring 504, 505
DSCCS, installing 501-503
setting up 500
validation errors, resolving 506
GetFirst() method 481
get first statement 210
GetLast() method 481
Get() method 477
GetSystemDatabaseName() function 519
get, table operations 198
global field properties 176, 177
global fields 174
global resources
changing, Modifier used 325
data type, changing 332, 333
formats, changing 328-332
message resources , changing 333, 334
native pictures, changing 326
pictures, changing 325, 327
string resource, changing 328
Go To button
creating 350-352
Go Tos
about 387, 410, 411
Item Maintenance window 411, 412
Item Transaction Inquiry window 411-414
GP 2013 considerations, VBA customizations 524
GpAddin.cs file 472
GPWindow
about 543-545
URL 543
Graphical forms designer 16
graphic elements, General Entry window modifications
about 319
adding or changing 319, 320
new picture, adding 321-324
picture, changing 324, 325
Grid methods
Hide 343
Move 343
Show 343
Grids 361
GroupBox control
about 464
properties 464

H

Height property 339
Help icon button, Sales Transaction Entry window 53, 54
History phase, transaction tables
about 39
URL 39
horizontal add-on
about 11
sampling 11
horizontal list box, Dexterity control types 553
HTTP 80
Hyperspace property 135, 265

I

IDE (Integrated Development Environment) 288
IG.chm 131
IMAP (Internet Message Access Protocol) 80
import 430

Import.chm 131
information icon, Sales Transaction Entry window 51, 52
INITIALIZE procedure 227
installation, VS Tools 448
running 450
integer, Dexterity control types 553
integer fields 402
Integrated Development Environment (IDE) 60
Integration Manager
about 73
capabilities 73
developer skills requisites 74
features 73
integrations, types
about 59
database-level integrations 59
user interface level integrations 59
invalid direct reference 268
invalid transitive reference 269
IsLoaded property 341
Item Maintenance window
about 412
Item Number field 412
Item Number field 412, 415
Item Transaction Inquiry window
(L) Display By field 416
(L) Display Options field 415
(L) End Location field 417
(L) Start Location field 416
about 413, 414
Item Number field 415

K

Knowledge Base (KB)
about 509
URL 509

L

Label control, properties
LinkField on dexLabelProvider 459
Label controls
about 459-462
properties 461, 462
Labels, VS Tools controls 86
launch file (Dynamics.set)
about 19
components 19
elements 20, 21
Left property 339
limitations, Dexterity 63
LineChange events
about 229, 363
BeforeLineChange 363
LineDelete event 230
Line Events, editable scrolling window
about 229
LineChange 229
LineDelete 230
LineFill 229
LineInsert 230
LinePost 229
LinePre 229
LineFill event 229
LineGotFocus events
about 362
AfterLineGotFocus 362
BeforeLineGotFocus 362
LineInsert event 230
line item note, Extender 443
LineLostFocus events
about 363
AfterLineLostFocus 363
BeforeLineLostFocus 363
Line navigation events 104
LinePost event 229
LinePre event 229
LinkedLookup property 135
LinkedPrompt property 135
LinkField on dexLabelProvider property 459
LinkField property 86
LinkTableKey property 145
LinkTable property 144
list box, Dexterity control types 554
ListBoxes, VS Tools controls 86
List of On Hand QTY field 406, 407
list view, Dexterity control types 554
Local Field Definition window 179
local fields 178, 179

Locked property 340
long integer, Dexterity control types 554
long integer fields 402
Lookup 430
lookup button, Vendor Maintenance window 46
lookup windows
about 126, 225
double-click, defaulting 228
lookup form, calling 226-228

M

macro language
URL 283
Main table abbreviations 42, 43
map link button, Sales Transaction Entry window 49
Master tables 36
menus, Visual Studio Tools for Dynamics
reference link 538
message resources
changing, Modifier used 333, 334
methods
about 342
using 354-356
Microsoft 445
Microsoft BizTalk AIC (Application Integration Component) service 76
Microsoft.Dexterity.Bridge.dll 447
Microsoft.Dexterity.Shell.UI.dll 446
Microsoft Dynamics GP. *See* **Dynamics GP**
Microsoft Dynamics GP Architecture
reference link 538
Microsoft Office 2010 522
Microsoft SQL Server 10
MIME (Multipurpose Internet Mail Extensions) 80
MiniDex.chm 131
Miscellaneous tables 38
Modal Dialog events, VBA
about 108
AfterModalDialog 109
BeforeModalDialog 109
Modifier
about 65, 233, 287
overview 287
Modifier customizations 520
Modifier & VBA articles and links 548
Modifier/VBA customizations
about 378, 519
hazards 519
issues, with Windows 7 381
Modifier 520
package files, creating 378, 379
package files, editing 380
package files, limitations 380
VBA 520
Modifier with VBA
about 65, 107, 287, 288
capabilities 67
developer skills requisites 67
end user prerequisites 67
features 65, 66
used, for modifying UI 87
Modifier with VBA sample applications
reference link 539
modifying tools, Dynamics GP
Dexterity 81
Extender / eXtender Enterprise 88
Modifier with VBA 87
VS Tools 83
MSDN
about 546
URL 546
MSMQ (Microsoft Message Queuing) service 76
MSTR 394
multicurrency icon, Sales Transaction Entry window 50
multidictionary environment, Dexterity 61
multi-select list box, Dexterity control types 555

N

name property 141, 145, 291, 340
namespace
referencing 467, 468
naming conventions, tables 164
native user interface, Dynamics GP
about 10
Horizontal add-on 11
Vertical add-on 11

navigation list 62
NET API 64
non-native list box, Dexterity control types 555
note button (record level), Vendor Maintenance window 45
note button (window level) Vendor Maintenance window 47
note window 430
numbering convention, Dynamics GP data tables
00000 - Master tables 36
10000 - Work transaction tables 38
20000 - History transaction tables 39
20000 - Open transaction tables 39
40000 - Setup tables 36
50000 - Temp tables 36
60000 - Relation or Cross Reference tables 36
70000 - Report Options tables 37
80000 - Posting Journal Reprint tables 37
90000 - Miscellaneous tables 38

O

Object Linking and Embedding Automation. *See* **OLE Automation**
object properties, Field
AutoCopy 134
Cancel 134
DataType 134
Default 134
Editable 135
Field 135
Hyperspace 135
LinkedLookup 135
LinkedPrompt 135
Required 135
object properties, Scrolling window properties
about 144
DefaultDblClick 144
LinkTable 144
LinkTableKey 145
Name 145
WindowType 145
object properties, window properties
AutoLinkTable 140
AutoOpen 140
CloseBox 141
Name 141
Title 141
objects
about 338, 339
adding, to VBA project 352
object types, Extender
detail form 430
detail window 430
dialog 430
form 429
import 430
Lookup 430
menu 430
note window 430
window 430
window group 430
ODBC
about 71
capabilities 72
developer skills requisites 72
features 71
OKtoDeleteVendor variable 376
OLE Automation
about 72
capabilities 72
developer skills requisites 72
features 72
OLEPath switch 25, 26
OLE Providers 71
online training, Dynamics GP 536
Open Database Connectivity. *See* **ODBC**
Open phase, transaction tables 39
Optimistic Concurrency Control (OCC) 165
Options window
accessing 348

P

package files
creating 378, 379
editing 380
limitations 380

parameters, database trigger registration
anonymous 236
form_name 236
script processing_procedure 236
table_operations 236
tag 236
Trigger_RegisterDatabase 236
parameters, Focus trigger registration
anonymous 234
attach_type 235
focus_type 234
script processing_procedure 235
tag 235
Trigger_RegisterFocus 234
parameters, Form trigger registration
accelerator_key 232
form_name 232
menu_item_name 232
script processing_procedure 233
tag 233
Trigger_RegisterForm 232
parameters, function trigger registration
attach_type 239
function_name 239
function processing_function 239
tag 239
Trigger_RegisterFunction 239
parameters, procedure trigger registration
about 237
attach_type 238
script 238
script processing_procedure 238
tag 238
Trigger_RegisterProcedure 238
passive locking 165
PHONE 134
Phonefmt 134
physicalname() function 219
physical tables 166
picture, Dexterity control types 556
picture resources
changing, Modifier used 325
PM Paid Transaction History File (PM30200) 481
Point of Sale (POS) systems 58
Position-Left property 136
Position-Top property 136
posted transactions 394
Posting Journal Reprint tables 37
Preferences file (Dex.ini)
about 21
ExportOneLineBody switch 28-31
OLEPath switch 25
RememberUser switch 26
ShowAdvancedMacroMenu switch 27, 28
SQLLogAllODBCMessages 23
SQLLogODBCMessages 23
SQLLogSQLStmt switch 22
Synchronize switch 23
Workstation2 switch 24
Workstation switch 23
printer icon, Vendor Maintenance window 45
Procedure events, VS Tools
InvokeAfterOriginal 106
InvokeBeforeOriginal 106
procedure or function changes 516
procedure trigger
about 98, 237
considerations 238
cross-dictionary considerations 239
registration procedure 237
processing procedure, Form trigger
creating 240
Product Information window
Compatibility ID 281
Compatibility Message 281
dictionary 280
Forms Dictionary 280
launch file 280
launch ID 280
product ID 280
product name 280
Reports Dictionary 281
progress indicator, Dexterity control types 556
project
additional data, storing 58, 59
current functionality, changing 58
data, exchanging between systems 58
defining 56
functionality, creating 58
integrations, types 59
scrolling window, adding to 362

windows behavior, changing 56
windows look, changing 56
property 339
push button, Dexterity control types 556
push buttons, Vendor Maintenance window 44

Q

QTY Available for Sale field 405, 406
quantity alert icon, Sales Transaction Entry window 50

R

RadioButton control
about 464
properties 464
radio button, Dexterity control types 556
Radio groups, Dexterity control types 556
range
about 210, 480
establishing 211, 212
range where statement 218, 219
setting 210-213
virtual key, creating 214-218
well-behaved range 211
working with 481, 482
RangeClear() method 482
RangeEnd() method 482
range end statement 216
RangeStart() method 482
range start statement 216
range where statement 218
record
creating 475
deleting 478
filtering 363-367
retrieving 477
updating 478
record-level notes 45
Record Note index 402
record, table operations
creating 200, 201
deleting 209
example, for retrieving 202
retrieving 202
updating 209
reference, Dexterity control types 556
referential diagnostics 268
RejectLine parameter 364
Relation or Cross Reference tables 36
release table, table operations 199
RememberUser switch 26
remove, table operations 199
Rental Company field 435
Rental Location field 435
report changes, VBA customizations 524
Report events, VBA
End 112
Start 112
Report Options tables 37
reports, Dexterity applications
about 267
linked prompts 269, 270
referential diagnostics 268
table relationships validation 271
Required property 135, 340, 341
ResDesc.chm 131
resource 118
Resource Explorer
about 13, 14, 122, 148
navigating 148
Resources List 149
Resources Tree 149
Status Area 149
toolbar 149
Workset 149
Resource Explorer toolbar 149
Resource Explorer window 181
resources
Dexterity documentation 283
macro language 283
Support Debugging Tool 284
Resources List 149
Resources Tree 149
restriction
adding, to SmartList object 408-410
creating 408, 409
RM_Customer_Address form 167
RM_Customer_Contact_Maintenance window 178
RM_Customer_Maintenance window 82

runtime components, VS Tools
AddIns folder 447
application assembly 447
Dexterity Bridge assembly 447
Dexterity Shell assembly 446
runtime environment, Dynamics GP application
Application dictionary (Dynamics.dic) 19
Dexterity Runtime engine (Dynamics.exe) 19
Forms dictionary (Forms.dic) 19
launch file (Dynamics.set) 19
Microsoft SQL Server databases 19
Preferences file (Dex.ini) 19
Reports dictionary (Reports.dic) 19
Runtime.exe 131

S

Sales Transaction Entry window
about 48
e-mail link button 48
expansion arrows 52
Help icon button 53, 54
information icon 51
map link button 49
multicurrency icon 50
quantity alert button 50
show details button 50
window elements 48
sanScript 62
about 14, 83, 94, 192
ranges 210
script analysis 203
Script Editor 202
scripts 192
table operations 196
Save() method 478
save table, table operations 199
script analysis 203
script changes, Dynamics GP SDK 489-491
script flow, sanScript
Change event 194
Post event 194
Pre event 194
script naming conventions 195
scripts, sanScript
about 192, 193
naming conventions 195
script flow 194
syntax rules 194
scrolling window
about 142, 180, 185, 220, 221, 361
adding, to project 362
AddsAllowed scrolling window 230
big line item 222
BrowseOnly scrolling window 224
creating 143, 180-188
editable scrolling window 228
fields, adding 146, 147
properties 188
small line item 222
types 221
scrolling window events, Dynamic GP
about 345, 362
After Line Changed 345
After Line Got Focus 345
After Line Lost Focus 345
Before Line Changed 345
Before Line Got Focus 345
Before Line Lost Focus 345
Before Line Populate 345
scrolling window events, Dexterity
about 96
ContextMenu 96
LineChange 96
LineDelete 96
LineFill 96
LineInsert 96
LinePost 96
LinePre 96
scrolling window events, VBA
AfterLineChange 111
AfterLineGofFocus 110
AfterLineLostFocus 111
AfterLinePopulate 111
BeforeLineChange 111
BeforeLineGofFocus 110
BeforeLineLostFocus 111
BeforeLinePopulate 111
scrolling window events, VS Tools
LineChangeAfterOriginal 103

LineChangeBeforeOriginal 103
LineDeleteAfterOriginal 104
LineDeleteBeforeOriginal 104
LineEnterAfterOriginal 102
LineEnterBeforeOriginal 102
LineFillAfterOriginal 102
LineFillBeforeOriginal 102
LineInsertAfterOriginal 104
LineInsertBeforeOriginal 104
LineLeaveAfterOriginal 103
LineLeaveBeforeOriginal 103
scrolling window fields 187
Scrolling window properties
about 144
object properties 144
visual properties 145
scrolling windows
SDK for eConnect
reference link 538
SDT
about 17, 284
installing 130
URL 130, 284, 549
security
granting, to SmartList Builder object 417-419
Service Reference 78
Setup tables 36
ShowAdvancedMacroMenu switch 27, 28
Show Details button, Sales Transaction Entry window 50
Size-Height property 136
Size-Width property 136
Small Line Item 146
small line item, scrolling windows
about 222
marking 223
SmartConnect 78
SmartList Builder
about 386, 387, 526
features 393
field options 400, 401
field types 401
introducing 388-390
templates, importing 390-392
SmartList Builder object
security, granting to 417-419
SmartList drill down 422
SmartList favorite 388
SmartList object
about 386, 392, 394
calculated fields 403-405 creating 393
fields, displaying 397, 399, 400
Go Tos 410, 411
restriction, adding to 408-410
tables, adding to 394-397
SMTP (Simple Mail Transfer Protocol) 80
SOAP (Simple Object Access Protocol) 79
Software development kit (SDK) for Visual Studio Tools
reference link 538
software manuals, Dynamics GP
Dexterity integration guide 537
training manuals 537
sort-by-list field, Vendor Maintenance window 46
source code control program 15
SQLLogAllODBCMessages switch 23
SQLLogODBCMessages switch 23
SQLLogSQLStmt switch 22
standard buttons 84
Standard button type 456
Standard Code Modules 346
standard resources, Dexterity 155
static text
adding, to window fields 188, 189
column headings 189
static text, General Entry window modifications
checkbox text, changing 317
line, adding 318
modifying 313
text prompt, changing 316
tool tip, adding 319
window title, changing 315
Status Area, Resource Explorer window 149
StatusArea button 85, 456
StatusArea property 446 84
stored procedure 39, 40
string, Dexterity control types 556
string fields 402, 403
string resource
changing, Modifier used 328

string terminators 219
subsidiary ledger
about 306
exmaples 306
Summary button
creating 349, 350
Support Debugging Tool. *See* **SDT**
Synchronize switch 23
syntax rules, sanScript 194
system requisites, Dexterity application deployment
feature specific requisites 247
general requisites 246, 247

T

Table 139
table buffer 196, 197
table changes 516
table changes, Dynamics GP SDK
about 495, 496
detail report 497, 498
summary 496
table creation routines, Dexterity application
about 253, 254
SQL Maintenance window, using 255
tables, creating automatically 263
utilities window, building 257-262
Table Import
about 74
capabilities 75
developer skills requisites 76
features 74
Table Lookup window 170
table operations
about 474
record, creating 475
record, deleting 478
record, retrieving 477
record, updating 478
table operations, sanScript
about 196
change 198
copy from table 200
copy to table 199
CRUD 196
get 198
record, creating 200, 201
record, deleting 209
record, retrieving 202
record, updating 209
release table 199
remove 199
save table 199
table options
about 165
Allow Active Locking 165
Use Row Timestamp 165
table relationships validation 271
tables
adding, to SmartList object 394-397
attaching, to form and windows 169
tables and keys, Dexterity
creating 159
tables, Dexterity
about 160, 267
Contact Phone Master 163
Customer Contact Master 160
naming conventions 164, 165
options 165
types 166
table types
creating 166
physical tables 166
temporary tables 166
virtual tables 166
tab sequence, window properties 295
Tab stop property 340
tags 79
Technical Name 41
Tek-Tips forum
about 541
URL 541
templates
importing 390-392
temporary tables 166
Temp tables 36
test mode, development environment
about 124, 125
Dynamics GP desktop 126
TextBox control
about 457-461
properties 460, 461

TextBox control, properties
AutoSetDexColors 459
TextBox control, VS Tools controls 86
TextChanged event 477
text, Dexterity control types 556
third-party dictionary 107
third-party resource 83
time, Dexterity control types 556
tiny integer, Dexterity control types 557
Title property 141, 291
Toolbar button 85, 456
ToolbarWithSeparator button 85, 456
tools
ADO 70
Continuum 68
DDE 70
Dexterity 60
eConnect 76
Extender 69
eXtender Enterprise 69
Integration Manager 73
Modifier 65
ODBC 70
OLE Automation 70
Table Import 74
VBA 65
VS Tools 63
Web services 79
tools documentation, Continuum API
reference link 538
tools documentation, Modifier with VBA
reference link 538
tools, Dynamics GP
Drill-Down Builder 387, 422-428
Excel Report Builder 386, 419-422
Extender 387, 429
SmartList Builder 386, 387
Top property 340
training manuals
about 537
reference links 537
transaction tables
about 38
History phase 39
Open phase 39
Work phase 39
Transact SQL (T-SQL) 404
transfer dictionary module 275, 277
tree view, Dexterity control types 557
Trigger Processing Procedure 94
Trigger_RegisterDatabaseName() function 237
Trigger_RegisterFocusByName() function 235
Trigger_RegisterFormByName() function 233
Trigger_RegisterFunctionByName() function 240
Trigger_RegisterProcedureByName() function 239
triggers, Dynamics GP
about 96, 97, 230, 231
database 235
Focus 233
Form 232
function 239
procedure 237
startup procedure 231
txtVendorID field 477

U

UDDI 79, 80
Universal Description, Discovery, and Integration. *See* **UDDI**
unposted transactions 394
update chunk
building 519
upgrade checklist, Dynamics GP data model
alternate forms and reports, creating 518
data, converting 517
forms and report dictionaries, updating 518
GP 2013 considerations 518
UserForm
about 345
creating 345, 346
User Interface, Dexterity
base resources 155
building 151
forms and windows, creating 167
scrolling windows 185
tables and keys, creating 159
window fields, working 188

user interface events 233
user interface level integrations
about 59
tasks 59
User Interface (UI)
about 43, 44
modifying 81
modifying tools 81
scrolling windows 185
Vendor Maintenance window 44
user security
modifying 126, 127, 128
UTC (Coordinated Universal Time) 165

V

validation errors
resolving 506
Value property 340
ValueSeg property 340
VBA
about 65, 107, 288, 335, 385
components 336
debugging 346, 347
Modules 346
online tutorials and sample applications 336
Options window, accessing 348
overview 336
UserForm 345
VBA customizations
about 520
environment changes 521
key changes 521
report changes 521
Window changes 521
VBA Developer's Guide 115
VBA Editor
about 336
launching 336, 338
VBA events
about 107
band 112
Field 109
Modal Dialog 108
report 112
scrolling window 110
window 108
VBA project
objects, adding to 352
Vendor Maintenance window, adding to 353
window fields, adding 354
windows, adding 354
VB.NET 445
VBScript 73
Vendor Extra window 88
VendorID_Changed() event 376
Vendor Maintenance window
about 44, 353
adding, to VBA project 353
browse buttons 46
lookup button 46
note button (record level) 45
note button (window level) 47
printer icon 45
push buttons 44
sort-by-list field 46, 47
zoom fields 45
Vendor Quick Entry project
about 451
creating, steps 451, 452
form, adding 453
window, creating 453, 454
Vendor Quick Entry window
assembly, building 474
assembly, testing 474
clearing 480
code, adding 472, 473
creating 453, 454
opening 471
window controls, adding 460
version and build, Dexterity application
minding 251, 252
vertical add-on
about 11
sampling 12
virtual key
about 214
creating 214-218
virtual tables 166
Visible property 136, 340

Visual Basic for Applications. *See* **VBA**
visual properties, Field
about 135
AltLineColor 135
Appearance 136
Border 136
Position-Left 136
Position-Top 136
Size-Height 136
Size-Width 136
Visible 136
WordWrap 136
Zoom 136
visual properties, Scrolling window properties
AltLineColor 145
visual properties, window properties
ControlArea 142
Visual Studio Tools. *See* **VS Tools 386**
Visual Studio Tools sample applications
reference link 539
visual switch, Dexterity control types 557
VS Tools
about 63, 98, 445, 526
architecture 446
capabilities 64
developer skills requisites 64
dictionary resources 99
downloading 448, 449
end user prerequisites 64
features 63
installation, running 450
installing 448
URL, for downloading 448
used, for modifying UI 83
WinForm control properties 84
WinForm properties 84
VS Tools controls
about 84
Buttons 84
ComboBoxes 86
Labels 86
ListBoxes 86
TextBoxes 86
VS Tools events
about 99
Field 105
Form 100
Function 106
Procedure 106
scrolling window 102
window 100
VS Tools integration, upgrading
about 526
application assemblies, rebuilding 529- 531
application, downloading 526
code, fixing 532
GP 2013 considerations 533
new solution, building 532, 533
references to assemblies, updating 531, 532
solution, opening 527, 528

W

WanttoDeleteVendor variable 376
WCF (Windows Communication Foundation) service 76
Web services
about 79
capabilities 80
developer skills requisites 80
end user prerequisites 81
Web Services Description Language. *See* **WSDL**
Web services sample applications
reference link 539
Web services SDK for Dynamics GP
reference link 538
well-behaved range 211
Width property 340
window
about 348, 349, 430
adding, to VBA project 354
clearing 480
field values, setting 356, 357
methods, using 354,-356
properties, using 354-356
window controls
about 454
button 455

ComboBox 460
GroupBox 460
Label 459
RadioButton 460
TextBox 457, 458
window events, Dynamics GP
about 344
After Activate 344
After Close 344
After Modal Dialog 344
After Open 344
Before Activate 344
Before Close 344
Before Modal Dialog 344
Before Open 344
window events, Dexterity
about 95
Activate 95
ContextMenu 95
Post 95
Pre 95
Print 95
window events, VBA
AfterActivate 108
AfterClose 108
AfterOpen 108
BeforeActivate 108
BeforeClose 108
BeforeOpen 108
window events, VS Tools
ActivateAfterOriginal 100
ActivateBeforeOriginal 100
AfterModalDialog 101
BeforeModalDialog 101
CloseAfterOriginal 100
CloseBeforeOriginal 100
OpenAfterOriginal 100
OpenBeforeOriginal 100
PrintAfterOriginal 101
PrintBeforeOriginal 100
window fields
about 182, 183, 348, 349
adding, to VBA project 354
static text, adding 188, 189
working with 188
window group 430
window layouts
modifying 298-303
window level notes 47
window methods
Activate 342
Close 342
Hide 342
Move 342
Open 342
PullFocus 342
Show 342
window object properties 17
window objects
big text fields 83
Comboboxes 83
currency field 83
date field 83
Dropdown lists 83
Horizontal listboxes 83
integer field 83
Lines and shapes 83
Listboxes 83
List views 83
Multi-select listboxes 83
Pictures 83
Progress indicators 83
Push buttons 83
Radio groups 83
Scrolling windows 83
Static text 83
String field 83
time field 83
Tree views 83
Visual switches 83
window properties
about 140, 291
AutoLinkTable 291
AutoOpen 291
Caption 341
Changed 341
CloseBox 291
EventMode 341
IsLoaded 341
Name 291
object properties 140
opening position, changing 292-295
Required 341

size, setting 292
tab sequence, setting 296-298
Title 291
visual properties 142
WINDOWS 23
windows and window fields
modifying 288
windows and window fields modifications
about 288
General Entry window, modifying 304
Modifier, launching 289, 290
window layouts 298
window properties 291
Windows Communication Foundation (WCF) 518
Windows Installer services 285
Windows Messaging Layer functionality 71
Windows Scheduler 74
window text properties 175
WindowType property
about 145
AddsAllowed 145
BrowseOnly 145
Editable 145
setting 224
WinForm control properties, VS Tools 84
WinForm properties, VS Tools
about 84
AutoSetDexColors 84
ControlArea 84
StatusArea 84
Winthrop Dexterity Consultants (WDC) 35
WordWrap property 136
Work phase, transaction tables 39
workset
about 153
creating 153, 154
Workset 149
Workstation2 switch 24
Workstation switch 23
WSDL 79, 80

X

XML (Extensible Markup Language) 77
XML tag 79

Z

zoom fields 203
zoom fields, Vendor Maintenance window 45
Zoom property 136

Thank you for buying

Developing Microsoft Dynamics GP Business Applications

About Packt Publishing

Packt, pronounced 'packed', published its first book "Mastering phpMyAdmin for Effective MySQL Management" in April 2004 and subsequently continued to specialize in publishing highly focused books on specific technologies and solutions.

Our books and publications share the experiences of your fellow IT professionals in adapting and customizing today's systems, applications, and frameworks. Our solution based books give you the knowledge and power to customize the software and technologies you're using to get the job done. Packt books are more specific and less general than the IT books you have seen in the past. Our unique business model allows us to bring you more focused information, giving you more of what you need to know, and less of what you don't.

Packt is a modern, yet unique publishing company, which focuses on producing quality, cutting-edge books for communities of developers, administrators, and newbies alike. For more information, please visit our website: `www.packtpub.com`.

About Packt Enterprise

In 2010, Packt launched two new brands, Packt Enterprise and Packt Open Source, in order to continue its focus on specialization. This book is part of the Packt Enterprise brand, home to books published on enterprise software - software created by major vendors, including (but not limited to) IBM, Microsoft and Oracle, often for use in other corporations. Its titles will offer information relevant to a range of users of this software, including administrators, developers, architects, and end users.

Writing for Packt

We welcome all inquiries from people who are interested in authoring. Book proposals should be sent to author@packtpub.com. If your book idea is still at an early stage and you would like to discuss it first before writing a formal book proposal, contact us; one of our commissioning editors will get in touch with you.

We're not just looking for published authors; if you have strong technical skills but no writing experience, our experienced editors can help you develop a writing career, or simply get some additional reward for your expertise.

[PACKT] enterprise
professional expertise distilled
PUBLISHING

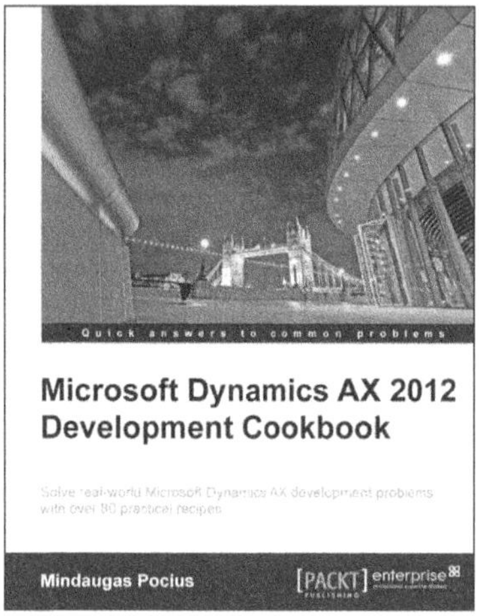

Microsoft Dynamics AX 2012 Development Cookbook

ISBN: 978-1-849684-64-4 Paperback: 372 pages

Solve real-world Microsoft Dynamics AX development problems with over 80 practical recipes.

1. Develop powerful, successful Dynamics AX projects with efficient X++ code with this book and eBook
2. Proven recipes that can be reused in numerous successful Dynamics AX projects
3. Covers general ledger, accounts payable, accounts receivable, project modules and general functionality of Dynamics AX

Microsoft Dynamics GP 2010 Reporting

ISBN: 978-1-849682-18-3 Paperback: 360 pages

Create and manage business reports with Dynamics GP

1. Identify the major reporting challenges facing your organization and select the most effective reporting tool to meet those challenges
2. Empower users from top to bottom in your organization to create their own reports
3. Discover how to use reporting tools to mine and analyze your Dynamics GP data for maximum competitive advantage

Please check **www.PacktPub.com** for information on our titles

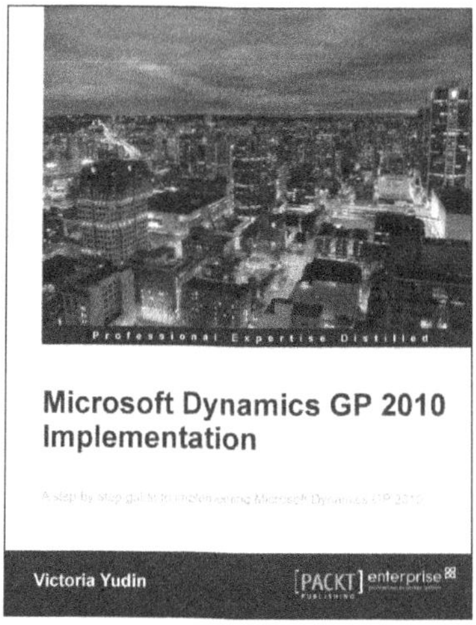

Microsoft Dynamics GP 2010 Implementation

ISBN: 978-1-849680-32-5 Paperback: 376 pages

A step-by-step guide to implementing Mircosoft Dynamics GP 2010

1. Master how to implement Microsoft Dynamics GP 2010 with real world examples and guidance from a Microsoft Dynamics GP MVP
2. Understand how to install Microsoft Dynamics GP 2010 and related applications, following detailed, step-by-step instructions
3. Discover the additional tools available from Microsoft for Dynamics GP

Microsoft Dynamics GP 2010 Cookbook

ISBN: 978-1-849680-42-4 Paperback: 324 pages

Solve real-world Dynamics GP problems with over 100 immediately usable and incredibly effective recipes

1. Discover how to solve real-world Dynamics GP problems with immediately useable recipes
2. Understand the various tips and tricks to master Dynamics GP, improve your system's stability, and enable you to get work done faster
3. Covers the new features in Dynamics GP 2010

Please check **www.PacktPub.com** for information on our titles